Springer Series in
MATERIALS SCIENCE

The Springer Series in Materials Science covers the complete spectrum of materials physics, including fundamental principles, physical properties, materials theory and design. Recognizing the increasing importance of materials science in future device technologies, the book titles in this series reflect the state-of-the-art in understanding and controlling the structure and properties of all important classes of materials.

26 **Gas Source Molecular Beam Epitaxy**
Growth and Properties of Phosphorus Containing III–V Heterostructures
By M. B. Panish and H. Temkin

27 **Physics of New Materials**
Editor: F. E. Fujita   2nd Edition

28 **Laser Ablation**
Principles and Applications
Editor: J. C. Miller

29 **Elements of Rapid Solidification**
Fundamentals and Applications
Editor: M. A. Otooni

30 **Process Technology for Semiconductor Lasers**
Crystal Growth and Microprocesses
By K. Iga and S. Kinoshita

31 **Nanostructures and Quantum Effects**
By H. Sakaki and H. Noge

32 **Nitride Semiconductors and Devices**
By H. Morkoç

33 **Supercarbon**
Synthesis, Properties and Applications
Editors: S. Yoshimura and R. P. H. Chang

34 **Computational Materials Design**
Editor: T. Saito

35 **Macromolecular Science and Engineering**
New Aspects
Editor: Y. Tanabe

36 **Ceramics**
Mechanical Properties, Failure Behaviour, Materials Selection
By D. Munz and T. Fett

37 **Technology and Applications of Amorphous Silicon**
Editor: R. A. Street

38 **Fullerene Polymers and Fullerene Polymer Composites**
Editors: P. C. Eklund and A. M. Rao

Series homepage – http://www.springer.de/phys/books/ssms/

Volumes 1–25 are listed at the end of the book.

Springer Series in
MATERIALS SCIENCE 38

Springer
Berlin
Heidelberg
New York
Barcelona
Hong Kong
London
Milan
Paris
Singapore
Tokyo

Peter C. Eklund Apparao M. Rao (Eds.)

Fullerene Polymers and Fullerene Polymer Composites

With 224 Figures and 19 Tables

Springer

Professor Dr. Peter C. Eklund
Dr. Apparao M. Rao
Department of Physics and Astronomy
and Center for Applied Energy Research
University of Kentucky
Lexington, KY 40506-0055
USA

Series Editors:

Prof. Robert Hull
University of Virginia
Dept. of Materials Science and Engineering
Thornton Hall
Charlottesville, VA 22903-2442, USA

Prof. H. Sakaki
Institute of Industrial Science
University of Tokyo
7-22-1 Roppongi, Minato-ku
Tokyo 106, Japan

Prof. R. M. Osgood, Jr.
Microelectronics Science Laboratory
Department of Electrical Engineering
Columbia University
Seeley W. Mudd Building
New York, NY 10027, USA

Prof. Alex Zunger
NREL
National Renewable Energy Laboratory
1617 Cole Boulevard
Golden Colorado 80401-3393, USA

Library of Congress Cataloging-in-Publication Data

Fullerene polymers and fullerene polymer composites / [edited by] Peter C. Eklund and Apparao M. Rao.
p. cm. -- (Springer series in materials science, ISSN 0933-033X ; 38)
Includes bibliographical references and index.
ISBN 3540648941 (hardcover : alk. paper)
1. Fullerene polymers. 2. Polymeric composites. I. Eklund, P. C. II. Rao, Apparao M., 1961- III. Springer series in materials science ; v. 38.

QD383.F84 F85 1999
620.1'92--dc21

99-050347

ISSN 0933-033x

ISBN 3-540-64894-1 Springer-Verlag Berlin Heidelberg New York

Springer-Verlag is a company in the specialist publishing group BertelsmanSpringer

Printed in Germany

Typesetting: EDV-Beratung F. Herweg, Hirschberg
Cover concept: eStudio Calamar Steinen
Cover production: *design & production* GmbH, Heidelberg

Printed on acid-free paper SPIN: 10688559 57/3144/mf - 5 4 3 2 1 0

Preface

The discovery of carbon cage molecules C_{2n} (or "fullerenes"), by Smalley, Kroto and coworkers in 1985, launched an intensive interdisciplinary inquiry into the fundamental properties of these highly symmetric molecules. The subsequent observation by Kratschmer, Huffman and coworkers that C_{60} and C_{70} could be produced in large enough quantities to be studied in the solid state opened up this research area to physicists, chemists and materials scientists, and the field of fullerene-based materials was born. In addition to the interesting chemical, physical and photophysical properties of the isolated molecules, many exciting materials properties were discovered when they were assembled in the solid state.

In this research monograph, we have attempted to collect important results in the particular sub-field of fullerenes devoted to polymeric materials containing, or built from, C_{60} and C_{70}. These materials can be sub-divided into three main classes: (1) "all-fullerene" polymers formed by inducing covalent connections (bonds) between the cages, (2) polymers formed by the chemical addition of C_{60} to the backbone of a conventional polymer, and (3) composite materials formed by the physical mixture of C_{60} and conventional polymers.

Herein will be found eleven chapters written by internationally recognized researchers in this sub-field of fullerene research. The first two chapters provide introductory material for the non-experts. Chapter 1, "Fundamental Properties of Fullerenes", provides an overview of the structure, bonding, electronic and vibrational states, and physical properties of fullerene molecules and fullerene solids. Chapter 2, "Introduction to Electronic, Optical and Transport properties of Conducting Polymers and Fullerene Polymer Composites", serves as a second cornerstone for the book, introducing some of the key photophysical and transport concepts from the perspective of polymer scientists. Of the remaining nine chapters, three are contributions from prominent theorists. They have applied molecular dynamics and state-of-the-art electronic energy band calculations to explain the structural, electronic and vibrational properties of C_{60} dimers, trimers and various C_{60} polymer systems. This information has proven useful in identifying the various polymeric phases obtained under different experimental conditions. Theoretical discussion of the polymerization mechanism(s) is also presented in these chapters. Three chapters are devoted to the observed properties of "all-fullerene" polymers, where polymerization (or covalent bonding between molecules) was driven either by the application of light, heat and pressure, or charge transfer doping. The remaining three chapters deal with the unusual physical and photophysical properties of composites of C_{60} – polymer systems, formed by the physical mixture of C_{60} with various conventional polymers; photodevice applications based on these materials are also discussed.

The authors are indebted to Dr. G. Sumanasekera for help with the preparation of the final electronic file of the text, and to the United States National Science Foundation for partial support during the preparation and editing of this manuscript. We gratefully acknowledge many helpful discussions with Prof. M.S. Dresselhaus and Dr. G. Dresselhaus yielding further insight into the properties of these novel materials.

November 1999

Peter C. Eklund
Apparao M. Rao

Contents

Contributors

G.B. Adams
Department of Physics and Astronomy, Arizona State University, Tempe, AZ 85287, USA

H. Araki
Department of Electronic Engineering, Faculty of Engineering, Osaka University, 2-1 Yamada-Oka, Suita, Osaka 565, Japan

G. Dresselhaus
Francis Bitter Magnet Laboratory, Massachusetts Institute of Technology, Cambridge, Massachusetts, 02139, USA

M.S. Dresselhaus
Department of Electrical Engineering and Computer Science and Department of Physics, Massachusetts Institute of Technology, Cambridge, Massachusetts 02139, USA

P.C. Eklund
Department of Physics and Astronomy and Center for Applied Energy Research, University of Kentucky, Lexington, KY 40506, USA

K. Harigaya
Physical Science Division, Electrotechnical Laboratory, Umezono 1-1-4, Tsukuba, Ibaraki 305, Japan

J.L. Hodeau
Laboratoire de Cristallographie, MNRS, BP166 Cedex 09, 38042 Grenoble, France

H. Kuzmany
Universität Wien, Institut für Materialphysik, Strudlhofgasse 4, A–1090 Wien, Austria
and
Ludwig Boltzmann Institut für Festkörperphysik Wien, Kopernikusgasse 15, A–1060 Wien, Austria

P.A. Lane
Department of Physics and Astronomy, University of Sheffield, United Kingdom

L. Marques
Laboratoire de Cristallographie, MNRS, BP166 Cedex 09, 38042 Grenoble, France

M.S. Meier
Department of Chemistry, University of Kentucky, Lexington, KY 40506-0055, USA

M. Menon
Department of Physics & Astronomy, University of Kentucky, Lexington, KY 40506-0055, USA
and
Center for Computational Sciences, University of Kentucky, Lexington, KY 40506-0045, USA

M. Núñez-Regueiro
EPM-Matformag, CNRS, BP166 Cedex 09, 38042 Grenoble, France

J.B. Page
Department of Physics and Astronomy, Arizona State University, Tempe, AZ 85287, USA

A.M. Rao
Department of Physics and Astronomy and Center for Applied Energy Research, University of Kentucky, Lexington, KY 40506, USA

N.S. Sariciftci
Institute for Polymers & Organic Solids University of California, Santa Barbara CA 93106, USA

G.E. Scuseria
Dept. of Chemistry and Rice Quantum Institute, Houston, TX77251-1892, USA

K.R. Subbaswamy
Department of Physics, University of Miami, P.O. Box 248004, Coral Gables, FL 33124-4620, USA

Z.V. Vardeny
Department of Physics, University of Utah, Salt Lake City, UT 84112, USA

J. Winter
Universität Wien, Institut für Materialphysik, Strudlhofgasse 4, A–1090 Wien, Austria
and
Ludwig Boltzmann Institut für Festkörperphysik Wien, Kopernikusgasse 15, A–1060 Wien, Austria

C.H. Xu
Dept. of Chemistry and Rice Quantum Institute, Houston, TX77251-1892, USA

K. Yoshino
Department of Electronic Engineering, Faculty of Engineering, Osaka University, 2-1 Yamada-Oka, Suita, Osaka 565, Japan

A.A. Zakhidov
Department of Electronic Engineering, Faculty of Engineering, Osaka University, 2-1 Yamada-Oka, Suita, Osaka 565, Japan
and
Permanent address: Department of Thermophysics, Uzbekistan Academy of Sciences, Katartal 28, Tashkent, 700135, Uzbekistan

1. Fundamental Properties of Fullerenes

M.S. Dresselhaus, G. Dresselhaus

1.1 Introduction to Fullerenes as Polymeric Materials and Composites

Fullerenes are cage-like all-carbon molecules which under special circumstances, such as exposure to high intensity photon or electron irradiation, ion plasma excitation, pressure, or doping, may form polymeric solids. This chapter reviews the properties of the fullerene molecules, the types of solids they normally form and their potential for forming polymeric solids. Fullerenes are also used to form fullerene-polymer composites, where their special molecular properties are exploited for specific applications.

Fullerenes, whether forming polymeric solids or fullerene-polymer composites, feature the fullerene molecule as the fundamental building block of the polymeric phase or polymeric composite. The unique structural features of these building blocks result in unique electronic structures and properties which are summarized in this review. Another attractive feature of fullerenes is their ready accessibility to doping, charge transfer, and the resulting control of their electronic properties. As a result of their unique structures and properties and potential for controlled synthesis, polymeric fullerenes and fullerene-polymer composites have become interesting for their potential for electronic and optical applications, which forms the focus of the present volume.

1.1.1 Historical Background to Fullerenes

In early gas phase work, a molecule with 60 carbon atoms was established experimentally by mass spectrographic analysis [1] as a relatively stable form compared to other carbon clusters. Based on stability arguments, it was conjectured that C_{60} is a closed cage molecule with icosahedral symmetry. The name “fullerene” was given to this family of gas phase molecules by Kroto and Smalley [1] because of resemblance of these cage molecules to the geodesic domes designed and built by R. Buckminster Fuller [2]. The name “buckminsterfullerene” or simply “buckyball” was given specifically to the C_{60} molecule. In the early gas phase work, the fullerene molecules were produced by the laser vaporization of carbon from a graphite target in a pulsed jet of helium [1, 3].

Definitive verification that the C_{60} molecule was indeed the shape of a regular truncated icosahedron came from spectroscopic studies, such as nu-

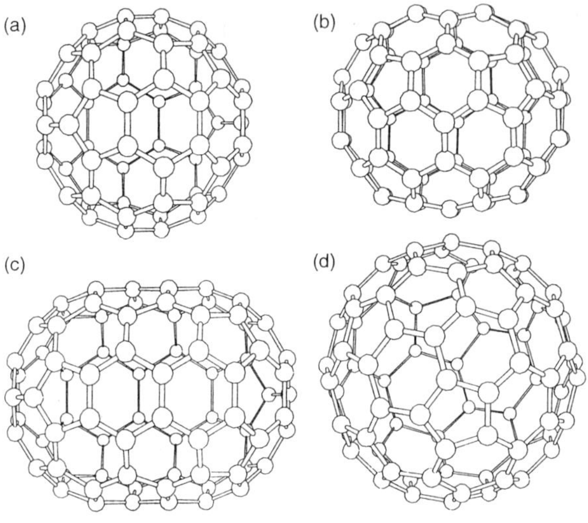

Fig. 1.1. Examples of closed shell fullerene configurations which avoid dangling bonds: (**a**) C_{60}, (**b**) C_{70}, (**c**) C_{80} with D_{5d} symmetry and (**d**) C_{80} with I_h symmetry. Also indicated on the figure are the point group symmetries of the various structures. We note that D_{5d} is a subgroup of the icosahedral group I_h, exhibiting inversion symmetry [6]

clear magnetic resonance (NMR), showing a single NMR line associated with the spin 1/2 ^{13}C nucleus [4, 5]. Since every carbon atom of the C_{60} molecule is in an identical site location, only one NMR line is expected for the 60 carbon atoms located at the vertices of a regular truncated icosahedron (see Fig. 1.1a). The identification of C_{60} as a molecule with icosahedral symmetry was further strengthened by observation of the simple four-line infrared absorption spectrum [7, 8], characteristic of this closed cage molecule with many (174) vibrational degrees of freedom but with very high (icosahedral) symmetry [6].

In 1990 a new type of condensed matter, based on C_{60}, was synthesized for the first time by Krätschmer et al. [9], who discovered a simple carbon arc method for preparing gram quantities of C_{60}, which had previously only been available in trace quantities in the gas phase [1, 10]. The availability of much larger quantities of fullerenes provided a great stimulus to this research field. It was soon found [11] that the intercalation of alkali metals into C_{60} to a stoichiometry M_3C_{60} (where M = K, Rb, Cs) greatly modifies the electronic properties of the host fullerenes and yields a conducting material with a relatively high superconducting transition temperature [12]. The first reported superconductor in the fullerene family was K_3C_{60} with $T_c \sim 19\,\mathrm{K}$ [12], and subsequent work has revealed superconductivity with $T_c \sim 40\,\mathrm{K}$ in Cs_3C_{60} under a pressure of 12 kbar [13]. The discovery of superconductivity [13, 14] in these compounds further spurred research activity in the field of C_{60}-related materials. The possibility for observing superconductivity in fullerene–polymer composites is discussed in Chap. 8.8.

1.1.2 Historical Background to Fullerene Polymers and Fullerene-Polymer Composites

The unusual structure of molecular fullerenes leads to unusual optical, and transport properties. As the structure-properties studies of fullerene-based films proceeded, it was soon demonstrated [15] that, by excitation through photons [15–17], electron beams [18], plasma discharge [19], pressure [20], and alkali metal doping [21], the films, initially in the crystalline phase, could be transformed into a polymeric phase, with major modifications to their optical, vibrational, and transport properties. Since the early discovery of these effects, extensive studies have been made of the polymer-induced modifications to the structure and properties of these films [6].

Early demonstration of the potential for the use of small additions of fullerenes to polymers to promote specific optoelectronic applications [22] stimulated much interest and activity, leading to a host of interesting potential applications and to an increased understanding of the transport and optical properties of this fascinating new class of optoelectronic materials.

1.2 Structural Properties of Fullerene Molecules

In the preparation of fullerenes by any of the conventional synthesis methods (by carbon arc discharge, laser pyrolysis, or combustion flames) the fullerene species of greatest abundance by far is C_{60}, the most stable of the fullerenes, and the fullerene molecule with the greatest symmetry. Every C_{60} molecule is identical, except for the possible presence of the ^{13}C isotope, with a 1.1% natural abundance which can substitute randomly for ^{12}C in the caged molecule [6]. C_{60} molecules thus form a very small (0.71 nm diameter) nanostructure of high (icosahedral I_h) symmetry.

Since the structure and properties of fullerene solids are strongly dependent on the structure and properties of the constituent fullerene molecules, we first review the structure of the molecular building blocks, which is followed by a review of the structure of the corresponding molecular solids. Because of the simplicity of C_{60} relative to other fullerenes and its relatively high abundance, most fullerene studies have focussed on the structure and properties of C_{60}.

1.2.1 Structure of C_{60}

To a good approximation, the 60 carbon atoms in C_{60} are located at the vertices of a regular truncated icosahedron. As mentioned above, every carbon site on the C_{60} molecule is equivalent to every other site (see Fig. 1.1a), consistent with a single sharp line in the NMR spectrum [4, 5], and with the simple four-line infrared spectrum (see Sect. 1.6.1) [7, 8]. The average nearest-neighbor carbon-carbon (C–C) distance a_{C-C} in the C_{60} molecule is very small (0.144 nm) and is almost identical to that in graphite (0.142 nm). Each carbon atom in C_{60} (and also in graphite) is trigonally bonded to three other carbon atoms, and 20 of the 32 faces on the regular truncated icosahedron are hexagons, the remaining 12 being pentagons. Thus C_{60} forms a closed shell molecule in keeping with Euler's theorem, which states that a closed surface consisting of hexagons and pentagons has exactly 12 pentagons and an arbitrary number of hexagons [23]. The introduction of pentagons gives rise to the curvature necessary to form a closed surface. To minimize local curvature, the pentagons are separated from each other in the self-assembly process, giving rise to the isolated pentagon rule, an important rule for stabilizing C_{60} clusters. The high reproducibility of C_{60} in the self-assembly process relates to the fact that the smallest cluster to obey the isolated pentagon rule is C_{60} and that there is only one way to assemble 60 carbon atoms in a closed cage configuration which obeys the isolated pentagon rule, and that structure is shown in Fig. 1.1a [6].

The symmetry operations of the icosahedron consists of the identity operation, 6 five-fold axes through the centers of the pentagonal faces giving rise to 24 independent symmetry operations, 10 three-fold axes through the centers of the hexagonal faces resulting in 20 independent symmetry operations, and 15 two-fold axes through centers of the edges joining two hexagons. Each of the 60 rotational symmetry operations is then compounded with the inversion operation, resulting in 120 symmetry operations in the icosahedral point group I_h [24] (see Fig. 1.1a). I_h is the highest possible point group symmetry for any molecule. The diameter of the C_{60} molecule is 0.710 nm (see Table 1.1), treating the carbon atoms as points [1, 25, 26].

From Euler's theorem it follows that the smallest possible fullerene is C_{20} which would form a regular dodecahedron with 12 pentagonal faces [6], but this structure is energetically unfavorable in accordance with the isolated pentagon rule, because of its high local curvature, giving rise to a high degree of local strain. Since the addition of a single hexagon adds two carbon atoms, all fullerenes C_{n_C} must have an even number of carbon atoms n_C, in agreement with the observed mass spectra for fullerenes (see Fig. 1.2 form $c > 36$) [3]. An estimate for the diameter of a fullerene can be found from the relation for an icosahedral fullerene $d_i = a_{C-C}\sqrt{15 n_C}/2\pi$ where $a_{C-C} = 0.144$ nm is the average nearest-neighbor carbon–carbon distance.

Although each carbon atom in C_{60} is equivalent to every other carbon atom, the three bonds emanating from each atom are not equivalent (see Fig. 1.1a). Each of the four valence electrons of each carbon atom is engaged in covalent bonds, so that each of the five atoms on the pentagon perimeter has two single bonds that are electron-poor and the bond between

Table 1.1. Physical constants for C_{60} molecules and for crystalline C_{60}

Quantity	Value	Reference
Average C–C distance	0.144 nm	[27]
C–C bond length on a pentagon	0.146 nm	[5, 28]
C–C bond length on a hexagon	0.140 nm	[5, 28]
C_{60} mean molecule diameter	0.710 nm	[26]
Moment of inertia I	1.0×10^{-43} kg m^2	[29]
Volume per C_{60}	1.87×10^{-22} cm^3	–
Number of distinct C sites	1	–
Number of distinct C–C bonds	2	–
Binding energy per atom[a]	7.40 eV	[30]
Heat of formation (per g C atom)	10.16 kcal	[31]
Electron affinity	2.65 ±0.05 eV	[32]
Cohesive energy per C atom	1.4 eV/atom	[33]
Spin-orbit splitting of C(2p)	0.00022 eV	[34]
Ionization potential	7.58 eV	[35]
Optical gap[b]	1.9 eV	[30]
fcc lattice constant	1.417 nm	[36]
C_{60}–C_{60} distance	1.002 nm	[25]
C_{60}–C_{60} cohesive energy	1.6 eV	[37]
Mass density	1.72 g/cm^3	[25]
Molecular density	1.44×10^{21}/cm^3	[25]
Transition temperature (T_{01})[c]	261 K	[38]
Vol. coeff. of thermal expansion	6.1×10^{-5}/K	[39]
Band gap (HOMO–LUMO)	1.7 eV	[15]
Work function	4.7 ±0.1 eV	[40]
Thermal conductivity (300 K)	0.4 W/mK	[41]
Electrical conductivity (300 K)	1.7×10^{-7} S/cm	[42]
Static dielectric constant	4.0–4.5	[43, 44]
Melting temperature	1180°C	[45]
Sublimation temperature	434°C	[46]
Heat of sublimation	40.1 kcal/mol	[46]
Latent heat	1.65 eV/C_{60}	[47]

[a] The binding energy for C_{60} is believed to be ~0.7 eV/C atom less than for graphite, although literature values for both are given as 7.4 eV/C atom. The reason for the apparent inconsistency is attributed to differences in calculational techniques.
[b] Calculated value for the optical band gap for the free C_{60} molecule.
[c] Structural phase transition associated with rotational order–disorder.

two adjacent hexagons (see Fig. 1.1a) which is an electron-rich double bond. Consistent with NMR and x-ray diffraction evidence, the structure of C_{60} is further stabilized by introducing a small distortion of the bond lengths to form the Kekulé structure of alternating single and double bonds around the hexagonal face. The single bonds that define the pentagonal faces are increased from the average bond length of 0.144 nm to a length $a_5 = 0.146$ nm,

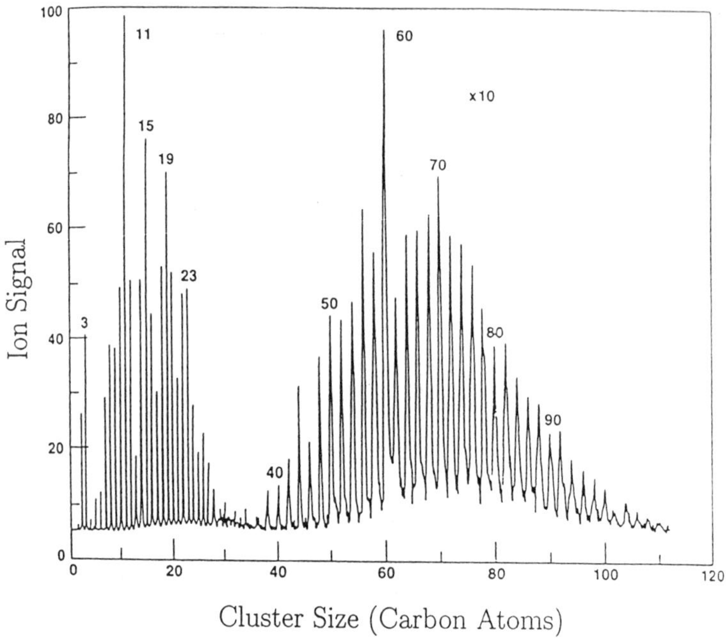

Fig. 1.2. Time-of-flight mass spectrum of carbon clusters produced in a supersonic beam by laser vaporization of a graphite target [3]

while the double bonds between adjacent hexagons are decreased in length to $a_6 = 0.140\,\text{nm}$ [48, 49]. We note that the icosahedral I_h symmetry is preserved under these distortions. Since each carbon atom has its valence requirements fully satisfied, a solid composed of C_{60} molecules is expected to form a van der Waals-bonded crystal which is nonconducting (an insulator or a semiconductor).

1.2.2 Structure of C_{70} and Higher Mass Fullerenes

In the synthesis of C_{60}, larger molecular weight fullerenes C_{n_C} ($n_C > 60$) are also formed, by far the most abundant being C_{70}. However, significant quantities of C_{76}, C_{78}, C_{84}, and higher mass fullerenes have also been isolated and studied in some detail [6]. Both C_{60} and C_{70}, either alone or in C_{60}–C_{70} mixtures, have been used in fullerene-polymer composites and applications.

C_{70} has been found to exhibit a rugby-ball shape (see Fig. 1.1b) [50], and its form can be envisioned either by adding a ring of 10 carbon atoms or a belt of 5 hexagons around the equatorial plane of the C_{60} molecule normal to one of the five-fold axes (see Fig. 1.1b). In contrast to the C_{60} molecule with I_h symmetry, the C_{70} molecule has the lower D_{5h} symmetry, which is a subgroup of I (but lacking inversion symmetry). Fullerenes often form isomers, since a given number n_C of carbon atoms can correspond to closed cage molecules C_{n_C} with different geometrical structures [50, 51], each distinct structure referring to a different isomer. Whereas C_{70}, like C_{60}, has only one isomer that obeys the isolated pentagon rule (see Fig. 1.1b), C_{80} might be formed in the shape of an elongated rugby ball prepared by adding two rows of 5 hexagons normal to a five-fold axis of C_{60} at the equator (see Fig. 1.1c); an icosahedral form of C_{80} can also be constructed (see Fig. 1.1d) as another of the 7 distinct isomers of C_{80}, satisfying the isolated pentagon rule. The symmetries of these distinct isomers of C_{80} are D_{5d}, D_2, $2C_{2v}$, D_3, D_{5h}, and I_h [52]. The three fullerenes obeying the isolated pentagon rule, and having just a single isomer are C_{60} (I_h), C_{70} (D_{5h}), and C_{76} (D_2) [6].

1.2.3 Structure of Metallofullerenes

In addition to the nanostructures formed purely from carbon-containing fullerenes, as described above, there are metallofullerenes which contain a metal ion within the fullerene cage [53].

Many species can be inserted into the interior hollow core of the C_{60} molecule to form an endohedrally-doped molecular unit, commonly called a metallofullerene for metal dopants. The insertion of one, two, or three metal species inside a single fullerene cage is common [54]. The endohedral fullerene configuration is denoted by La@C_{60} for one endohedral lanthanum in C_{60}, or Y_2@C_{82} for two Y atoms inside a C_{82} fullerene [55]. Slow progress with the isolation and purification of sufficient quantities of metallofullerene materials has, however, delayed study of the structure and properties of metallofullerenes in the solid state [6]. Although metallofullerenes have potential for forming interesting polymer compounds or fullerene-polymer composites, due to the difficulty in obtaining significant quantities of metallofullerenes, the literature on this topic is limited at the present time.

1.3 Synthesis

Synthesis methods for fullerenes are rapidly improving, thereby increasing yields and lowering costs. Likewise, separation and purification methods are also improving rapidly, making available C_{60} samples with purities well beyond 99%. Furthermore, sufficient quantities of purified C_{70} are now available for many physical measurements. Substantial progress has also been made with the separation of higher mass fullerenes, and of metallofullerenes so that physical measurements of the structure and properties of these nanoclusters is now becoming active. Through the use of scanning tunneling microscopy and electron energy loss spectroscopy, physical measurements on periodic arrays of higher mass fullerenes and of metallofullerenes, as occur in crystal lattices have now been carried out [6, 56].

Fullerenes are usually synthesized using an arc discharge between graphite electrodes (20 V, 60 A) in approximately 200 torr of He gas. The heat generated at the contact point between the electrodes evaporates carbon to form soot and fullerenes, which condense on the water-cooled walls of the reactor. This discharge produces a carbon soot which can contain up to ~15% fullerenes: C_{60} (~13%) and C_{70} (~2%). The fullerenes are next separated from the soot according to their mass (which is proportional to the number of carbon atoms in the fullerene molecule) using liquid chromatography and a solvent such as toluene for the chromatography column. Extraction and purification steps follow the separation to yield powder samples of specific fullerenes [6].

Metallofullerenes are prepared by endohedral doping of guest species such as rare earth, alkaline earth or alkali metal ions into the interior of the fullerene molecule. The synthesis is carried out by impregnating the positive electrode of the arc discharge apparatus with graphite powder mixed with the desired dopant. During the arc discharge process, the dopant species is released into the plasma gas and becomes entrained within the fullerenes. Sophisticated liquid chromatography techniques are used for concentrating the minute amounts of a given metallofullerene that are prepared in the synthesis process, which is followed by further separation and purification steps [53].

Property measurements of fullerenes are made in the gas phase, in solution or in the solid phase, using powder samples, films or single crystals. C_{60} powder is obtained by vacuum evaporation of the solvent from a solution during the synthesis process discussed above. Single crystals and polycrystalline films are then prepared from these purified powders. Fullerene films are prepared by vacuum sublimation of the fullerenes on substrates selected for the specific use of the films [6]. Single crystals are best grown by the vacuum sublimation technique [57], since growth from solution yields crystals of lower chemical purity because of the incorporation of solvent during crystallization.

1.4 Structure in Condensed Phases

In this section we discuss the structure of fullerenes and related materials in the crystalline phase and the degree of ordering that is provided by single crystal growth, film growth on a substrate, and intercalation. The deposition of C_{60} on a substrate will in general result in a disordered array of fullerene molecules, but if the substrate is chosen for lattice matching [e.g., GeS (001)] and the growth conditions are appropriately controlled, epitaxial fullerene growth can occur [58]. In all cases, whether single crystal growth, film growth or intercalation is involved, the fullerene building blocks retain their molecular structural integrity in forming these crystalline phases [6].

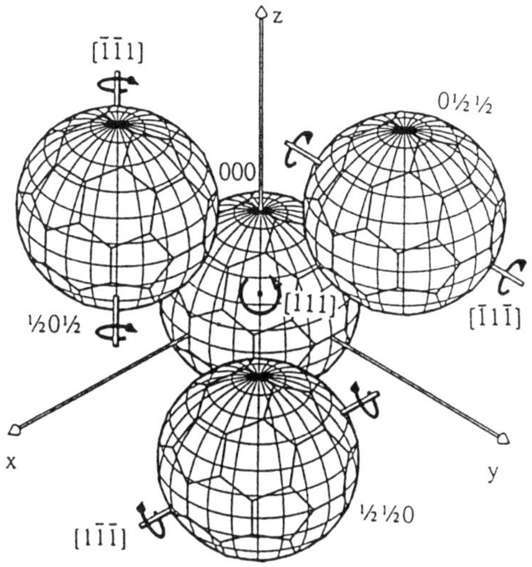

Fig. 1.3. The four molecules in the unit cell of C_{60} showing the same standard orientation, with two-fold axes aligned parallel to the cube edges. Starting from this orientation, molecules at (0,0,0), ($\frac{1}{2}$,$\frac{1}{2}$,0), ($\frac{1}{2}$,0,$\frac{1}{2}$), (0,$\frac{1}{2}$,$\frac{1}{2}$) are rotated by the same angle about local axes $\langle 111\rangle$, $\langle 1\bar{1}\bar{1}\rangle$, $\langle \bar{1}\bar{1}1\rangle$, and $\langle \bar{1}1\bar{1}\rangle$, respectively. The sense of rotation about each $\{111\}$ axis is indicated in the figure [59]

1.4.1 Crystalline C_{60}

In the crystalline phase, the C_{60} molecules crystallize into a cubic structure with a lattice constant of 1.417 nm, a nearest neighbor C_{60}–C_{60} distance of 1.002 nm [25] and a density of 1.72 g/cm^3 (corresponding to 1.44×10^{21} C_{60} molecules/cm^3) (see Table 1.1). At room temperature, the molecules are rotating rapidly with full rotational freedom, and the centers of the molecules are arranged on a face-centered cubic (fcc) lattice with one C_{60} molecule per primitive fcc unit cell, or 4 molecules per simple cubic unit cell. The pertinent space group is O_h^5 or $Fm\bar{3}m$, which has been established directly by x-ray and neutron diffraction [39, 60, 61]. Because of their rapid rotation at room temperature relative to typical measurement probes, all the C_{60} molecules in the array are equivalent, so that an ordered phase for the fullerenes is achieved upon crystallization.

Below an ordering temperature of $T_{01} = 260$ K, the C_{60} molecules lose most of their rotational degrees of freedom, so that the residual rotational motion in the low temperature phase below T_{01} occurs only along the four $\langle 111\rangle$ axes [39, 61, 64], each molecule rotating about a different $\langle 111\rangle$ axis (see Fig. 1.3). The structure of solid C_{60} below T_{01} thus becomes simple cubic with a space group T_h^6 or $Pa\bar{3}$, whereby the four distinct molecules shown in Fig. 1.3 are inequivalent in the low temperature phase (see also Fig. 1.4a) [39, 61, 65], thereby resulting in a lowering of the symmetry. As the temperature is lowered further below T_{01}, additional ordering of the C_{60} molecules occurs as the rotations about the $\langle 111\rangle$ axes become hindered. Adjacent C_{60} molecules thus develop correlated orientations, so that an electron-rich double bond of one molecule tends to be opposite an electron-poor pentagonal face on an adjacent molecule [6].

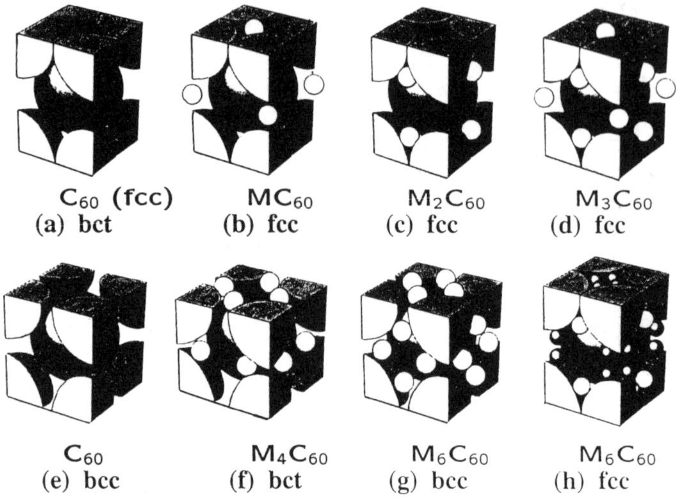

Fig. 1.4. Crystal structures for the alkali metal fullerides (**a**) undoped fcc C_{60}, (**b**) MC_{60}, (**c**) M_2C_{60}, (**d**) M_3C_{60}, (**e**) C_{60} in a bcc structure, (**f**) M_4C_{60}, and two structures for M_6C_{60}, (**g**) M_6C_{60} (bcc) for M = K, Rb, Cs, and (**h**) M_6C_{60} (fcc), which is appropriate for M = Na [62]. The large balls denote C_{60} molecules and the small balls denote alkali metal ions. For fcc M_3C_{60}, which has four C_{60} molecules per cubic unit cell, the M atoms can be on either octahedral or tetrahedral sites. Undoped solid C_{60} exhibits the fcc crystal structure at room temperature, but in this case all tetrahedral and octahedral sites are unoccupied. For (**f**) bct (body-centered tetragonal) M_4C_{60} and (**g**) bcc M_6C_{60}, all the M atoms are on distorted tetrahedral sites. For (**h**) we see that four Na ions can occupy an octahedral site of this fcc lattice. The M_1C_{60} compounds (for M = Na, K, Rb, Cs) crystallize in the rocksalt structure [6, 62, 63]

In the low temperature phase, the [100] cubic axes pass through three mutually orthogonal 2-fold molecular axes on the icosahedron (the centers of the electron-rich hexagon-hexagon edges), and four $\langle 111 \rangle$ cubic axes pass through centers of the hexagonal molecular faces (see Fig. 1.3). Since the fullerene molecule has no four-fold axis, the molecules are ordered on the cubic lattice relative to their two-fold axis. There are, however, two standard orientations of the icosahedral C_{60} molecule about its two-fold axes, which can be obtained from one another by a rotation about a three-fold axis. Thus, when the C_{60} molecules crystallize in the simple cubic $Pa\overline{3}$ structure for temperatures below T_{01}, they will have equal probability for orienting their two-fold axes along either of the two standard orientations, therefore introducing a type of disorder to the system which is called *merohedral disorder* [45]. Although intermolecular interactions result in preferred orientations between adjacent fullerenes and some partial molecular alignment occurs, orientational (merohedral) disorder persists down to the lowest temperatures because of the basic incompatibility of the icosahedral symmetry and cubic symmetry with regard to the four-fold axes [6].

1.4.2 Crystalline C_{70}

Because of the rugby-ball shape of the C_{70} molecule (see Fig. 1.1b), the structure for crystalline C_{70} is more complex than that for C_{60} [66–68], evolving through several distinct crystal structures as a function of temperature. At high temperature ($T \gg 340\,\mathrm{K}$), the fcc phase (lattice constant $a = 1.501\,\mathrm{nm}$) with freely rotating molecules is most stable, but since the ideal hexagonal close packed (hcp) phase with $c/a = 1.63$ is almost equally stable, fcc crystals of C_{70} tend to be severely twinned and show many stacking faults. As the temperature is lowered, the fcc structure is continuously transformed by deformation into a rhombohedral structure with the long diagonal 3-fold axis aligned parallel to the $\langle 111 \rangle$ direction of the fcc structure [67]. At lower temperatures, a hexagonal phase (space group $P6_3/mmc$) with the C_{70} molecules rotating with full rotational symmetry is stabilized. This C_{70} crystal phase has lattice constants $a = b = 1.056\,\mathrm{nm}$ and $c = 1.718\,\mathrm{nm}$ and a nearly ideal c/a ratio of 1.63. A transition to another hcp occurs at $\sim 337\,\mathrm{K}$, but with $a = b = 1.011\,\mathrm{nm}$ and a larger c/a ratio of 1.82 for the lower temperature phase. This larger c/a ratio is associated with the orientation of the C_{70} molecules along their long axis, as the free molecular rotation (full rotational symmetry) that is prevalent in the higher temperature phase freezes into a rotation about the 5-fold axis of the C_{70} molecule [67]. As the temperature is further lowered to $\sim 270\,\mathrm{K}$, the free rotation about the c-axis also becomes frozen, resulting in a monoclinic structure with the unique axis along the c-axis of the hcp structure, and the monoclinic angle β is close to $120°$. In the low temperature phases ($T < 337\,\mathrm{K}$), the main 5-fold axes for the C_{70} molecules are aligned, and the two-fold axes assume some degree of correlation. But because of the incompatibility of the D_{5h} point group operations with those of the $P6_3/mmc$ space group, orientational misalignment with respect to the three-fold and two-fold axes persists down to the lowest temperatures [6].

1.4.3 Crystalline Phases for Higher Mass Fullerenes

Because of the small sample quantities currently available for the higher mass fullerenes C_{n_C} ($n_C > 70$), structural studies of their crystalline phases by conventional methods (such as x-ray diffraction and neutron scattering) are difficult. The most direct structural measurements for higher mass fullerenes thus far have come from scanning tunneling microscopy (STM) studies [69, 70] and selected area electron diffraction studies using electron energy loss spectroscopy (EELS) [56].

STM measurements have been done on C_{76}, C_{80}, C_{82}, and C_{84} fullerenes adsorbed on Si(100) 2×1 and GaAs(110) surfaces [69, 70], showing that all of these higher mass fullerenes crystallize in an fcc structure with C_{n_C}–C_{n_C} nearest-neighbor distances of 1.13 nm, 1.10 nm, 1.174 nm, 1.21 nm for C_{76}, C_{78}, C_{82}, and C_{84}, respectively, and fcc lattice constants are obtained from these distances by multiplication by $\sqrt{2}$. The more massive fullerenes are less mobile rotationally, so that the orientational phase transition at T_{01}, discussed above for crystalline C_{60} and C_{70}, is driven to higher temperatures (above room temperature) in the higher mass fullerenes, if the phase transition occurs at all. Since higher mass fullerenes generally have multiple isomers, the fullerenes on the fcc lattice sites are not expected to be identical. Therefore disorder occurs due to both the presence of multiple isomers and orientational misalignment (as discussed above for crystalline C_{60} and C_{70} for their low temperature phases). Novertheless, electron diffraction measurements have provided quantitative values for the lattice constants and grain sizes of very small crystalline samples of the higher fullerenes, yielding a plot of lattice constants a_{fcc} as a function of $\sqrt{n_C}$, as shown in Fig. 1.5 [56], where n_C is the number of carbon atoms in C_{n_C}.

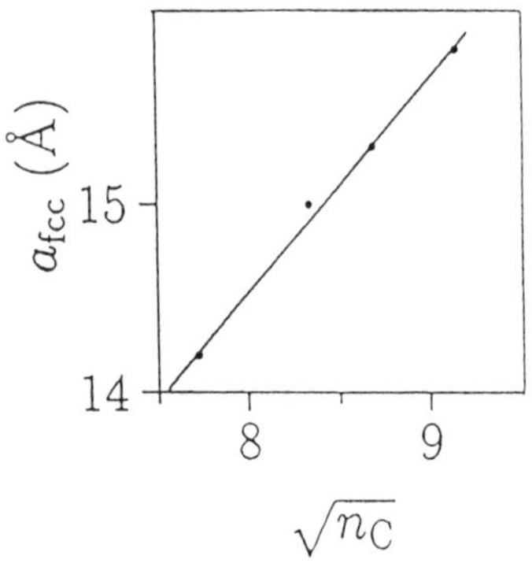

Fig. 1.5. The lattice constant for the higher-mass fullerenes C_{n_C} as a function of $\sqrt{n_C}$, where $60 \leq n_C \leq 84$, as obtained by electron diffraction measurements [56]

1.4.4 Intercalated Fullerene Structures

Fullerene solids are unique insofar as they can be doped in several different ways, including endohedral doping (where the dopant is inside the fullerene shell), substitutional doping (where the dopant is included in the fullerene shell), and exohedral doping (where the dopant is outside or between fullerene shells). Doped fullerenes in the crystalline phase are often called fullerides, in contrast to the term fullerite, which refers to the crystalline phase prior to doping. In many instances, the doping of fullerenes with guest species leads to charge transfer between the guest species and the host, while in clathrate materials, such charge transfer does not occur. Several stable crystalline phases for exohedrally doped or intercalated C_{60} have been identified. Much less information is available for the crystalline phases of endohedrally doped fullerenes or for substitutionally doped fullerenes, such as the fullerene-derived BC_{59} [6], although such information is now being assembled by a number of groups wordlwide.

Almost all studies of intercalated fullerene structures have employed alkali metal guest species as exohedral dopants (see Fig. 1.4), though some studies have also been carried out on alkaline earth dopants, and a very few on rare earth dopants [6]. In the intercalation process [71], the guest species is introduced into the interstitial positions between adjacent molecules (exohedral locations) (see Fig. 1.4). The guest species in Fig. 1.4 tend to be small (relative to the fullerene anions) and they fit into the interstitial spaces shown in the figure with only small increases in lattice constant of the crystal upon doping. Charge transfer can take place between the M atoms and the fullerene molecules, so that the M atoms become positively charged cations and the fullerene molecules become negatively charged anions with predominately delocalized electrons. With exohedral doping, the conductivity of fullerene solids can be increased by many orders of magnitude [6].

Among the alkali metals, Li, Na, K, Rb, and Cs and their alloys have been used as exohedral dopants [72, 73], and stable M_xC_{60} compounds have been achieved for $x = 1, 3, 4$, and 6. The stable structure for crystalline M_1C_{60} is the rocksalt structure (Fig. 1.4b) consisting of two interpenetrating fcc lattices and this phase is typically stable above 425 K. The most intensively studied of the M_xC_{60} compounds are the M_3C_{60} compounds (M = K, Rb) which form face-centered cubic (fcc) structures. The addition of higher concentrations of dopants destabilizes the close packed fcc structure, with the M_4C_{60} (M = K, Rb, Cs) phases forming body-centered tetragonal (bct) structures and the M_6C_{60} (M = K, Rb, Cs) phases forming bcc structures.

Of particular interest to the literature on polymeric fullerenes are the M_1C_{60} (M = K, Rb) compounds which crystallize in the fcc rocksalt structure above ~400 K. Upon slow cooling from elevated temperatures (e.g., 400 K), K_1C_{60} forms an orthorhombic polymerized phase with relatively strong intermolecular coupling along one crystalline axis for which the lattice constant is ~10% shorter than for the two orthogonal directions [74–76]. Some authors have referred to this crystalline phase as a polymeric phase because of its chain-like character (see Fig. 1.6). Evidence for this phase transition is also found in optical [77–81] and electron spin resonance [21, 75, 80] studies.

Some success has also been achieved with the intercalation of alkaline earth dopants, such as Ca, Sr, and Ba [36, 82, 83] into the C_{60} lattice, but in this case, two electrons per metal atom M are transferred to the C_{60} molecules for low concentrations of metal atoms, while for high concentrations of metal atoms the charge transfer is less than two electrons per alkaline earth ion. In general, the alkaline earth ions are smaller than alkali metal ions in the same row of the periodic table. For this reason, the crystal structures formed with alkaline earth doping are often different from those for alkali metal dopants. Doping fullerenes with acceptors has been more difficult than with donors because of the high electron affinity of C_{60} (see Table 1.1). Although a

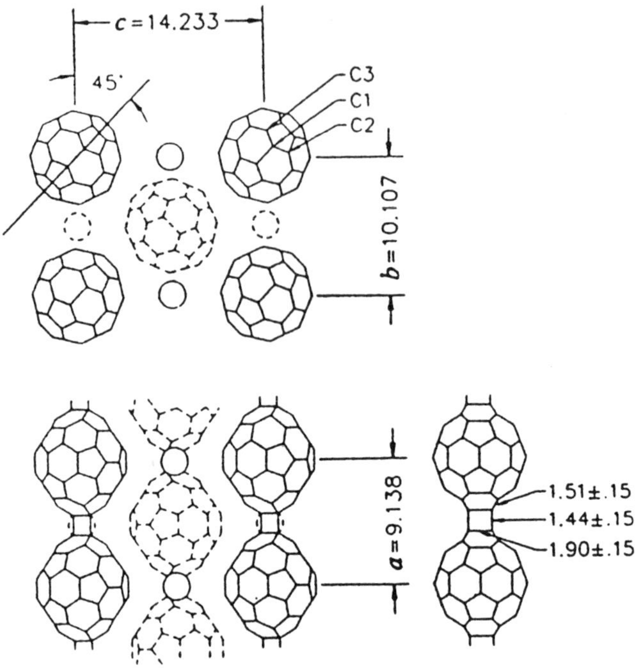

Fig. 1.6. Atomic structure of the polymeric Rb_1C_{60} phase determined from x-ray diffraction patterns [74]. Solid (broken) lines show fullerene anions and metal cations centered at $x = 0$ ($x = 1/2$) in the b–c plane at the top, and at $y = 0$ ($y = 1/2$) in the a–c plane at the bottom. To the right is a projection of one chain perpendicular to the molecular mirror plane

very few stable compounds with acceptor-type dopants have been synthesized [84, 85], acceptor doping has not been important for polymer-related fullerene materials.

1.5 Polymerized Fullerenes

Polymerization has been induced in C_{60} films through excitation by photons (Sect. 1.5.1 and Chap. 3.5) [15–17, 86], electrons (Sect. 1.5.2) [18], in a plasma discharge (Sect. 1.5.4), as a result of hydrostatic pressure (Sect. 1.5.3) [20], and by chemical means, such as in $(\text{-}C_{60}Pd\text{-})_n$ chains or $(\text{-}K_1C_{60}\text{-})_n$ "pearl necklaces." Of the various excitation processes, the phototransformation process has been studied most extensively (Sect. 1.5.1) and has also been studied in C_{70} (Sect. 1.5.5) [87]. The electron beam-induced polymerization (Sect. 1.5.2) was observed for both a 3 eV electron beam in an STM instrument and a 1500 eV external electron beam [18]. Pressure-induced polymerization of C_{60} (Sect. 1.5.3) has been reported using a diamond anvil cell [20]. Phototransformation of fullerenes is treated in detail in Chap. 3.5 of this volume.

1.5.1 Photopolymerization of C_{60}

In the absence of diatomic oxygen, solid C_{60} undergoes a phototransformation when exposed to visible or ultraviolet light of sufficient intensity and with photon energies greater than the optical absorption edge (~1.7 eV) [15, 86, 88]. Phototransformation renders the C_{60} films insoluble in toluene, a property which may be exploitable in photoresist applications, since the unexposed portion of the C_{60} films remain soluble in toluene. Because the optical penetration depth for C_{60} at visible or ultraviolet frequencies is on the order of ~0.1–1 μm, small particle-size powders or thin films can undergo phototransformation over a significant fractional volume of the sample. Heating the phototransformed films to ~100°C has been reported to return the system to the pristine state [89].

On the basis of various probes, such as Raman and infrared spectroscopy, luminescence, and optical absorption, it has been proposed [15, 90] that in the phototransformation process, covalent C–C bonds are formed between adjacent C_{60} molecules to produce a polymerically-linked solid. Direct evidence for the formation of covalent intermolecular bonds is provided by laser desorption mass spectroscopy (LDMS) measurements [15, 91], as shown in Fig. 1.7 for a phototransformed C_{60} film. Using a pulsed nitrogen laser (337 nm wavelength, ~8 ns pulse, and 10 mJ/pulse energy) to desorb molecules from the film, mass peaks were observed at 720 atomic mass units (amu), corresponding to the C_{60} molecule, and at all integer multiples of this mass

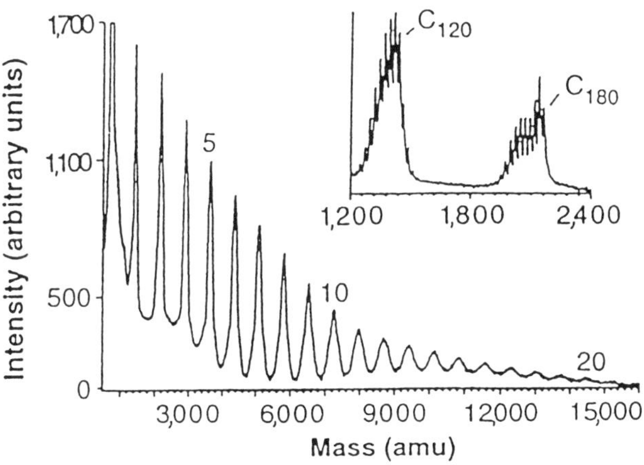

Fig. 1.7. Laser desorption mass spectrum (LDMS) of a C_{60} film (thickness ~2000 Å) that was phototransformed by exposure to UV–visible radiation from a 300 W Hg arc lamp for 12 h. The material was desorbed from the film with a pulsed nitrogen laser (337 nm wavelength, ~8 ns pulse, and 10 mJ/pulse energy) under vacuum (~ 10^{-7} torr) conditions. A series of 20 peaks have been identified with N cross-linked fullerenes of mass ~720 N atomic mass units (amu). Inset: LDMS spectrum indicating C_2 loss or gain in the desorption process [15]

out to 20 × 720 amu, consistent with the desorption of a distribution of covalently-linked clusters of C_{60} molecules from the film surface, including dimers, trimers, and higher oligomers, as shown in Fig. 1.7. The pulsed nitrogen laser used in the laser desorption mass spectrometer, which yielded the spectrum in Fig. 1.7, was shown to also contribute some weak cross-linking of the fullerenes in the C_{60} films.

Consistent with a one-photon excitation of fullerene molecules, the rate of transformation to the polymeric state at $T = 300\,\mathrm{K}$ is proportional to the incident light intensity [92]. As mentioned above, the presence of dioxygen in the film appears to prevent (or considerably retard) the photopolymerization process [15, 93], indicating that the lowest excited triplet state (see Sect. 1.8.1) is involved in the photochemical reaction, since O_2 is known to quench the population of the triplet state effectively.

The photopolymerization process has been found to occur only at temperatures $T > T_{01}$, where $T_{01} \simeq 260\,\mathrm{K}$ is the orientational ordering temperature [93] discussed in Sect. 1.4.1. It has been proposed that molecular rotations occurring above this transition temperature allow adjacent C_{60} molecules to get close to each other, thereby forming a four-membered ring linking adjacent molecules, [called a "2+2 cycloaddition" (see Fig. 1.8)]. Photopolymerization of C_{60} seems to be most effectively carried out near room temperature. At elevated temperatures (~100°C), the thermal scission rate of the covalent

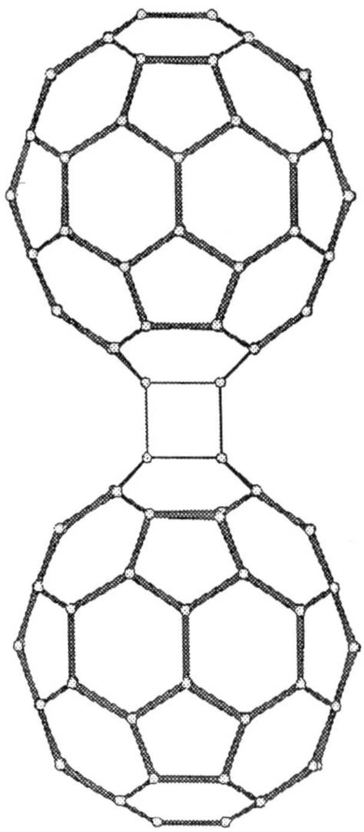

Fig. 1.8. Structure of two adjacent phototransformed C_{60} molecules forming a dimer with a dumbbell shape. The dimer structure shows a four-membered ring bridging adjacent fullerenes [15]

bonds between the molecules is high enough so that the film reverts to the pristine monomer form, as observed by Raman spectroscopy [89]. Thus the stability range of the phototransformed C_{60} phase is between T_{01} (~260 K) and T_p (~420 K) (see Sect. 1.6.6) [6].

1.5.2 Electron Beam–Induced Polymerization of C_{60}

Electron beam-induced polymerization of C_{60} has been reported both for ~3 eV electrons from a scanning tunneling microscope (STM) at nanometer dimensions and for 1500 eV electrons from an electron gun on a larger spatial scale [18]. Electron beams can be used to write patterns on C_{60} surfaces, either very fine lines on a nanometer scale using an STM tip or thicker lines using an electron gun [18].

The STM-induced modifications to solid C_{60} were carried out both for monolayer coverage of C_{60} on a GaAs(110) substrate and for a 10-monolayer C_{60} film for 3 eV electron irradiation. Typical conditions used to induce polymerization in the C_{60} film were a flux of $\sim 1 \times 10^8$ electrons per second per C_{60} molecule [18]. The same STM technique, but operating at lower voltages (< 2 eV), was used to probe the polymerization effect of the electron irradiation, which was carried out by repeated scanning (for approximately 30 min) of a small area of the image (e.g., $80 \times 80\,\text{Å}^2$) at a higher voltage (~3 eV). The electron irradiation-induced modification to the solid C_{60} was identified as the appearance of a modified "speckled" region in the STM scan. Annealing to 470 K restored the original ordered surface, as it appeared prior to electron irradiation [18]. These results thus imply operating ranges where STM techniques can be used for the investigation of C_{60} surfaces, without introducing surface damage due to electron irradiation.

Electron irradiation–induced polymerization on a larger spatial scale was demonstrated using 1500 eV electrons from a 100 μA beam incident on a $2 \times 3\,\text{mm}^2$ C_{60} surface, corresponding to ~100 primary electrons/s per C_{60} molecule, but also including many secondary electrons associated with electron scattering processes. Electron irradiation effects thus produced were also probed by STM techniques [18]. The observation that annealing at 470 K restores a polymerized film to its original monomer form supports the basic model for C_{60} polymerization discussed in connection with the phototransformation studies (Sect. 1.5.1).

1.5.3 Pressure-Induced Polymerization of C_{60}

Infrared (IR) spectroscopy studies of C_{60} in a diamond anvil cell have shown that hydrostatic pressure can induce polymerization of solid C_{60} at room temperature [20, 94]. In experiments up to 10 GPa, a continuous decrease in the intensity of the four F_{1u} IR-active lines was observed (see Sect. 1.6.1). After compression to 6 GPa, an ambient pressure spectrum revealed additional

IR lines, many of which are in good agreement with frequencies reported for photopolymerized C_{60} [15]. Thus, these IR results were interpreted [20] as evidence for a pressure-induced polymerization of C_{60}.

1.5.4 Plasma-Induced Polymerization of C_{60}

The polymerization of mixed C_{60}/C_{70} films, induced by the interaction of the film with a radio frequency (rf) Ar ion plasma (driven at 13.6 MHz) during film deposition, has been reported using laser desorption mass spectroscopy detection [19]. A pulsed nitrogen laser was used to desorb material from the fullerite film and a series of six close peaks in the mass spectrum was observed at masses $m = Nm_0$ where $N = 1, 2, \ldots, 6$, is the number of monomers in the oligomer and $m_0 = 720$ amu is the mass of a single C_{60} molecule. Furthermore, the UV-visible spectra for the plasma-transformed films [19] are similar to the absorption bands seen for the phototransformed films [95] discussed in Sect. 1.5.1.

The properties of the plasma-induced polymerized films were characterized by various physical property measurements including temperature-dependent conductivity (10^{-7} S/cm under ambient conditions), molecular weight distribution using LDMS, surface morphology using scanning electron microscopy (SEM), surface wetting properties using contact angle measurements, electron paramagnetic resonance (EPR) measurements to determine the unpaired spin concentrations ($10^{17}/\text{cm}^3$), Fourier transform infrared (FTIR), and UV-visible spectroscopy measurements to document polymerization-induced spectral changes, among others [19].

1.5.5 Photopolymerization of C_{70} Films

Experimental evidence for the UV-visible photopolymerization of C_{70} films in the absence of dioxygen has also been reported [87, 96]. Results from laser desorption mass spectroscopy, vibrational spectroscopy, and the insolubility of thin irradiated films in toluene all have been interpreted [87] as evidence for the phototransformation of C_{70} to a polymeric solid. However, the cross-section for the phototransformation in C_{70} is found to be considerably smaller than that observed for C_{60} [15], and this reduced cross-section was attributed to a reduction in the number of reactive double bonds from 30 in C_{60} to only 5 in each polar cap of a C_{70} molecule.

The tendency of C_{70} to dimerize via the formation of a four-membered ring between C_{70} shells (see Fig. 1.8) was also shown theoretically [87], using a generalized, tight binding molecular dynamics (GTBMD) simulation [96–98]. These molecular dynamics simulations find that the lowest-energy dimer configuration contains a four-membered ring between monomers, the same dimeric bonding unit as is found theoretically for the C_{60} dimer [99]. Experimentally, laser desorption mass spectroscopy has been used to show the mass distribution of the oligomers coming from a phototransformed C_{70} film. A comparison of the LDMS for phototransformed C_{60} and C_{70} mass spectra clearly show that the phototransformation of C_{70} has a much lower cross-section than that observed for solid C_{60}. In contrast to photopolymerized C_{60} [15, 91], no clear signature for the formation of intermolecular bonds between C_{70} shells has yet been observed in either the Raman spectrum or the UV-visible electronic absorption spectrum [87], though evidence for photoinduced changes in the vibrational mode spectra of the C_{70} films was detected using FTIR transmission.

1.6 Vibrational Properties

The normal modes for solid C_{60} can be clearly subdivided into two main categories: "intramolecular" and "intermolecular" modes, because of the apparent weak coupling between molecules. The former are often simply called "molecular" modes, since their frequencies and eigenvectors closely resemble those of an isolated molecule. The latter are also called lattice modes, and can be further subdivided into librational, acoustic and optic modes. All of these types of modes, except the acoustic lattice modes, appear schematically in Fig. 1.9 [14] where (a) and (b) refer, respectively, to the librational and optical lattice modes. The schematic density of states shown for these modes represents a continuum of mode frequencies, which are found throughout the entire Brillouin zone. The frequencies for the intermolecular modes are low, reflecting the weak van der Waals bonds between fullerene molecules. In the limit that the molecule is treated as a "point", the molecular moment of inertia I approaches zero, and the librational modes are lost from the spectrum. The optic modes (c) in the figure are associated with metal-doped C_{60}, where electrons are transferred from the metal atom sublattice (e.g., Na, K, Rb or Cs) to the C_{60} sublattice to form fullerene anions and alkali metal cations. Finally, (d) and (e) refer, respectively, to particular molecular modes (with H_g symmetry), that are Raman-active. The arrows indicate the C-atom displacements in each mode. The schematic density of states for all the molecular vibrations exhibits two broad peaks, with the lower and upper peaks associated, respectively, with the modes having predominantly radial and tangential molecular displacements, as discussed below. It should be noted that the overall schematic density of vibrational states in Fig. 1.9 is generally applicable to any fullerene solid.

Because of the high symmetry of the C_{60} molecule (point group I_h), there are only 46 distinct molecular mode frequencies corresponding to the $180 - 6 = 174$ degrees of freedom for the isolated C_{60} molecule, and of these only 4 are infrared-active (all with T_{1u} symmetry) and 10 are Raman-active (2 with A_g symmetry and 8 with H_g symmetry). Mode displacements associated with the two A_g modes are shown in Fig. 1.10 [6].

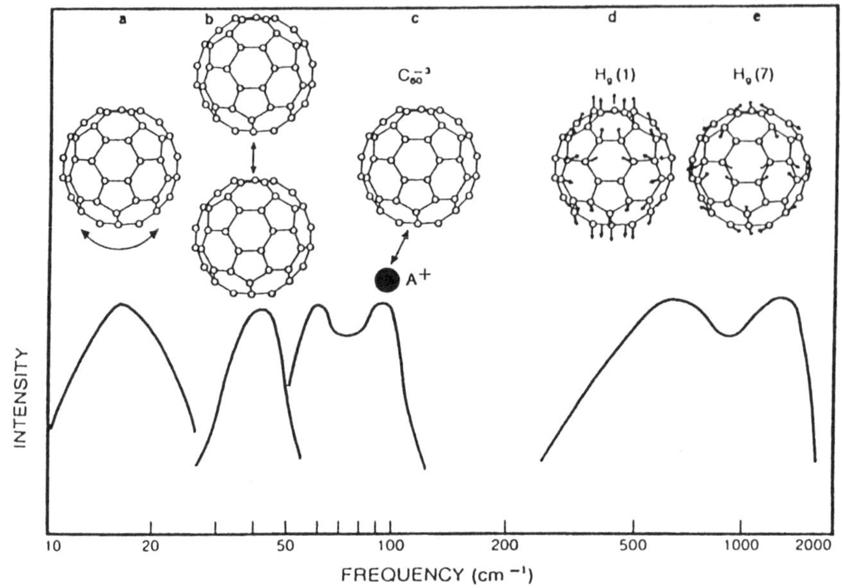

Fig. 1.9. Various vibrations in the M_3C_{60} compounds. At lower energies the compounds exhibit librational modes of individual C_{60} molecules (**a**), intermolecular modes (**b**), and optic modes between the C_{60}^{-3} anions and alkali metal cations (**c**). At higher energies the intramolecular modes are dominant. These "on-ball" modes, have radial character (**d**) at relatively lower energies (for example the $H_g(1)$ mode) and tangential character (**e**) at higher energies (for example, the $H_g(7)$ mode) [14]

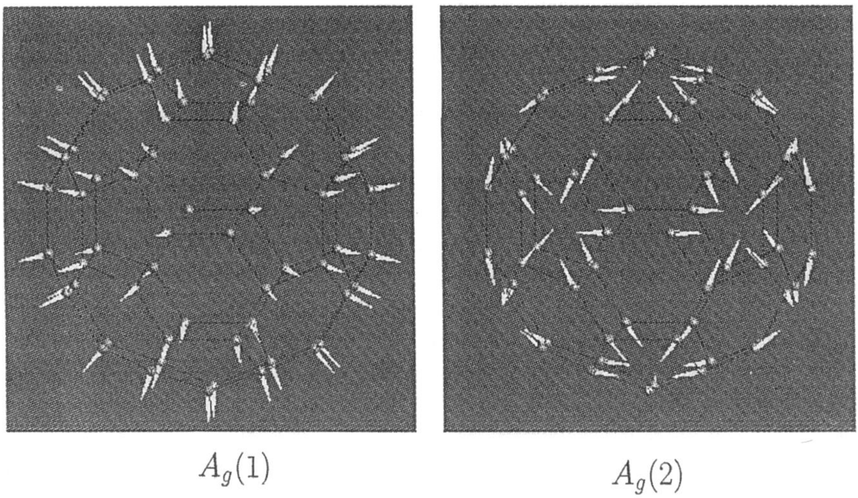

Fig. 1.10. Schematic diagram for the displacements corresponding to the "radial breathing" mode $A_g(1)$ shown on the *left* and the "pentagonal pinch" mode $A_g(2)$ shown on the *right*. For the radial $A_g(1)$ mode, the displacements for all 60 carbon atoms are in the radial direction and all are of equal magnitude and in phase. The figure was kindly supplied by Dr M. Grabow

The remaining 32 eigenfrequencies correspond to silent modes. Raman and infrared spectroscopy provide sensitive methods for determining the mode frequencies. The first-order spectra yield the 14 optically-active modes, while the combination modes and overtones of the second-order spectra can be used to obtain mode frequencies for the 32 silent modes [100–103]. Raman and infrared spectroscopy can also be used to distinguish C_{60} from higher molecular weight fullerenes with lower symmetry (e.g., C_{70} has D_{5h} symmetry). Since most of the higher molecular weight fullerenes have lower symmetry, more isomers, as well as more degrees of freedom, they have many more infrared- and Raman-active modes, as shown in Fig. 1.11 [6].

1.6.1 Infrared-active Modes in C_{60}

The simplicity of the infrared (IR) spectrum of solid C_{60} (see Fig. 1.11a), which shows four prominent lines at 527, 576, 1183, 1428 cm^{-1} [9], provides a convenient method for characterizing the compositional purity of C_{60} samples, especially with regard to their contamination with C_{70} which has 53 Raman-active and 31 infrared-active modes (see Fig. 1.11) [9, 104]. The IR spectrum of solid C_{60} remains almost unchanged relative to the isolated C_{60} molecule, with the most prominent addition being the weak feature at 1539 cm^{-1} associated with a combination mode in the second-order spectrum [105–107]. The strong correspondence between the solution and/or gas phase IR spectrum and the solid state IR spectrum is indicative of the highly molecular nature of solid C_{60}. The infrared spectra shown in Fig. 1.11 are for thin C_{60} and C_{70} films on alkali halide substrates. Figure 1.11 also shows the positions of the toluene bands indicating how IR spectroscopy might be used to monitor the presence of residual solvent (at the 1–2% level) in fullerene-based solid state samples.

1.6.2 Raman-active Modes in C_{60}

Raman spectroscopy (see Fig. 1.12) [108, 109] provides an especially valuable tool for investigating the intramolecular and intermolecular bonding in solid C_{60} and C_{60}-related compounds [11, 108–114]. The Raman spectrum in Fig. 1.12 for solid C_{60} shows 10 strong features for the Raman-allowed intramolecular modes of the free molecule. The normal modes in molecular C_{60} above about 1000 cm^{-1} involve carbon atom displacements that are predominantly tangential to the C_{60} ball surface, while the modes below ~800 cm^{-1} involve predominantly radial motion [115]. Numerous calculations of the vibrational spectra for C_{60} have been carried out [115–121], and from these calculations the symmetries and eigenvectors for the normal mode displacements for all the vibrational modes are available [114]. The low frequency acoustic and rotational-librational intermolecular modes have been studied in some detail [6, 110, 122].

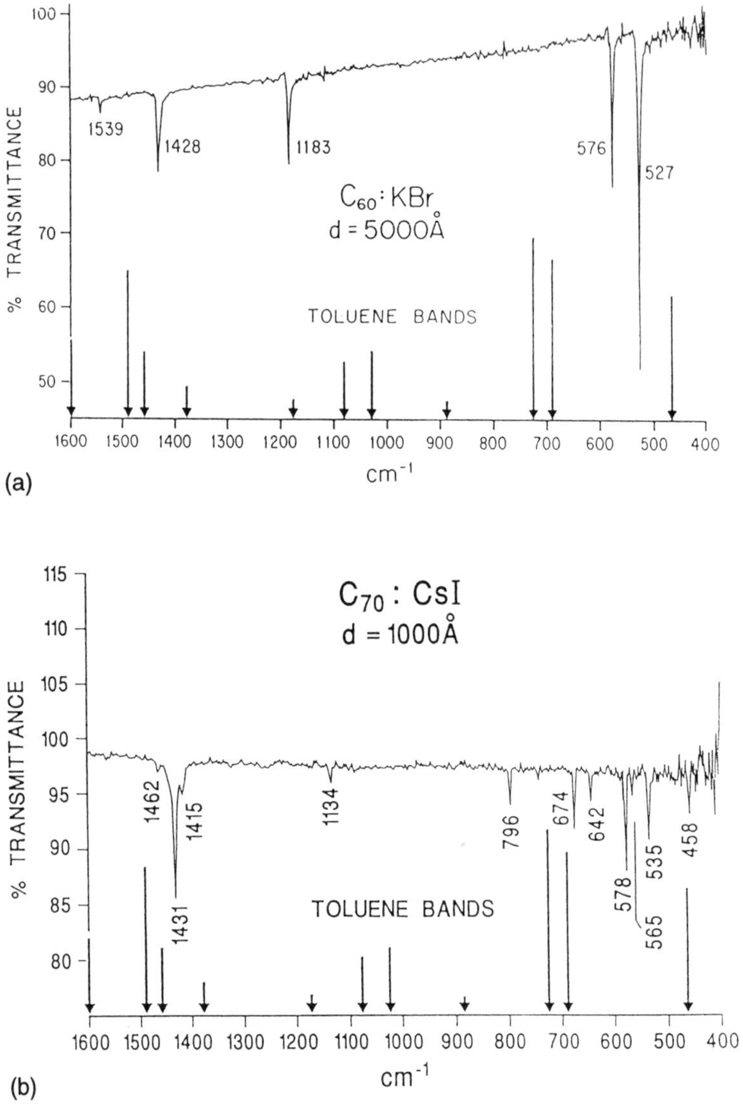

Fig. 1.11. The infrared spectra of (**a**) C_{60} and (**b**) C_{70}. Also shown schematically are the IR bands for toluene, a common solvent for fullerenes

The $A_g(2)$ mode ($1469\,\mathrm{cm}^{-1}$) shown in Fig. 1.10b [88] corresponds to tangential displacements of the 5 carbon atoms around each of the 12 pentagons and is called the "pentagonal pinch" mode. Under high laser flux from an Ar ion laser, this mode frequency downshifts to $1458\,\mathrm{cm}^{-1}$, and this downshift has been interpreted as a signature of a photoinduced structural transfor-

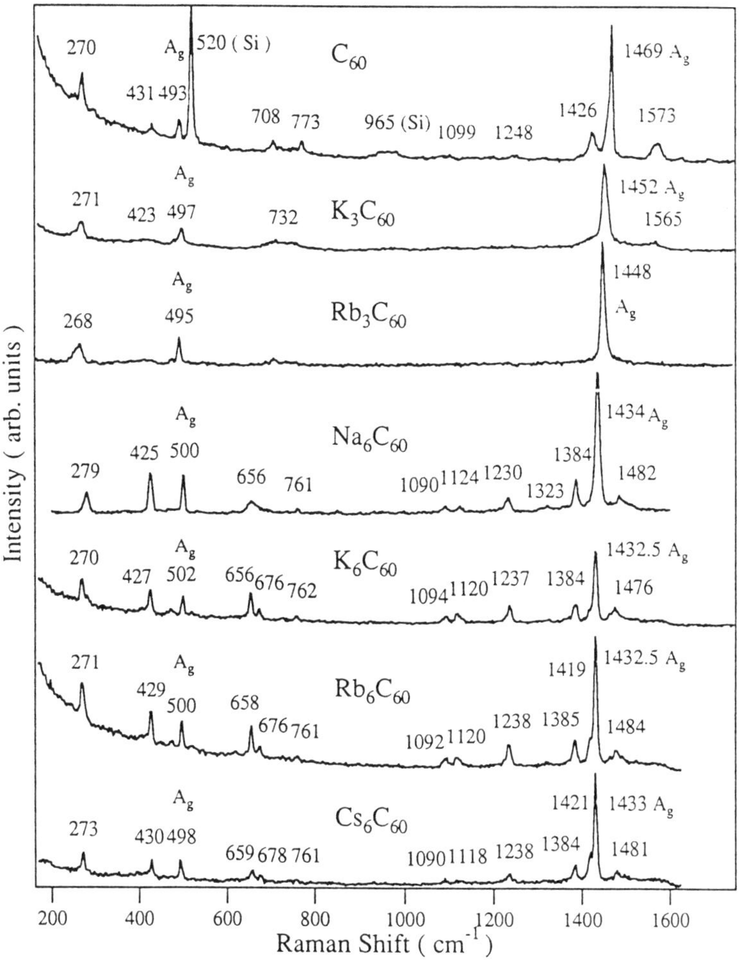

Fig. 1.12. Unpolarized Raman spectra ($T = 300\,\mathrm{K}$) for solid C_{60}, K_3C_{60}, Rb_3C_{60}, Na_6C_{60}, K_6C_{60}, Rb_6C_{60} and Cs_6C_{60} [108, 109]. The tangential and radial modes of A_g symmetry are identified, as is the feature associated with the Si substrate in the C_{60} spectrum. From these spectra it is concluded that the intermolecular interactions between the C_{60} molecules are weak, as are also the interactions between the C_{60} anions and the alkali metal cations in the case of the M_3C_{60} and M_6C_{60} compounds

mation [88]. The effect of phototransformation on the vibrational spectra of solid C_{60} is discussed in Sect. 1.6.6.

Because of the metastability of the lowest-lying triplet electronic state above the Fermi level (see Sect. 1.8), it is possible to observe the Raman spectrum in this excited triplet state. The frequency of the pentagonal pinch mode is found to be downshifted by $\sim 11\,\mathrm{cm}^{-1}$ relative to that for the vibration in the electronic ground state [110].

In the solid phase, cubic crystal field interactions are expected to lower the symmetry of the lattice modes and to cause a splitting of the five-fold degenerate H_g-derived molecular modes, into 3-fold (T_g) and 2-fold (E_g) lattice modes. The study of line splittings and their symmetry in solid C_{60} requires polarization studies and the use of single crystals [24, 105]. Studies of the temperature dependence of the Raman spectra have been used to monitor the effect of the 260 K phase transition on the line shifts, linewidths, line splittings and their symmetries [110, 114].

1.6.3 Silent Modes in C_{60}

A number of experimental techniques, including inelastic neutron scattering, electron energy loss spectroscopy, and singlet oxygen photoluminescence (PL) have provided valuable information about the silent modes of C_{60} [6]. Because of the highly molecular nature of fullerene solids, it is possible to obtain sharp higher-order Raman and infrared spectral features associated with combination modes, and, in the case of the Raman spectrum, features associated with overtones are also observed [6, 100, 101]. When symmetry selection rules and a force constant model is then used, an identification of the 32 silent modes can be made [6, 100, 101]. Since phototransformation broadens the infrared and Raman spectra, these higher-order features cannot be readily followed in phototransformed samples, so that spectroscopic studies for phototransformed films are predominately confined to observation of the first-order spectra.

1.6.4 Vibrational Spectra for C_{70}

The Raman and infrared spectra for C_{70} are much more complicated than for C_{60} because of the lower symmetry and the large number of Raman-active modes (53) and infrared-active modes (31), out of a total of 122 possible vibrational mode frequencies. Nevertheless, well-resolved infrared spectra [104, 123] and Raman spectra for C_{70} have been observed (see Fig. 1.11) [111, 123, 124]. Making use of a force constant model based on C_{60} and a small perturbation to account for the weakening of the force constants for the 10 extra carbon atoms in the belt region of the C_{70} molecule, qualitative agreement between the model calculation and the experimentally determined lattice modes [123, 125] has been achieved [6].

1.6.5 Vibrational Modes in Doped Fullerene Solids

The intercalation of alkali metal dopants into C_{60} has a relatively minor effect on the Raman spectra, as shown in Fig. 1.12 where the Raman spectrum for solid C_{60} are shown in comparison to various M_3C_{60} (M = K, Rb) and M_6C_{60} (M = K, Rb, Cs) spectra [108, 109]. One can, in fact, identify each of the lines in the M_xC_{60} spectra with those of pristine C_{60}, and very little change in the mode frequencies is found from one alkali metal dopant to another [9]. The small magnitude of the perturbation of the Raman spectrum by alkali metal doping and the insensitivity of the M_6C_{60} spectra to the specific alkali metal species indicates a very weak coupling between the C_{60} anions and the M^+ cations. In the case of the superconducting M_3C_{60} phase (M= K, Rb), the spectra (see Fig. 1.12) are again quite similar to that of C_{60}, except for a major broadening of the H_g-derived spectral features and the apparent absence in the M_3C_{60} spectra of several of the Raman lines derived from the H_g modes in C_{60}. Since the M_6C_{60} (M = K, Rb, Cs) compounds are semiconductors, the broadening effects dominate in the M_3C_{60} compounds (due to the large electron-phonon coupling) are not present in the M_6C_{60} compounds (see Fig. 1.12).

As a result of alkali metal doping, electrons are transferred to the π-electron orbitals on the surface of the C_{60} molecules, elongating the C–C bonds and down-shifting the intramolecular tangential modes. Referring to Fig. 1.13, the dependence of the Raman frequencies on M concentration x for three dominant vibrational modes in M_xC_{60} (M = K, Rb, Cs) is presented. The data in Fig. 1.13 come from several publications of the Kentucky, Vienna, and AT&T research groups [105, 126, 127]. Two of the three mode eigenvectors ($A_g(2)$, tangential: $1469\,cm^{-1}$) ($H_g(1)$, radial: $270\,cm^{-1}$) are also shown in the figure. The eigenvector for the breathing mode ($A_g(1)$ at $493\,cm^{-1}$, with radial displacements) is not shown.

The softening of the $1469\,cm^{-1}$ tangential $A_g(2)$ mode by alkali metal doping has been used as a convenient method to characterize the stoichiometry x of stable K_xC_{60} samples. This mode in M_xC_{60} is downshifted by $\sim 6\,cm^{-1}$/M atom, based on direct mode frequency measurements on the stable compounds K_3C_{60} and K_6C_{60} [105]. Since all end-point M_6C_{60} compounds (M = K, Rb, Cs) exhibit approximately the same value for this mode frequency ($1432\,cm^{-1}$), it is reasonable to expect that a downshift of $\sim 6\,cm^{-1}$/M atom is approximately applicable also for Rb and Cs. The radial $A_g(1)$ mode, on the other hand, stiffens slightly due to competing effects associated with a mode softening arising from the charge transfer effect (similar to the situation for the tangential modes) and a larger mode stiffening effect due to electrostatic interactions between the charged C_{60} anions and the surrounding charged alkali cations as their atomic separations change during a normal mode displacement [128].

The Raman spectra for Rb_1C_{60} have been investigated by several groups [76, 129] and the spectra show some unusual features, not present in the corresponding spectra for the other doped M_xC_{60} compounds (see Fig. 1.14). These unusual features are related to the special structural phases of Rb_1C_{60} (see Sect. 1.4.4). At 125°C, where Rb_1C_{60} assumes a rock salt structure con-

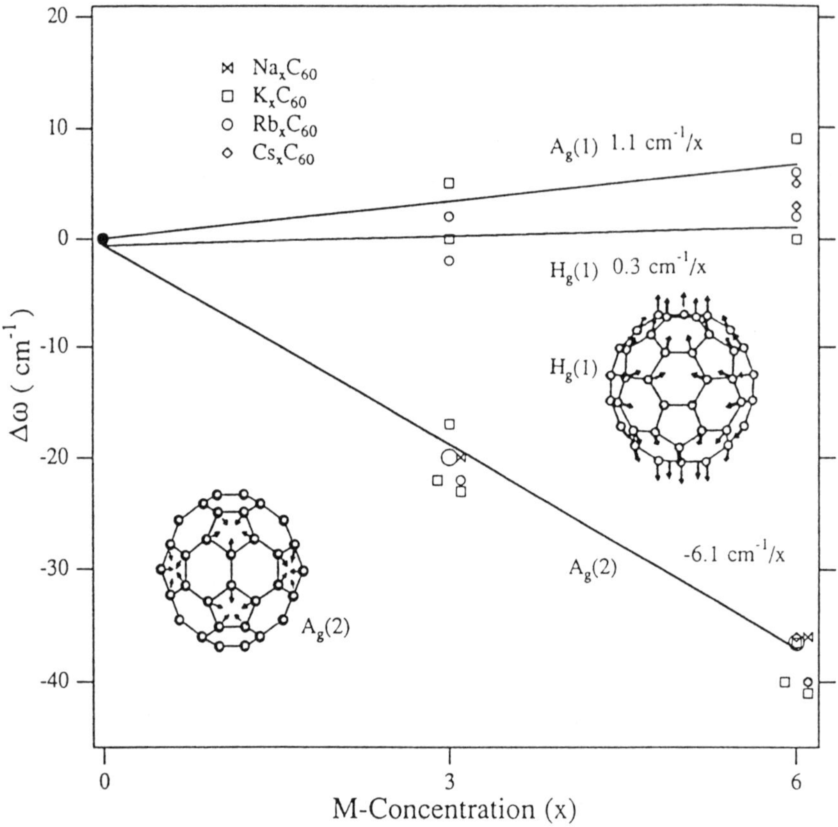

Fig. 1.13. Dependence of the frequencies of the $A_g(1)$, $A_g(2)$, and $H_g(1)$ modes on alkali metal concentration x in M_xC_{60}, where M= Na, K, Rb, Cs. The frequency shifts of these Raman-active modes are plotted relative to the frequencies $\omega(A_g(1)) = 493\,\text{cm}^{-1}$, $\omega(A_g(2)) = 1469\,\text{cm}^{-1}$ and $\omega(H_g(1)) = 270\,\text{cm}^{-1}$ at $T = 300\,\text{K}$ C_{60}. Schematic of the displacements for the eigenvectors for $A_g(2)$ "pentagonal pinch" and $H_g(1)$ modes are shown. All the C atom displacements for the $A_g(1)$ modes are radial and of equal magnitude. Data from [126] and [127] are artificially displaced to the left, and right, respectively. The centered data is from [105]

sisting of two interpenetrating fcc structures [76], the 10 Raman-active modes of C_{60} are clearly seen with little line broadening relative to the C_{60} spectrum (see Fig. 1.14). If the sample is slowly cooled to 25°C, it is believed that a polymerized phase is formed (see Sect. 1.4.4); the lower symmetry of this polymerized phase causes many more modes to become Raman-active, although the observed spectrum suggests that inversion symmetry is conserved in the 25°C spectrum (see Fig. 1.14). If instead, the Rb_1C_{60} is rapidly quenched to −130°C, additional lines then appear in the Raman spectrum

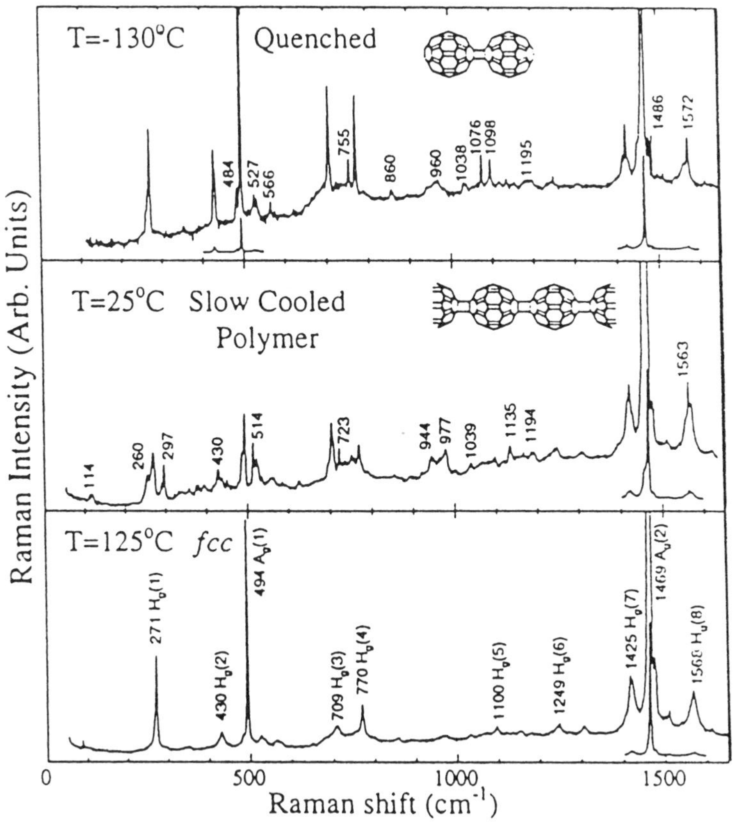

Fig. 1.14. Raman spectra of a Rb_1C_{60} sample in its rocksalt fcc structure (*bottom panel*), its polymer state (*middle*), and its "quenched" state (*top*). The respective temperatures are indicated on the curves. The 10 Raman-active modes of pristine C_{60} and their identification are labeled in the *bottom panel*; the frequencies of modes activated by the reduction in symmetry are labeled for the quenched and polymer state spectra in the *upper two panels*. The inset sketches represent the dimer and polymer structures of the C_{60} molecules in the quenched and slow-cooled states of Rb_1C_{60}, respectively [130]

(see Fig. 1.14) and these additional lines have been identified with the breakdown of inversion symmetry [130].

The introduction of alkali metal dopants into the lattice is expected to give rise to low frequency intermolecular vibrational modes, whereby the alkali metal ions vibrate relative to the large fullerene balls.

1.6.6 Vibrational Properties of Phototransformed Fullerenes

As an introduction to Chap. 4 of this volume, we discuss here the use of Raman and infrared spectroscopy (see Fig. 1.15) to probe the extent of the

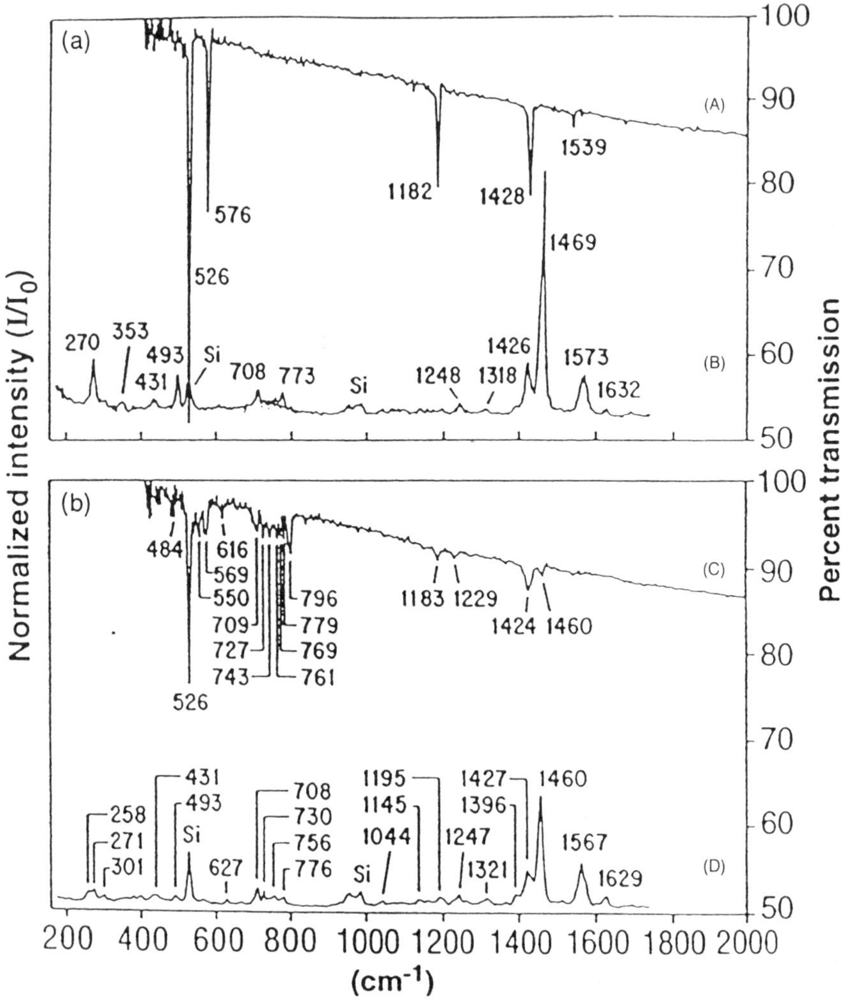

Fig. 1.15. Effect of photoinduced dimerization and polymerization on the infrared and Raman spectra of C_{60}. The *upper panel* shows infrared (A) and Raman (B) spectra for C_{60} taken with low incident optical power levels ($< 50\,\mathrm{mW/mm^2}$). The *lower panels* show the corresponding infrared (C) and Raman (D) spectra taken after exposure of the C_{60} film to intense optical power levels for several hours, until the Raman line at $1469\,\mathrm{cm}^{-1}$ disappeared. After the intense optical irradiation, many more infrared and Raman lines appear in these spectra, due to a lowering of the symmetry caused by the photoinduced covalent bonding between the fullerene molecules [15]

phototransformation that has occurred in the sample. Photopolymerization, discussed in Chap. 4 is most easily detected by Raman scattering using 488 or 514 nm excitation, as evidenced by a decrease in the intensity of the polarized $A_g(2)$ Raman line at $1469\,\mathrm{cm}^{-1}$ (intrinsic pentagonal pinch mode), and

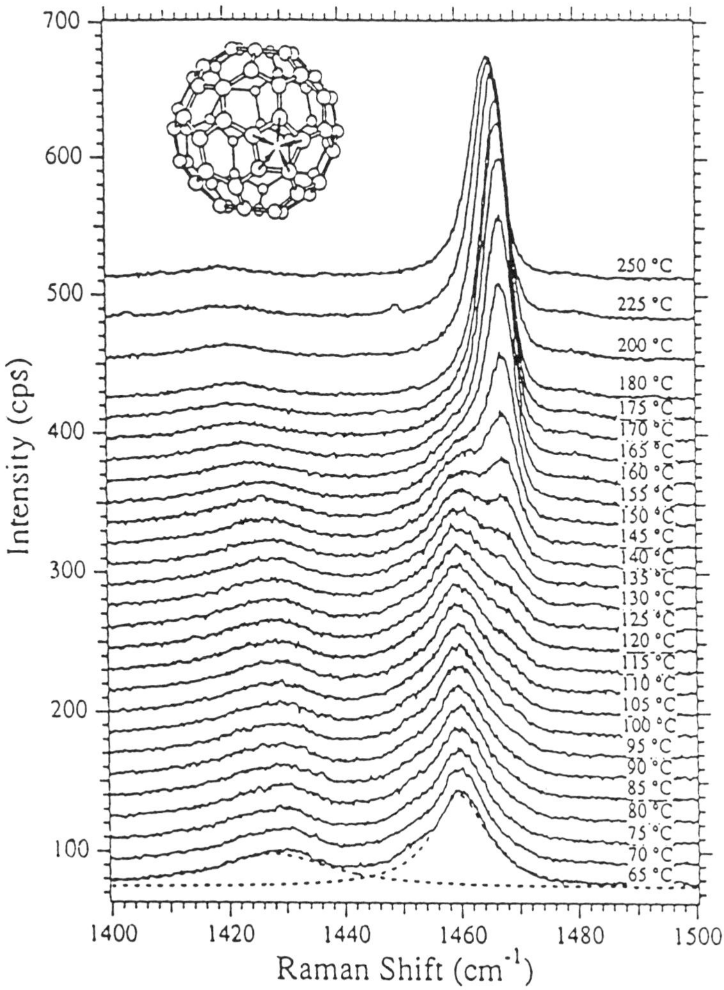

Fig. 1.16. Raman spectra in the vicinity of the pentagonal pinch mode for a photochemically polymerized C_{60} film ($d \approx 4500$ Å) on a Suprasil (fused silica) substrate for various temperatures. The *dashed curves* in the bottom spectrum represent a Lorentzian lineshape fit to each of the Raman lines in the spectrum at 65°C. As the temperature is increased, the features associated with the polymerized phase are quenched

the concomitant emergence of a broader, unpolarized peak (with shoulders) at $1458\,\mathrm{cm}^{-1}$ identified with photochemically produced dimers and higher oligomers (see Fig. 1.16). The $1458\,\mathrm{cm}^{-1}$ line continues to grow at the expense of the $1469\,\mathrm{cm}^{-1}$ line as the phototransformation proceeds to completion.

Because of the symmetry lowering of the phototransformed fullerene phases, Raman and infrared spectroscopies provide powerful tools for the characterization of the phototransformation process (see Fig. 1.15). Although Brillouin scattering can in principle be used for characterizing phototransformed C_{60}, the power levels required to carry out Brillouin scattering are so large that the Brillouin scattering probe itself introduces phototransformation of C_{60} [131].

The overall effect of phototransformation on the vibrational spectra of solid C_{60} is shown in Fig. 1.15 [15], where the infrared transmission and Raman spectra of solid C_{60} films can be compared for the pristine fcc phase (upper panel) and the phototransformed phase (lower panel). Phototransformation occurs at a moderate optical flux ($> 5\,W/cm^2$) in the visible and in the ultraviolet (UV) regions of the spectrum when light is incident on thin solid films [15]. The various effects observed in the Raman and infrared spectra in Fig. 1.15 have been interpreted [15] to be due to a photopolymerization of the lattice in which the C_{60} molecules become cross-linked by covalent bonds, rather than by the weaker van der Waals forces that normally bond C_{60} monomer molecules in the solid phase.

In Fig. 1.16 we see the Raman spectra for polymerized C_{60} in the vicinity of the pentagonal pinch mode for various temperatures. At the lowest temperature shown (65°C), the trace exhibits the characteristic spectral features of the phototransformed film, showing the broad Raman line at $1459\,cm^{-1}$ associated with the $A_g(2)$ mode of C_{60}, perturbed by the formation of dimers, trimers, and higher oligomers [92]. As the temperature is increased, the narrow $A_g(2)$ line at $1469\,cm^{-1}$ grows at the expense of the broad $1459\,cm^{-1}$ line, indicating a transformation back to the monomeric solid by a thermal detachment of C_{60} monomers from the oligomers. The competition between the photoinduced attachment of C_{60} monomers to form oligomers and the thermal detachment process to form monomers has been studied by plotting the temperature dependence of the integrated intensity of the $1459\,cm^{-1}$ feature associated with the polymerized phase and the corresponding intensity of the $1469\,cm^{-1}$ sharp Raman line associated with the C_{60} monomer [92, 132].

Additional radial and tangential molecular modes are activated (Fig. 1.15) by the apparent breaking of the icosahedral symmetry, resulting from bonds that cross-link adjacent molecules (see Fig. 1.8). A new Raman-active mode is also observed at $118\,cm^{-1}$ in phototransformed C_{60} [15], and this new mode is identified with a stretching of the cross-linking bonds *between* molecules (see Fig. 1.17a) [92]. This frequency falls in the gap between the lattice and molecular modes of pristine C_{60} in Fig. 1.9, and the magnitude of this mode frequency is in good agreement with theoretical predictions for this mode at $104\,cm^{-1}$ [99]. Figure 1.17b shows the temperature dependence of the intensity of the $118\,cm^{-1}$ mode and the suppression of this mode as the thermal detachment of C_{60} monomers from oligomer clusters proceeds with increasing temperature. The decrease in intensity of the $118\,cm^{-1}$ mode with increasing temperature is closely correlated with the attenuation of the $1459\,cm^{-1}$ Raman feature, discussed above and shown in Fig. 1.16 [92].

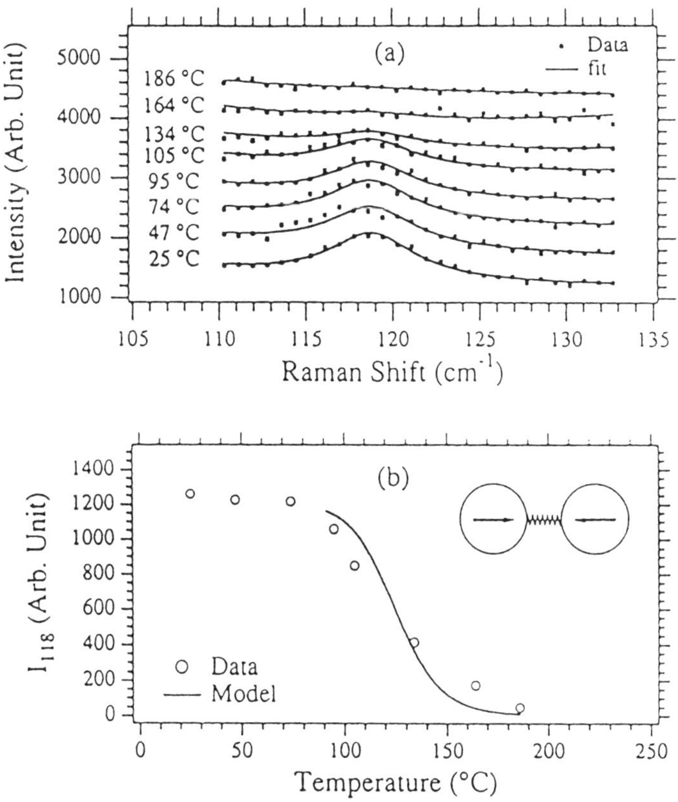

Fig. 1.17. Temperature dependence of the Raman-active $118\,\mathrm{cm}^{-1}$ intermolecular mode in a photopolymerized C_{60} film ($d \approx 4500$ Å, Si (100) substrate). (**a**) Raman spectra taken in the 110–133 cm^{-1} range at the indicated temperatures. The solid lines are the results of a Lorentzian lineshape analysis in which the peak frequency and width were fixed at the values obtained at 25°C. (**b**) Integrated intensity I_{118} of the photoinduced intermolecular mode at $118\,\mathrm{cm}^{-1}$ vs. temperature. The circles are obtained from a Lorentzian lineshape analysis and the solid curve is calculated from a rate equation model. The inset shows the intermolecular dimer C_{60}–C_{60} mode schematically [92]

1.7 Electronic Properties

On the basis of vibrational (Sect. 1.6) and optical (Sect. 1.8) spectroscopy studies, it is concluded that fullerenes form highly molecular solids. Thus, their electronic structures are expected to be closely related to the electronic levels of the isolated fullerene molecules. The electronic properties of polymers and polymer composites are reviewed in Chap. 2 and also in Chaps. 5 and 9 in this volume.

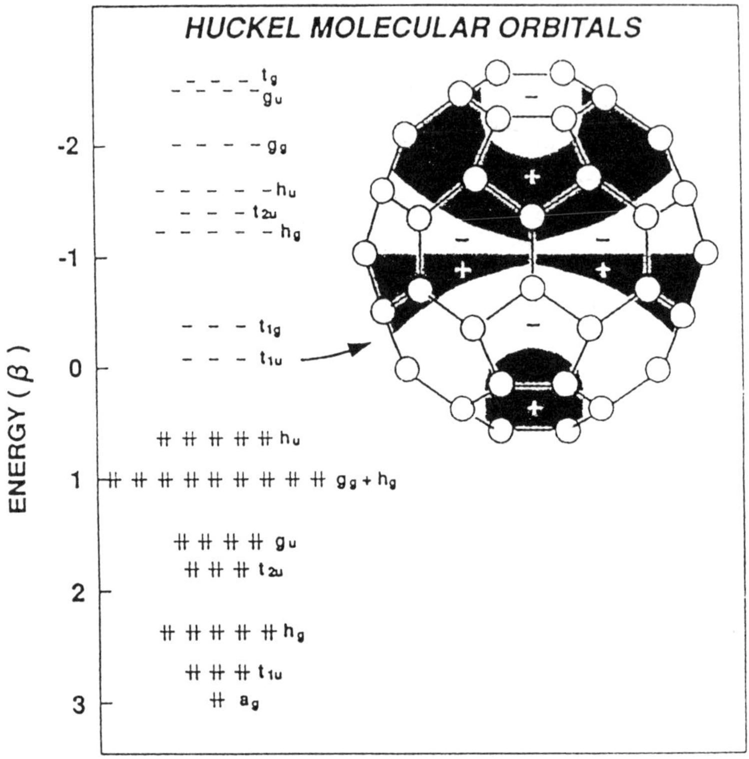

Fig. 1.18. Molecular orbital levels for C_{60} from a Hückel calculation. The filled states are labeled '+' and the empty states '−'. The shading on the icosahedron on the right-hand side of the figure pertains to whether the wave function is negative or positive, and the figure schematically shows three nodes and odd parity for the wave function for the t_{1u} (f_{1u}) triply-degenerate conduction states defining the lowest unoccupied molecular orbital (LUMO) for C_{60} [133]

The most extensive calculations of the electronic structure of fullerenes so far have been done for C_{60}. The ordering of the molecular energy levels and their icosahedral symmetry identifications based on a Hückel calculation [133] are shown in Fig. 1.18, where the filled levels are indicated as '+' and the unoccupied states by '−'. Figure 1.18, and the results of a local density calculation of the molecular energy levels for the C_{60} molecule shown in Fig. 1.19a are both widely used in the literature for discussion of the electronic structure of the C_{60} molecule. For example, the molecular orbital calculation in Fig. 1.19a implies that the bonding σ electrons lie more than 7 eV below the highest occupied molecular orbital (HOMO) of C_{60} and the antibonding σ electron states lie more than 6 eV above the Fermi level. The bandwidth of the molecular levels in Fig. 1.19 is large ($\sim$25 eV) and of similar magnitude to the corresponding states in graphite.

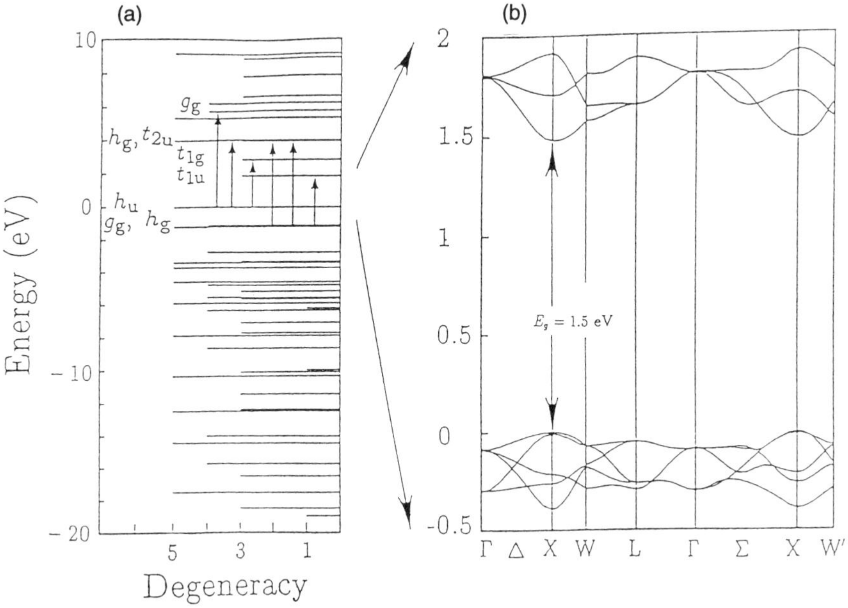

Fig. 1.19. Calculated electronic structure of an (**a**) isolated C_{60} molecule and (**b**) a face-centered cubic (fcc) solid C_{60}, where the calculated direct band gap at the X point is 1.5 eV [134]

A number of calculations of the electronic states for the free C_{60} molecule have been carried out using a variety of calculational methods [135–138], and general agreement is found with the level ordering shown in Figs. 1.18 and 1.19a. Some models have even been successful in obtaining bond lengths of 1.45 Å for a_5 and 1.40 Å for a_6 [139], in good agreement with experimental values (see Sect. 1.2.1) [140, 141]. Another relevant energy is the ionization potential for the free molecule (7.6 eV), denoting the energy needed to remove an electron from neutral C_{60} to create the C_{60}^+ cation. This energy is to be compared with the much smaller electron affinity (2.65 eV [142]), the energy needed to add one electron, giving rise to the C_{60}^- anion. Whereas the electron affinity for C_{60} is similar to that for other acceptor molecules (e.g., TCNQ, for which the electron affinity is 2.82 eV), the large ionization potential for C_{60} makes it unlikely for C_{60} to become an electron donor.

Representative results for the electronic levels of the free C_{60} molecule are shown in Fig. 1.19(a) where it is seen that because of the molecular nature of solid C_{60}, the electronic structure for the solid phase is expected to be closely related to that of the free molecule [143]. It has further been shown that a molecular approach can explain a large body of experimental observations pertinent to the optical properties of C_{60}, which are difficult to explain on the basis of a one-electron energy band picture.

According to the molecular approach, many-electron-orbitals for the "valence" and "conduction" electron states are constructed for the π-electrons on the C_{60} molecule, since the lower-lying σ orbitals are filled. The number of π-electrons on the C_{60} molecule is 60. In relating the levels of icosahedral C_{60} to those with full rotational symmetry, we note that 50 π-electrons fully occupy the angular momentum states through $\ell = 4$, and the remaining 10 electrons are available to start filling the $\ell = 5$ state.

In full spherical symmetry, the $\ell = 5$ state can accommodate 22 electrons, and the splitting of the $\ell = 5$ state in icosahedral symmetry is into the $h_u + f_{1u} + f_{2u}$ irreducible representations. The level of lowest energy by Hund's rule is the 5-fold h_u level which is completely filled by the 10 available electrons. Neglecting any thermal excitation, the two 3-fold f_{1u} and f_{2u} (many authors refer to these levels as t_{1u} and t_{2u}) levels in icosahedral symmetry are empty. Whether C_{60} is considered in terms of a spherical approximation or its icosahedral symmetry, the ground state is non-degenerate with a total angular momentum of $J = 0$. This special circumstance has been suggested as being in part responsible for the high stability of the C_{60} molecule [144]. Since the h_u highest occupied molecular orbital (HOMO) for C_{60} is fully occupied and the f_{1u} lowest unoccupied molecular orbital (LUMO) is completely empty, C_{60} is a semiconductor.

Transport measurements show that C_{60} is a semiconductor with a very high resistivity ($> 10^8 \Omega$-cm), though the measured value of the resistivity is strongly dependent of the oxygen uptake [6]. However, by doping C_{60} with alkali metals, alkaline earths, or other donor species, the resulting charge transfer greatly reduces the resistivity and in some cases can yield metallic conduction. Because of the similar parity for the HOMO and LUMO levels (see Fig. 1.18), electric dipole transitions across the band gap are symmetry forbidden. Thus, the optical absorption at the absorption edge is weak in intensity and involves vibronic excitations to excitonic states both for the free molecules and in the solid state.

Assuming one electron to be transferred to the C_{60} molecule per alkali-metal atom dopant, the LUMO levels are expected to become half-occupied at the alkali metal stoichiometry M_3C_{60} and totally occupied at M_6C_{60}, leading to a filled-shell configuration for the $f_{1u}(t_{1u})$ band (see Fig. 1.19). Thus, M_6C_{60} would be expected to be semiconducting with a band gap between the f_{1u} and f_{1g} (t_{1u} and t_{1g}) levels (see Fig. 1.19), while M_3C_{60} should be metallic, provided that no band gap is introduced at the Fermi level by a Peierls distortion. A large increase in optical absorptivity in M_xC_{60} with alkali metal doping is observed as $x \to 3$, and this observation supports the metallic nature of M_3C_{60}, in agreement with the large increase in electrical conductivity observed at M_3C_{60} for M = K, Rb. Below ~0.5 eV, the free electron contribution to the optical properties of M_3C_{60} is dominant, and above ~0.5 eV, interband transitions from the partially occupied f_{1u} level to the higher lying f_{1g} level take place. It should be emphasized that the level-filling arguments for M_xC_{60} pertain both to the free molecule and to the levels in the solid state.

Fermi surface calculations for K_3C_{60} indicate a hole Fermi surface around the Γ-point of the Brillouin zone, and this hole surface contributes about 12% to the density of states at the Fermi surface [145]. These calculations also indicate a larger and more complicated Fermi surface consisting of both electron and hole orbits which contribute the remaining 88% to the density of states. Calculated values for m^* based on a band model yield $1.3m_e$ (where m_e is the free electron mass) for the conduction band carriers of C_{60}, and $1.5m_e$ and $3.4m_e$ for the valence band carriers.

Because of the weak van der Waals interaction of the fullerene molecules with each other and with the alkali metal dopants, solid M_3C_{60} can be described as a molecular solid, having energy levels with little dispersion, and thus giving rise to a very high density of states near the Fermi level. This property is important in understanding superconductivity in the K_3C_{60}, Rb_3C_{60}, and related compounds (see Sect. 1.10).

1.8 Optical Properties

The optical properties of C_{60} have a number of interesting features of particular importance to polymerized fullerenes and fullerene-polymer composites (see Chap.7.4). Since the optical transitions near the HOMO-LUMO gap are symmetry-forbidden, their absorption strengths are consequently very low, and the absorption process thus involves excitonic and vibronic states. Further, the high photosensitivity of C_{60} leads to the phototransformation of fullerite films (see Sect. 1.5.1).

1.8.1 Introduction to Molecular Photophysics

In this section we discuss the photophysical properties of isolated C_{60} molecules. Isolation of the C_{60} molecules can be realized in the gas phase and can also be achieved approximately for molecules in solution. Since C_{60} is soluble in a wide variety of solvents [6], numerous optical studies on solutions have been carried out to investigate the "photophysics" of this unusual molecule. By a "photophysical" process, it is usually meant that an electronic excitation is achieved through the absorption of a photon, without a resulting chemical change (e.g., fragmentation, bond rearrangement). Photophysical processes can either be unimolecular or bimolecular (i.e., involve one or multiple molecules), and they can involve the absorption or emission of one or several photons. In the case of C_{60}, and to some degree C_{70}, photochemical transformation to a dimer or oligomer has been reported for fullerenes in the solid state, where the molecules are in closer contact [15, 16, 87, 93, 146]. These photochemical changes are, however, much less likely (but also possible) in solution.

Photoexcited electronic states of molecules, in general, include spin triplet ($S = 1$) and spin singlet ($S = 0$) states arising from the excited electron and the hole that is left behind. The conventional notation denoting singlet and triplet states in photophysical processes is as follows [147]:

S_0		ground singlet state
S_1		first excited singlet state
S_p	$(p > 1)$	higher excited singlet states
T_1		first excited triplet state
T_q	$(q > 1)$	higher excited triplet states.

Because of the coupling of vibrational and electronic levels by the electron–molecular vibration (EMV) interaction, a ladder of "vibronic" states with approximate energy

$$E_n = E_i + (n + 1/2)\hbar\omega_j \tag{1.1}$$

is realized, where E_i and $\hbar\omega_j$ are, respectively, particular (renormalized) electronic and vibrational energies, and n is an integer $n = 0, 1, 2, \ldots$. In Fig. 1.20, we show schematically optical transitions between the ground state vibronic ladder and an excited state vibronic ladder for a generic molecule. The transition $1 \rightarrow 0$ is commonly called a "hot band" because elevated temperatures are needed to populate the $n = 1$ vibrational level in the ground electronic

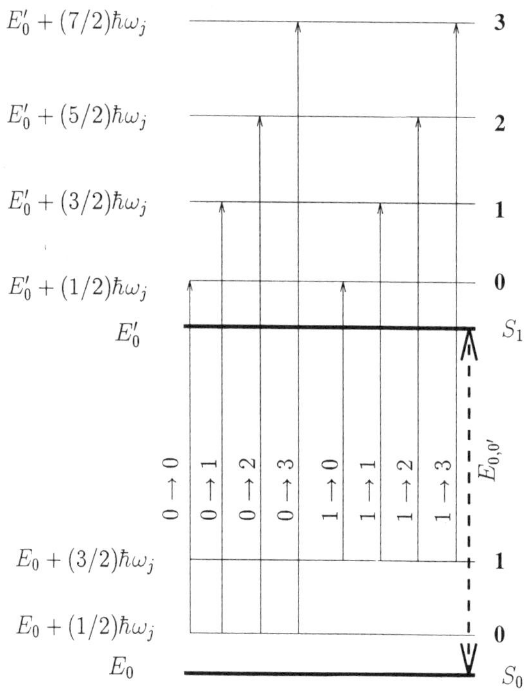

Fig. 1.20. Schematic level diagram for vibronic absorption transitions from the singlet S_0 electronic ground state to the first excited singlet electronic state S_1. The symbols $n \rightarrow m$ denote the optical transitions between vibronic levels associated with the initial and final electronic states E_0 and E_0', respectively

state (e.g., a $250\,\mathrm{cm}^{-1}$ molecular vibration corresponds to a temperature of 360 K). Note that the vibrational frequency of a particular vibrational mode in the electronic ground state should differ slightly from the value of this mode in the excited electronic states, because of a softening of the harmonic potential due to the excited state electronic configuration. The energy $E_{0,0'}$ (see Fig. 1.20) denotes the energy difference between the excited and ground electronic zero vibrational states, while $E_{n,m'}$ denotes the energy difference between the vibrational states of E_n and E'_m, where the prime notation is used to denote the excited electronic state (as shown in Fig. 1.20). These vibronic ladders can be generated from either triplet or singlet electronic states. Such vibronic states are required in the case of C_{60} to understand the low-lying weak, optical absorption, near the absorption edge.

In Fig. 1.21, we display a generic energy level diagram for an isolated molecule, showing only the $n = 0$ vibronic states (i.e., only the base of each vibronic ladder of Fig. 1.20). For clarity, the singlet and triplet states in Fig. 1.21 are drawn to the left and right, respectively. The molecule depicted in Fig. 1.21 has a singlet ground state (S_0), as does C_{60}. Of particular interest to us here are the following transitions:

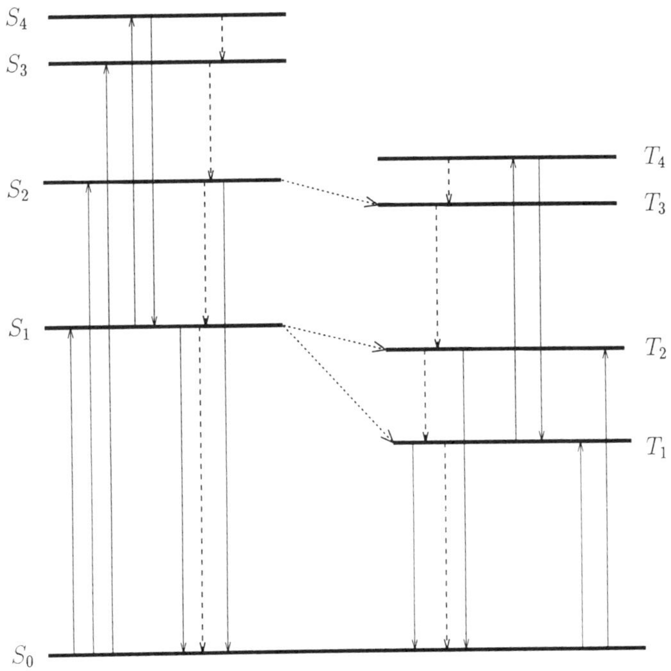

Fig. 1.21. Level diagram for optical absorption and emission transitions between singlet and triplet molecular states. Solid lines denote optical transitions, and dashed lines denote radiationless transitions. The dotted lines denote intersystem crossings [6]

1. Excited state absorption ($S_1 \to S_p$ and $T_1 \to T_p$; $p > 1$), which are spin-conserving transitions, and therefore can be quite strong (shown in Fig. 1.21 as solid lines).
2. Fluorescence ($S_1 \to S_0$),
3. Phosphorescence ($T_1 \to S_0$).
4. Internal conversion ($S_p \to S_{p-1}$ and $T_p \to T_{p-1}$), which are radiationless transitions (shown as dashed vertical lines in Fig. 1.21).
5. Intersystem crossing ($S_p \to T_q$ for $p > 0$, or vice versa), which requires the spin-orbit interaction to mix levels with differing spin (the selection rule is that there is no symmetry change for the electronic wave function in an intersystem crossing (dotted lines)).

Typically, internal conversion is a rapid, spin-conserving process which returns the photoexcited molecule to either an S_1 or T_1 state. From these states, luminescence, either fluorescence (S_1) or phosphorescence (T_1), can occur. Because phosphorescence is not a spin-conserving transition (and the spin–orbit interaction is needed to couple the T_1 and S_0 levels), the radiative lifetime of the T_1 state is long. Duration times for phosphorescent emission from a collection of molecules are typically $\sim$ 1–10^4 ms, while typical duration times for fluorescent luminescence are considerably shorter, $\sim$ 1–10^3 ns [6].

The most sensitive measurement of the optical absorption spectra is to probe the spectral properties of the optical density (or absorbance) which is defined as $\log_{10}(1/\text{transmittance})$. Results for the optical absorbance shown as the optical density vs wavelength in the ultraviolet (UV)-visible range (200 – 700 nm) for C_{60} and C_{70} in hexane [148] are depicted in Figs. 1.22a and b, respectively. The strong absorption bands in the figure (below $\sim$400 nm in wavelength or above $\sim$2.9 eV in energy) are identified with electric dipole-allowed transitions between occupied (bonding) and empty (antibonding) molecular orbitals, denoted in Fig. 1.19 by either the $h_u \to t_{1g}$ (or f_{1g}) transition or the $h_g \to t_{1u}$ (or f_{1u}) transition. At longer wavelengths $490 < \lambda < 640$ nm (or $1.9 < E < 2.5$ eV), Fig. 1.22 show that weak absorption takes place, and is associated with electric dipole-forbidden transitions between the one-electron HOMO level with h_u symmetry and the one-electron LUMO level with t_{1u} (or f_{1u}) symmetry (see Fig. 1.19). Using many-electron molecular states [149], the fully occupied ground HOMO state is identified with a singlet S_0 state (having quantum numbers $L = 0$, $S = 0$, $J = 0$ and A_g symmetry).

The lowest parity-forbidden HOMO→LUMO optical transition is denoted by the configurational transition $h_u^{10} \to h_u^9 f_{1u}^1$. The ground state ($h_u^{10}$) is a totally filled h_u multiplet which can accommodate 10 electrons, and by Hund's rule, has ^1A_g symmetry and is a singlet state ($S = 0$). In the lowest excited state, one electron is promoted to an excited state (f_{1u}), leaving one hole behind (h_u^9), each with spin 1/2, so that the interacting electron and hole can either form a singlet S_i state ($S = 0$), or a triplet T_i state ($S = 1$). Since

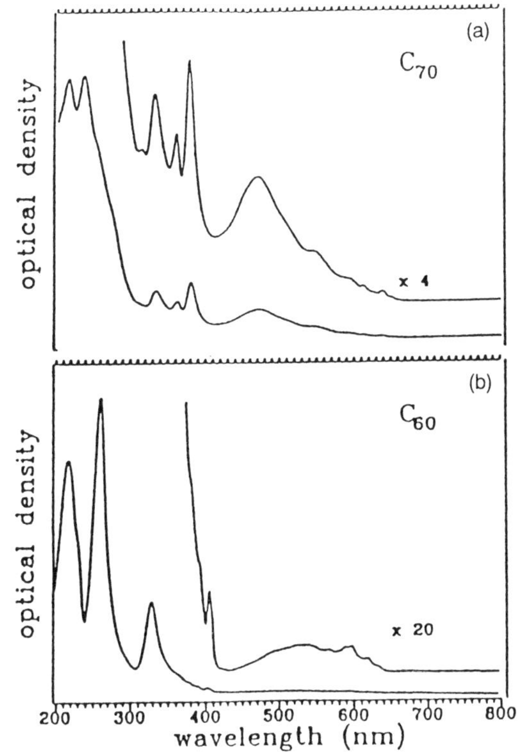

Fig. 1.22. Optical density vs wavelength for (**a**) C_{60} and (**b**) C_{70} in hexane solutions and the characteristic absorption bands are identified [148]

the electric dipole matrix element for optical transitions $\mathcal{H}'_{\rm em} = (\mathbf{p} \cdot \mathbf{A})/(mc)$ does not depend on spin, there is a strong spin selection rule ($\Delta S = 0$) for optical electric dipole transitions, so that the strong optical transitions are either between singlet states or between triplet states. This strong selection rule arises from the very weak spin-orbit interaction for carbon, which has a low atomic number ($Z = 6$) [6].

Having satisfied the spin selection rule, the orbital selection rule must now be considered. To satisfy the electric dipole selection rule, appropriate odd-parity vibrational modes must be admixed with the initial and (or) final electronic states for the optical transition (see Fig. 1.23). Thus the weak absorption below 2.9 eV involves optical transitions between appropriate vibronic levels. In group theoretical language, the symmetries for the various states in the $h_u^9 f_{1u}^1$ excitonic state configuration are found by taking the direct product of $H_u \otimes F_{1u}$ which requires the final excited state to have one of the following symmetries: F_{1g}, F_{2g}, G_g or H_g. Each of these four excited state configurations must be combined with appropriate odd-parity vibrations (see Sect. 1.6) to satisfy the electric dipole selection rule and allow coupling to the A_g ground state [150–154].

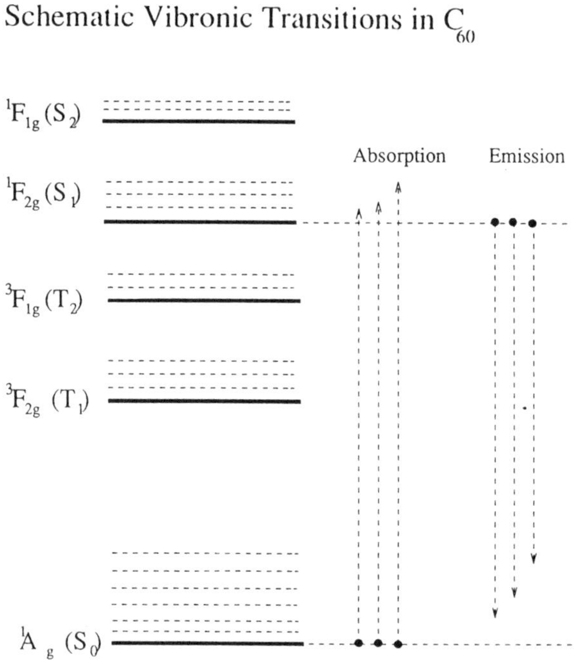

Fig. 1.23. Schematic diagram of the excitonic vibronic transitions in C_{60}^0 between an h_u^{10} ground state with ${}^1A_g$ symmetry and an $h_u^9 f_{1u}^1$ excited state on absorption and on emission

On the basis of these arguments, the weak absorption observed between 490 and 640 nm for C_{60} in solution (Fig. 1.22a) [148] is assigned to transitions between the singlet ground state S_0 (also denoted as ${}^1A_g$) and the lowest excited singlet state S_1. Figure 1.23 also shows the vibronic manifold associated with the lowest excitonic state. The lowest allowed vibronic transition is to a state with ${}^1F_{2g}$ symmetry. From the long wavelength limit (~640 nm) of this weak absorption in solution, the energy of the S_1 state relative to the HOMO ground state is estimated to be ~1.9 eV.

For C_{70}, molecular orbital calculations [134] reveal a large number of closely-spaced orbitals both above and below the HOMO-LUMO gap $E_{HL} = 1.65\,\text{eV}$ [134]. The large number of orbitals make it difficult to assign particular groups of transitions to structure observed in the solution spectra of C_{70}. UV-visible solution spectra for higher fullerenes (C_n; $n = 76, 78, 82, 84, 90, 96$) have also been reported [50, 51, 155] and molar extinction coefficients for C_{60} and C_{70} have been obtained at several wavelengths [156].

Further insight into the electronic structure is provided by pulsed laser studies of C_{60} and C_{70}. Such time-resolved studies of fullerenes in solution have been used to probe the photodynamics of the optical excitation/luminescence spectra. The importance of these dynamic studies is to show that photoexcitation in the long-wavelength portion of the UV-visible spectrum leads to the promotion of C_{60} from the singlet S_0 ground state (^1A_g) into a singlet S_1 excited state, which decays quickly with a nearly 100% efficiency [147, 157] via an intersystem crossing to the lowest excited triplet state T_1 (see Fig. 1.21), which has the same orbital quantum numbers as the excited S_1 state. This rapid singlet-triplet decay (~33 ps [157]) is fostered by the overlap in energy between the vibronic manifold of electronic states associated with the lowest singlet S_1 state and the corresponding vibronic manifold for the triplet T_1 state.

Once in the T_1 triplet excitonic manifold, a very rapid transition to the lowest level of the T_1 triplet manifold occurs (denoted in Fig. 1.23 as $^3F_{2g}$). The energy difference between the lowest excited singlet (S_1) and triplet (T_1) states is the exchange energy which for C_{60} has been found to be 0.29 eV [6, 157]. Also of significance is the long life of the lowest triplet state ($> 2.8 \times 10^{-4}$ s) relative to the lowest singlet state (1.2 ns) in the room temperature solution spectra [158]. The value of the T_1 triplet lifetime varies widely from one study to another, presumably because of the high sensitivity of this lifetime to the presence of oxygen, which has been shown for different solutions to reduce the triplet lifetime from < 40 μs to ~300 ps [157, 159]. Since the triplet T_1 lifetime is much longer than the singlet S_1 lifetime, a metastable T_1 state with a relatively high population can be established by optical pumping, leading to interesting nonlinear optical properties such as optical limiting [6, 160–163] and photo-induced excited state optical absorption [6, 160–163].

Further evidence for the existence of a metastable T_1 excited state comes from observations of a downshifted $A_g(2)$ Raman-active pentagonal pinch mode at 1458 cm^{-1} identified with vibrations from a C_{60} molecule in the excited T_1 state [110]. The quenching of the T_1 state by the presence of oxygen can be explained by spin-orbit coupling that is induced by the dioxygen (which has a triplet ground state), allowing a rapid nonradiative decay of the T_1 excited state for C_{60} to its ground state [164].

Most optical studies on pristine fullerene solids have been carried out in transmission on thin solid films deposited on various substrates (such as quartz, Si, KBr). As mentioned above for the spectra near the optical absorption edge, the UV-visible transmission spectra for C_{60} and C_{70} solid films are observed to be remarkably similar to the respective solution spectra, thus providing further evidence for the molecular nature of fullerene solids. This is shown for example in Fig. 1.24 in which the transmittance for solid C_{60} is compared to C_{60} dissolved in decalin [43]. The optical properties of solid C_{60} and C_{70} have been studied over a wide frequency range using the variable angle spectroscopic ellipsometry (VASE) technique [43, 165]. From the VASE measurements, the complex refractive index $\tilde{n}(\omega) = n(\omega) + ik(\omega)$ is determined, which is related to the optical dielectric function $\epsilon(\omega) = \epsilon_1(\omega) + i\epsilon_2(\omega) = \tilde{n}^2(\omega)$. The results for $\epsilon_1(\omega)$ and

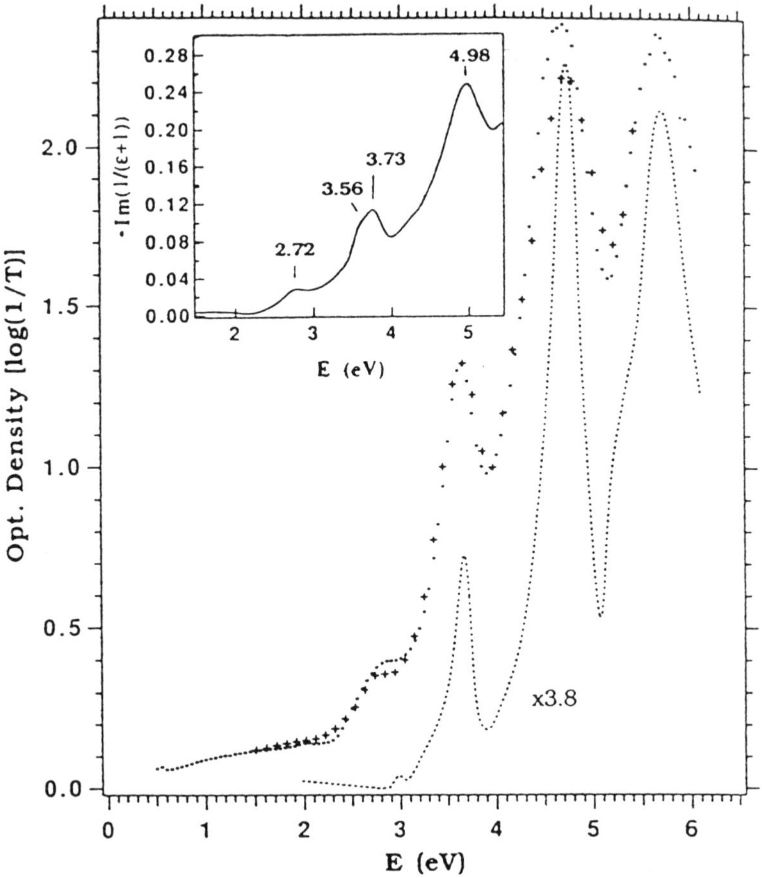

Fig. 1.24. Optical density of solid C_{60} on a Suprasil substrate derived from ellipsometry measurements (+) together with the frequency dependent optical density measured directly for C_{60} on Suprasil by normal incidence transmission spectroscopy (•). For comparison, the solution spectrum for C_{60} dissolved in decalin (dotted curve) is shown below the spectra for the films [43]. The inset is a plot of $-\mathrm{Im}[(1+\epsilon)^{-1}]$ vs photon energy E for comparison of the peaks in the optical data with the peaks in the high-resolution electron energy loss spectrum (HREELS) shown in the inset

$\epsilon_2(\omega)$ obtained from VASE and transmission-FTIR studies on thin solid films of C_{60} at $T = 300\,\mathrm{K}$ are shown on a logarithmic frequency scale in Fig. 1.25 [166]. In Fig. 1.25 is shown the frequency dependence of the dispersive part [$\epsilon_1(\omega)$; upper panel] and the absorptive part [$\epsilon_2(\omega)$; lower panel] of $\epsilon(\omega)$ for solid C_{60} films at room temperature, obtained from a variety of optical experiments, as discussed below [15]. The data are plotted on a semilog frequency scale and cover the frequency range from the IR to the ultraviolet (~0.05 to 30 eV). Above 7 eV, data for $\epsilon_1(\omega)$ obtained from electron energy loss (EELS) studies is included. The strong, sharp structure at low energy is identified with

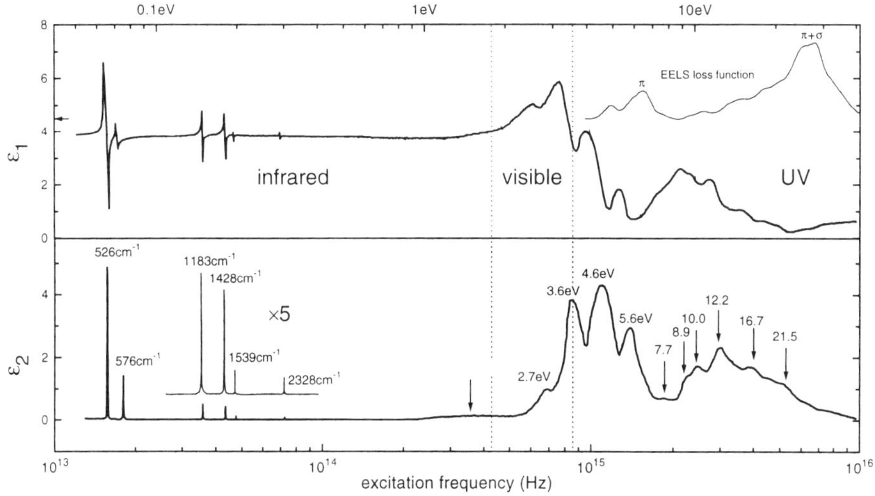

Fig. 1.25. Summary of real $\epsilon_1(\omega)$ and imaginary $\epsilon_2(\omega)$ parts of the dielectric function for C_{60} vacuum-sublimed solid films at room temperature over a wide frequency range. The data between 0.05 and 0.5 eV (mid- to near-infrared) were collected using the Fourier transform infrared (FTIR) transmission technique[15]. The visible-UV range was investigated by variable angle spectroscopic ellipsometry (VASE) [43] and near-normal-incidence reflection and transmission experiments[44]. UV data above ~7 eV were obtained using electron energy loss spectroscopy (EELS)[167] by Kramers–Kronig analysis of the EELS loss function (inset). The arrow at the left axis points to $\epsilon_1 = 4.4$, the low-frequency value of the dielectric constant determined by capacitance measurements [168]

infrared-active optic phonons and at higher energies the structure is due to electronic transitions.

From the VASE studies of C_{60} and C_{70}, values for the static dielectric function $\epsilon_1(0) = [n(0)]^2$ are obtained: $\epsilon_1(0)$ =3.61 (for C_{60}) [43] and $\epsilon_1(0)$=3.94 (for C_{70}) [165], to be compared with the value $\epsilon_1(0)$=4.4 ±0.2 for C_{60} from capacitance measurements [168] and the value $\epsilon_1(0)$=4.1 from fringe structure in the optical transmission data of C_{60} solid films.

For alkali metal-saturated C_{60} solid films (e.g., M_6C_{60}: M = K, Rb, Cs), similar transmission and reflection spectra are obtained, largely insensitive to the dopant or intercalate species (M) [44]. These results give strong evidence for only weak hybridization between M and C_{60} states. The optical spectra are thus consistent with a molecular picture in which complete charge transfer of the alkali metal s-electrons takes place to fill a lower lying, six-fold degenerate C_{60} band (F_{1u} symmetry). The narrow spectral features in M_6C_{60} observed in $\epsilon_2(\omega)$ for $\omega < 3$ eV have been assigned to transitions between narrow energy bands (bandwidth between 0.3 and 0.5 eV) or broadened molecular states whose spacing is in general agreement with theory [145, 169]. Photo-induced absorption [6, 170] and optical absorption between molecular anion states [6] show that the energy separation between the F_{1u}-derived and F_{1g}-derived states is ~1.1 eV.

Because of the high delocalization of electrons on the shell of fullerenes and the relatively high isolation between molecules, fullerene-based materials are expected to exhibit large nonlinear optical effects such as third-order nonlinear optical susceptibility, comparable to those for π-conjugated organic polymers [171]. Since the nonlinear polarizability for polymers depends on the fourth power of the length of the polymer chain, C_{70} would be expected to exhibit an even larger nonlinear response than C_{60} [172], in agreement with experiment. These nonlinear effects have been extensively studied by degenerate four-wave mixing (DFWM) [172–175], third harmonic generation [176–179], and electric field–induced second harmonic generation [180], both in solution and in high-quality films [6].

1.8.2 Optical Transitions in Photopolymerized C_{60}

In this section we summarize results for the optical absorption (OA) and photoluminescence (PL) spectra of photopolymerized C_{60} solid films. The features in the PL spectra of C_{60} films are observed to broaden and downshift by ~330 cm^{-1} upon phototransformation, whereas the OA features upshift and broaden considerably. These results are consistent with the reduced symmetry imposed by the introduction of covalent bonds between C_{60} molecules. Photopolymerization by the exciting source may also be the cause for the previously reported [181] problems in the spot-to-spot reproducibility in the photoluminescence spectra of C_{60}.

In Fig. 1.26(a) we compare the room temperature absorbance ($\mathcal{A} = -\log_{10}(\mathcal{T})$ where $\mathcal{T}$ is the transmission) of a C_{60} film ($d \sim 500$ Å) on a Suprasil substrate in the pristine (solid curve) and phototransformed (dashed curve) phases. The four prominent absorption peaks in Fig. 1.26a at ~2.7, 3.6, 4.7, and 5.6 eV for the pristine phase are identified with predominantly "dipole-allowed" electronic transitions [44].

After photopolymerization of the sample, the optical bands associated with these optical transitions are noticeably broadened and reduced in peak intensity (Fig. 1.26a). This broadening (by ~0.3 eV) is attributed to a random photochemical cross-linking of C_{60} molecules, giving rise to inhomogeneous broadening. Cross-linking the molecular shells completely removes the degeneracies of the electronic energy levels in the C_{60} monomers. Also, the phototransformation generates a distribution of C_{60} oligomeric units. Both of these changes in the system should contribute to line broadening.

In addition to the photoinduced broadening of the absorption bands in polyfullerene, the peaks at ~3.6, 4.7, and 5.6 eV in pristine solid C_{60} are noticeably blue-shifted in the polymer. It is difficult to judge whether the ~2.7 eV feature is shifted because the absorption band in the polymer is too broad to locate its center. To visualize the effect of photopolymerization on the lowest energy states of crystalline C_{60}, we refer to the schematic diagram

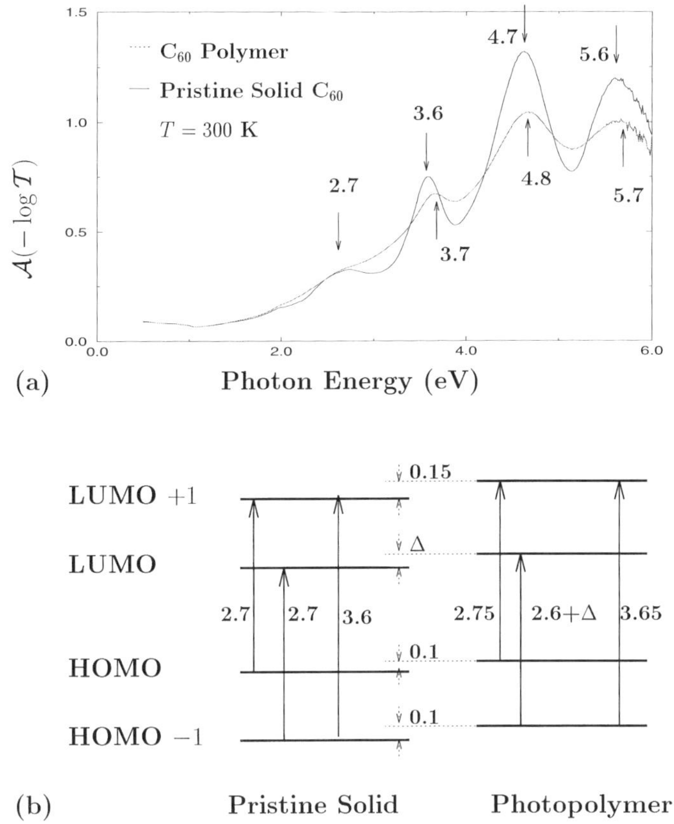

Fig. 1.26. (**a**) Optical absorbance $\mathcal{A}$ for pristine and photopolymerized C_{60} solid films (thickness ~500 Å) on a Suprasil (fused silica) substrate in the photon energy range 0.5–6.0 eV. The same C_{60} film was used for both spectra. (**b**) Schematic diagram of the electronic states and optical transitions of pristine and photopolymerized solid C_{60} near the Fermi level using simple molecular orbital notation. The numbers denote the transition energies (in eV) at the center of the optical absorption bands [44]

in Fig. 1.26b that was used to interpret the absorption spectra in Fig. 1.26a. The solid horizontal lines on the left in Fig. 1.26b represent the peak positions of the calculated Gaussian density of state (DOS) for the contributing band of electronic states [169], and allowed optical transitions and transition energies (in eV) are indicated in Fig. 1.26b. To explain the observed OA blue shift upon phototransformation, and to be consistent with the photoemission data on the photopolymer [182], it is concluded that the electronic levels of the phototransformed C_{60} are upshifted with respect to the electronic levels of pristine C_{60}, as shown in Fig. 1.26b.

1.9 Transport and Photoconductivity

The transport properties of fullerene solids that have been studied include electrical conductivity (dc up through microwave frequencies), Hall effect, magnetoresistance, photoconductivity, and thermoelectric power. Most attention has focused on studies of the electrical conductivity and photoconductivity. This section provides background for Chap. 8 and 10 in this volume which reviews the transport properties of polymers and fullerene-polymer composites.

1.9.1 Electrical Conductivity

The room temperature resistivity of the undoped fullerene solids is high with $\rho \sim 10^{14}\,\Omega$-cm reported for undoped mixtures of C_{60} and C_{70} films exposed to air [183] and $\rho \sim 10^{8}\,\Omega$-cm for oxygen-free C_{60} films [184]. Almost no carriers are available for transport in C_{60} unless they are thermally or optically excited, or more importantly if carriers are introduced by donor dopants, such as the alkali metals, which are efficient in providing electron charge transfer to the fullerenes and in creating carriers near the Fermi level. Because of the degenerate ground states of the fullerene anions (see Fig. 1.20), Jahn–Teller distortions of the molecule may occur, thereby contributing to the overlap of the wave functions between adjacent fullerenes and to the enhancement of their transfer integral.

Because of the high-resistivity of undoped C_{60}, the doping of C_{60} with alkali metals decreases the electrical resistivity ρ of C_{60} by many orders of magnitude. As x in M_xC_{60} increases, the resistivity $\rho(x)$ decreases and eventually approaches a minimum at $x = 3.0 \pm 0.05$ [185, 186], corresponding to a half-filled f_{1u}-derived (t_{1u}-derived) conduction band. Then, upon further increase in x from 3 to 6, ρ again increases, as is shown in Fig. 1.27 [184]. It should be noted that stable crystallographic K_xC_{60} phases occur only for $x = 0, 1, 3, 4$, and 6 (see Sect. 1.4.4). The compounds corresponding to filled molecular levels (C_{60} and M_6C_{60}) are the most stable and exhibit maxima in the resistivity of M_xC_{60} (M = K, Rb) as a function of x, consistent with a filled band.

At the minimum resistivity in M_xC_{60}, the value of ρ found in Fig. 1.27 for K_3C_{60} ($2.5 \times 10^{-3}\,\Omega$-cm) is high, typical of a high-resistivity metal [186]. Conduction is believed to occur by charge transfer from one C_{60} molecular ion to another along the (110) directions through the weak overlap of the wave functions on adjacent C_{60} anions, perhaps enhanced by a Jahn–Teller distortion of the molecule to lower the degeneracy of the ground state. Consistent with the band structure studies [135], nuclear magnetic resonance (NMR) Knight shift studies of ^{39}K [187] indicate that the charge transfer from the alkali metal dopants of the C_{60} is complete, so that the alkali metal ions do not participate in electrical conduction. From the charge transfer and the stoichiometry of the compound, the carrier density can be inferred. Then

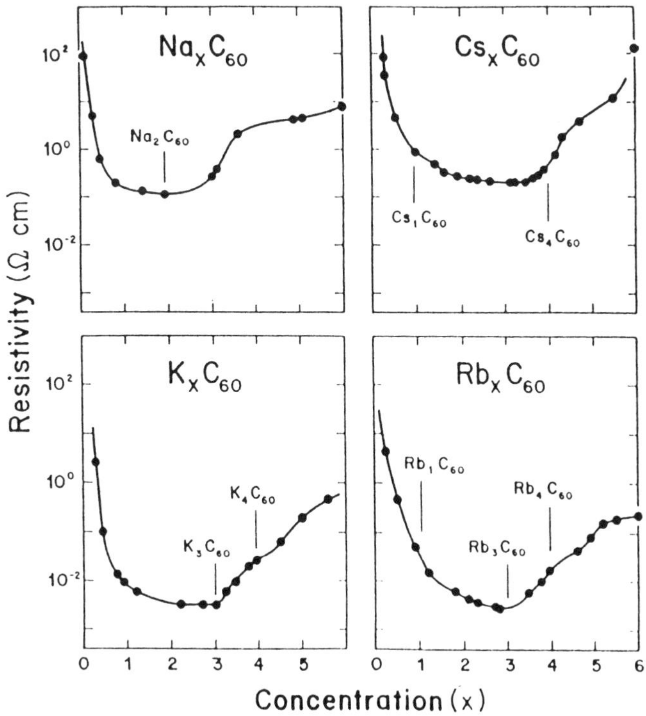

Fig. 1.27. Dependence of resistivity $\rho(x)$ on alkali metal concentration for thick films of C_{60} doped with Na, K, Rb, and Cs. Points indicate where exposure to the alkali-metal source was stopped and where x-ray and ultraviolet photoemission spectra were acquired to determine the concentration x. The labels indicate the known fulleride phases at 300 K. The minima in $\rho(x)$ occur for stoichiometries corresponding to Na_2C_{60}, K_3C_{60}, Rb_3C_{60}, and $Cs_{3.5}C_{60}$. Structure in $\rho(x)$ can be associated with the development of different stable phases [184]

assuming three electrons per C_{60}^{3-} anion in K_3C_{60}, the carrier density is expected to be $\sim 4 \times 10^{21}/\text{cm}^3$, which is quite low for conducting systems. Of the various stoichiometries and alkali metal species for the bulk compounds indicated on Fig. 1.27, only the compounds close to the K_3C_{60} and Rb_3C_{60} stoichiometry exhibit metallic temperature coefficients of the resistivity.

The high value of ρ_{min} arises in large part from the merohedral disorder which is present because icosahedral symmetry has no fourfold axis, thus giving rise to the random occurrence of the two "standard" orientations for the 2-fold axes of the C_{60} molecules (see Sect. 1.4.1). Since the intermolecular hopping matrix element depends strongly on relative intermolecular orientations of adjacent fullerene molecules, merohedral disorder significantly reduces the intermolecular hopping matrix element, and provides an important scattering mechanism for doped C_{60} [188–190].

The $\rho_{\min}$ values for Na_xC_{60} and Cs_xC_{60} in Fig. 1.27 are 0.11 Ω-cm and 0.20 Ω-cm, respectively, much higher than $\rho_{\min}$ for K_xC_{60} and Rb_xC_{60}, and the different behavior for Na_3C_{60} and Cs_3C_{60} relative to K_3C_{60} and Rb_3C_{60} has been related to the size of the Na^+ and Cs^+ cations [13, 94]. All compounds and phases in Fig. 1.27 show a local maximum in resistivity at $x = 6$, corresponding to the filling of the t_{1u} band, if one electron per alkali metal dopant is transferred to the C_{60} anions.

Since the alkali metal–doped fullerenes are chemically unstable in air and are reactive with other chemical species, great care needs to be exercised in sample handling and in the execution of transport measurements. In general, the electrodes are attached prior to alkali metal doping, and the doping procedure is carried out under the constraints of maintaining the integrity of the electrodes. Because of the small magnitude of the mean free path ℓ (~2–31 Å) in the M_3C_{60} compounds and because ℓ tends to be comparable to the lattice constant a_0 (14 Å) and to the superconducting coherence length ξ_0 (~25 Å), the effects of crystal defects and grain boundaries are very important, so that film and single-crystal samples tend to yield somewhat different detailed transport behavior, especially regarding the magnitude of $\rho_{\min}$ and ℓ and the temperature dependence of $\rho(T)$.

There have been reports of the transport in the M_1C_{60} polymerized materials [191, 192]. The temperature dependence of the polymerized K_1C_{60} shown in Fig. 1.28 provides evidence for a metal-to-semiconductor phase transition near 50 K. The same metal-to-semiconductor phase transition has been observed in polymerized Rb_1C_{60} and Cs_1C_{60} [191]. Between 400 and 50 K, the temperature dependence of $\rho(T)$ is metal-like and $\rho(T)$ decreases smoothly with decreasing temperature. Under applied hydrostatic pressure, ρ decreases, and the 50 K transition is suppressed. No anomalous magnetoresistance is observed up to 7 T and no evidence for superconductivity is found above 1.9 K [192]. The transport and optical properties of the polymeric phase of Rb_1C_{60} is reviewed in detail in Chap. 6.3.4 of this volume.

1.9.2 Photoconductivity

Many measurements of the photoconductivity of undoped C_{60} films have been published [183, 193–197], and a few measurements have also been reported on undoped C_{70} films [196, 198] and potassium-doped C_{70} films [196]. In addition, photoconductivity studies have been carried out on polymers that have been doped with small amounts of fullerenes [22, 199–206]. Wide disagreement is, however, found in the published photoconductivity results because of four dominant reasons. First, the magnitude of the photoconductivity signal is reduced by several orders of magnitude by the presence of oxygen, which acts as a trapping center for carriers; it is therefore important to work with oxygen-free fullerene films for measurements of the intrinsic photoconductivity [207–209]. Second, the persistent photoconductivity effect [210–212] implies that photocarrier relaxation times in fullerenes can extend to days.

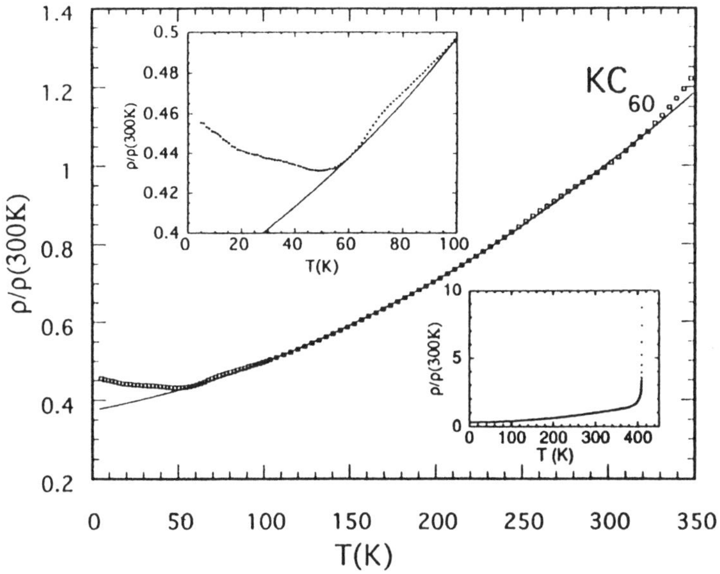

Fig. 1.28. Normalized dc electrical resistivity $\rho(T)$ for KC_{60}. The solid line is the empirical form $\rho(T) = a + bT + cT^2$, fitted to the data between 300 and 100 K and extended throughout the range of the data. The inset in the *lower right* shows resistivity data over an extended temperature range. The inset in the *upper left* shows the low-temperature resistivity: the material shows a transition from a metallic to a semiconducting phase near 50 K [192]

These extremely long relaxation times must be taken into account when using light pulses for photoexcitation and detection. Third, excessive incident light intensities can lead to photopolymerization of the fullerene film [15], thereby modifying the material during the measurement process. Finally, photoconductivity measurements show sensitivity to the degree of crystallinity and to the concentration of defects [6].

Oxygen-free C_{60} has a very high quantum efficiency (~55%), which is defined as the number of photogenerated carriers per absorbed photon [195, 197]. This high quantum efficiency, which is attributed to the long lifetimes of excited carriers in C_{60}, can be reduced by several orders of magnitude by oxygen uptake. Electrons are easily trapped at the interstitial octahedral sites where the oxygen molecules collect [213], giving rise to shallow traps for electron–hole recombination [197]. Electrons have been shown to be the dominant photocarriers in C_{60} films [209, 214]. The photoconductivity shows a wide range of rise times and decay times ranging from hundreds of picoseconds to days, with different mechanisms involved as the time scale is varied. As the different timescales are probed with various experimental techniques, different photocarrier excitation mechanisms come into play. It is interesting that the transient photoconductivity is very large as compared to the steady-state photoconductivity (by up to nine orders of magnitude) [215]. The transient photoconductivity of a pristine C_{60} film consists of short- and longer-lived components with relaxation times of 693 ps and 7.2 ns, respectively [216].

In contrast to the short relaxation times obtained in transient photoconductivity experiments, a persistent photoconductivity (PPC) effect has been observed in C_{60} films corresponding to photogenerated carrier decay times as long as $\sim 10^6$ s [211]. The PPC effect is observed as a metastable increase in dark conductivity by up to an order of magnitude upon exposure of the sample to light (see Fig. 1.29). The PPC effect in C_{60} persists to 260°C (which is the highest recorded temperature for this effect for any materials system), and it is excited by photons with energies above the weak absorption edge (see Sect. 1.8.1) [211].

Photoconductivity can be a very sensitive probe of changes in the density of gap states in a semiconductor. To determine these changes and the recombination process, the photoconductivity is studied as function of tem-

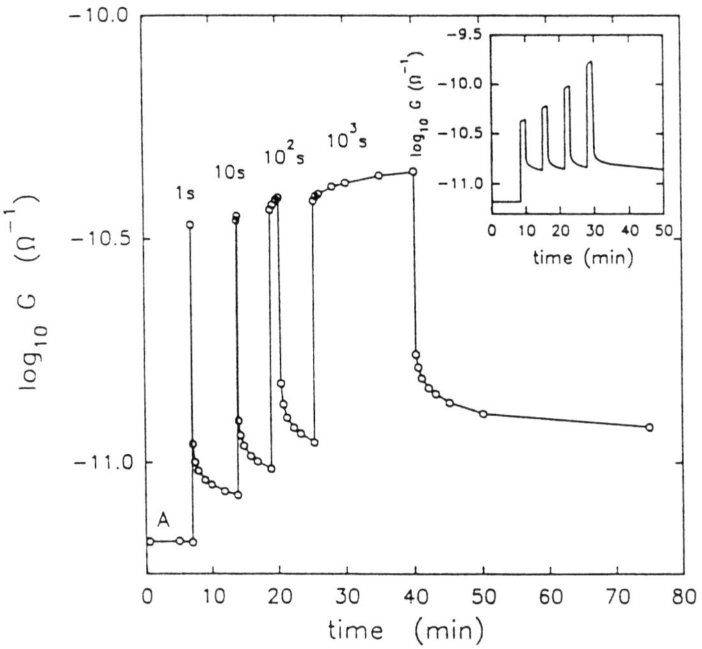

Fig. 1.29. Conductance G (on a log scale) as a function of time of a C_{60} thin film before, during, and after exposure to 2 mW/cm^2 heat-filtered white light for increasing exposure times from 1 s to 1000 s. The inset shows the effect on the conductance (on a log scale) of illuminating the film for the same duration (1 min) with increasing light intensities of 2, 6, 20, and 60 mW/cm^2 [211]

perature and light intensity. Since the presence of oxygen decreases the dark conductivity of C_{60} films by four orders of magnitude and decreases the photocurrent by a factor of $\sim$400 [209], the relative importance of dark carriers and photocarriers in the recombination process will sensitively depend on oxygen content [196, 217, 218].

It appears that photoconductivity in C_{60} is related to carrier generation via a direct absorption process. In the presence of oxygen, rapid carrier electron–hole recombination occurs at shallow oxygen trap levels, lowering the free carrier lifetime and yielding a low quantum efficiency. When oxygen is rigidly excluded from the sample, the quantum efficiency is high and carrier transport between C_{60} molecules occurs via a hopping mechanism, consistent with the dominant mechanism for dark carrier transport. The explanation of why the spectral dependence of the photoconductivity follows that of the absorption coefficient presents a challenge, since no energy seems to be required to overcome the exciton binding energy which lowers the energy of the absorption edge relative to the HOMO–LUMO gap in an optical absorption or luminescence process. One explanation for the absence of the exciton binding energy from the photoconductivity process relates to the rapid formation of a self-trapped polaron by the photoexcited hole, consistent with the interpretation of optically detected magnetic resonance (ODMR) and electron paramagnetic resonance (EPR) measurements [6]. This self-trapped polaron is strongly localized on a single fullerene molecule and in the polaron state the hole is screened from its photo-generated electron by a charged cloud from the π electrons on the C_{60} molecule. Thus the photogenerated electron can participate in electrical conduction using the same transport mechanism as for the dark carriers.

Much less has been reported about the photoconductivity of C_{70} relative to C_{60} [196, 198]. The photoconductivity for C_{70} may, however, be significant for applications because of the onset of the photoconductivity effect at lower photon energies in the technologically important longer-wavelength regime [196, 219].

Photocarrier excitation plays an important role for charge transfer in fullerene doped polymers. In this process photoinduced charge transfer creates C_{60} anions, leaving holes on the polymer chains for conduction. A schematic band diagram useful for understanding the mechanism for the photoinduced electron transfer reaction between a polymer semiconductor and C_{60} is shown in Fig. 1.30. As proposed [199, 203], the incident photon is absorbed across the semiconducting gap of the polymer. This excitation drives a rapid structural relaxation of the polymer ($< 10^{-13}$ s) due to electron–phonon coupling, thereby creating self-trapped polarons whose energy levels are in the semiconducting gap. The upper polaron level strongly couples to the LUMO of C_{60}, and electron transfer on a time scale of $\sim 10^{-12}$ s produces a metastable charge separation. The hole (positive polaron) then is free to drift away from the C_{60} anion that was produced, provided that the

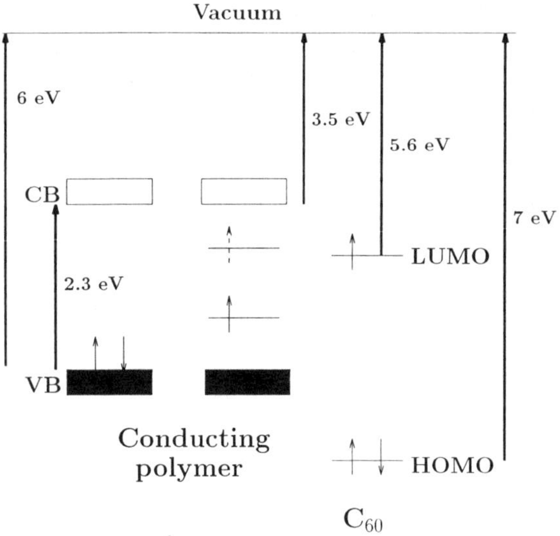

Fig. 1.30. Schematic energy level diagram for the photoinduced electron transfer from a semiconducting polymer on to C_{60} [203]

interaction between the hole and the C_{60} anion is sufficiently screened. Thus the process shown in Fig. 1.30 represents a metastable photodoping process. These concepts are discussed in detail in Chap. 3.

1.10 Superconductivity

This section reviews superconductivity in doped fullerenes. The superconductiviting properties of fullerene composites is covered in Chap. 8.8.

Perhaps the most striking property of the C_{60}-related materials is the observation of relatively high temperature superconductivity ($T_c \sim 40\,K$). The first observation of superconductivity in an alkali metal-doped carbon material was in 1965, when superconductivity was observed in the first stage alkali metal graphite intercalation compound (GIC) C_8K [220]. Except for the novelty of observing superconductivity in a compound having no superconducting constituents, this observation did not attract a great deal of attention, since the T_c was very low ($\sim 140\,mK$) [221]. Later, higher T_c values were observed in GICs using superconducting intercalants (e.g., $KHgC_8$, for which $T_c = 1.9\,K$ [222]), and in subjecting the alkali metal GICs to pressure (e.g., NaC_2, for which $T_c \sim 5\,K$) [223].

The early observation of superconductivity in doped fullerenes was in K_3C_{60} (with $T_c \sim 18\,K$) [11] and this work was soon followed by observations of superconductivity at even higher temperatures: in Rb_3C_{60} ($T_c = 30\,K$) [224, 225], $Rb_xCs_yC_{60}$ ($T_c = 33\,K$) [73], and Cs_3C_{60} ($T_c = 40\,K$) [13]. As the quality of the materials have improved, the reported transition temperatures for a given compound have risen slightly and the widths of the transition (ΔT_c) have decreased in some cases. An increase in T_c from 18 K to 40 K was achieved in going to compounds with larger intercalate atoms resulting in unit cells of larger size and with larger lattice constants (see Fig. 1.31) [60, 72, 139, 226]. As the lattice constant increases, the coupling between adjacent C_{60}^{3-} anions decreases, narrowing the bandwidth of the LUMO level, and thereby increasing the corresponding density of states. Several experiments [229] and calculations [230] provide supporting evidence for an increased density of states at the Fermi level resulting from an increase in lattice constant. Figure 1.31 shows an empirical, nearly linear, relation between T_c and the lattice constant a_0 for superconducting alkali-metal doped samples with a_0 greater than that for undoped C_{60} (see Table 1.2) [88]. This correlation includes compounds derived from alkali-metal dopants, alloys of different alkali metals [226] and samples under pressure [228, 235, 236]. Because of the close connection between the electronic density of states at the Fermi level $N(E_F)$ and the lattice constant a_0, plots of T_c vs $N(E_F)$ similar to Fig. 1.31 have been made [230]. Thus, the increase in T_c is consistent with the BCS expression relating the transition temperature to the density of states $N(E_F)$

$$T_c \sim \omega_{ph} \exp(-1/VN(E_F)), \tag{1.2}$$

where V is the electron–phonon coupling energy. For the donor C_{60} compounds intercalated with Li, Na and Ca for which a_0 falls below that for C_{60} itself (see Table 1.2), the correlation between T_c and a_0 breaks down [72, 226, 237], because of the attractive electrostatic interaction introduced by the presence of the alkali metal ions in the lattice. The effect of pressure on K_3C_{60} and Rb_3C_{60} [13, 228, 235] reduces T_c in accordance with an approximately linear dependence of T_c on a_0, as shown by the open points in Fig. 1.31. It has also been noted that Li_3C_{60}, Na_3C_{60} and Cs_3C_{60} do not show superconducting properties [226].

The reason why the T_c is so much higher in the M_3C_{60} materials relative to other carbon-based materials appears to be closely related to the high density of states that can be achieved at the Fermi level when the t_{1u} (or f_{1u}) LUMO molecular level is half filled with carriers. However, several authors have found an enhancement in the electron-phonon coupling in C_{60}-related materials [238–241]. If the electron-phonon interaction provides the pairing mechanism for superconductivity in C_{60}-based materials, it is believed [238, 240] that it is the H_g-derived intramolecular phonons that play the dominant role in the coupling. The observation of broad H_g-derived Raman lines [105, 113] in M_3C_{60} is consistent with a strong electron-phonon coupling. Calculations of the contributions of the various phonon modes to the electron-phonon interaction strength [239] indicate that the high frequency intra-molecular phonons may be able to account for the high T_c values that are observed experimentally. A variety of experimental techniques including infrared reflectivity [242], NMR [243, 244], and muon spin relaxation [245]

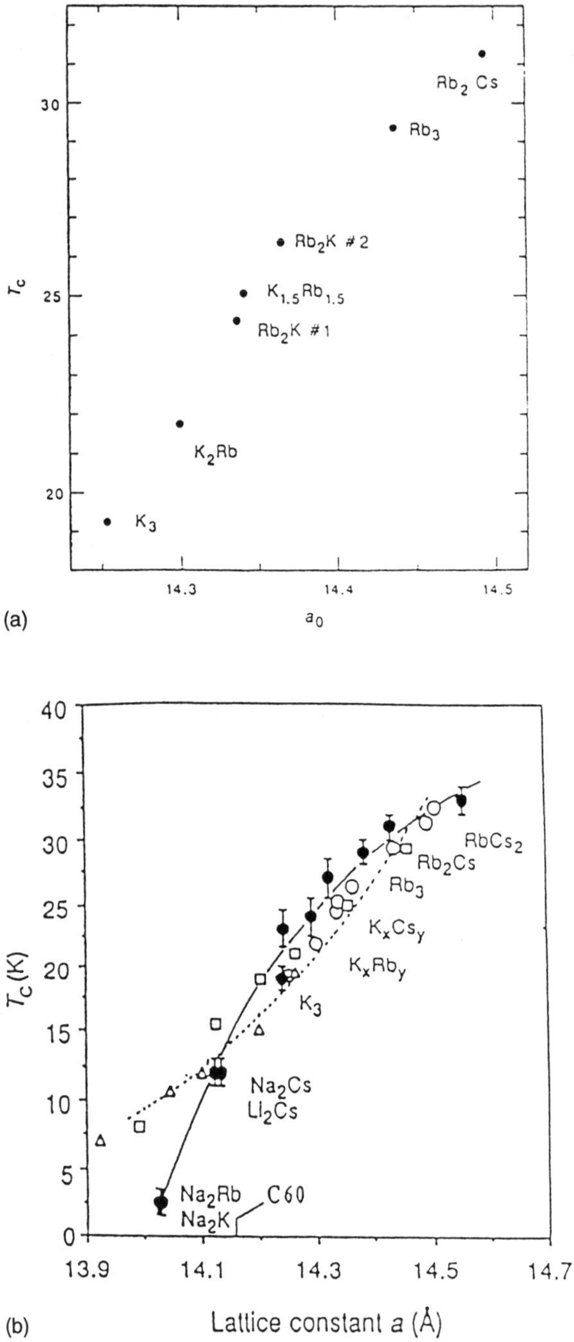

Fig. 1.31. (**a**) Early reports on the dependence of T_c for various $M_xM'_{3-x}C_{60}$ compounds on the fcc lattice constant a_0 [60]. (**b**) More complete summary of the dependence of T_c on a_0 [227], including points provided by pressure-dependent studies of T_c [228]

Table 1.2. Lattice constants and superconducting T_c values for alkali metal M_3C_{60}, C_{60}, and related compounds[a]

M_3C_{60}	Dopant site sym.[b]	a (Å)	T_c (K)	References
C_{60}	–	14.161	–	[60]
Li_2RbC_{60}	Li(T)Li(T)Rb(O)	13.896	<2.0	[227]
Li_2CsC_{60}	Li(T)Li(T)Cs(O)	14.120	12.0	[226]
Na_2KC_{60}	Na(T)Na(T)K(O)	14.025	–	[231]
Na_2RbC_{60}	Na(T)Na(T)Rb(O)	14.028	3.5	[231]
Na_2CsC_{60}	Na(T)Na(T)Cs(O)	14.134	12.5	[231]
KRb_2C_{60}	K(T)Rb(T)Rb(O)	14.243	23.0	[227]
$K_{1.5}Rb_{1.5}C_{60}$	O and T	14.253	22.2	[60]
K_2RbC_{60}	K(T)K(T)Rb(O)	14.243	23.0	[227, 232]
K_2CsC_{60}	K(T)K(T)Cs(O)	14.292	24.0	[227]
K_3C_{60}	O and T	14.240	19.3	[227]
$RbCs_2C_{60}$	Rb(T)Cs(T)Cs(O)	14.555	33.0	[227]
Rb_2KC_{60}	O and T	14.323	27	[227]
Rb_2CsC_{60}	Rb(T)Rb(T)Cs(O)	14.431	31.3	[60]
Rb_3C_{60}	O and T	14.384	29	[227]
Cs_3C_{60}[d]	O and T	[c]	40.0	[13]
$Na_2Cs(NH_3)_4C_{60}$	O and T	14.473	29.6	[93, 233]
$K_3(NH_3)C_{60}$[d]	O and T	14.520	–	[234]

[a] All superconducting compounds in this table crystallize in the fcc structure (except for Cs_3C_{60}), neglecting the orientation of molecules with respect to crystal axes (see text). However, the compounds LiK_2C_{60}, Li_2KC_{60}, and NaK_2C_{60} are multiphase, and KCs_2C_{60} is unstable [231].

[b] The octahedral and tetrahedral site symmetries are denoted by O and T, respectively.

[c] The T_c for Cs_3C_{60} increases under application of pressure, with a maximum T_c of ~40 K achieved at ~12 kbar [13]. However, the crystal structure of the superconducting phase for Cs_3C_{60} is not believed to be fcc [13].

[d] Orthorhombic space group $P2221$, which is distorted from the fcc structure common to M_3C_{60} compounds. The value of 14.52 Å given in the table is an average of the a, b, c lattice constants of the orthorhombic structure, $a = 14.971$ Å, $b = 14.895$ Å, and $c = 13.687$ Å [234].

indicate that the zero temperature bandgap $2\Delta(0)$ behaves like a weak coupling BCS superconductor with $2\Delta(0)/kT_c \sim 3.5$.

There have been several measurements of the isotope effect in superconducting M_3C_{60} to provide insight into whether or not the mechanism responsible for superconductivity in the doped fullerenes involves the coupling between electrons and phonons. In the experiments where the isotopic enrichment has involved ^{13}C [246–248], the experiments all suggest that $T_c \propto M^{-\alpha}$ where the observed α values are compared with the BCS prediction of $\alpha = 0.5$. Magnetization experiments using a SQUID magnetometer on Rb_3C_{60}, where up to 75% of the carbon was the ^{13}C isotope, gave $\alpha = 0.37 \pm 0.05$ [248]. The result of Chen and Lieber [246] on K_3C_{60} prepared from 99% ^{13}C powder show a value for $\alpha = 0.30 \pm 0.06$ which is consistent with a somewhat smaller α value than for the BCS theory. Similar experiments on samples with only a 33% enrichment of ^{13}C gave a much larger value of $\alpha = 1.4 \pm 0.5$ [247, 249]. On the basis of the superconducting gap equation (1.2), all of the reported values of the exponent α suggest that phonons are involved in the pairing mechanism for superconductivity and that the electron–phonon coupling constant is relatively large [247, 248]. Future work is needed to clarify the experimental picture of the isotope effect in the M_3C_{60} compounds. Of significance to the pairing mechanism for superconductivity in the doped fullerenes is the muon spin resonance (μSR) experiment which shows an isotropic superconducting gap [245], as is also found in BCS superconductors.

Of interest also is the dependence of the superconducting parameters on pressure and magnetic field. Closely related to the high compressibility of C_{60} [48] and M_3C_{60} (M = K, Rb) [228] is the large linear decrease in T_c with pressure observed in K_3C_{60} [228] and Rb_3C_{60} [228] (see Table 1.3) for measurements up to a pressure of 1.9 GPa. By attributing the smaller size of the K^+ ion relative to Rb^+ (0.186Å) to an effective "chemical pressure" of 1.06 GPa, the pressure dependence of T_c for both K_3C_{60} and Rb_3C_{60} can be made to coincide, and the result can be fit to the same functional form

$$T_c(P) = T_c(0)\exp(-\gamma P) \tag{1.3}$$

with the same value of $\gamma = 0.44 \pm 0.03\,\mathrm{GPa}^{-1}$ [228].

Table 1.3. Superconductivity Parameters in M_3C_{60} (M = K, Rb)

Parameter	K_3C_{60}	Rb_3C_{60}	Ref.
fcc a_0 (Å)	14.253	14.436	[36]
T_c (K)	19.3	29.6	[11]
$2\Delta(0)/kT_c$	3.6	3.6	[251–253]
$(dT_c/dP)_{p=0}$ (K/GPa)	-7.8	-9.7	[228]
$H_{c1}(0)$ mT)	13	12	[228]
$H_{c2}(0)$ (T)	49	78	[228]
$H_c(0)$ (T)	0.38	0.44	[228]
ξ_0 (nm)	2.6 ± 0.3	2.0 ± 0.25	[228]
λ_L (nm)	480 ± 15	460 ± 15	[245, 254]
$\kappa = \lambda_L/\xi_0$	92	124	[228]

Also listed in Table 1.3 are values for characteristic superconducting parameters relevant to their magnetic field behavior [228, 235, 255]. Based on measurements of the linear temperature dependence of the upper critical field H_{c2} near T_c, the high value for the zero temperature $H_{c2}(0)$ has been estimated [228, 235, 256] using the Werthamer–Helfand–Hohenberg (WHH) formula [257]

$$H_{c2}(0) = 0.69 \left[\frac{\partial H_{c2}}{\partial T} \right]_{T_c} T_c, \qquad (1.4)$$

and high values of $H_{c2}(0)$, exceeding the Pauli-limiting field of 35 T, have been obtained for both K_3C_{60} and Rb_3C_{60} (see Table 1.3). Measurements of the lower critical field H_{c1}, where the magnetic flux penetration initially occurs [228], show that $H_{c1}(0)$ is very low and that $H_{c1}(T)$ follows the simple empirical formula

$$[H_{c1}(T)/H_{c1}(0)] = 1 - (T/T_c)^2. \qquad (1.5)$$

Values for $H_{c1}(0)$ are also given in Table 1.3. Based on the relations

$$H_{c2}(0) = \frac{\phi_0}{2\pi\xi_0^2}, \qquad (1.6)$$

where ϕ_0 is the magnetic flux quantum and

$$H_{c1}(0) = \frac{\phi_0}{4\pi\lambda_L^2} \ln(\lambda_L/\xi_0), \qquad (1.7)$$

the zero temperature superconducting coherence length ξ_0 and the London penetration depth λ_L have been estimated (see Table 1.3) [228], showing values of ξ_0 only slightly larger than a lattice constant for the fcc unit cell. The thermodynamic critical field listed in Table 1.3 is found from the relation

$$H_c^2(0) = H_{c1}(0) H_{c2}(0) / \ln\kappa, \qquad (1.8)$$

where $\kappa = \lambda_L/\xi_0$ [228]. The large values of κ listed in Table 1.3 confirm that the M_3C_{60} compounds are strongly type II superconductors.

The possibility of the occurrence of superconductivity in fullerene polymers and fullerene-polymer composites is discussed in Chap. 8.8.

1.11 Concluding Remarks

The review of the structure and properties of fullerenes presented in this chapter is focused on providing background for the other chapters in this volume, without overlapping excessively with the content of these chapters. Because of the diversity of the articles in this volume, it is not possible to cover all the pertinent background in adequate depth. Therefore the reader is referred to a number of books and review articles on the structure and properties of fullerenes [6, 114, 258].

Acknowledgement. We gratefully acknowledge the helpful discussions with Professor P.C. Eklund and Dr A.M. Rao and with many other colleagues regarding their seminal contributions to the field. The research at MIT is funded by NSF grant DMR-98-04734.

References

1. H. W. Kroto, J. R. Heath, S. C. O'Brien, R. F. Curl, R. E. Smalley: Nature (London) **318**, 162 (1985).
2. R. B. Fuller: in W. Marlin (Ed.), The Artifacts of R. Buckminster Fuller: A Comprehensive Collection of His Designs and Drawings, New York (1984). Garland Publishing.
3. E. A. Rohlfing, D. M. Cox, A. Kaldor: J. Chem. Phys. **81**, 3322 (1984).
4. R. Taylor, J. P. Hare, A. K. Abdul-Sada, H. W. Kroto: J. Chem. Soc. Chem. Commun. **20**, 1423 (1990).
5. R. D. Johnson, G. Meijer, D. S. Bethune: J. Amer. Chem. Soc. **112**, 8983 (1990).
6. M. S. Dresselhaus, G. Dresselhaus, P. C. Eklund: Science of Fullerenes and Carbon Nanotubes (Academic Press, New York, NY 1996).
7. W. Krätschmer, K. Fostiropoulos, D. R. Huffman: in E. Bussolettii and A. A. Vittone (eds.), *Dusty Objects in the Universe*, (Kluwer Academic Publishers, Dordrecht 1989). Proceedings of the 4th Inter. Workshop of the Astronomical Observatory of Capodimonte, Capri, Italy.
8. W. Krätschmer, K. Fostiropoulos, D. R. Huffman: Chem. Phys. Lett. **170**, 167 (1990).
9. W. Krätschmer, L. D. Lamb, K. Fostiropoulos, D. R. Huffman: Nature (London) **347**, 354 (1990).
10. R. F. Curl and R. E. Smalley: Science **242**, 1017 (1988).
11. R. C. Haddon, A. F. Hebard, M. J. Rosseinsky, D. W. Murphy, S. J. Duclos, K. B. Lyons, B. Miller, J. M. Rosamilia, R. M. Fleming, A. R. Kortan, S. H. Glarum, A. V. Makhija, A. J. Muller, R. H. Eick, S. M. Zahurak, R. Tycko, G. Dabbagh, F. A. Thiel: Nature (London) **350**, 320 (1991).
12. A. F. Hebard, M. J. Rosseinsky, R. C. Haddon, D. W. Murphy, S. H. Glarum, T. T. M. Palstra, A. P. Ramirez, A. R. Kortan: Nature (London) **350**, 600 (1991).
13. T. T. M. Palstra, O. Zhou, Y. Iwasa, P. Sulewski, R. Fleming, B. Zegarski: Solid State Commun. **93**, 327 (1995).
14. A. F. Hebard: Physics Today **45**(Nov.), 26 (1992).
15. A. M. Rao, P. Zhou, K.-A. Wang, G. T. Hager, J. M. Holden, Y. Wang, W. T. Lee, X.-X. Bi, P. C. Eklund, D. S. Cornett, M. A. Duncan, I. J. Amster: Science **259**, 955 (1993).
16. C. Yeretzian, K. Hansen, F. N. Diederich, R. L. Whetten: Nature (London) **359**, 44 (1992).
17. M. Matus, J. Winter, H. Kuzmany: in H. Kuzmany, J. Fink, M. Mehring, S. Roth (eds.), *Electronic Properties of Fullerenes*, vol. 117, pp. 255–258, Berlin (1993). Springer-Verlag: Springer Series in Solid-State Sciences.
18. Y. B. Zhao, D. M. Poirier, R. J. Pechman, J. H. Weaver: Appl. Phys. Lett. **64**, 577 (1994).
19. N. Takahashi, H. Dock, N. Matsuzawa, M. Ata: J. Appl. Phys. **74**, 5790 (1993).
20. H. Yamawaki, M. Yoshida, Y. Kakudate, S. Usuba, H. Yokoi, S. Fujiwara, K. Aoki, R. Ruoff, R. Malhotra, D. C. Lorents: J. Phys. Chem. **97**, 11161 (1993).
21. S. Pekker, A. Jànossy, L. Mihaly, O. Chauver, M. Carrard, L. Forró: Science **265**, 1077 (1994).
22. Y. Wang: Nature (London) **356**, 585 (1992).
23. M. S. Dresselhaus, G. Dresselhaus, P. C. Eklund: J. Mater. Res. **8**, 2054 (1993).
24. G. Dresselhaus, M. S. Dresselhaus, P. C. Eklund: Phys. Rev. B **45**, 6923 (1992).
25. P. W. Stephens, L. Mihaly, P. L. Lee, R. L. Whetten, S. M. Huang, R. Kaner, F. Diederich, K. Holczer: Nature (London) **351**, 632 (1991).

26. R.D. Johnson, D.S. Bethune, C.S. Yannoni: Acc. Chem. Res. **25**, 169 (1992).
27. W.I.F. David, R.M. Ibberson, J.C. Matthewman, K. Prassides, T.J.S. Dennis, J.P. Hare, H.W. Kroto, R. Taylor, D.R.M. Walton: Nature (London) **353**, 147 (1991).
28. R.D. Johnson, N. Yannoni, G. Meijer, D.S. Bethune: in R.M. Fleming et al. (eds.), *Videotape of Late News Session on Buckyballs, MRS, Boston.* (Materials Research Society Press, Pittsburgh, PA, 1990).
29. C. Christides, T.J.S. Dennis, K. Prassides, R.L. Cappelletti, D.A. Neumann, J.R.D. Copley: Phys. Rev. B **49**, 2897 (1994).
30. S. Saito and A. Oshiyama: Phys. Rev. B **44**, 11532 (1991).
31. H.D. Beckhaus, C. Rüchart, M. Kas, F. Diederich, C.S. Foote: Angew. Chem. Int. Edn. Engl **31**, 63 (1992).
32. D.L. Lichtenberger, K.W. Nebesny, C.D. Ray, D.R. Huffman, L.D. Lamb: Chem. Phys. Lett. **176**, 203 (1991).
33. A. Tokmakoff, D.R. Haynes, S.M. George: Chem. Phys. Lett. **186**, 450 (1991).
34. G. Dresselhaus, M.S. Dresselhaus, J.G. Mavroides: Carbon **4**, 433 (1966).
35. J. de Vries, H. Steger, B. Kamke, C. Menzel, B. Weisser, W. Kamke, I.V. Hertel: Chem. Phys. Lett. **188**, 159 (1992).
36. A.R. Kortan, N. Kopylov, S.H. Glarum, E.M. Gyorgy, A.P. Ramirez, R.M. Fleming, F.A. Thiel, R.C. Haddon: Nature (London) **355**, 529 (1992).
37. M.B. Jost, N. Troullier, D.M. Poirier, J.L. Martins, J.H. Weaver, L.P.F. Chibante, R.E. Smalley: Phys. Rev. B **44**, 1966 (1991).
38. J.E. Fischer: Mater. Sci. Eng. **B19**, 90 (1993).
39. P.A. Heiney, G.B.M. Vaughan, J.E. Fischer, N. Coustel, D.E. Cox, J.R.D. Copley, D.A. Neumann, W.A. Kamitakahara, K.M. Creegan, D.M. Cox, J.P. McCauley, Jr., A.B. Smith, III: Phys. Rev. B **45**, 4544 (1992).
40. G. Gensterblum, K. Hevesi, B.Y. Han, L.M. Yu, J.J. Pireaux, P.A. Thiry, D. Bernaerts, S. Amelinckx, G. Van Tendeloo, G. Bendele, T. Buslaps, R.L. Johnson, M. Foss, R. Feidenhans'l, G. LeLay: Phys. Rev. B **50**, 11981 (1994).
41. R.C. Yu, N. Tea, M.B. Salamon, D. Lorens, R. Malhotra: Phys. Rev. Lett. **68**, 2050 (1992).
42. C. Wen, J. Li, K. Kitazawa, T. Aida, I. Honma, H. Komiyama, K. Yamada: Appl. Phys. Lett. **61**, 2162 (1992).
43. S.L. Ren, Y. Wang, A.M. Rao, E. McRae, G.T. Hager, K.A. Wang, W.T. Lee, H.F. Ni, J. Selegue, P.C. Eklund: Appl. Phys. Lett. **59**, 2678 (1991).
44. Y. Wang, J.M. Holden, A.M. Rao, W.T. Lee, X.X. Bi, S.L. Ren, G.W. Lehman, G.T. Hager, P.C. Eklund: Phys. Rev. B **45**, 14396 (1992).
45. J.E. Fischer and P.A. Heiney: J. Phys. Chem. Solids **54**, 1725 (1993).
46. C. Pan, M.P. Sampson, Y. Chai, R.H. Hauge, J.L. Margrave: J. Phys. Chem. **95**, 2944 (1991).
47. E. Kim, Y.H. Lee, J.Y. Lee: Phys. Rev. B **48**, 18230 (1993).
48. D.S. Fischer: Phys. Rev. B **43**, 130 (1991).
49. J.E. Fischer, P.A. Heiney, A.B. Smith, III: Accounts Chem. Res. **25**, 112 (1992).
50. F. Diederich and R.L. Whetten: Acc. Chem. Res. **25**, 119 (1992).
51. K. Kikuchi, N. Nakahara, T. Wakabayashi, S. Suzuki, H. Shiramaru, Y. Miyake, K. Saito, I. Ikemoto, M. Kainosho, Y. Achiba: Nature (London) **357**, 142 (1992).
52. D.E. Manolopoulos and P.W. Fowler: Chem. Phys. Lett. **187**, 1 (1991).
53. D.S. Bethune, R.D. Johnson, J.R. Salem, M.S. de Vries, C.S. Yannoni: Nature (London) **366**, 123 (1993).
54. R.E. Smalley: Mater. Sci. Eng. **B19**, 1 (1993).
55. R.E. Smalley: Accounts Chem. Res. **25**, 98 (1992).

56. J.F. Armbruster, H.A. Romberg, P. Schweiss, P. Adelmann, M. Knupfer, J. Fink, R.H. Michel, J. Rockenberger, F. Hennrich, H. Schreiber, M.M. Kappes: Z. Phys. B **95**, 469 (1994).
57. R.L. Meng, D. Ramirez, X. Jiang, P.C. Chow, C. Diaz, K. Matsuishi, S.C. Moss, P.H. Hor, C.W. Chu: Appl. Phys. Lett. **59**, 3402 (1991).
58. G. Gensterblum, L.M. Yu, J.J. Pireaux, P.A. Thiry, R. Caudano, J.M. Themlin, S. Bouzidi, F. Coletti, J.M. Debever: Appl. Phys. A **56**, 175 (1993).
59. J.R.D. Copley, W.I.F. David, D.A. Neumann: Neutron News **4**, 20 (1993).
60. R.M. Fleming, A.P. Ramirez, M.J. Rosseinsky, D.W. Murphy, R.C. Haddon, S.M. Zahurak, A.V. Makhija: Nature (London) **352**, 787 (1991).
61. P.A. Heiney, J.E. Fischer, A.R. McGhie, W.J. Romanow, A.M. Denenstein, J.P. McCauley, Jr., A.B. Smith, III, D.E. Cox: Phys. Rev. Lett. **67**, 1468 (1991).
62. J.H. Weaver and D.M. Poirier: in H. Ehrenreich and F. Spaepen (Eds.), *Solid State Physics*, vol. 48, p. 1, (Academic Press, New York, 1994). Chap. 1.
63. D.W. Murphy, M.J. Rosseinsky, R.M. Fleming, R. Tycko, A.P. Ramirez, R.C. Haddon, T. Siegrist, G. Dabbagh, J.C. Tully, R.E. Walstedt: J. Phys. Chem. Solids **53**, 1321 (1992).
64. W.I.F. David, R.M. Ibberson, T.J.S. Dennis, J.P. Hare, K. Prassides: Europhys. Lett. **18**, 219 (1992).
65. R.M. Fleming, M.J. Rosseinsky, A.P. Ramirez, D.W. Murphy, J.C. Tully, R.C. Haddon, T. Siegrist, R. Tycko, S.H. Glarum, P.M. Marsh, G. Dabbagh, S.M. Zahurak, A.V. Makhija, C. Hampton: Nature (London) **352**, 701 (1991).
66. G.B. Vaughan, P.A. Heiney, J.E. Fischer, D.E. Luzzi, D.A. Ricketts-Foot, A.R. McGhie, Y.W. Hui, A.L. Smith, D.E. Cox, W.J. Romanow, B.H. Allen, N. Coustel, J.P. McCauley, Jr., A.B. Smith, III: Science **254**, 1350 (1991).
67. M.A. Verheijen, H. Meekes, G. Meijer, P. Bennema, J.L. de Boer, S. van Smaalen, G.V. Tendeloo, S. Amelinckx, S. Muto, J. van Landuyt: Chem. Phys. **166**, 287 (1992).
68. G. Van Tendeloo, S. Amelinckx, M.A. Verheijen, P.H.M. van Loosdrecht, G. Meijer: Phys. Rev. Lett. **69**, 1065 (1992).
69. X.-D. Wang, T. Hashizume, H. Shinohara, Y. Saito, Y. Nishina, T. Sakurai: Phys. Rev. B **47**, 15923 (1993).
70. Y.Z. Li, J.C. Patrin, M. Chander, J.H. Weaver, K. Kikuchi, Y. Achiba: Phys. Rev. B **47**, 10867 (1993).
71. M.S. Dresselhaus and G. Dresselhaus: Adv. Phys. **30**, 139 (1981).
72. M.J. Rosseinsky, D.W. Murphy, R.M. Fleming, R. Tycko, A.P. Ramirez, T. Siegrist, G. Dabbagh, S.E. Barrett: Nature (London) **356**, 416 (1992).
73. K. Tanigaki, T.W. Ebbesen, S. Saito, J. Mizuki, J.S. Tsai, Y. Kubo, S. Kuroshima: Nature (London) **352**, 222 (1991).
74. P.W. Stephens, G. Bortel, G. Falgel, M. Tegze, A. Jànossy, S. Pekker, G. Oszànyi, L. Forró: Nature (London) **370**, 636 (1994).
75. O. Chauvet, G. Oszànyi, L. Forró, P.W. Stephens, M. Tegze, G. Faigel, A. Jànossy: Phys. Rev. Lett. **72**, 2721 (1994).
76. Q. Zhu, O. Zhou, J.E. Fischer, A.R. McGhie, W.J. Romanow, R.M. Strongin, M.A. Cichy, A.B. Smith, III: Phys. Rev. B **47**, 13948 (1993).
77. M.C. Martin, D. Koller, A. Rosenberg, C. Kendziora, L. Mihaly: Phys. Rev. B **51**, 3210 (1995).
78. J. Winter and H. Kuzmany: Solid State Commun. **83**, 1321 (1992).
79. A. Jànossy, O. Chauvet, S. Pekker, J.R. Cooper, L. Forró: Phys. Rev. Lett. **71**, 1091 (1993).

80. L. Mihaly, D. Koller, M. C. Martin: in H. Kuzmany, J. Fink, M. Mehring, S. Roth (Eds.), *Progress in Fullerene Research: International Winter School on Electronic Properties of Novel Materials*, Kirchberg Winter School, (World Scientific Publishing Co., Ltd., Singapore, 1994) p. 265.
81. M. C. Martin, D. Koller, X. Du, P. W. Stephens, L. Mihaly: Phys. Rev. B **49**, 10818 (1994).
82. A. R. Kortan, N. Kopylov, E. Özdas, A. P. Ramirez, R. M. Fleming, R. C. Haddon: Chem. Phys. Lett. **223**, 501 (1994).
83. Y. Chen, F. Stepniak, J. H. Weaver, L. P. F. Chibante, R. E. Smalley: Phys. Rev. B **45**, 8845 (1992).
84. W. R. Datars, P. K. Ummat, T. Olech, R. K. Nkum: Solid. State Commun. **86**, 579 (1993).
85. W. R. Datars, T. R. Chien, R. K. Nkum, P. K. Ummat: Phys. Rev. B **50**, 4937 (1994).
86. H. Kuzmany: in J. Shinar, Z. V. Vardeny, Z. H. Kafafi (eds.), *Optical and Electronic Properties of Fullerenes and Fullerene-Based Materials*, (Marcel Dekker, Inc, New York 1999).
87. A. M. Rao, M. Menon, K. A. Wang, P. C. Eklund, K. R. Subbaswamy, D. S. Cornett, M. A. Duncan, I. J. Amster: Chem. Phys. Lett. **224**, 106 (1994).
88. P. Zhou, A. M. Rao, K. A. Wang, J. D. Robertson, C. Eloi, M. S. Meier, S. L. Ren, X. X. Bi, P. C. Eklund, M. S. Dresselhaus: Appl. Phys. Lett. **60**, 2871 (1992).
89. Y. Wang, J. M. Holden, X. X. Bi, P. C. Eklund: Chem. Phys. Lett. **217**, 413 (1994).
90. M. S. Dresselhaus, G. Dresselhaus, M. A. Pimenta, P. C. Eklund: *Analytical Applications of Raman Spectroscopy*, (Blackwell Science Ltd., Oxford UK, 1999) p. 367.
91. D. S. Cornett, I. J. Amster, M. A. Duncan, A. M. Rao, P. C. Eklund: J. Phys. Chem. **97**, 5036 (1993).
92. Y. Wang, J. M. Holden, Z. H. Dong, X. X. Bi, P. C. Eklund: Chem. Phys. Lett. **211**, 341 (1993).
93. P. Zhou, Z. H. Dong, A. M. Rao, P. C. Eklund: Chem. Phys. Lett. **211**, 337 (1993).
94. Y. Iwasa, T. Arima, R. M. Fleming, T. Siegrist, O. Zhou, R. C. Haddon, L. J. Rothberg, K. B. Lyons, H. L. Carter, Jr., A. F. Hebard, R. Tycko, G. Dabbagh, J. J. Krajewski, G. A. Thomas, T. Yagi: Science **264**, 1570 (1994).
95. Y. Wang, J. M. Holden, A. M. Rao, P. C. Eklund, U. Venkateswaran, D. Eastwood, R. L. Lidberg, G. Dresselhaus, M. S. Dresselhaus: Phys. Rev. B **51**, 4547 (1995).
96. M. Menon, A. M. Rao, K. R. Subbaswamy, P. C. Eklund: Phys. Rev. B **51**, 800 (1995).
97. M. Menon and K. R. Subbaswamy: Phys. Rev. Lett. **67**, 3487 (1991).
98. M. Menon, K. R. Subbaswamy, M. Sawtarie: Phys. Rev. B **48**, 8398 (1993).
99. M. Menon, K. R. Subbaswamy, M. Sawtarie: Phys. Rev. B **49**, 13966 (1994).
100. Z. H. Dong, P. Zhou, J. M. Holden, P. C. Eklund, M. S. Dresselhaus, G. Dresselhaus: Phys. Rev. B **48**, 2862 (1993).
101. K. A. Wang, A. M. Rao, P. C. Eklund, M. S. Dresselhaus, G. Dresselhaus: Phys. Rev. B **48**, 11375 (1993).
102. M. C. Martin, D. Koller, L. Mihaly: Phys. Rev. B **50**, 6538 (1994).
103. M. C. Martin, X. Du, L. Mihaly: in H. Kuzmany, J. Fink, M. Mehring, S. Roth (eds.), *Progress in Fullerene Research: International Winter School on Electronic Properties of Novel Materials*, Kirchberg Winter School, (World Scientific Publishing Co., Ltd., Singapore, 1994) p. 353.

104. D.S. Bethune, G. Meijer, W.C. Tang, H.J. Rosen: Chem. Phys. Lett. **174**, 219 (1990).
105. P.C. Eklund, P. Zhou, K.-A. Wang, G. Dresselhaus, M.S. Dresselhaus: J. Phys. Chem. Solids **53**, 1391 (1992).
106. A.M. Vassallo, L.S. Pang, P.A. Cole-Clark, M.A. Wilson: J. Am. Chem. Soc. **113**, 7820 (1991).
107. B. Chase, N. Herron, E. Holler: J. Phys. Chem. **96**, 4262 (1992).
108. K.A. Wang, Y. Wang, P. Zhou, J.M. Holden, S.L. Ren, G.T. Hager, H.F. Ni, P.C. Eklund, G. Dresselhaus, M.S. Dresselhaus: Phys. Rev. B **45**, 1955 (1992).
109. P. Zhou, K.-A. Wang, Y. Wang, P.C. Eklund, M.S. Dresselhaus, G. Dresselhaus, R.A. Jishi: Phys. Rev. B **46**, 2595 (1992).
110. P.H.M. van Loosdrecht, P.J.M. van Bentum, G. Meijer: Phys. Rev. Lett. **68**, 1176 (1992).
111. D.S. Bethune, G. Meijer, W.C. Tang, H.J. Rosen, W.G. Golden, H. Seki, C.A. Brown, M.S. de Vries: Chem. Phys. Lett. **179**, 181 (1991).
112. G. Meijer, D.S. Bethune, W.C. Tang, H.J. Rosen, R.D. Johnson, R.J. Wilson, D.D. Chambliss, W.G. Golden, H. Seki, M.S. de Vries, C.A. Brown, J.R. Salem, H.E. Hunziker, H.R. Wendt: in R.S. Averback, J. Bernholc, D.L. Nelson (Eds.), *Clusters and Cluster-Assembled Materials, MRS Symposia Proceedings, Boston*, vol. 206, (Materials Research Society Press, Pittsburgh, PA, 1991) p. 619.
113. S.J. Duclos, R.C. Haddon, S.H. Glarum, A.F. Hebard, K.B. Lyons: Science **254**, 1625 (1991).
114. M.S. Dresselhaus, G. Dresselhaus, P.C. Eklund: J. Raman Spectrosc. **27**, 351 (1995).
115. R.E. Stanton and M.D. Newton: J. Phys. Chem. **92**, 2141 (1988).
116. F. Negri, G. Orlandi, F. Zerbetto: Chem. Phys. Lett. **144**, 31 (1988).
117. M. Häser, J. Almlöf, G.E. Scuseria: Chem. Phys. Lett. **181**, 497 (1991).
118. Z.C. Wu, D.A. Jelski, T.F. George: Chem. Phys. Lett. **137**, 291 (1987).
119. S.J. Cyvin, E. Brendsdal, B.N. Cyvin, J. Brunvoll: Chem. Phys. Lett. **143**, 377 (1988).
120. D.E. Weeks and W.G. Harter: J. Chem. Phys. **90**, 4744 (1989).
121. R.A. Jishi, R.M. Mirie, M.S. Dresselhaus: Phys. Rev. B **45**, 13685 (1992).
122. Y. Wang, D. Tománek, G.F. Bertsch: Phys. Rev. B **44**, 6562 (1991).
123. R.A. Jishi, M.S. Dresselhaus, G. Dresselhaus, K.A. Wang, P. Zhou, A.M. Rao, P.C. Eklund: Chem. Phys. Lett. **206**, 187 (1993).
124. D.W. Snoke, Y.S. Raptis, K. Syassen: Phys. Rev. B **45**, 14419 (1992).
125. R.A. Jishi, R.M. Mirie, M.S. Dresselhaus, G. Dresselhaus, P.C. Eklund: Phys. Rev. B **48**, 5634 (1993).
126. T. Pichler, M. Matus, J. Kürti, H. Kuzmany: Phys. Rev. B **45**, 13841 (1992).
127. S.H. Glarum, S.J. Duclos, R.C. Haddon: J. Am. Chem. Soc. **114**, 1996 (1992).
128. R.A. Jishi and M.S. Dresselhaus: Phys. Rev. B **45**, 6914 (1992).
129. J. Winter and H. Kuzmany: in H. Kuzmany, J. Fink, M. Mehring, S. Roth (eds.), *Progress in Fullerene Research: International Winter School on Electronic Properties of Novel Materials*, Kirchberg Winter School, (World Scientific Publishing Co., Ltd., Singapore, 1994) p. 271.
130. M.C. Martin, D. Koller, A. Rosenberg, C. Kendziora, L. Mihaly: in P. Bernier, T.W. Ebbesen, D.S. Bethune, R.M. Metzger, L.Y. Chiang, J.W. Mintmire (Eds.), *Science and Technology of Fullerene Materials, MRS Symposia Proceedings, Boston*, vol. 359, (Materials Research Society Press, Pittsburgh, PA, 1995) p. 319.
131. M. Manfredini, C.E. Bottani, P. Milani: Chem. Phys. Lett. **226**, 600 (1994).

132. J.D. Lorentzen, S. Guha, J. Menendez, P. Giannozzi, S. Baroni: Chem. Phys. Lett. **270**, 129 (1997).
133. R.C. Haddon: Acc. Chem. Res. **25**, 127 (1992).
134. S. Saito and A. Oshiyama: Phys. Rev. Lett. **66**, 2637 (1991).
135. W.E. Pickett: in H. Ehrenreich and F. Spaepen (Eds.), *Solid State Physics*, vol. 48, (Academic Press, New York (1994) p. 225.
136. R.C. Haddon, L.E. Brus, K. Raghavachari: Chem. Phys. Lett. **125**, 459 (1986).
137. S.G. Louie and E.L. Shirley: J. Phys. Chem. Solids **54**, 1767 (1993).
138. B.I. Dunlap, D.W. Brenner, J.W. Mintmire, R.C. Mowrey, C.T. White: J. Phys. Chem. **95**, 5763 (1991).
139. Z. Zhang, C.-C. Chen, S.P. Kelty, H. Dai, C.M. Lieber: Nature (London) **353**, 333 (1991).
140. Q.M. Zhang, J.Y. Yi, J. Bernholc: Phys. Rev. Lett. **66**, 2633 (1991).
141. P. Damay and F. Leclercq: Phys. Rev. B **49**, 7790 (1994).
142. L.-S. Wang: Chem. Phys. Lett. **182**, 5 (1992).
143. F. Negri, G. Orlandi, F. Zerbetto: J. Am. Chem. Soc. **114**, 2910 (1992).
144. R. Saito, G. Dresselhaus, M.S. Dresselhaus: Phys. Rev. B **46**, 9906 (1992).
145. S.C. Erwin and M.R. Pederson: Phys. Rev. Lett. **67**, 1610 (1991).
146. P.C. Eklund, A.M. Rao, P. Zhou, Y. Wang, J.M. Holden: Thin Solid Films **257**, 185 (1995).
147. J.B. Birks: *A general review of the molecular spectroscopy of aromatic molecules.* (Wiley and Sons, London, 1970)
148. H. Ajie, M.M. Alvarez, S.J. Anz, R.D. Beck, F. Diederich, K. Fostiropoulos, D.R. Huffman, W. Krätschmer, Y. Rubin, K.E. Schriver, D. Sensharma, R.L. Whetten: J. Phys. Chem. **94**, 8630 (1990).
149. G. Herzberg: *Molecular Spectra and Molecular Structure*, (Van Nostrand, Princeton, NJ, 1945).
150. T. Pichler, M. Matus, J. Kürti, H. Kuzmany: Solid State Commun. **81**, 859 (1992).
151. S.P. Sibley, S.M. Argentine, A.H. Francis: Chem. Phys. Lett. **188**, 187 (1992).
152. A. Nakayama, K. Suzuki, T. Enoki, S.L. di Vittorio, M.S. Dresselhaus, K. Koga, M. Endo, N. Shindo: Syn. Met. **55-57**, 3736 (1993).
153. M. Matus, H. Kuzmany, E. Sohmen: Phys. Rev. Lett. **68**, 2822 (1992).
154. M.S. Dresselhaus, G. Dresselhaus, R. Saito: Mater. Sci. Eng. **B19**, 122 (1993).
155. F. Diederich, R. Ettl, Y. Rubin, R.L. Whetten, R. Beck, M. Alvarez, S. Anz, D. Sensharma, F. Wudl, K.C. Khemani, A. Koch: Science **252**, 548 (1991).
156. P.-M. Allemand, K.C. Khemani, A. Koch, F. Wudl, K. Holczer, S. Donovan, G. Grüner, J.D. Thompson: Science **253**, 301 (1991).
157. R.R. Hung and J.J. Grabowski: J. Phys. Chem. **95**, 6073 (1991).
158. T.W. Ebbesen, K. Tanigaki, S. Kuroshima: Chem. Phys. Lett. **181**, 501 (1991).
159. J.W. Arbogast, A.P. Darmanyan, C.S. Foote, Y. Rubin, F.N. Diederich, M.M. Alvarez, S.J. Anz, R.L. Whetten: J. Phys. Chem. **95**, 11 (1991).
160. B.L. Justus, Z.H. Kafafi, A.L. Huston: Opt. Lett. **18**, 1603 (1993).
161. M.P. Joshi, S.R. Mishra, H.S. Rawat, S.C. Mehendale, K.C. Rustagi: Appl. Phys. Lett. **62**, 1763 (1993).
162. A. Kost, L. Tutt, M.B. Klein, T.K. Dougherty, W.E. Elias: Opt. Lett. **18**, 334 (1993).
163. F. Henari, J. Callaghan, H. Stiel, W. Blau, D.J. Cardin: Chem. Phys. Lett. **199**, 144 (1992).
164. M.K. Nissen, S.M. Wilson, M.L.W. Thewalt: Phys. Rev. Lett. **69**, 2423 (1992).

165. S. L. Ren, K. A. Wang, P. Zhou, Y. Wang, A. M. Rao, M. S. Meier, J. Selegue, P. C. Eklund: Appl. Phys. Lett. **61**, 124 (1992).
166. A. M. Rao, K. A. Wang, P. Zhou, Y. Wang, S. L. Ren, P. C. Eklund: *Springer Series in Solid-State Sciences.* vol. 113 (Springer-Verlag, New York, Berlin, 1992).
167. E. Sohmen, J. Fink, W. Krätschmer: Z. Phys. B **86**, 87 (1992).
168. A. F. Hebard, R. C. Haddon, R. M. Fleming, A. R. Kortan: Appl. Phys. Lett. **59**, 2109 (1991).
169. W. Y. Ching, M.-Z. Huang, Y.-N. Xu, W. G. Harter, F. T. Chan: Phys. Rev. Lett. **67**, 2045 (1991).
170. D. Dick, X. Wei, S. Jeglinski, R. E. Benner, Z. V. Vardeny, D. Moses, V. I. Srdanov, F. Wudl: Phys. Rev. Lett. **73**, 2760 (1994).
171. P. Prasad: *Nonlinear Optical Properties of Organic Materials*: (Plenum, New York, 1991).
172. N. Tang, J. P. Partanen, R. W. Hellwarth, R. J. Knize: Phys. Rev. B **48**, 8404 (1993).
173. Z. H. Kafafi, F. J. Bartoli, J. R. Lindle, R. G. S. Pong: Phys. Rev. Lett. **68**, 2705 (1992).
174. Q. H. Gong, Y. X. Sun, Z. J. Xia, Y. H. Zou, Z. N. Gu, X. H. Zhou, D. Qiang: J. Appl. Phys. **71**, 3025 (1992).
175. J. R. Lindle, R. G. S. Pong, F. J. Bartoli, Z. H. Kafafi: Phys. Rev. B **48**, 9447 (1993).
176. H. Hoshi, N. Nakamura, Y. Maruyama, T. Nakagawa, S. Suzuki, H. Shiromaru, Y. Achiba: Jpn. J. Appl. Phys. **30**, L1397 (1991).
177. J. S. Meth, H. Vanherzeele, Y. Wang: Chem. Phys. Lett. **197**, 26 (1992).
178. F. Kajzar, C. Taliani, R. Zamboni, S. Rosini, R. Danieli: Synth. Met. **54**, 21 (1993).
179. D. Neher, G. I. Stegeman, F. A. Tinker: Opt. Lett. **17**, 1491 (1992).
180. Y. Wang and L. T. Cheng: J. Phys. Chem. **96**, 1530 (1992).
181. C. Reber, L. Yee, J. McKiernan, J. I. Zink, R. S. Williams, W. M. Tong, D. A. A. Ohlberg, R. L. Whetten, F. Diederich: J. Phys. Chem. **95**, 2127 (1991).
182. A. Ito, T. Morihawa, T. Takahashi: Chem. Phys. Lett. **211**, 333 (1993).
183. J. Mort, R. Ziolo, M. Machonkin, D. R. Huffman, M. I. Ferguson: Chem. Phys. Lett. **186**, 284 (1991).
184. F. Stepniak, P. J. Benning, D. M. Poirier, J. H. Weaver: Phys. Rev. B **48**, 1899 (1993).
185. A. F. Hebard: in P. Jena, S. N. Khanna, B. K. Rao (Eds.), *Physics and Chemistry of Finite Systems: From Clusters to Crystals*, vol. 374, NATO ASI Series C: Mathematical and Physical Sciences. (Kluwer Academic Publishers, Dordrecht, 1991) pp. 1213–1220.
186. G. P. Kochanski, A. F. Hebard, R. C. Haddon, A. T. Fiory: Science **255**, 184 (1992).
187. K. Holczer and R. L. Whetten: Carbon **30**, 1261 (1992).
188. M. P. Gelfand, J. P. Lu: Phys. Rev. Lett. **68**, 1050 (1992).
189. M. P. Gelfand, J. P. Lu: Phys. Rev. B **46**, 4367 (1992): Erratum: Phys. Rev. B **47**, 4149 (1992).
190. M. P. Gelfand and J. P. Lu: Appl. Phys. A **56**, 215 (1993).
191. F. Bommeli, L. Degiorgi, P. Wachter, O. Legeza, A. Jànossy, G. Oszlànyi, O. Chauvet, L. Forró: Phys. Rev. B **51**, 14794 (1995).
192. J. Hone, M. S. Fuhrer, K. Khazeni, A. Zettl: Phys. Rev. B **52**, R8700 (1995).
193. J. Mort, M. Machonkin, R. Ziolo, D. R. Huffman, M. I. Ferguson: Appl. Phys. Lett. **60**, 1735 (1992).

194. M. Kaiser, J. Reichenbach, H. J. Byrne, J. Anders, W. Maser, S. Roth, A. Zahab, P. Bernier: Solid State Commun. **81**, 261 (1992).
195. H. Yonehara and C. Pac: Appl. Phys. Lett. **61**, 575 (1992).
196. M. Hosoya, K. Ichimura, Z. H. Wang, G. Dresselhaus, M. S. Dresselhaus, P. C. Eklund: Phys. Rev. B **49**, 4981 (1994).
197. N. Minami and M. Sato: Synth. Met. **56**, 3092 (1993).
198. A. Hamed, H. Rasmussen, P. H. Hor: Appl. Phys. Lett. **64**, 526 (1994).
199. N. S. Sariciftci, L. Smilowitz, A. J. Heeger, F. Wudl: Science **258**, 1474 (1992).
200. N. S. Sariciftci, D. Braun, C. Zhang, V. I. Srdanov, A. J. Heeger, C. Stucky, F. Wudl: Appl. Phys. Lett. **62**, 585 (1993).
201. K. Lee, R. A. J. Janssen, N. S. Sariciftci, A. J. Heeger: Phys. Rev. B **49**, 5781 (1994).
202. C. H. Lee, G. Yu, D. Moses, K. Pakbaz, C. Zhang, N. S. Sariciftci, A. J. Heeger, F. Wudl: Phys. Rev. B **48**, 15425 (1993).
203. N. S. Sariciftci and A. J. Heeger: Int. J. Mod. Phys. B **8**, 237 (1994).
204. N. S. Sariciftci, B. Kraabel, C. H. Lee, K. Pakbaz, A. J. Heeger: Phys. Rev. B **50**, 12044 (1994).
205. N. S. Sariciftci, L. Smilowitz, C. Zhang, V. I. Srdanov, A. J. Heeger, F. Wudl: Proc. SPIE – Int. Soc. Opt. Eng. **1852**, 297 (1993).
206. B. Kraabel, D. W. McBranch, K. H. Lee, N. S. Sariciftci: in Z. H. Kafafi (Ed.), *Proceedings of the International Society for Optical Engineering (SPIE). "Fullerenes and Photonics I"*; San Diego, CA, July 24-29, 1994. *Proc. vol. 2284*, pp. 194–207,
207. M. Kaiser, W. K. Maser, H. J. Byrne, A. Mittelbach, S. Roth: Solid State Commun. **87**, 281 (1993).
208. R. Könenkamp, J. Erxmeyer, A. Weidinger: Appl. Phys. Lett. **65**, 758 (1994).
209. S. Kazaoui, R. Ross, N. Minami: Solid State Commun. **90**, 623 (1994).
210. A. Hamed, Y. Y. Sun, Y. K. Tao, R. L. Meng, P. H. Hor: Phys. Rev. B **47**, 10873 (1993).
211. A. Hamed, H. Rasmussen, P. H. Hor: Phys. Rev. B **48**, 14760 (1993).
212. A. Hamed, D. Y. Xing, C. C. Chow, P. H. Hor: Phys. Rev. B **50**, 11219 (1994).
213. R. A. Assink, J. E. Schirber, D. A. Loy, B. Morosin, G. A. Carlson: J. Mater. Res. **7**, 2136 (1992).
214. D. Sarkar and N. J. Halas: Solid State Commun. **90**, 261 (1994).
215. C. H. Lee, G. Yu, D. Moses, V. I. Srdanov, X. Wei, Z. V. Vardeny: Phys. Rev. B **48**(11), 8506 (1993).
216. C. H. Lee, G. Yu, B. Kraabel, D. Moses, V. I. Srdanov: Phys. Rev. B **49**, 10572 (1994).
217. H. Yonehara and C. Pac: Appl. Phys. Lett. **61**, 575 (1992).
218. A. Hamed, R. Escalante, P. H. Hor: Phys. Rev. B **50**, 8050 (1994).
219. A. Hirao, H. Nishizawa, H. Miyamoto, M. Sugiuchi, M. Hosoya: in Z. H. Kafafi (Ed.), vol 2526, (San Diego, CA, July 13, 1995) pp. 71–80.
220. N. B. Hannay, T. H. Geballe, B. T. Matthias, K. Andres, P. Schmidt, D. MacNair: Phys. Rev. Lett. **14**, 225 (1965).
221. Y. Koike, K. Higuchi, S. Tanuma: Solid State Commun. **27**, 623 (1978).
222. A. Chaiken, M. S. Dresselhaus, T. P. Orlando, G. Dresselhaus, P. Tedrow, D. A. Neumann, W. A. Kamitakahara: Phys. Rev. B **41**, 71 (1990).
223. I. T. Belash, A. D. Bronnikov, O. V. Zharikov, A. V. Pal'nichenko: Synth. Met. **36**, 283 (1990).
224. A. F. Hebard, T. T. M. Palstra, R. C. Haddon, R. M. Fleming: Phys. Rev. B **48**, 9945 (1993).
225. K. Holczer, O. Klein, G. Grüner, S.-M. Huang, R. B. Kaner, K.-J. Fu, R. L. Whetten, F. Diederich: Science **252**, 1154 (1991).

226. K. Tanigaki, I. Hirosawa, T. W. Ebbesen, J. Mizuki, Y. Shimakawa, Y. Kubo, J. S. Tsai, S. Kuroshima: Nature (London) **356**, 419 (1992).
227. K. Tanigaki, S. Kuroshima, J. Fujita, T. W. Ebbesen: Appl. Phys. Lett. **63**, 2351 (1993).
228. G. Sparn, J. D. Thompson, R. L. Whetten, S.-M. Huang, R. B. Kaner, F. Diederich, G. Grüner, K. Holczer: Phys. Rev. Lett. **68**, 1228 (1992).
229. T. Inabe, H. Ogata, Y. Maruyama, Y. Achiba, S. Suzuki, K. Kikuchi, I. Ikemoto: Phys. Rev. Lett. **69**, 3797 (1992).
230. A. Oshiyama, S. Saito, N. Hamada, Y. Miyamoto: J. Phys. Chem. Solids **53**, 1457 (1992).
231. K. Tanigaki, I. Hirosawa, T. W. Ebbesen, J. Mizuki, S. Kuroshima: Chem. Phys. Lett. **203**, 33 (1993).
232. I. Hirosawa, J. Mizuki, K. Tanigaki, H. Kumura: Solid State Commun. **89**, 55 (1994).
233. O. Zhou, R. M. Fleming, D. W. Murphy, M. J. Rosseinsky, A. P. Ramirez, R. B. van Dover, R. C. Haddon: Nature (London) **362**, 433 (1993).
234. M. J. Rosseinsky, D. W. Murphy, R. M. Fleming, O. Zhou: Nature (London) **364**, 425 (1993).
235. G. Sparn, J. D. Thompson, S.-M. Huang, R. B. Kaner, F. Diederich, R. L. Whetten, G. Grüner, K. Holczer: Science **252**, 1829 (1991).
236. J. E. Schirber, D. L. Overmyer, H. H. Wang, J. M. Wang, K. D. Carlson, A. M. Kini, M. J. Pellin, U. Welp, W.-K. Kwok: Physica C **178**, 137 (1991).
237. P. W. Stephens: Nature (London) **356**, 383 (1992).
238. C. M. Varma, J. Zaanen, K. Raghavachari: Science **254**, 989 (1991).
239. M. A. Schlüter, M. Lannoo, M. Needels, G. A. Baraff, D. Tománek: Phys. Rev. Lett. **68**, 526 (1992).
240. K. H. Johnson, M. E. McHenry, D. P. Clougherty: Physica C **183**, 319 (1991).
241. R. A. Jishi and M. S. Dresselhaus: Phys. Rev. B **45**, 2597 (1992).
242. L. Degiorgi, E. J. Nicol, O. Klein, G. Grüner, P. Wachter, S. M. Huang, J. Wiley, R. B. Kaner: Phys. Rev. B **49**, 7012 (1994).
243. V. A. Stenger, C. H. Pennington, D. R. Buffinger, R. P. Ziebarth: Phys. Rev. Lett. **74**, 1649 (1995).
244. V. A. Stenger, C. Recchia, J. Vance, C. H. Pennington, D. R. Buffinger, R. P. Ziebarth: Phys. Rev. B **48**, 9942 (1993).
245. Y. J. Uemura, A. Keren, L. P. Le, G. M. Luke, B. J. Sternlieb, W. D. Wu, J. H. Brewer, R. L. Whetten, S. M. Huand, S. Lin, R. B. Kaner, F. Diederich, S. Donovan, G. Grüner: Nature (London) **352**, 605 (1991).
246. C.-C. Chen and C. M. Lieber: J. Am. Chem. Soc. **114**, 3141 (1992).
247. T. W. Ebbesen, J. S. Tsai, K. Tanigaki, J. Tabuchi, Y. Shimakawa, Y. Kubo, I. Hirosawa, J. Mizuki: Nature (London) **355**, 620 (1992).
248. A. P. Ramirez, A. R. Kortan, M. J. Rosseinsky, S. J. Duclos, A. M. Mujsce, R. C. Haddon, D. W. Murphy, A. V. Makhija, S. M. Zahurak, K. B. Lyons: Phys. Rev. Lett. **68**, 1058 (1992).
249. A. A. Zakhidov, K. Imaeda, D. M. Petty, K. Yakushi, H. Inokuchi, K. Kikuchi, I. Ikemoto, S. Suzuki, Y. Achiba: Phys. Lett. A **164**, 355 (1992).
250. Z. Zhang, C. C. Chen, C. M. Lieber: Science **254**, 1619 (1991).
251. R. F. Kiefl, W. A. MacFarlane, K. H. Chow, S. Dunsiger, T. L. Duty, T. M. S. Johnston, J. W. Schneider, J. Sonier, L. Brard, R. M. Strongin, J. E. Fischer, A. B. Smith, III: Phys. Rev. Lett. **70**, 3987 (1993).
252. L. D. Rotter, Z. Schlesinger, J. P. McCauley, Jr., N. Coustel, J. E. Fischer, A. B. Smith, III: Nature (London) **355**, 532 (1992).

253. L. Degiorgi, G. Grüner, P. Wachter, S. M. Huang, J. Wiley, R. L. Whetten, R. B. Kaner, K. Holczer, F. Diederich: Phys. Rev. B **46**, 11250 (1992).
254. R. Tycko, G. Dabbagh, M. J. Rosseinsky, D. W. Murphy, A. P. Ramirez, R. M. Fleming: Phys. Rev. Lett. **68**, 1912 (1992).
255. K. Holczer, O. Klein, G. Grüner, J. D. Thompson, F. Diederich, R. L. Whetten: Phys. Rev. Lett. **67**, 271 (1991).
256. S. Foner, E. J. McNiff, Jr., D. Heiman, S. M. Huang, R. B. Kaner: Phys. Rev. B **46**, 14936 (1992).
257. N. R. Werthamer, E. Helfand, P. C. Hohenberg: Phys. Rev. **147**, 295 (1966).
258. N. S. Sariciftci: Prog. Quantum Electron. **19**, 131 (1995).

2. Introduction to Electronic, Optical, and Transport Properties of Conducting Polymers and Fullerene-Polymer Composites

P.A. Lane, Z.V. Vardeny

2.1 Introduction

Conducting polymers are a remarkable new class of organic materials with potential applications such as light emitting diodes (LEDs), thin film transistors, solid state batteries and optical switches. Serious investigations into conducting polymers were spurred by the discovery in 1977 that polyacetylene $(CH)_x$ could be chemically doped to 1000 S/cm at room temperature [1, 2]. Subsequent improvements in polymerization methods raised this value to 145 000 S/cm [3–6]. These materials have a highly anisotropic quasi-one-dimensional electronic structure which is fundamentally different from conventional inorganic semiconductors. The chain-like structure makes conducting polymers susceptible to strong coupling of the electronic states to conformational distortions leading to excitations peculiar to the 1D system. In addition, the relatively weak interchain binding allows diffusion of dopant molecules into the structure, while strong carbon–carbon intrachain binding maintains the integrity of the polymer.

2.2 Electronic States in Conducting Polymers

The simplest example of conducting polymers is polyacetylene $(CH)_x$, depicted in Fig. 2.1. It consists of weakly coupled chains of CH units forming a pseudo-1D lattice. The stable isomer is trans-$(CH)_x$, whereas cis-$(CH)_x$ is unstable at high temperatures or under strong illumination. Three of the four carbon valence electrons form σ-bonds from sp^2-hybridized orbitals, resulting in a planar structure, while the fourth valence electron is in a π-orbital perpindicular to the plane of the chain. The π-electron forms the highest occupied (HOMO) and lowest occupied (LUMO) molecular orbitals, which together span an energy range of about 10 eV. If all bond lengths were equal, polyacetylene would be a quasi-1D metal with a half-filled π-band. However, this structure is unstable with respect to a dimerization distortion, the so-called Peierls instability [7, 8]. Adjacent CH groups move towards each other, forming alternating single and double bonds, and an energy gap of 1.7 eV develops between the HOMO and LUMO.

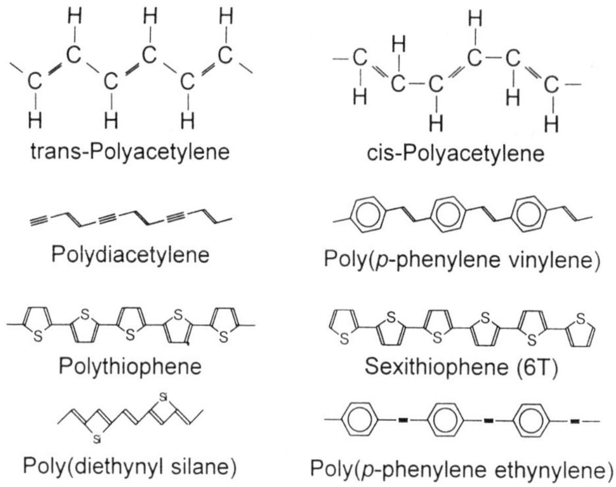

Fig. 2.1. Chemical structure of conducting polymers under discussion

Trans-$(CH)_x$ is especially interesting in that it has a two-fold degenerate ground state. From symmetry considerations, one can exchange the single and double bonds without changing the energy of the system. Hence, there are two degenerate energy states with distinct bonding structures. This two-fold degeneracy allows the existence of topological defects (solitons), which are domain walls between the two degenerate structures and appear to be responsible for many of the remarkable properties of t-$(CH)_x$ [9, 10]. On the other hand, if single and double bonds are interchanged in cis-$(CH)_x$, adjacent neighbors differ by the relative position of their third nearest neighbor, lifting the degeneracy. This is also true of the other non-degenerate ground state (NDGS) conducting polymers depicted in Fig. 2.1, polydiacetylene (PDA), polythiophene (PT) and its oligomer precursor 6T, and poly(*p*-phenylenevinylene) (PPV).

An optical transition across the Peierls gap in a degenerate ground state (DGS) polymer forms an entirely different excitation than the traditional electron–hole pairs in conventional inorganic semiconductors. The proper description of quasi-particles in DGS polymers is of 1D domain walls or solitons, which separate the two degenerate ground state structures. One Hamiltonian describing DGS polymers was proposed by Su et al. [11, 12]:

$$H_{SSH} = \sum_n t_{n,n+1}(a_n^+ a_{n+1} + a_{n+1}^+ a_n) + \frac{1}{2}\sum_n K(u_n - u_{n+1})^2, \qquad (2.1)$$

where $t_{n,n+1}$ is the intersite transfer integral, a_n creates an electron on the nth state, and u_n measures the displacement of the nth carbon atom from its equilibrium position. Within the framework of H_{SSH}, which includes electron–phonon interactions but not electron–electron and 3D interactions, a photoexcited e–h pair is unstable towards the formation of a soliton–antisoliton pair.

Furthermore, as a consequence of the Pauli principle and charge conjugation symmetry, the photogenerated pair of solitons must be oppositely charged. Within 100 fs after photoexcitation, the lattice spontaneously deforms and the half-occupied orbitals shift symmetrically into the gap. At later times, the localized deformation splits into a pair of charged solitons [12].

The soliton excitation in t-$(CH)_x$ is an amphoteric defect which can accomodate 0, 1, or 2 electrons (Fig. 2.2). The neutral soliton (S^0, spin=1/2) has one electron; positively and negatively charged solitons ($S^{\pm}$, spin-0) have zero and two electrons, respectively. The study of photoexcited t-$(CH)_x$, however, has revealed several unexpected phenomena not predicted by the SSH model. Most important is the observation of an overall neutral excitation which has been identified as due to $\bar{S}^{\circ}S^{\circ}$ pairs. Furthermore, transitions of neutral and charged solitons in t-$(CH)_x$ do not appear exactly at mid-gap as predicted (Fig. 2.2). Thus, any complete desciption of excited state dynamics in DGS polymers must include electron–electron (e–e) interactions. This implies that the electronic gap is partially due to e–e correlations, rather than entirely due to electron–phonon (e–p) interactions as in the SSH model. These findings have also stimulated research into other conducting polymers, most with NDGS [13–15]. The previous picture of bound soliton-antisoliton pairs has been relaxed. In fact, photoexcitation of singlet and triplet excitons has been demonstrated in many conducting polymers. Their existence can be theoretically justified only when e–e interaction is added to the model Hamiltonian. In a noninteracting model, charged solitons are degenerate with neutral solitons. The degeneracy is lifted by e–e interactions which must be included in any quantitative treatment. One approach is to supplement the SSH Hamiltonian with a Hubbard term including on-site correlation effects [16]:

$$H = H_{SSH} + U \sum \rho_{n,\uparrow}\rho_{n,\downarrow}, \tag{2.2}$$

where $\rho_{n,\uparrow}$ is the number of electrons with spin σ on the nth carbon atom. In this model, the degeneracy of charged and neutral solitons is lifted. The various soliton charge states are depicted in Fig. 2.2, using experimentally determined values to locate states in the gap for t-$(CH)_x$ [10, 12].

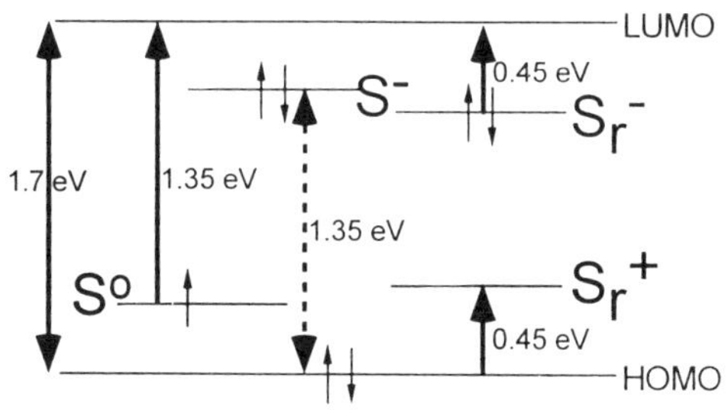

Fig. 2.2. Energy levels of charged and neutral solitons and their transitions, [41]

The most striking property of many conducting polymers is a bright photoluminescence (PL), with quantum yields exceeding 50% in thin films and 80% in solution [17]. This efficient PL as well as a chemically tuneable gap has led to intensive investigations of conducting polymers with respect to their applications as active elements in LEDs. On the other hand, many other conducting polymers exhibit little or no PL. Key to understanding this fundamental difference between luminescent and nonluminescent polymers is the parity (or symmetry) of the ground and excited states. Note, this takes into account the symmetry of all occupied levels. The ground state is of even parity and thus written as the 1A$_g$ state in this notation. Optical transitions involving absorption or emission of one photon are allowed only between states of opposite parity. Conversely, two photon absorption (2PA) occurs between states of the same parity. Kasha's rule [18] for molecules requires that fluorescence come from the lowest excited singlet of whatever symmetry. Hence, the relative location of the lowest odd (1B$_u$) and even (2A$_g$) states will determine whether a polymer is luminescent or nonluminescent. If the 2A$_g$ state lies below the 1B$_u$, as in t-CH$_x$ and PDA, emission from the lowest excited state is dipole-forbidden and the polymer will be nonluminescent. On the other hand, if the 1B$_u$ lies below the 2A$_g$, as for polythiophene and PPV (Fig. 2.1), emission is dipole-allowed and the polymer is luminescent. Comparison of absorption and 2PA studies of a variety of conducting polymers have confirmed this picture [19–21].

The relative ordering of the 2A$_g$ and 1B$_u$ states is a consequence of e–e interactions and demonstrates the fundamental effects on various properties such as PL, electroabsorption (EA), and third-order nonlinear optical susceptibilities. In cases where the bond alternation is relatively large, the ordering of states is such that $E(2A_g) > E(1B_u)$, resulting in an intense PL [19]. When the bond alternation is relatively small, the ordering of these states is reversed and luminescence is quite weak [22]. For instance, the benzene ring in the backbone structure of PPV gives rise to a large bond alternation for the extended π-electrons, and therefore to strong PL [19]. The Coulomb interaction among the π-electrons, even when it is not dominant, leads to behavior qualitatively different from the predictions of the single particle SSH model. Theoretical calculations within a Pariser-Parr-Pople (PPP) type Hamiltonian have provided excellent agreement between experiment and theory [23, 24]:

$$\begin{aligned} H_{PPP} = &\sum_{n,\sigma} t_{n,n+1}(a^+_{n,\sigma}a_{n+1,\sigma} + a^+_{n+1,\sigma}a_{n,\sigma}) \\ &+ U\sum_i n_{i,\uparrow}n_{i,\downarrow} + \frac{1}{2}\sum_{i,u} V_{i,j}(n_i - 1)(n_j - 1). \end{aligned} \tag{2.3}$$

Here $a^+_{i,\sigma}$, creates a π-electron of spin σ on carbon atom n, $t_{n,n+1}$ is the hopping integral between n and $n+1$, $n_{i,\sigma} = a^+_{i,\sigma}a_{i,\sigma}$, and U and V$_{ij}$ are the onsite and intersite Coulomb interactions, respectively.

Whether absorption in conducting polymers is band-like or excitonic is a subject of great debate in the field [15]. The conventional picture [12] has been that these materials behave like quasi-1D semiconductors whose behavior can be described by the SSH Hamiltonian, (2.1) and (2.2). Within this picture, the elementary excitations in NDGS polymers are bound soliton-antisoliton pairs, which can be described as polarons and bipolarons [12]. Recombination between oppositely-charged polarons is responsible for PL. The SSH model has been successful in explaining a variety of optical experiments on conducting polymers. On the other hand, PPP and molecular exciton models [19, 22, 24] focus on e–e interactions and predict that the lowest excited state is an exciton with a binding energy of at least 0.5 eV [24]. Supporting this view is the fact that absorption spectra of ordered conducting polymers, such as stretch-oriented PPV, reveal vibronic structure and closely resemble that of their corresponding oligomers, save for a red-shift [25, 26]. Similarities in the photoinduced absorption (PA) spectra of α-substituted sexithiophene (α-6T) and polythiophene (PT), which indicate that photoexcitations in oligomers and polymers are also similar, will be detailed below. Regardless as to the ultimate resolution of this controversy, it is clear that both e–h and e–e interactions must be considered in any complete model of conducting polymers.

Studies of excited states in conducting polymers have revealed a rich photophysics with interesting and important states above the gap. In addition to the low-lying $1B_u$ and $2A_g$ excitons, 2-photon absorption and electroabsorption studies have revealed an even-parity exciton with particularly strong dipole coupling to the $1B_u$, the mA_g. The existence of biexcitons (BX) has been theorized [23], but not yet definitively identified by experiment. We consider the mA_g state to be an excited state of the $1B_u$ level, whereas the BX level is a bound state of two excitons [23]. The existence of a conducting, or continuum, band above $E(1B_u) + E_{binding}$ has been proposed and tentatively identified by photoconductivity measurements [24]. These singlet states are shown in Fig. 2.3a. Figure 2.3b contains the most important states in the triplet manifold. The lowest triplet level is 1^3B_u, which is lower than the $1B_u$ by the singlet-triplet energy splitting Δ_{S-T}. In principle, the 1^3B_u can directly recombine to the ground state by emission of a photon (phosphorescence). However, this transition is dipole-forbidden and, if present in conducting polymers, extremely weak. Also shown are the m^3A_g, which is equivalent to the singlet mA_g, and the TX level, composed of bound singlet–triplet pairs.

The proper description of the charged excitations in NDGS polymers has been [12] the spin 1/2 polaron ($P^{\pm}$) and the spinless bipolaron ($BP^{2\pm}$). Upon removing of charge from the polymer chain by chemical doping or photoexcitation, a structural relaxation shifts the HOMO and the now singly-occupied SOMO level intrudes into the gap, resulting in the energy diagram of Fig. 2.4a. In oligomers, calculations [27] have shown that the parity of these levels alternates between even (g) and odd (u). Thus, a positive po-

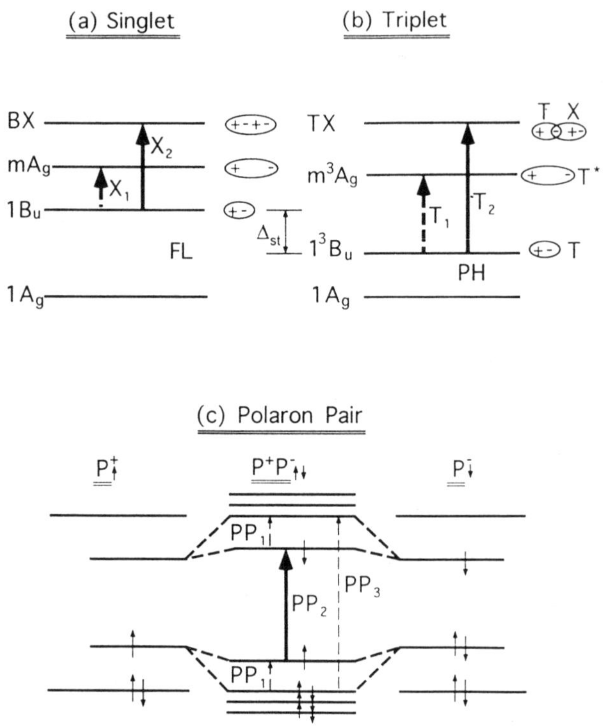

Fig. 2.3 a–c. Energy levels and transitions of neutral photoexcitations in conducting polymers with nondegenerate ground state

laron will have two strong optical transitions P_1 and P_2, as the third possible transition (P_3) is dipole forbidden (Fig. 2.4a). Similarly, a negative polaron will have two similar transitions, though P_1 may be slightly shifted due to differences between positive and negative charges (charge conjugation symmetry violation). Similarities between the photoinduced absorption of α-6T and PT, to be discussed later, lead us to adopt this model for polarons in the polymer. Upon removing a second charge from the same level, there is a further structural relaxation, resulting in the energy diagram of Fig. 2.4b for a positive bipolaron (BP^{2+}). The parity of levels alternates, resulting in only one strong absorption band (BP_1) red-shifted with respect to the P_2 band of polarons with the second transition BP_2 being dipole-forbidden.

When two polarons come together on the same chain, theory predicts that they are unstable with respect to formation of a bipolaron [12]. The validity of this model in real polymers is still an open question. Nevertheless, BPs must appear in heavily doped conducting polymers, whether or not their formation is exothermic. However, another excitation can appear in polymers: a π-dimer (PD) in which like-charged polarons on different chains (or different segments on the same chain) are coupled. PD excitations are

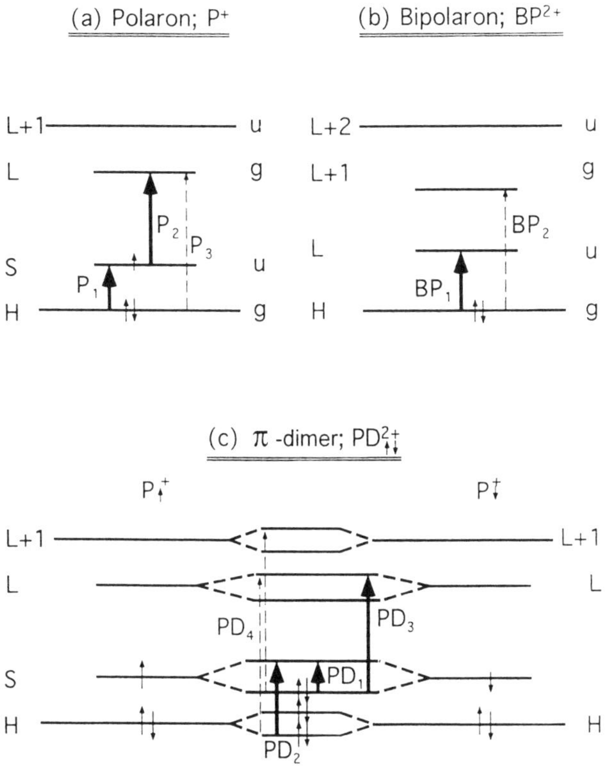

Fig. 2.4 a–c. Energy levels of charged excitations in NDGS polymers

spinless, and their formation has been recently proposed [28] to explain the dramatic decrease in unpaired spins observed in α-6T doped by $FeCl_3^-$. The possibility that PDs can be photogenerated merits consideration. The energy levels of a PD^{2+} are shown in Fig. 2.4c.

We conclude our survey of excited states with bound, oppositely-charged polaron pairs (PP), which have also been called charge transfer excitons or exciplexes. The PL quantum yield (η) of luminescent polymers dramatically drops upon going to film from solution and time-resolved PA studies have revealed a state photogenerated within picoseconds [17]. As η in films is roughly 10–20%, this suggests that the primary photoexcitations in films may be interchain polaron pairs. These excitations cannot be polarons or bipolarons, as the former is photogenerated with a yield of $\sim$1% whereas the latter is not expected to form within ps. The binding energy of a PP should be primarily Coulombic, in contrast to lattice relaxation for a π-dimer. The stronger overlap leads to a larger splitting of the P^+ and P^- levels than for the PD, as shown in Fig. 2.3c. Following the same parity arguments as above, we expect three strong transitions PP_1–PP_3. For a loosely bound pair, these transitions should be similar to the corresponding polaronic transitions. However, for a tightly bound pair, we expect a single transition, PP_2, to dominate the spectrum because PP_1 is considered to be interband with traditional low intensity and PP_3 is close to the energy gap and therefore difficult to distinguish from the HOMO–LUMO transition. We note that it would be difficult to distinguish a tightly bound polaron pair from a deeply trapped exciton.

2.3 Absorption Studies of Conducting Polymers

Perhaps the best way to detect and characterize various photoexcitations of conducting polymers is to study their optical absorption. We begin by discussing the linear and electric field-modulated absorption (EA) spectra. The absorption spectrum of t-$(CH)_x$, shown in Fig. 2.5a, consists of a single featureless band which can be fit by both single-particle SSH [12] and excitonic models [24]. In contrast to $(CH)_x$ and PDA, phenylene-based conducting polymers such as MEH-PPV (Fig. 2.5b) exhibit multiple absorption features extending well into the ultraviolet. Recent studies by several groups [24, 29–31] have emphasized the importance of fitting the whole absorption spectrum of PPV. Such models can provide direct information as to the strength of e–e interactions as well as the role of chemical substitution, finite chain length, and CCS violation.

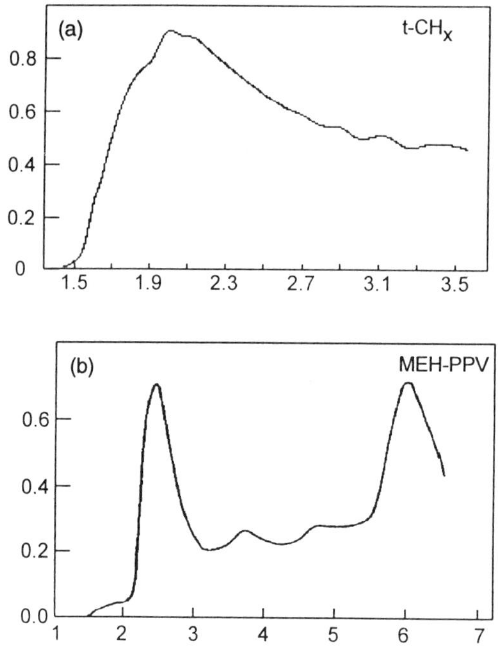

Fig. 2.5 a,b. Absorption spectra of t-CH_x and MEH-PPV, [10, 24]

EA probes changes in optical absorption due to an external electric field. Due to transfer of oscillator strength between states, previously dipole-forbidden transitions can become allowed. Typically, the sample is deposited onto a transparent substrate with a series of interpenetrating metal fingers. An AC electric field is applied, and lockin-detected changes in transmission through the sample are monitored. The optical constants of these polymers are such that to a good approximation, $-\Delta T/T = \Delta \alpha d$, where T is the transmission through the sample, ΔT is the field-induced change in T, and αd is the absorbance of the sample. EA spectra of conducting polymers follow either the first or second derivative of the linear absorption. The EA spectra of PDES and PPE films are shown as the solid lines in Fig. 2.6, whereas the corresponding first and second derivatives of the absorption are shown as dashed lines [32].

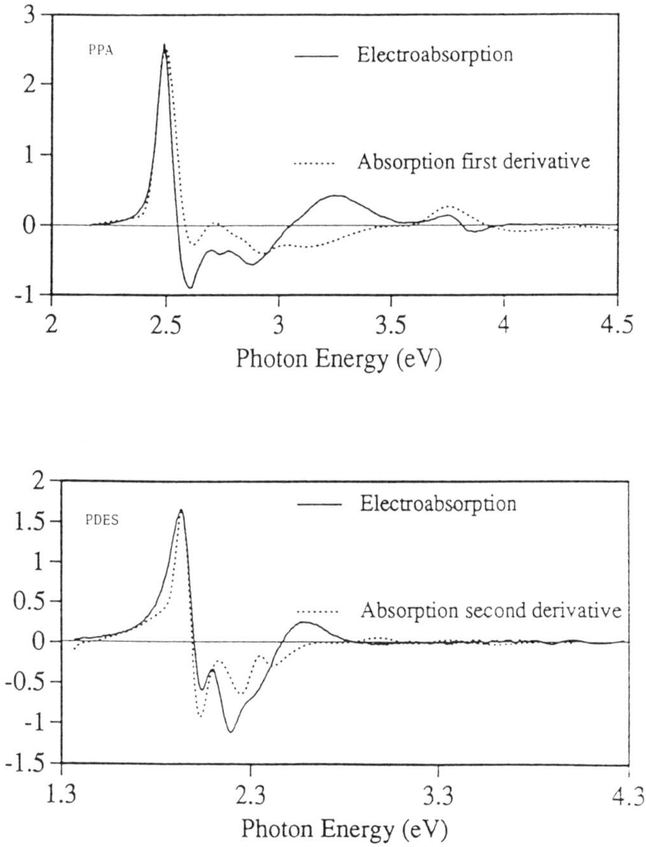

Fig. 2.6. Electroabsorption and absorption derivatives for PPE and PDES, [32]

The quadratic Stark shift of an energy level in response to an electric field is given by $\Delta E = -\frac{1}{2}pF^2$, where p is the polarizability of the energy state and F the field strength [32]. This shift produces an EA spectrum that follows the first derivative of the absorption. Assuming a dipole moment only along the chain leads to the following relation [33]:

$$\Delta\alpha d = \frac{7}{60}\Delta p F^2 \frac{\mathrm{d}(\alpha d)}{\mathrm{d}E}, \tag{2.4}$$

where Δp is the polarizability difference between the $1B_u$ exciton and the ground state.

The second derivative nature of the EA spectrum of some polymers precludes direct application of (2.4). One possibility which does yield a second derivative lineshape is some sort of charge transfer mechanism. However, the absorption and EA spectra of various polymers are not sufficiently different that one can reasonably invoke a Stark shift for one class and neglect it for the other. One alternative explanation is that coupling between close-lying allowed and forbidden transitions in an electric field produces a second derivative-like lineshape. This effect has long been known for F-centers in alkali-halide crystals. However, theory predicts that when the $2A_g$ and $1B_u$ states are separated by more than 0.2 eV, coupling between these levels is negligible. Thus, the generality of this explanation is suspect. The deviation from a first derivative lineshape can be explained by considering the nature of real polymer films. Any individual polymer is made up of a large number of conjugated segements (chromophores). The energy of any segment varies inversely with the conjugation length of each segment and the strength of its contribution to the electroabsorption spectrum is proportional the cube of the conjugation legnth. Hence, the electroabsorption spectrum will be dominated by the longest segments, leading to a spectrum which will more closely resemble the second derivative of the absorption spectrum with respect to energy. A sum-over-states model was recently developed which correctly simulated the absorption and electroabsorption spectra. This model incorporated three essential states and their strongly coupled vibrations, and a conjugation length distribution of the excited states energies [32]. These essential states are the ground state ($1A_g$), the lowest odd-parity excited state ($1B_u$) and an even parity state strongly coupled to the $1B_u$ (mA_g). The mA_g state can be seen in the EA spectra of both PPA and PDES (Fig. 2.6) as an induced absorption band entirely unrelated to the derivative absorption spectra. According to current theories of polymer energy states [23, 24, 34] the mA_g would also mark a lower limit to the location of the continuum band. Therefore, EA has a direct bearing on the binding energy of the 1Bu exciton. These results imply a binding energy in conducting polymers of order 0.7 eV.

2.4 Excited State Optical Studies

Next, we consider photoluminescence (PL) and photoinduced absorption (PA) studies of conducting polymers. Techniques discussed include transient spectroscopy with picosecond (ps) resolution, steady state (cw), and optically detected magnetic resonance (ODMR). Typical experimental setups for transient and cw measurements are shown in Fig. 2.7. In ps transient spectroscopy [35], the sample is photoexcited by picosecond pulses of the pump

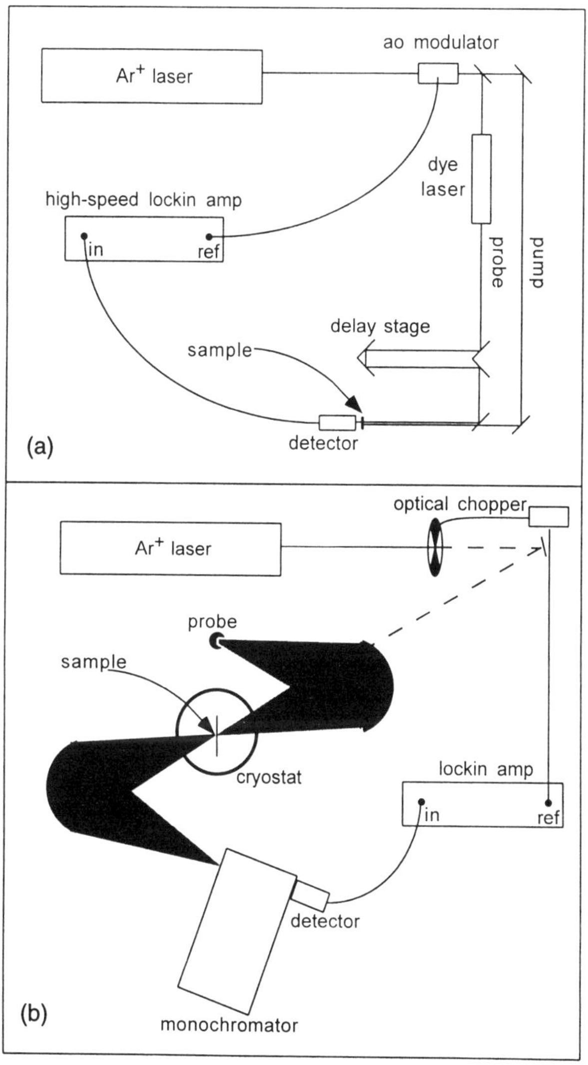

Fig. 2.7. Schematic diagrams of (**a**) ps transient and (**b**) cw PA spectrometers

beam with PL measurements made by a streak camera or multichannel diode array. Transient spectra of photoinduced changes ΔT in transmission T of a time-correlated IR probe beam are studied by varying the mechanically introduced time-delay and the probe wavelength. The spectral range of the ps pump-probe correlation measurements has been 0.7 to 2.8 eV. For cw measurments (Fig. 2.7b) [36], the pump beam is amplitude-modulated and the probe beam is dispersed through a monochromator onto a photodiode. The use of lockin-amplification permits signal resolution $\Delta T/T$ of better than 10^{-6}. By use of multiple detectors and diffraction gratings, PA can be measured in the spectral range of 0.1 to 2.4 eV.

Since its inception in 1967, ODMR has proven to be a highly sensitive method for detection of paramagnetic excited states in a variety of organic and inorganic materials. This technique relies on the fact that the recombination of photoexcitations is often spin-dependent. Hence, population redistribution among magnetic sublevels will produce a change in either the absorbed or emitted light associated with the excitation. The change in light is thus used as a detector of magnetic resonance, replacing direct observation of microwave absorption by the paramagnetic species (ESR). This scaling up from the microwave to the optical range makes ODMR methods extremely sensitive; up to 10^5 times more sensitive than conventional ESR. An ODMR spectrometer resembles a cw PL and PA spectrometer, with the sample placed in a microwave cavity between the pole pieces of an electromagnet [36]. The sample is constantly illuminated by both pump and probe beams; amplitude-modulated microwaves are introduced into the cavity through a waveguide. Changes in PL δI (PLDMR) or transmission δT (ADMR) are detected by lockin-amplification of the photodiode signal. Two types of ADMR spectra are usually obtained: H-ADMR spectra in which δT is measured at a fixed probe wavelength λ while sweeping the magnetic field H, and λ-ADMR where δT is measured at fixed H on resonance, while λ-probe is varied. By ODMR, it is possible to identify the spin of various PA bands as well as their effects on the PL. We briefly discuss models for spin 1/2 and spin 1 (triplet) resonances [36].

We take the case of an electron–hole pair formed from two initially uncorrelated, i.e. non-geminate, charge carriers. In conducting polymers, the pair would be formed from the combination of a negative polaron and a positive polaron. There are four possible spin states, two with anti-parallel spins and two with parallel spins. As recombination to the ground state requires that the pair be in an anti-parallel spin configuration, the steady state population of parallel pairs will be greater. Resonant absorption of a microwave photon will flip the spin of one carrier, which converts the spin configuration from parallel to an anti-parallel and thereby increases the recombination rate. The g values of positive and negative charges can be sufficiently different from one another that an ODMR signal may contain two resonances. However, only a single resonance at $g = 2$ has been observed by ODMR in conjugated polymers. It has been shown that magnetic dipole interactions between polarons composing the pair broaden the resonance to the point where distinct contributions from positive and negative polarons cannot be resolved.

The spin Hamiltonian for triplet excitons states is more complex than that of doublets, due to spin–spin interactions:

$$H = H_{elec} + g\mu_B \mathbf{H}.\mathbf{S} + D\left(S_z^2 - \frac{1}{3}S^2\right) + E(S_x^2 - S_y^2) \tag{2.5}$$

where

$$D = \frac{3g^2\mu_B}{4}\left\langle\frac{r^2 - 3z^2}{r^5}\right\rangle;$$
$$E = \frac{3g^2\mu_B}{4}\left\langle\frac{y^2 - x^2}{r^5}\right\rangle \tag{2.6}$$

H$_{elec}$ is the electronic portion of the Hamiltonian and x, y, z and r all refer to the reduced coordinates of the spins in the triplet. In principle, there will be three microwave induced transitions for a single triplet, two $\Delta m_s = 1$ transitions and one $\Delta m_s = 2$ transition. For polycrystalline or amorphous samples, a powder pattern is formed due to the random orientation of the various triplets with respect to the applied field. While analysis is somewhat more complex than for a single crystal, it is not difficult to extract the ZFS parameters D and E from a powder pattern. A useful approximation for the triplet wavefunction extent R can be derived from (6):

$$R^3 = \frac{2.78 \times 10^4}{D} \tag{2.7}$$

where R is measured in Å and D is in Gauss.

2.5 Optical Studies of t-(CH)$_x$

The cw PA spectrum of t-(CH)$_x$ at 4 K is shown in Fig. 2.8; [37] the t-(CH)$_x$ sample used was a thin film ($d \sim 100$ nm) polymerized on a sapphire substrate by the Shirakawa method. The PA spectrum consists of two main PA bands with $\Delta\alpha > 0$, and photobleaching ($\Delta\alpha < 0$) for $\hbar\omega > 1.65$ eV. The low energy (LE) PA band peaks at 0.43 eV and is correlated with photoinduced infrared active vibrations (IRAVs) [38]. As neutral photoexcitations are not correlated with IRAVs, the band must originate from photoinduced charged defects. The intensity of the LE band easily saturates with increased excitation power and is sample-dependent [38]. As was demonstrated in oriented (CH)$_x$ films, the LE band is induced preferentially with light polarized perpendicular to the chain's direction. Therefore, it seems that the LE band is stabilized by extrinsic defects of the polymer chain. The LE band is now considered to be due to transitions of charged solitons [12]; the transitions are not found at midgap due to e–e interactions [38, 39], as discussed previously (see Fig. 2.2).

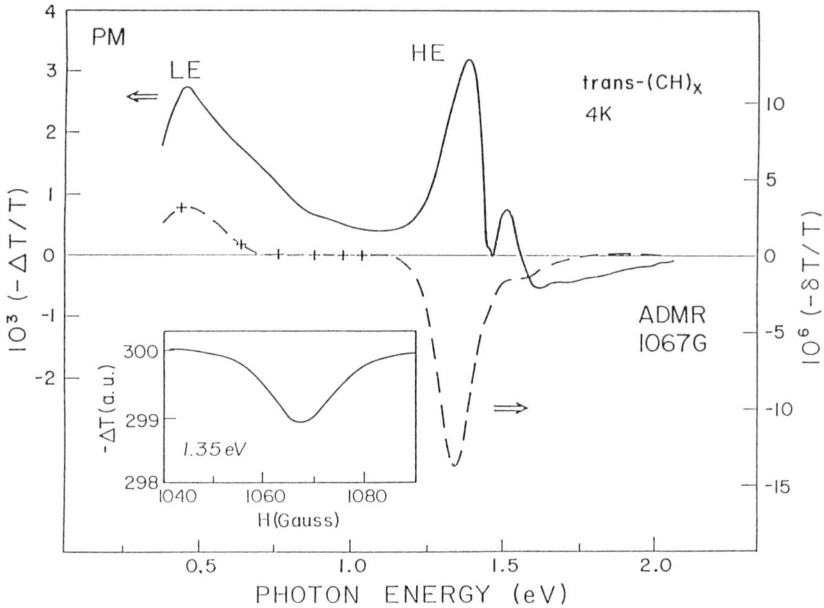

Fig. 2.8. CW PA and PADMR of t-CH$_x$ (Inset: H-PADMR spectrum), [45]

The high energy (HE) PA band peaks at 1.35 eV (Fig. 2.8) and is due to a neutral state which has been tentatively identified as a bound pair of neutral solitons $\bar{S}^\circ S^\circ$ [40], where S is an antisoliton. This PA band does not depend on sample treatment nor does it saturate at high laser illumination. Therefore, it is intrinsic to the $(CH)_x$ chain [38]. The bleaching spectrum is in reverse to $\alpha(\omega)$, indicating bleaching of the interband transitions. However, the bleaching spectrum associated with the LE band peaks at about 1.45 eV and is negligibly small at 2 eV. Therefore, it has been proposed that the long lived $S^{\pm}$ are stabilized by neutral defects that are not intrinsic to the polymerchains. Later ADMR studies [37] proved that the HE photoexcitation has spin 1/2; therefore, it must be correlated with photogenerated neutral solitons.

The picosecond to nanosecond dynamics of the photoinduced bleaching at 2 eV is shown in Fig. 2.9 for both parallel and perpendicular pump-probe beam polarizations [41]. Starting from $t = 0$, $\Delta\alpha_{\parallel} \neq \Delta\alpha_{\perp}$. This photoinduced dichroism implies that the electronic structure is quasi 1D and that the photoexcitations are immediately localized [41]. The induced anisotropy is lost in the nanosecond time range with an exponential time dependence. This is shown in the inset of Fig. 2.9 where the polarization memory P, where:

$$P = (\Delta\alpha_{\parallel} - \Delta\alpha_{\perp})/(\Delta\alpha_{\parallel} - \Delta\alpha_{\perp}), \qquad (2.8)$$

is plotted versus time in semilog scale; P decays with a time constant of about 1.2 ns. However, $\Delta\alpha > 0$ persists out to even longer times, implying a

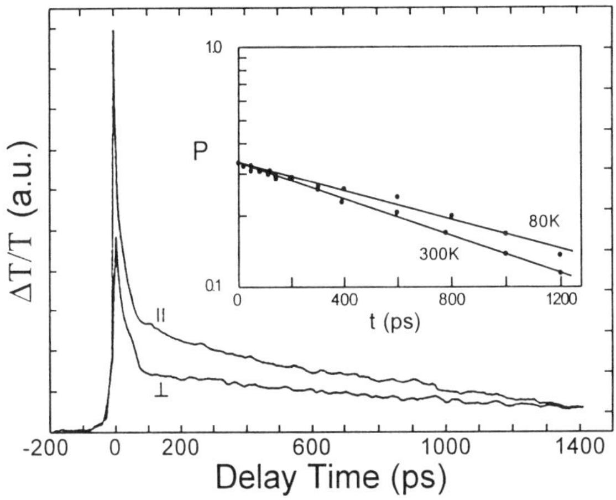

Fig. 2.9. Transient PA dynamics of t-CH_x, [41]

relatively long lifetime for the HE PA band. The polarization memory results from the polarization of photoexcited $\bar{S}^\circ S^\circ$ pairs in regions of the polymer where the $(CH)_x$ chains are parallel to the pump beam polarization. The eventual loss of P results from the diffusion of $\bar{S}^\circ S^\circ$ pairs over sufficiently long distances (100 nm) that the excited $(CH)_x$ are random with respect to the pump polarization. It was calculated [41]:

$$P(t) = P_0\, e^{-4Dt/R^2}, \tag{2.9}$$

where R is the radius of curvature of the $(CH)_x$ fibril and D is the pair diffusion coefficient. Taking $R \sim 100\,\text{nm}$, best fits to the data in Fig. 2.9 yielded $D = 2.2 \times 10^{-2}\,\text{cm}^2/\text{sec}$ at 300 K and $1.7 \times 10^{-2}\,\text{cm}^2/\text{sec}$ at 80 K. These values of D are comparable to those obtained for neutral soliton defects from dynamic magnetic resonance measurements [42].

The H-ADMR spectrum at a fixed probe energy of 1.35 eV (at the HE peak) is shown in Fig. 2.8 (inset) [37]. The spectrum was taken at 4 K, with a resonant microwave frequency of 2.99 GHz and microwave power of 100 mW, modulated at 500 Hz; pump wavelength at 488 nm with an intensity of 500 mW/cm^2. A reduction of PA (i.e $\delta N < 0$), with $\delta N/N \sim -3\times10^{-3}$, is observed at 1067 G, which corresponds to spin 1/2 with $g \sim 2.003$. The resonant lineshape is Lorentzian, with a full width at the half maximum (FWHM) of 16.3 G. The spin 1/2 resonance is the only ADMR signal which has been observed in t-$(CH)_x$. No microwave power saturation has been observed, and the microwave modulation frequency dependence of the ADMR signal was the same as the pump modulation frequency dependence of the HE band in the PM spectrum. Hence, ADMR shows that the HE PA band is correlated with spin-1/2 photoexcitations. Oppositely charged S^+/S^- soliton pairs cannot be the source as these should be spinless photoexcitations and, hence, not observable by ADMR. However, bound $\bar{S}^\circ S^\circ$ pairs are composed of spin-1/2 excitations and we thus identify these pairs, arising from dissociation of $2A_g$ excitons, as the source of the HE band in PA and ADMR.

2.6 Optical Studies of Polythiophene

We extend our review to an oligomer and its polymer counterpart of one important NDGS polymer, polythiophene (PT). Since the PA bands of the oligomer are more well defined compared to those of the polymer films, we discuss the spectroscopy of the oligomer first, followed by the results in PT. As explained above, NDGS polymers do not support soliton excitations. We therefore expect to observe PA related to polarons and bipolarons as the dominant charged excitations; whereas singlet and triplet excitons would be the dominant neutral excitations. The distinction between charged and neutral excitations in the PA spectra can be made relatively easily by studying a possible correlation between the PA bands and the photoinduced IRAVs, regarding their frequency, temperature and intensity dependencies [40]. PA bands of neutral excitations are not correlated, whereas PA bands of charged excitations are correlated with the IRAVs.

The backbone structure of α-sexithiophene, α-6T, was shown in Fig. 2.1. It consists of 6 thiophene rings, and thus can serve as a model oligomer for PT. The sample films ($d \sim 1000$) were prepared by vapor deposition of well purified, pure α-6T molecules on sapphire substrates. α-6T has been shown to support polaron and bipolaron-like excitations when doped. Fichou et al. [43] have shown that the absorption spectrum of the radical cation $6T^{\bullet +}$ consists mainly of two strong bands, each with a pronounced shoulder on its high energy side, which collapse into a single band with a high energy shoulder when the oligomer is further oxidized to its $6T^{++}$ spinless form. It therefore appears that radical ions (positive and negative polarons) have two electronic transitions, whereas the divalent ions (positive and negative bipolarons) show only a single absorption band. This is consistent with the theoretical calculations of Cornil et al. [24].

The PA spectrum of α-6T at 80 K is shown in Fig. 2.10a [44]. The spectrum exhibits PA bands at 0.80, 1.1, and 1.54 eV, respectively. Each of these three peaks is accompanied by a high energy phonon replica. The bands at 0.8 and 1.54 eV have the same dependence on temperature and modulation frequency (ν_c), whereas that at 1.1 eV falls more rapidly as ν_c increases and has a sharply different temperature dependence. We note that each of the main PA peaks in the PM spectrum corresponds to the respective bands in the doping-induced absorption mentioned above [43]. We therefore identify the bands at 0.8 and 1.54 eV as the transitions of photogenerated polarons (P_1 and P_2); the single band at 1.1 eV is related to photogenerated bipolarons (BP_1) in the film. This is in agreement with Figs. 2.4a and 2.4b for the origin of the various PA bands for $P^{\pm}$ and $BP^{\pm}$ excitations. The bleaching of the PA above 1.7 eV is due to a thermal modulation of the film PL [44].

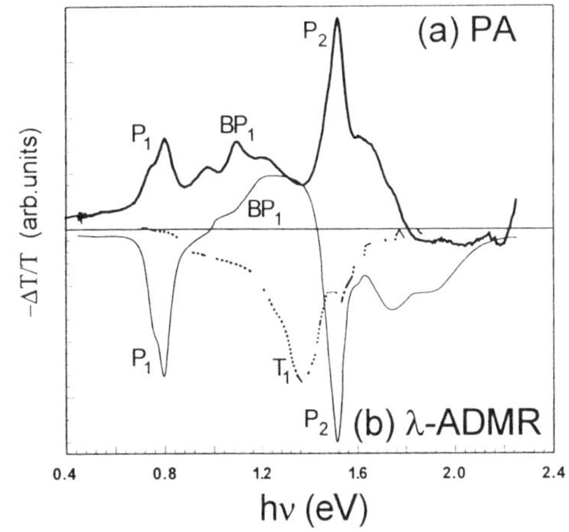

Fig. 2.10. CW PA and λ-PADMR of α-6T at 10K, [44]

The H-ADMR spectrum of α-6T at 4 K at a probe photon energy of 1.54 eV is shown in Fig. 2.11 [44]. The weak resonance at 420 G is due to a $\Delta m_s = 2$ transition of a triplet state and the narrower line at 1070 G is due to $S = 1/2$ photoexcited species. The λ-ADMR spectra of the $S = 1/2$ resonance at 1070 G (solid line), and the half-field triplet resonance at 420 G (dashed line), respectively, are shown in Fig. 2.10b. The $S = 1/2$ spectrum consists of two relatively narrow $\delta N < 0$ bands; P_1 at 0.8 and P_2 at 1.52 eV, respectively, and a broad $\delta N > 0$ band centered at ~ 1.2 eV. The two narrow

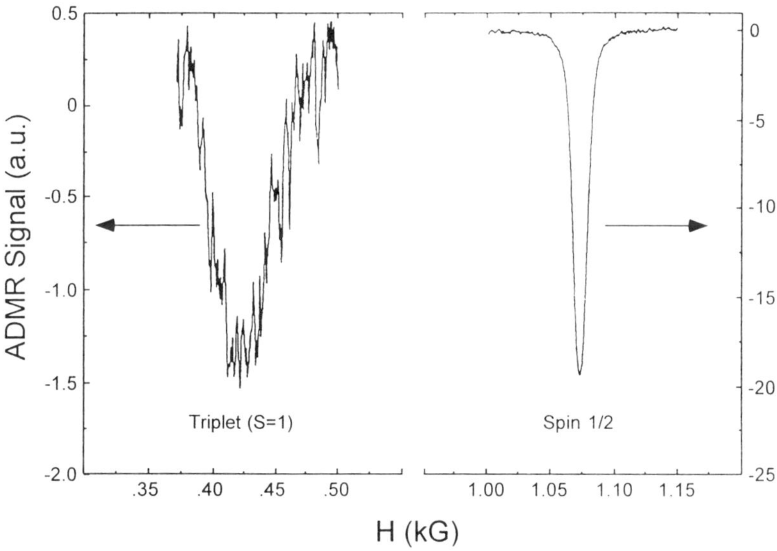

Fig. 2.11. H-PADMR of α-6T, [44]

bands coincide in energy with the polaron bands in the PA spectrum. We therefore conclude that P_1 and P_2 are due to spin 1/2 polarons, which recombine via a "distant pair" process. This is in agreement with the conclusion drawn above from the comparison of the PM with the doping induced spectra. We attribute the broad $\delta N > 0$ band to bipolarons. As bipolarons recombine by annihilating oppositely charged polarons, magnetic resonance conditions which decrease polaron populations will indirectly increase bipolaron populations, leading to a resonance of the opposite sign. A clear bipolaron peak cannot be seen as the high energy shoulder of the LE polaron band will interfere with the bipolaron resonance. If one assumes the both λ-ADMR polaron bands to have similar spectra, the remainder λ-ADMR spectrum is at least qualitatively similar to the BP_1 band observed in PA [36]. The spin 1/2 λ-ADMR shows an additional band at 1.9 eV; as it is correlated with spin 1/2 excitations but not due to polarons, we conjecture the possibility that this may be due to polaron pairs (Fig. 2.3) or a trapped exciton.

Following the discussion of photoexcitations in α-6T, we will discuss two types of PT films: an electrochemically polymerized PT (e-PT) film, where defects play a major role in its photophysics, and a soluble PT derivative, the chemically prepared poly(3-butylthiophene) [P3BT]. The PA spectrum of e-PT at 4 K is shown in the top half of Fig. 2.12 [45]. It contains three PA bands: P_1 at 0.45 eV, P_2 at 1.25 eV and PP_2 at 1.8 eV. These PA bands do not share a common origin. Studying their dependence on the laser excitation intensity, modulation frequency, temperature and decay time after laser pulsed excitation leads to the conclusion that the P_1 and P_2 bands are correlated with each other, whereas the PP_2 band is not. In particular the PP_2 band decays much faster than the two other PA bands and therefore is associated with excitations with much shorter lifetime [45].

The ps transient PA spectrum of e-PT at delay times of 0 (circles) and 400 ps (triangles) is shown in Fig. 2.13 [45]. This consists of a broad PA band that changes into photobleaching at 2.05 eV. The PA band is formed within 300 fs following the initial PB. The dynamics of the PA and PL from 10 ps to 3 ns are shown in Fig. 2.13 (inset). The decays are nonexponential, and up to 40 ps the PL signal decays in a form close to the derivative of the PA. Hence, at short times, the PA and PL share the same origin since $PA \sim N(t)$ and $PL \sim \frac{\mathrm{d}}{\mathrm{d}t}N(t)$. From this and the similarity of the broad PA spectrum to that of poly(diacetylene), in which self-trapped excitons were identified, it was conjectured that the dominant photoexcitations are self-trapped (polaronic) excitons. After 40 ps, when the separation between polarons increases, the recombination dynamics substantially alter and become nonradiative.

The H-ADMR spectra (Fig. 2.12, inset) measured at the peaks of these three PA bands, all show similar properties: The ADMR is negative, has a Lorentzian line shape, and peaks at 1067 G ($g \sim 2.003$) with FWHM of about 9 G. This indicates that all three PA bands are associated with spin 1/2 excitations; spin 1 ADMR has not been detected in e-PT films. The

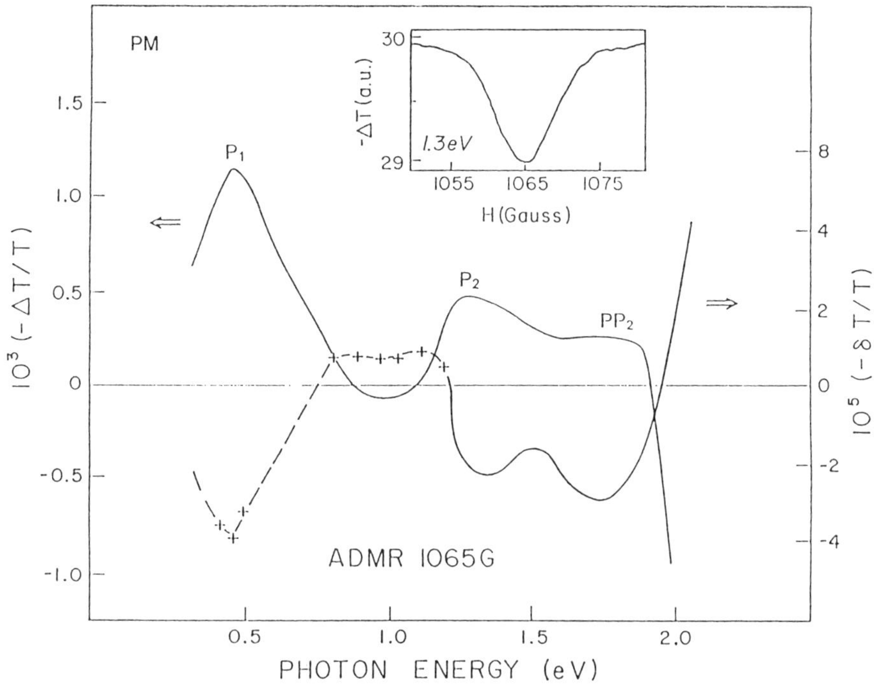

Fig. 2.12. Steady state PA and λ-PADMR of e-PT (inset H-PADMR at 1.3 eV), [45]

λ-ADMR spectrum at 1067 G (Fig. 2.12, lower) contains the same three bands as in the PM spectrum, but with different intensities. The relative microwave induced changes in the photoexcitation density for both P_1 and P_2 bands, whereas for the PP_2 band. As in α-6T, a positive spin 1/2 ADMR signal has been found around 1eV. A similar analysis as for the $\delta N > 0$ signal in α-6T leads to a conclusion that the $\delta N > 0$ band actually shows a peak at about 0.85 eV. From the similarity of the λ-ADMR spectra of e-PT and α-6T, we assign the P_1 and P_2 bands in e-PT as due to polarons P, the $\delta N > 0$ band (now BP_1) to bipolarons BP_2 and the PP_2 band is caused by polaron pairs P^+P^-. This differs somewhat with the original assignment of these PA bands [45], but is in agreement with the new results on α-6T [36, 44].

When we compare the positions of the bands labelled P_1, P_2 and BP_1 in e-PT (Fig. 2.12) and α-6T (Fig. 2.10b), we find a consistent red-shift of about 0.3 eV in e-PT. This is a strong argument in favor of the $P^{\pm}$ identification as the source of P_1 and P_2 bands and $BP^{2\pm}$ of BP_1 in e-PT. The red-shift of the charged excitation transitions in e-PT is presumably caused by the existence of chains longer than 6 rings in this film. We also note that the PP_2 transition stays the same (1.8 eV) in e-PT and α-6T. It seems, therefore, that PP_2 transition does not depend on chain length, in contrast to transitions

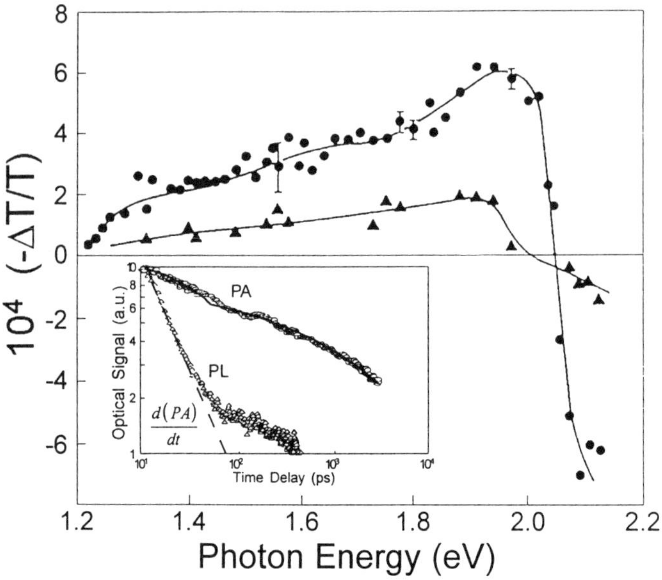

Fig. 2.13. The ps transient spectrum of e-PT at 0 ns (circles) and 400 ps (triangles). Inset: dynamics of Pl and PA from 10 ps to 3 ns, [45]

P_1, $P^{\pm}$ and BP_1 above. In agreement with our assignment, this indicates that the PP excitation is more tightly bound than the P or $BP^{2\pm}$ excitations in conducting polymers.

It is also important to study chemically substituted PT films as there is a strong PA band associated with triplet excitons in the PA spectra of these films. The strong PL emission in P3BT films implies that excitons live longer in these PT-derivatives than in e-PT films. In fact, the excitons in P3BT are sufficiently long-lived to allow intersystem crossing (ISC) to occur into the triplet manifold. This may explain why the PA spectrum in P3BT (as well as in P3OT [36] and PPV derivatives) is dominated by long-lived triplet excitations [46, 47]. The excitons in polymer films with a higher defect density such as e-PT quickly disassociate, before ISC can occur, explaining both the weak triplet PA band and strong PP band in the PM spectra of e-PT (Fig. 2.12). Intersystem crossing may also be affected by the presence of the sulfur atom with relatively strong spin-orbit coupling (the heavy atom effect).

As seen in Fig. 2.14, the PM spectrum of P3BT (upper half) is dominated by a PA band (T_1) at 1.45 eV; there is a second, weaker band (P_1) at 0.55 eV. From the correlation (or lack of) with the photoinduced IRAVs, seen in Fig. 14 at $\hbar\omega < 0.15eV$, we inferred that P_1 is due to charged excitations, whereas T_1 is caused by neutral excitations. Both $S = 1/2$ and $S = 1$ excitations have been identified by ADMR spectroscopy; their λ-ADMR is

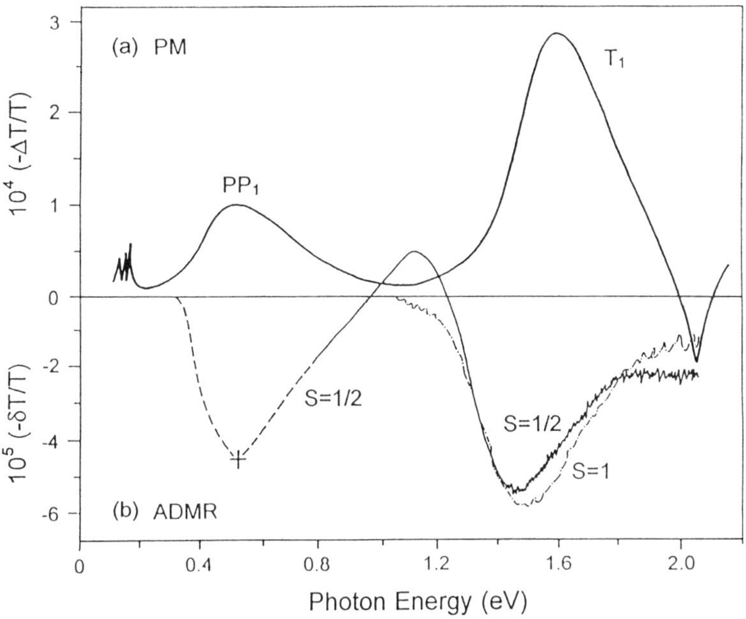

Fig. 2.14 a,b. CW PA and λ-PADMR of P3BT, [36]

also shown in Fig. 2.14 (lower half). There are two strong $S = 1/2\delta N < 0$ bands at 0.55 and 1.4 eV, respectively, a weaker shoulder at 1.8–1.9 eV, and a $S = 1/2\delta N > 0$ band at about 1 eV. The $S = 1$ λ-ADMR, in contrast, contains a single $\delta N < 0$ band at 1.5 eV (Fig. 2.14, lower). As this is quite similar to the results in α-6T, we identify the strong negative $S = 1/2$ ADMR bands as due to P_1 and P_2 transitions associated with P excitations, the $\delta N < 0$ shoulder at 1.9 eV as due to PP excitations, the δN¿0 band as BP_1 associated with $BP^{2\pm}$. It is worth noting the potential of the ADMR spectroscopy to separate these various PA bands from such a relative featureless PA spectrum (Fig. 2.14a). The PA bands associated with the charged excitations ($P^{\pm}$ and $BP^{2\pm}$) are red-shifted (~0.2 eV) compared to those in α-6T, but the PA bands due to the neutral excitations (T_1 and PP_2) are not, showing that they are tightly bound, independent of the PT chain length. The energy difference between the polaron bands is 0.8 eV, as in e-PT and α-6T films.

2.7 Optical Properties of C_{60}: Polymer Composites

The discovery of photoinduced charge transfer (PCT) from a π-conjugated polymer onto C_{60} has stirred widespread interest [48–50]. Persistent photoconductivity in fullerene-polymer complexes has drawn particular attention due to applications as low-cost photovoltaic devices [51, 52]. While there is no doping in the ground state, PA studies have demonstrated an excited-state absorption spectrum different from that of the polymer or C_{60}. [53] Light-induced ESR (LESR) experiments have clearly identified the spin resonance of positive polarons on PPV and negatively-charged C_{60} ions [53]. C_{60} doping not only quenches the PL, but sharply reduces the lifetime of the remnant emission [54, 55]. This behavior is consistent with nonradiative quenching of ^1B_u singlet excitons by the positive polaron injected onto the chain from C_{60}. It is also similar to the effect of photoinduced oxidation on PPV, where the formation of carbonyl centers both quenches the PL and sharply reduces the PL lifetime [56].

The ps PA spectra of freshly prepared DOO-PPV with pump-probe delay times of 0 and 1 ns, respectively, are shown in Fig. 2.15a [57]. The $t = 0$ ps spectrum shows a broad band centered at 1.5 eV and a structured photo-bleaching (PB) band above about 1.7 eV. As the ps PB is identical to the PL spectrum at 300 K, this signal was identified as due to stimulated emission (SE) of photogenerated excitons by the probe beam [57]. Since the decay of the PA and SE are correlated up to approximately 600 ps, the initial PA band at 1.5 eV must be due to excitons. At 1 ns, the PA band narrows and its decay is no longer correlated with the SE. Since the triplet PA band in DOO-PPV is at 1.5 eV, it is possible that this PA band is due to the formation of triplet excitons.

Figure 2.15b shows the ps transient PA in 10% C_{60}-doped DOO-PPV film at delay times of 0 and 2 ns, resp. In contrast to pristine DOO-PPV film, the transient PA in 10% C_{60}-doped films is much broader, quite stable in time, and contains very little SE for $t \leq 50$ ps. The lack of appreciable SE combined with the fact that ADMR measurements have shown that triplet resonance has been almost entirely quenched, lead us to conclude that singlet and triplet excitons play only a minor role in this transient PA spectrum. Furthermore, separated charges on the polymer chain and C_{60} cannot follow the the fast pump modulation frequency ($f \approx 4$ MHz), because these charges recombine in times much longer than $1/f$ (250 ns). Therefore, the contributions of PPV^+/C_{60}^- complexes to the measured transient PA spectra will be negligibly small. All of these factors lead us to conclude that the transient PA of Fig. 2.15b is due to polaron pairs on the *polymer chain*.

The steady-state PA spectrum of undoped MEH-PPV, shown in Fig. 2.16a, is dominated by a band with a maximum at 1.5 eV, labeled T. As it is not correlated with IRAVs, it must be due to neutral photoexcitations. -ADMR measurements [34] have shown that there is band with spin-1 with roughly the same spectrum. Hence, we identify this band as due to triplet excitons. Upon 50% doping by weight with C_{60}, the triplet band is almost entirely quenched, a pair of peaks arise at ≈ 1.3 eV as well as a dramatic enhancement of a band at 0.4 eV. These bands are accompanied by a series of sharp IRAVs, leading us to conclude that they arise from charged excitations. These results are consistent with polarons generated through charge transfer from

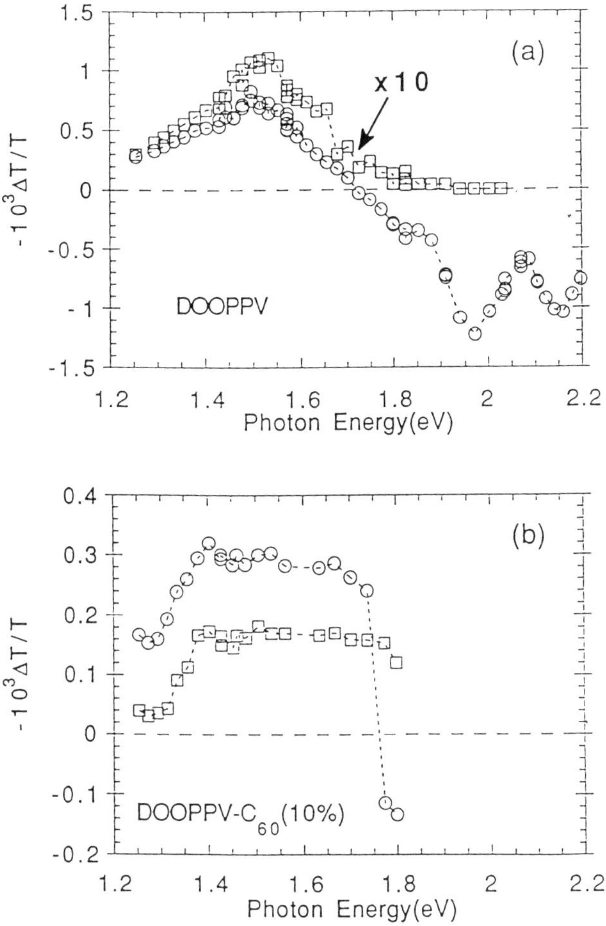

Fig. 2.15 a,b. Transient PA spectra of (**a**) undoped and (**b**) C_{60}-doped DOO-PPV. Circles are for a OPS delay time between and squares for 2 ns

PPV to C_{60}. We also note that these spectra indicate that the generated polarons have two subgap transitions, as with α-6T and polythiophene.

2.8 Transport Properties of Conducting Polymers

The conductivity of $(CH)_x$ varies from 10^{-13} to 10^5 S/cm, depending upon preparation conditions and doping [1–6]. Given this range of conductivity, it is not surprising that a variety of models have been proposed. In many polymeric systems, charge transport is believed to proceed via phonon-assisted hopping [58, 59]. A charge carrier moves between sites accompanied by a localized deformation of atomic positions. Hence, charge transport will be

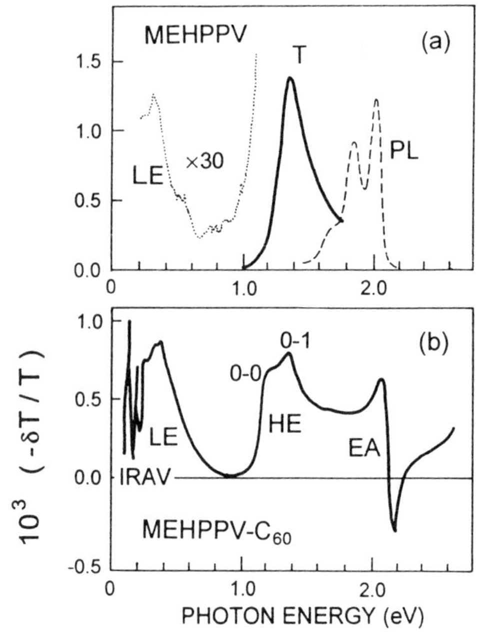

Fig. 2.16 a,b. CW PA spectra of (**a**) undoped and (**b**) C_{60}-doped MEH-PPV

dependent upon the strength of e–p interactions. A polaron can move only when atoms, whose displacements cause the self-trapping, alter their positions. If the polaron is deeply trapped, the atomic displacements will be substantial and its motion is severely impeded. On the other hand, a lightly trapped (or itinerant) polaron is associated with atomic displacements that are sufficiently small that its motion is only slightly impeded. This distinction between trapped and itenerant polarons is not surprising in view of two observations: PL and ODMR studies support the view that polarons recombine nonradiatively and may even be PL-quenching defects [47]. However, it is clear that the origin of electroluminescence in polymer diodes is from radiative recombination of charged carriers, presumably itinerant polarons. Examining the temperature dependence of the DC conductivity is a useful tool for distinguishing transport mechanisms. Results of early experiments on t-$(CH)_x$ prepared by the Shirakawa method are shown in Fig. 2.17. The data shown have an excellent fit to the following relationship:

$$log\sigma \sim (T_0/T)^{\frac{-1}{4}}, \tag{2.10}$$

where $T_0 = 1.9 \times 10^9\,\mathrm{K}$. This fits very well to a model of 3-dimensional variable range hopping, which predicts a weak temperature dependence of this form [39]. While a detailed discussion on various transport models in conducting polymers is beyond the scope of this chapter, the presence of defects, which break charge conjugation along the chains, requires some form of hopping-related transport.

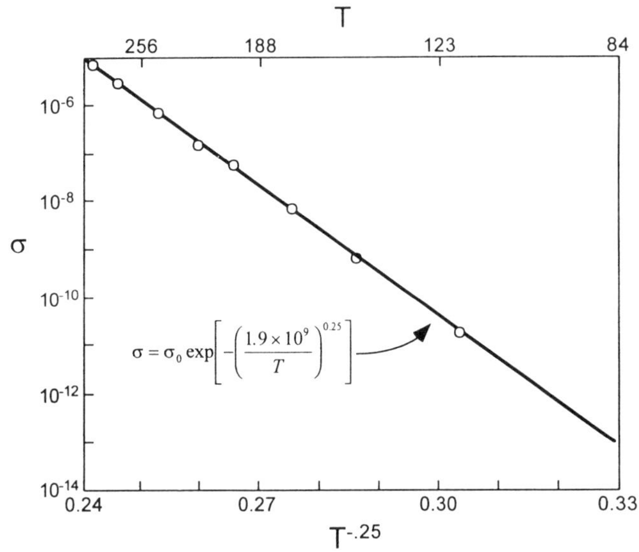

Fig. 2.17. Temperature dependence of conductivity in t-CH_x, [58]

Transport studies also have direct bearing on the fundamental issues in conducting polymer physics. The results of photoconductivity (PC) studies have proven to be controversial. The PC quantum efficiency of a conducting polymer-based photodiode with MEH-PPV as the active layer is shown in Fig. 2.18 [24], along with the absorption spectrum of MEH-PPV. PC was measured in a sandwich structure, with successive layers of indium tin oxide (ITO), MEH-PPV, and an aluminum electrode (Fig. 2.19a). Measurements

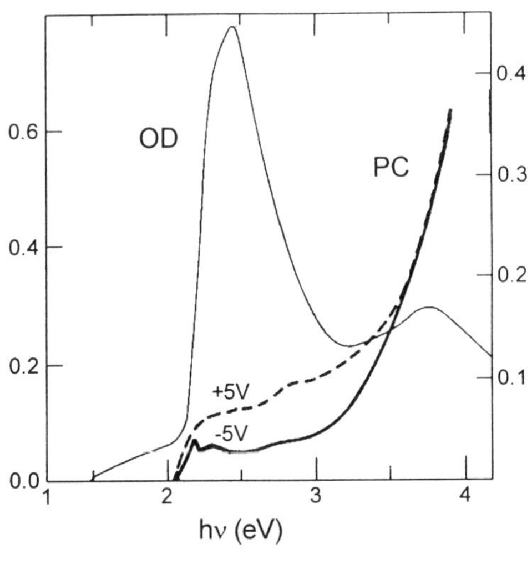

Fig. 2.18. The spectra of the optical density (left axis) and photoconductivity quantum yield (right axis) in a device configuration, [24]

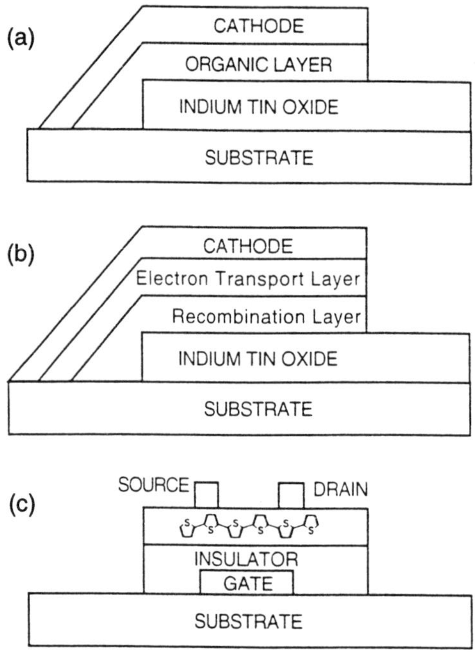

Fig. 2.19 a,b. Schematic diagrams of (**a**) single-layer and (**b**) bilayer LEDs and (**c**) an organic TFT

were preformed at room temperature in vacuum with applied biases of +5 V (solid line) and −5 V (dashed line). The onset of PC at the absorption edge has been explained as supporting a band-like picture [53]. However, there is a second onset of PC at 3.1 eV where there are no corresponding features in the absorption spectrum. An alternative explanation to the first onset that it arises from charged polarons created by exciton dissociation [24]. This is consistent with optical studies detailed previously. Then the second onset is due to excitation directly into the continuum band ~0.7 to 0.9 eV above the $1\,B_u$ exciton. This picture is consistent with the results of recent ODMR studies [60], where visible excitation of PPV yields a PL-enhancing resonance due to trapped polarons and UV excitation ($h\nu > 3\,\text{eV}$) yields a PL-quenching resonance presumably due to itenerant polarons.

2.9 Device Applications of Conducting Polymers

The potential of conducting polymers and oligomers for active devices such as light-emitting diodes (LEDs), solar cells, and thin film transistors has attracted much interest. Polymer LEDs in particular, have gained considerable attention since 1990, when a group at Cambridge University reported [61] that PPV devices exhibited significant electroluminescence (EL). In principle, displays built from polymer devices might be simpler to design and manufacture than active matrix liquid crystal displays. The advantages of polymer LEDs include high quantum yields, high thermal stability, flexibility and the ability to prepare large areas. A number of polymer LEDs also exhibit EL in both forward and reverse biases, permitting AC operation [62, 63]. As noted above, a sandwich geometry is generally used where ITO serves as a transparent anode and a metal with a low work function (ϕ) such as aluminum or calcium serves as the cathode. For conducting polymers such as PPV, electron injection has limited device efficiency when the anode is ITO. Lowering the work function by changing from aluminum ($\phi = 4.3\,\text{eV}$) to calcium ($\phi = 2.9\,\text{eV}$) has improved device efficiency from 0.002 to 0.1% [64], where efficiency is defined as the number of photons emitted per electron injected. As calcium is highly reactive, it is not a satisfactory solution to the problem of electron injection.

Several approaches to overcoming this problem have been undertaken. Insertion of an electron transporting (ET) layer between the cathode and PPV improves device efficiency by confining holes to the PPV layer and preventing unipolar current flow [65]. Furthermore, electron injection is facilitated, permitting the use of more stable metals without compromising device efficiency. An ET layer also diminishes exciton quenching near the metal electrode by acting as a spacer. One widely used material is 2-(4-biphenyl)-5-(4-tert-butylphenyl)-1,3,4-oxadiazole (PBD) dispersed in an inert polymer matrix such as poly(methyl methacrylate (PMMA) [66]. The geometry of a polymer LED incorporating an ET layer is shown in Fig. 2.19b. Substantial efforts are being devoted to the discovery and application of new materials as ET layers [66].

An alternative to the use of a distinct ET material is modification of the conducting polymer itself through incorporation of an electronegative acceptor group [55]. Cyano-substituted PPV (CN-PPV) is prepared by use of a precursor molecule which incorporates a CN unit replacing a hydrogen atom at the vinyl linkage. Substituted CN-PPV is both soluble and has a high electron affinity. Single-layer devices based on CN-PPV show quantum efficiencies up to 0.2% with either aluminum or calcium electrodes. Other single-layer devices show at least a ten-fold reduction in efficiency when changing from calcium to aluminum. Combining the above approaches by using MEH-PPV as the hole-transporting layer and CN-PPV as the electron-transporting layer in a bilayer configuration raised the internal quantum efficiency to $\sim$4% [55]. The spectrum of EL from these devices matches that of a CN-PPV LED, indicating recombination takes place in the CN-PPV layer, probably near the interface between the two polymer layers. Work on replacing the other electrode with the conducting polymer PANi has resulted in higher device efficiencies due to improved hole injection. Substantial progress has been made towards improving device efficiency and stability. Researchers at Phillips Electronics have reported large area polymer-based devices with projected lifetimes well over 10 000 hours with a light intensity of 200 cd/m^2 biased at 3.5 V and power efficiencies of 17 lumens/W [67].

Solar cells are another area of interest. The high cost of photovoltaics has limited their use to niche applications such as powering calculators. For a solar cell, the device must absorb photons and the resulting electron–hole pair separated. The excitonic nature of conducting polymers has made this a difficult task, but the incorporation of an ET material blended into the polymer layer has improved device efficiency by a factor of 100. For this application, both CN-PPV [68] and C_{60} have shown promise, with a combination of both materials proving most effective [69]. Unfortunately, these polymers are not stable when exposed to strong light, a serious liability for a solar cell. For that reason, these devices are more likely to succeed in less demanding applications such as photodetectors.

Thin film transistors (TFTs) made from conducting polymers are interesting from both commercial applications and scientific standpoints. In particular, field effect carrier mobilities can be directly measured and film conductivity can be extracted from a study of device characteristics. Studies of polymer TFTs have proven useful in learning about the effects of doping, structure, and materials preparation [70]. A typical polymer TFT is shown in Fig. 2.19c. The semiconducting layer is composed of a thin film of organic material to which two metal electrodes, the source and the drain, are attached forming ohmic contacts. The source-drain current is controlled by a third electrode, the gate, which is isolated from the polymer by an insulating layer. Carrier mobilities in polymer TFTs have typically been in the range of 10^{-4} to 10^{-5} cm^2/V/s, far below the levels of amorphous silicon ($\mu \approx 1$).

While polymers appear unlikely candidates for commercial applications of organic TFTs, oligomers of polythiophene (nT) have shown promise. Oligomers have well defined electronic properties as well as a crystalline, rather than an amorphous, microstructure. Table 2.1 summarizes the characteristics of TFTs made from nT and α-substituted nT with SiO_2 as the insulating layer [71]. The conductivity and carrier mobility increase as n increases, peaking for $n = 6$. This maximum at 6T is not surprising in view of spectroscopic data that shows octithiophene (8T) contains an appreciable number of $\alpha\beta'$-coupled 4T units. Alkyl-substitution at the site on the end of the chain markedly improves carrier mobility appears due to enhanced ordering of the chains. However, substitution at mid-chain β sites effectively destroys it, which may be due to spacing the chains from one another or a lower degree of order. These results underscore the importance of structural order on transport properties in conducting polymers and oligomers. With on-off ratios of 10^6 and field-effect mobilities as high as 10^{-1}, α-6T TFTs have reached levels where they can be considered competitive with a-Si [72, 73]. Improvements in crystallinity as well as purity of longer oligomers hold great promise.

2.10 Summary

We have reviewed electronic states in undoped conducting polymers and described the results of optical and transport studies. We focused on studies of photoexcitations in two systems of conducting polymers; trans-polyacetylene with its unique two-fold degeneracy, and in polythiophene, a representative NDGS polymer. The long lived charged and neutral photoexcitations in both systems have been identified. The charged photoexcitations in t-$(CH)_x$ are solitons with a single PA band in the gap, whereas in polythiophene they form polarons with two PA bands in the gap. However, we have also identified photogenerated bipolarons in NDGS polymers, which have a single PA band below the gap. BPs may play an important role in more ordered polymer films, where two like-charged polarons may easily recombine to form a bipolaron. The long-lived neutral photoexcitations in both polymer classes were also identified. In the singlet manifold, neutral photoexcitations form soliton pairs, $\bar{S}^{\circ}S^{\circ}$ in t-$(CH)_x$ and P^+P^- polaron pairs in polythiophene. Both pairs show a single, strong PA band below the gap, with an associated negative spin 1/2 ADMR signal, which is caused by the exchange coupling interaction between the spin 1/2 particles composing the pair. Both SS and PP pairs can be generated in the ps time domain directly from the allowed exciton bands of the polymers. From the ZFS parameters of triplet excitons, we have determined that the triplet excitations in NDGS polymers and oligomers are tightly localized.

Studies of transport properties in conducting polymers indicate a hopping behavior for charge transport, leading to a weak temperature dependence. Polymer devices have proveen to be quite promising; they are quite useful for studying the excited state properties of charged excitations in these materials and appear to be on the verge of commercial breakthroughs. Their combination of unique electronic states and potential applications make studies of conducting polymers a fascinating field.

Acknowledgement. We would like to thank our collaborators without whom this work would not be possible. They includes Profs T. Barton, R.E. Benner, E. Ehrenfreund, J. Shinar, B.C. Hess and K. Yoshino; Drs S. Jeglinski, G.S. Kanner, J.M. Leng, X. Wei, S. Frolov, and M. Liess.

This work was supported in part by the Office of Naval Research under grant ONR N00014-94-1-0853 and by the Department of Energy under grant DOE FG-03-93 ER 45490.

References

1. H. Shirakawa, E. J. Louis, A. G. MacDiarmid, C. K. Chiang, A. J. Heeger: JCS Chem. Comm. 578 (1977).
2. C. K. Chiang et al.: Phys. Rev. Lett. **39**, 1098 (1977).
3. N. Basescu et al.: Nature **327**, 403 (1987).
4. K. Araya, A. Mukoh, T. Narahara, H. Shirakawa: Chem. Lett. 1141 (1984).
5. H. Naarmann and N. Theophilou: Synth. Met. **22**, 1 (1987).
6. J. Tsukamoto, A. Takahashi, K. Kawasaki: Jpn. J. Appl. Phys. **29**, 125 (1990).
7. R. E. Peierls: *Quantum Theory of Solids*, (Oxford University Press, London, 1955) p. 108.
8. J. T. Devrees, R. P. Evrard, V. E. Van Doren (Eds): *Highly Conducting One-Dimensional Solids*, (Plenum, NY, 1976).
9. T. W. Hagler and A. J. Heeger: *Chem. Phys. Lett.* **189**, 333 (1992).
10. *Handbook of Conducting Polymers*, 2nd Edition, Eds: T. Skotheim, R. Elsenbaumer and R. L. Reynolds, (Dekker, New York, 1997).
11. W. P. Su, J. R. Schrieffer, A. J. Heeger: Phys. Rev. B **22**, 2099 (1981).
12. A. J. Heeger, S. Kivelson, J. R. Schrieffer, W. P. Su: Rev. Mod. Phys. **60**, 781 (1988).
13. Proc. Int. Conf. Synthetic Metals, Seoul, S. Korea, (1994), Synth. Met. **69-71** (1995).
14. Proc. Int. Conf. Synthetic Metals, Snowbird, Utah, (1996), Synth. Met. **84-86** (1997).
15. *Proc. Optical Probes of Conjugated Polymers III*, Eds L. J. Rothberg and Z. V. Vardeny, Proc. SPIE **3145**, 132 (1997).
16. W. P. Su: Mol. Cryst. Liq. Cryst. **256**, 45 (1994).
17. M. Yan, L. J. Rothberg, F. Papadimitrakopoulos, M. E. Galvin, T. M. Miller: Phys. Rev. Lett. **72**, 1104 (1994).
18. M. Kasha: Discuss. Faraday Soc. **9**, 14 (1950).
19. Z. G. Soos, S. Etemad, D. S. Galvao, S. Ramaseshu: Chem. Phys. Lett. **194**, 341 (1992).
20. C. J. Baker, O. M. Gelsen, D. D. C. Bradley: Chem. Phys. Lett. **201**, 127 (1993).
21. G. S. Kanner, S. Frolov, Z. V. Vardeny: Phys. Rev. Lett. **74**, 1609 (1993).
22. Z. G. Soos, R. Ramaseshon, D. S. Galvao: Phys. Rev. Lett. **71**, 1609 (1993).
23. F. Guo, M. Chandross, S. Mazumdar: Phys. Rev. Lett. **74**, 2068 (1995).
24. M. Chandross, S. Mazumdar, S. Jeglinski, X. Wei, T. Miller, Z. V. Vardeny: Phys. Rev. Lett. **50**, 14702 (1994).
25. K. Pichler, D. A. Halliday, D. D. C. Bradley, P. L. Burns, R. H. Friend, A. B. Holmes: J. Phys. Condens. Matter **5**, 7155 (1993).
26. R. Kersting, U. Lemmer, R. F. Mahrt, K. Leo, H. Kurz, H. Bssler, E. O. Gbel: Phys. Rev. Lett. **70**, 3820 (1993).
27. D. Beljonne, Z. Shuai, J. L. Brdas: J. Chem. Phys. **98**, 8819 (1993).
28. M. G. Hill, K. R. Mann, L. L. Miller, J.-F. Penneau: J. Am. Chem. Soc. **114**, 2728 (1992).
29. J. Cornil, D. Beljonne, R. H. Friend, J. L. Brdas: Chem. Phys. Lett. **223**, 82 (1994).
30. M. J. Rice and Y. N. Gartstein: Phys. Rev. Lett. **73**, 2504 (1994).
31. Y. Shimoi and S. Abe: Synth. Met. **91**, 363 (1997).
32. M. Liess, S. Jeglinski, Z. V. Vardeny, M. Ozaki, K. Yoshino, Y. Ding, T. J. Barton: Phys. Rev. B **56**, 5781 (1997).
33. R. G. Kepler and Z. G. Soos: Phys. Rev. B **43**, 12530 (1991).
34. J. M. Leng, S. Jeglinski, X. Wei, R. E. Benner, Z. V. Vardeny, F. Guo, S. Mazumdar: Phys. Rev. Lett. **69**, 668 (1992).

35. Z.V. Vardeny and J. Tauc: in *Semiconductor Processes Probed by Ultrafast Laser Spectroscopy*, (Ed.) R.R. Alfano, (Academic Press, NY 1985) pp. 91–137.
36. P.A. Lane, X. Wei, Z.V. Vardeny: Phys. Rev. B **56**, 4626 (1997).
37. X. Wei, B.C. Hess, Z.V. Vardeny, F. Wudl: Phys. Rev. Lett. **68**, 666 (1992).
38. J. Orenstein, in ref [10].
39. Z.V. Vardeny and J. Tauc: Phys. Rev. Lett. **54**, 1844 (1985); **56**, 1510 (1986); Philos. Mag. B **52**, 313 (1985).
40. J. Orenstein and G.L. Baker: Phys. Rev. Lett. **49**, 1043 (1982).
41. Z.V. Vardeny, J. Strait, D. Moses, T.C. Chung, A.J. Heeger: Phys. Rev. Lett. **49** (1982).
42. K. Holczer, J.P. Boucher, F. Devreu, M. Nechtschein: Phys. Rev. B **23**, 1051 (1981).
43. D. Fichou, G. Horowitz, B. Xu, F. Garnier: Synth. Met. **39**, 243 (1990).
44. P.A. Lane, X. Wei, Z.V. Vardeny, J. Poplawski, E. Ehrenfreund, M. Ibrahim, A.J. Frank: Chem. Phys. **210**, 229 (1996).
45. G.S. Kanner, X. Wei, B.C. Hess, L. Chen, Z.V. Vardeny: Phys. Rev. Lett. **69**, 538 (1992).
46. N.F. Colaneri, D.D.C. Bradley, R.H. Friend, P.L. Burn, A.B. Holmes, C.W. Spangler: Phys. Rev. B **42**, 11671 (1990).
47. L.S. Swanson, P.A. Lane, J. Shinar, F. Wudl: Phys. Rev. B **44**, 10617 (1991).
48. Y. Wang: Nature **356**, 585 (1992).
49. N.S. Sariciftci, L. Smilowitz, A.J. Heeger, F. Wudl: Science **258**, 1474 (1992).
50. S. Morita, A.A. Zakhidov, K. Yoshino: Solid State Comm. **82**, 249 (1992).
51. S. Morita, A.A. Zakhidov, K. Yoshino: Solid State Comm. **82**, 249 (1992).
52. N.S. Sariciftci, D. Braun, C. Zhang, V. Srdanov, A.J. Heeger, G. Stucky, F. Wudl: Appl. Phys. Lett. **62**, 585 (1993).
53. L. Smilowitz, N.S. Sariciftci, R. Wu, C. Gettinger, A.J. Heeger, F. Wudl: Phys. Rev. B **47**, 13835 (1993).
54. N.S. Sariciftci, L. Smilowitz, A.J. Heeger, F. Wudl: Science **258**, 174 (1992).
55. S. Morita et al.: *Jpn. J. Appl. Phys.* **31**, L890 (1992); K. Yoshino et al.: *ibid.* **32**, L140 (1993); *ibid.* L357.
56. M. Yan, L.J. Rothberg, F. Papadimitrakopoulos, M.E. Galvin, T.M. Miller: Phys. Rev. Lett. **73**, 744 (1993).
57. S.V. Frolov, Z.V. Vardeny, M. Ozaki, K. Yoshino: Bull. Amer. Phys. Soc. **40**, 103 (1995).
58. A. Epstein: *Handbook of Conducting Polymers*, T.A. Skotheim (Ed.) (Marcel Dekker, NY) Vol. 1 (1986).
59. S. Kivelson: Phys. Rev. Lett. **46**, 1344 (1981).
60. J. Shinar et al.: Mol. Cryst. Liq. Cryst. **256**, 691-696 (1994).
61. J.H. Burroughes et al.: Nature **347**, 539 (1990).
62. S.A. Jeglinski, O. Amir, X. Wei, Z.V. Vardeny, J. Shinar, T. Cerkvenik, W. Chen, T.J. Barton: Appl. Phys. Lett. 67, 3960 (1995).
63. S. Jeglinski, Ph.D. Thesis, University of Utah, 1995 (unpublished)
64. D. Braun and A.J. Heeger: Appl. Phys. Lett. **58**, 1982 (1991).
65. N.C. Greenham, S.C. Maratti, D.D.C. Bradley, A.B. Holmes, R.H. Friend: Nature **365**, 628 (1993).
66. M. Strukelj, F. Papadimitrakopoulos, T.M. Miller, L.J. Rothberg: Science **267**, 1969 (1995).
67. H.P. Schoo: Oral Communication, 1998 Spring Materials Research Society Meeting, San Francisco, CA.
68. J.J.M. Halls et al.: Nature **376**, 498 (1995).
69. R.F. Service: Science **269**, 920 (1995).

70. F. Garnier, G. Horowitz, X. Peng, D. Fichou: Adv. Mater. **2**, 592 (1990); J. Appl. Phys. **67**, 528 (1990).
71. G. Horowitz, F. Deloffe, F. Garnier, R. Hajlaovi, M. Hmyene, A. Yasser: Synth. Met. **54**, 435 (1993).
72. G. Horowitz, F. Garnier, A. Yassar, R. Hajlaoui, F. Kouki: Adv. Mater. **8**, 52 (1996).
73. F. Garnier, Philos. Trans. R. Soc. Lond. A, Mathe. Phys. Eng. Sci. **355**, 815 (1997).

3. Supramolecular Photophysics and Photovoltaic Devices of Fullerene-Conjugated Polymer Composites

N.S. Sariciftci

3.1 Introduction

3.1.1 Semiconducting, Conjugated Polymers as Photoexcited Electron Donors

Light harvesting polymer systems utilizing the photoinduced energy transfer and/or electron transfer mechanisms are attracting more and more attention from the scientific community [1]. The photophysics and photochemistry of polymers are particularly interesting because modern synthetic methods enable the chemist and material scientist to modify, functionalize and derivatize donor and acceptor units. Light sensitive polymer arrays are under consideration for photocatalysis, non-silver based imaging, all optical information storage, and electroresponsive systems for displays and sensors, among many other applications [2–6].

The discovery of semiconducting, conjugated polymers and the ability to dope these polymers over the full range from insulator to metal resulted in the creation of a class of new materials that combines the electronic and optical properties of semiconductors and metals with the attractive mechanical properties and processing advantages of polymers [7–9]. Moreover, the ability to control the energy gap and electronegativity through molecular design and engineering has enabled the synthesis of semiconducting polymers with a range of ionization potentials and electron affinities [10, 11]. The novel phenomena associated with nonlinear excitations (solitons, polarons and bipolarons), characteristic of these quasi-one-dimensional solids in which the electronic structure is strongly coupled to the lattice, have been a focal point of interest in the physics of conducting polymers [7, 8, 12]. In conjugated polymers with a degenerate ground state, solitons are the lowest energy charge storage configurations. The photogeneration of solitons has been studied in detail in polyacetylene [8]; transient spectroscopy measurements have demonstrated that soliton formation occurs at times less than 10^{-13} s following photoabsorption, in agreement with the predictions of Su and Schrieffer [13]. When the ground state degeneracy is lifted, the solitons are confined, and the nonlinear excitations are polarons and/or bipolarons [7, 8, 12]. Thus, quite generally in semiconducting polymers, photoexcitation is rapidly followed by structural relaxation (at times less than 10^{-13} s) due to the electron–phonon coupling, thereby creating new energy states within the π^* gap.

Conjugated polymers in their undoped, semiconducting state are electron donors upon photoexcitation (electrons promoted to the antibonding π^* band), as will be described below in great detail. The idea of using this property in conjunction with a molecular electron acceptor to achieve long living charge separation was based on the stability of the photoinduced nonlinear excitations (such as polarons) on the conjugated polymer backbone. Once the photoexcited electron is transferred to an acceptor unit, the resulting cation radical (positive polaron) species on the conjugated polymer backbone is known to be highly delocalized and stable as shown in electrochemical and/or chemical oxidative doping studies. Analogous to the chemical doping process, we will describe the photoinduced electron transfer from the conjugated polymer donor onto an acceptor moeity as "photodoping".

3.1.2 Buckminsterfullerene, C_{60}, as Electron Acceptor

As a new form of carbon, buckminsterfullerene, C_{60}, has generated considerable interest in the scientific community [14–16]. C_{60} is an excellent electron acceptor capable of taking on as many as six electrons [17]; C_{60}, therefore, forms charge transfer salts with a variety of strong donors. The discovery of important solid state properties in charge transfer salts of C_{60} has opened exciting new directions of research activity; examples include the metallic state and superconductivity in salts with alkali metal ions [18, 19], and soft ferromagnetism in an organic salt with (C_{60}) [20, 21]. Furthermore, buckminsterfullerene shows quite interesting nonlinear optical (NLO) properties which are important from the point of view of photonics applications. Third order NLO response and optical limiting have been reported for C_{60} [22–26]. Since a comprehensive introduction and review of the field can be found in the previous chapters, we will strictly focus on the supramolecular photophysics of conjugated polymer/buckminsterfullerene composites.

During the last three years, we have reported extensively on photoinduced electron transfer and metastable charge separation in semiconducting conjugated polymer/C_{60} composites [27–32]. A schematic description of this phenomenon is displayed in Fig. 3.1. The forward electron transfer from the semiconducting polymer (as donor) onto C_{60} (as acceptor) occurs in less than 1 ps, thereby quenching the luminescence by three orders of magnitude as well as any intersystem crossing to a triplet state [27, 33]. Using this molecular effect at the interface between bilayers consisting of semiconducting polymer (poly(2-methoxy,5-(2'-ethyl-hexoxy)-p-phenylene) vinylene, hereafter referred to as MEH-PPV, see, Abbreviations of Molecular Structures, below) and C_{60} films, diodes were demonstrated with rectification ratios on the order of 104 which exhibited a photovoltaic effect [34]. Significant improvement of the relatively low collection efficiency of the Donor/Acceptor D/A bilayer

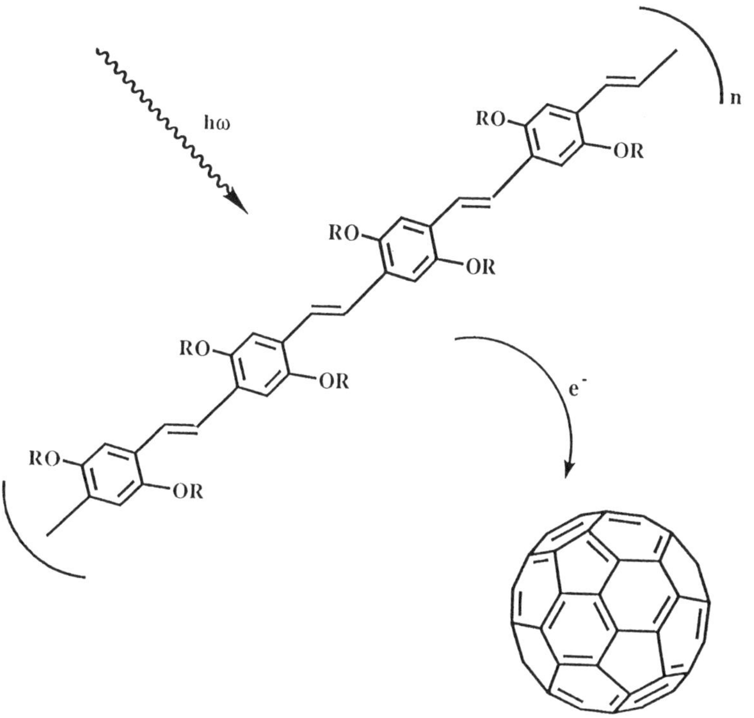

Fig. 3.1. Schematic illustration of the photoinduced electron transfer from conjugated, semiconducting polymers onto buckminsterfullerene, C_{60}

has been achieved by using phase separated composite materials; through control of the morphology of the phase separation into an interpenetrating network, one can achieve a high interfacial area within a bulk donor/acceptor composite. Using this approach, the power conversion efficiency of solar cells made from MEH-PPV/C_{60} composites was subsequently increased by two orders of magnitude to approximately 3% [35].

The stabilization of the charge separated state in these composites is assumed to result from the stability and delocalization of the radical cations (positive polarons) on the semiconducting polymer backbone and from structural relaxation of the fullerene following the photoexcitation. Thus, the possible sensitivity of the photoinduced electron transfer process (from D* to A) to the excited state dynamics of the semiconducting polymer (as D) is an important issue. Does electron transfer occur even faster than soliton, or exciton formation? We analyzed this question by comparative studies of different semiconducting polymers as donors and found that soliton formation on polyacetylenes and exciton formation on polydiacetylenes inhibits the photoinduced electron transfer [28, 36].

Morita et al. [37] reported independent studies on the mixtures of poly(3-alkylthiophene)/C_{60}. They observed a strong quenching of semiconducting polymer luminescence. The same group reported extended studies on the photoinduced electron transfer in conjugated polymer/buckminsterfullerene composites during recent years and their results will be summarized in a separate Chapter of this volume [37–44].

3.2 Experimental Results: Ultrafast, Reversible, Photoinduced Electron Transfer in Conjugated Polymer/Fullerene Composite Films

3.2.1 Linear Absorption and Photoluminescence

MEH-PPV/C_{60} Composites. Figure 3.2 shows the optical absorption spectrum of a MEH-PPV/C_{60} film with different C_{60} content compared to

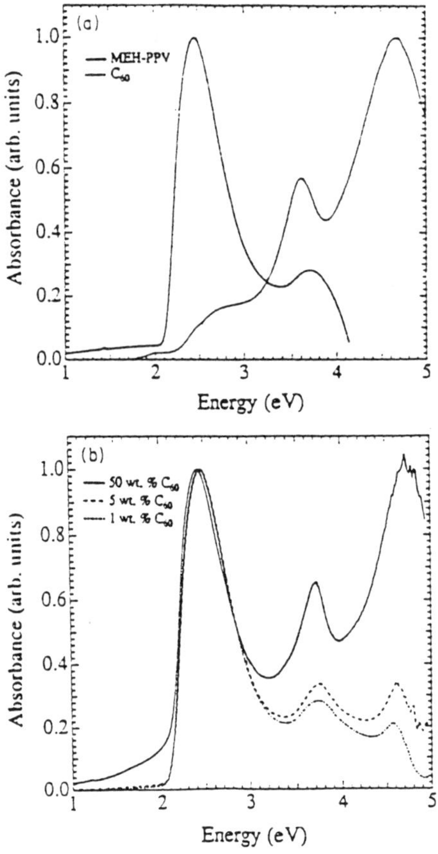

Fig. 3.2. (**a**) Optical absorption spectra of MEH-PPV and C_{60}. (**b**) Optical absorption spectra of MEH-PPV/C_{60} films with different concentrations of C_{60} (reproduced with permission of The American Physical Society from [29])

the optical absorption spectrum of the components alone. The π^* absorption of MEH-PPV (peak at 2.5 eV) is clearly observed along with the first dipole allowed transition in C_{60} (at 3.75 eV). The spectrum is a simple superposition of the two components without any indication of states below the π^* gap of the conducting polymer as might arise from interaction between the two materials in the ground state.

Figures 3.3 and 3.4 show the luminescence spectrum and the time-resolved emission decay of MEH-PPV/C_{60} composites compared to MEH-PPV alone. The strong photoluminescence of MEH-PPV is quenched by a factor in ex-

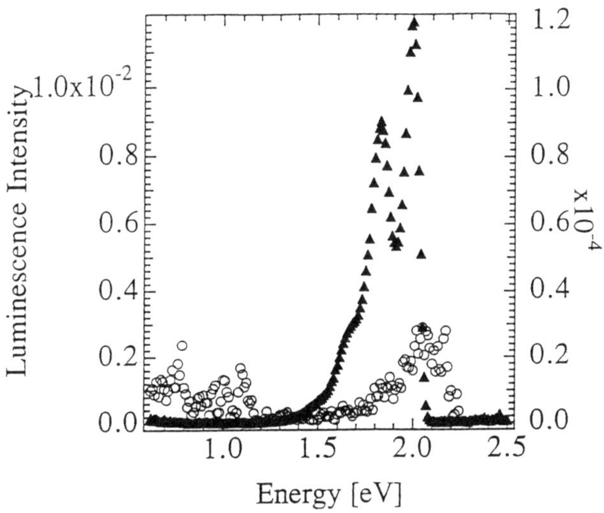

Fig. 3.3. Luminescence of MEH-PPV (*solid triangles, left-hand axis*) and MEH-PPV/C_{60} composite (*circles, right-hand axis*). Reproduced with permission of AAAS from [27]

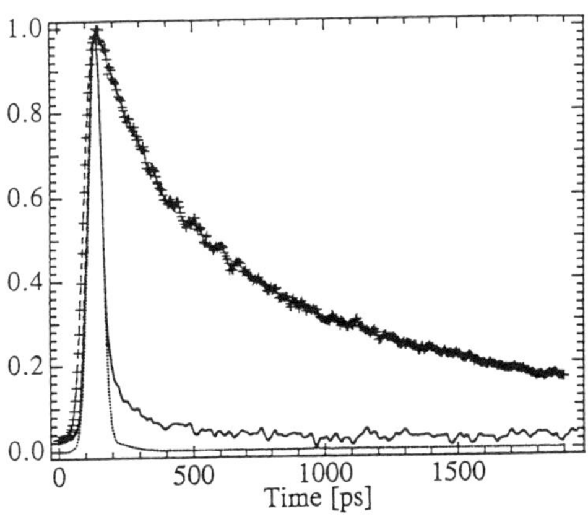

Fig. 3.4. The time-resolved emission decay of MEH-PPV/C_{60} composite (+) compared to MEH-PPV alone (*solid line*) and the system response (*dashed line*). Reproduced with permission of Elsevier from [102]

cess of 100, and the luminescence decay time is reduced from to 550 ps to trad $\leq$ 60 ps (the instrumental resolution) indicating the existence of a rapid quenching process [27, 28, 45]. An estimate of the transfer rate, $1/\tau_{ct}$, is given by decay rate of the photoluminescence in the MEH-PPV/C_{60} composite (charge transfer will cut-off the radiative decay).

$$\frac{1}{\tau_{ct}} = \frac{1}{\tau_{rad}} - \frac{1}{\tau_0} \tag{3.1}$$

Because the photoluminescence decay in h composite is uresolved within the 60 ps of the excitation pulse, we use the ratio of the intensities to estimate 1/tct;

$$\frac{1}{\tau_{ct}} = \frac{1}{\tau_{rad}}\frac{I_0}{I_{comp}} \tag{3.2}$$

where $1/\tau_0$ and $1/\tau_{rad}$ are the radiative decay rates, I_0 and I_{comp} are the integrated photoluminescence intensities of MEH-PPV and the MEH-PPV/C_{60} composite, respectively. The data imply, therefore, that $1/\tau_{ct}$ 10^{12}; i.e. electron transfer occurs on the picosecond time scale. Since the charge transfer rate is more than two orders of magnitude faster than competing radiative and non-radiative recombination processes, the quantum efficiency for charge transfer must be close to unity.

P3OT/C_{60} Composites. In earlier reports, a change in the P3OT absorption spectrum was reported upon adding the C_{60} to P3OT; the π^* absorption of P3OT is considerably broader with a tail extending into the gap. This observation was interpreted as indicative of significant mixing of the P3OT and C_{60} ground state electronic wave functions [28, 37, 46]. Morita et al. [37] proposed a weak ground state doping upon adding C_{60} to P3OT. This, however, would imply the presence of the C_{60} anion ESR signal in the ground state of the P3OT/C_{60} composite. As shown below, however, the C_{60} anion signal appears only in light induced ESR, ruling out doping-induced electron transfer in the ground state. Moreover, the electrical conductivity is not consistent with doping of P3OT by C_{60}. The spectral changes in the P3OT/C_{60} composite was proposed to originate from partial charge transfer in the ground state, from D + A $D^{\delta+} - A^{\delta-}$ (D for donor, A for acceptor and δ for fractional charge transferred) [28]. Alternatively, because C_{60} phase segregates within the P3OT matrix at high loading ratios, the changes in the optical absorption of P3OT upon mixing C_{60} can possibly be accounted for by the disorder, granular structure and related diffuse scattering and reflectance.

We have recently revisited the linear optical absorption of these composites using derivatized C_{60} molecules [6,6]PCBCa. Due to the high solubility and processability of these C_{60} derivatives, composite films can be fabricated with excellent optical quality. Even at very high loading ratios, the optical aborption of the composite is a simple superposition of the optical absorption spectra of the two components. There is no ground state mixing of the wavefunctions, and there is no ground state doping.

3.2.2 Near Steady State Photoinduced Absorption

MEH-PPV/C_{60} Composites. To clearly distinguish the spin multiplicity of the PIA peaks, photoinduced absorption detected magnetic resonance (PIADMR) experiments were performed [33]. The full spectral range photoinduced absorption spectra are displayed in Fig. 3.5 [33]. Results show that the PIA bands centered around 0.3 eV and 1.1 eV are significant only in the composite material. Pure MEH-PPV shows the single PIA band from the triplet-triplet absorption, as assigned earlier [46].

The PIA band at 1.35 eV in the pure MEH-PPV shows a strong response at the (forbidden) half field resonance of spin=1, indicating triplet character for this PIA band and dominance of the neutral photoexcitations in MEH-PPV (Fig. 3.6). There is a small, residual spin = 1/2 response even within the pure MEH-PPV. Upon adding C_{60}, the triplet signal for the 1.35 eV

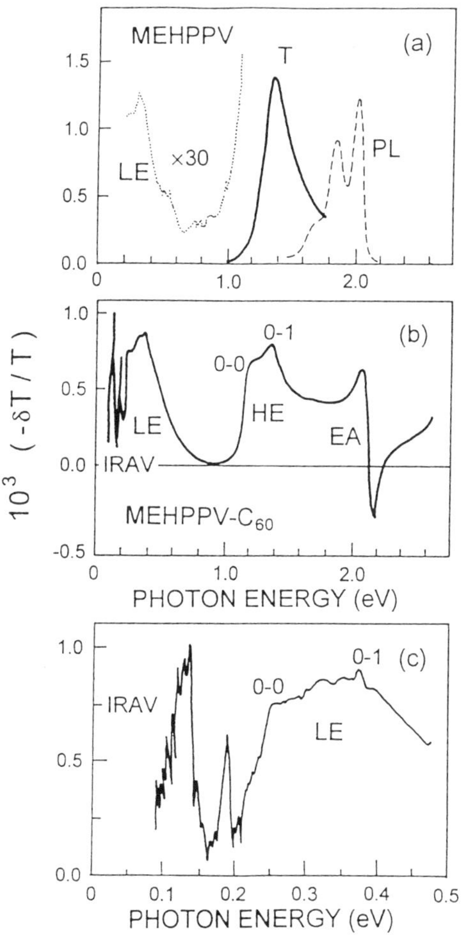

Fig. 3.5. (**a**) Photoinduced absorption and photoluminescence (PL) spectra of MEH-PPV at 80 K; (**b**) and (**c**) The photoinduced absorption spectra of MEH-PPV/C_{60} at 4 K with argon ion laser at 514.5 nm modulated at 200 Hz. The abbreviations T, LE, HE, EA stand for triplet–triplet absorption, low-energy band, high-energy band and electroabsorption features, respectively. IRAV stands for infrared activated modes (see Section 3.2.5) [33]

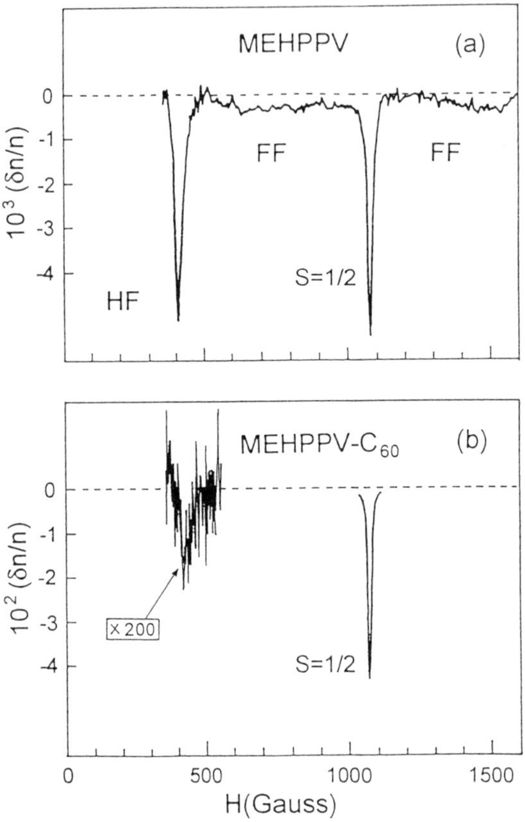

Fig. 3.6. (**a**) Photoinduced absorption detected magnetic resonance (ADMR) spectrum of MEH-PPV. HF and FF abbreviates the half field and full field powder pattern for the triplet (S=1) resonance, respectively. (**b**) ADMR spectrum of MEH-PPV/C_{60} composite film. Both spectra were measured at probe energy 1.35 eV, T = 4 K and 3 GHz resonant microwave frequency [33]

PIA band is completely quenched. Instead a strong spin = 1/2 signal dominates, indicating charged polarons as photoexcitations on the polymer donor (Fig. 3.6 [33]). This confirms that the photoinduced electron transfer occcurs on a time scale sufficiently fast to quench the intersystem crossing to the triplet state.

Bleaching of the MEH-PPV π^* absorption at 2.5 eV is observed in both the composite and pure material, but the luminescence is quenched only in the composite [28].

The relaxation of the PIA bands was measured by monitoring the dependence of the PIA band intensities on the chopping frequency [28]. The results are displayed for comparison in Fig. 3.7. The relaxation is similar for all PIA bands in the composite. We estimate the lifetime of the charge separated configuration to be of the order of milliseconds at 80 K [28]. The intensity of the PIA bands increase with the laser pump intensity as $I^{0.59}$, as shown in Fig. 3.8. To perform a final check to show that all PIA bands in the composite originate from excitation of the MEH-PPV donor, we determined the excitation profile by recording the PIA spectrum with different

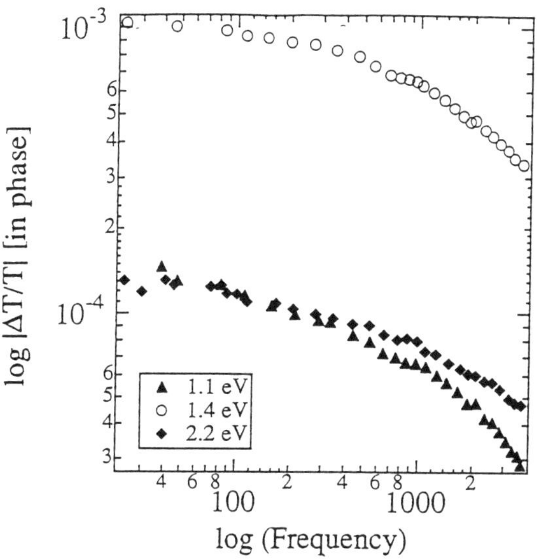

Fig. 3.7. Dependence of the photoinduced absorption features of MEH-PPV/C_{60} on the chopping frequency. Reproduced with permission of The American Physical Society from [28]

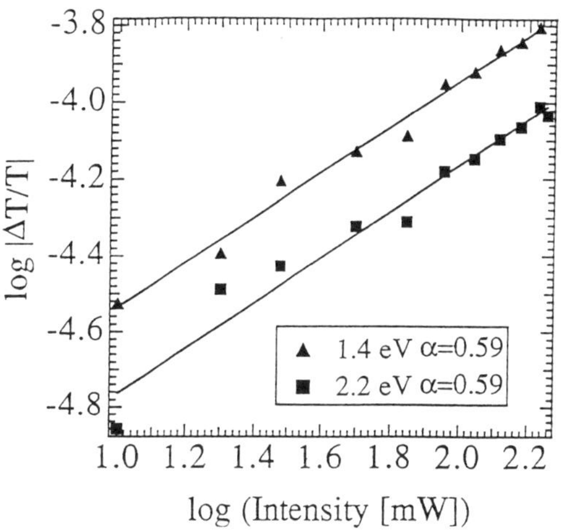

Fig. 3.8. Dependence of the photoinduced absorption features of MEH-PPV/C_{60} on the pump intensity. Reproduced with permission of The American Physical Society from [28]

excitation energies. The excitation profile displayed in Fig. 3.9 clearly shows the correlation of all the PIA bands in the composite with the π^* excitation of MEH-PPV component [27].

P3OT/C_{60} Composites. Figure 3.10 shows the photoinduced absorption spectra of P3OT alone and in a composite with C_{60}. The PIA spectrum of P3OT with a sharp PIA band at 1.05 eV and a broad PIA band centered at 1.25 eV is considerably different from the PIA spectrum reported earlier [47]. The 1.05 eV band has a different relaxation mechanism from the other bands. The chopping frequency dependency of the 1.05 eV band has been fitted to

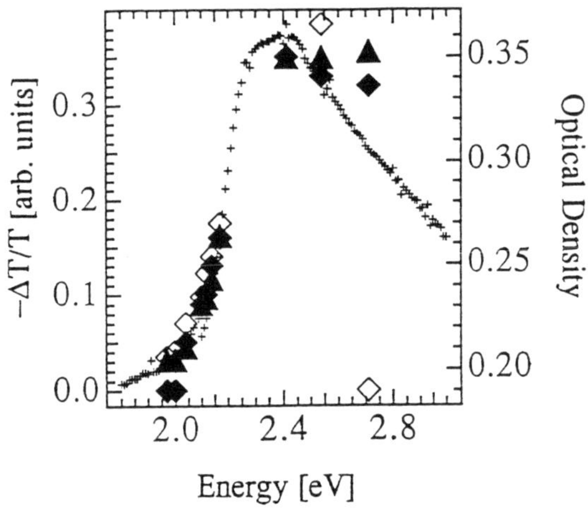

Fig. 3.9. Excitation profiles (*left axis*) at 1.33 eV (*solid triangles*), at 1.1 eV (*solid squares*) and at 1.8 eV (*empty squares*) of MEH-PPV/C_{60} for photoinduced absorption peaks diplayed in Fig. 3.5. Reproduced with permission of Elsevier from [102]

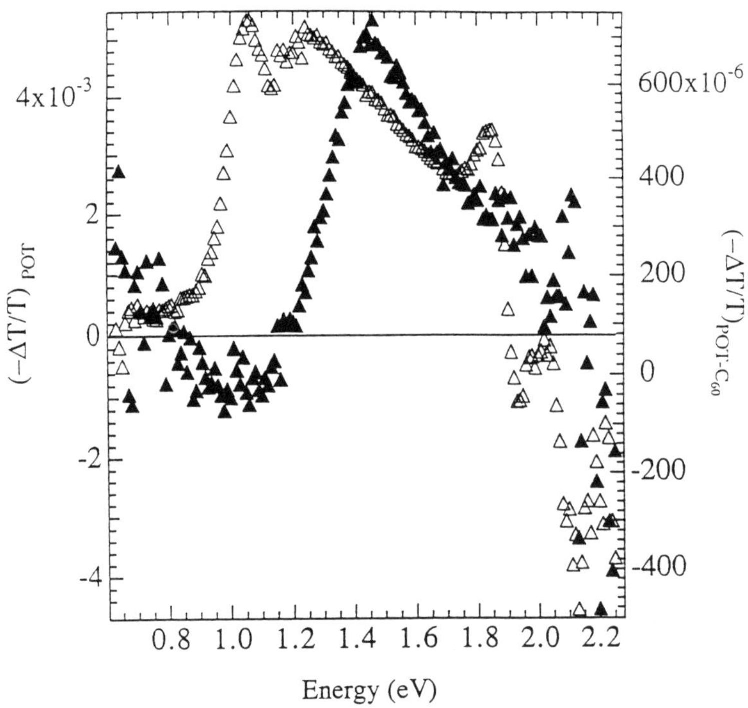

Fig. 3.10. Photoinduced absorption spectra of P3OT (*open triangles*) and P3OT/C_{60} composite (*solid triangles*) at 80 K obtained by pumping with argon-ion laser at 2.41 eV with 50 mW. Reproduced with permission of The American Physical Society from [28]

a monomolecular decay mechanism [28]. The lifetime derived from the best fit is $\tau = 175\,\mathrm{ms}$. This specific evidence of monomolecular decay behavior is indicative of an excitation with a well defined lifetime in contrast to the power law dependence typically found for trapped charged excitations [28, 46]. Because of this monomolecular decay and the long lifetime, we assign this 1.05 eV peak to a triplet excitation. A similar shift in PIA states from trapped charged excitations to metastable triplet excitons has been observed in MEH-PPV on going from a disordered film to an oriented blend and in PPV by using an improved synthesis technique [28, 46, 48]. To clarify this assignment unambigously, spin dependent PIA experiments are needed, as carried out for the MEH-PPV/C_{60} case (see previous section) [28].

Figure 3.11 demonstrates that the other PIA bands follow more closely a power law dependence on chopping frequency with $\Delta T/T$ increasing as ω decreases, even at the lowest chopping frequencies. The dependency of the PIA features on pump intensity has also been investigated; all show a power law dependence on pump intensity with exponents between 0.6 and 0.8 [28].

Upon adding C_{60} to P3OT, there are significant changes in the photoinduced absorption (see Fig. 3.10). In the composite, only one PIA is observed, centered at 1.45 eV, along with an onset of photoinduced infrared absorption at low energies. The PIA spectrum of the P3OT/C_{60} is similar to that which was observed previously for trapped charged excitations on P3OT [47] and to the absorption spectrum of the doped P3OT. This supports the identification of the PIA features in the P3OT/C_{60} composite as arising from a charged excitation [28]. An explicit absorption band of C_{60}- is not observed in the P3OT/C_{60} composite in these experiments due to spectral dominance of the polaronic absorptions from the conjugated polymer backbone. The absorption band of C_{60}- has been clearly identified, however, in the photoinduced FTIR experiments, to be described below.

3.2.3 Sub-Picosecond Photoinduced Absorption

MEH-PPV, BCHA-PPV and Their Composites With C_{60}. The sub-picosecond PIA spectrum and its time evolution after photoexcitation are altered significantly upon addition of C_{60} into MEH-PPV and BCHA-PPV, consistent with photoinduced electron transfer [30]. As shown in Fig. 3.12b and 3.13b, a broad absorption band in the PIA spectrum centered near 1.45 eV is observed in both BCHA-PPV/C_{60} (1%) and MEH-PPV/C_{60} (1%). In both cases, the spectra are different from that observed in the pristine polymer; i.e. without the C_{60}. The PIA spectrum for MEH-PPV/C_{60} (1%) (Fig. 3.13b) at 1 ps is similar to the PIA spectrum observed in the ms time regime, with a shoulder on the low energy side of the peak, and a plateau above 1.8 eV. The observation of the similar spectrum at 1 ps that is observed in the millisecond time regime provides direct evidence that the photoinduced electron transfer occurs on a subpicosecond time scale and persists into the millisecond regime, showing an assymmetry of 10 orders of magnitude.

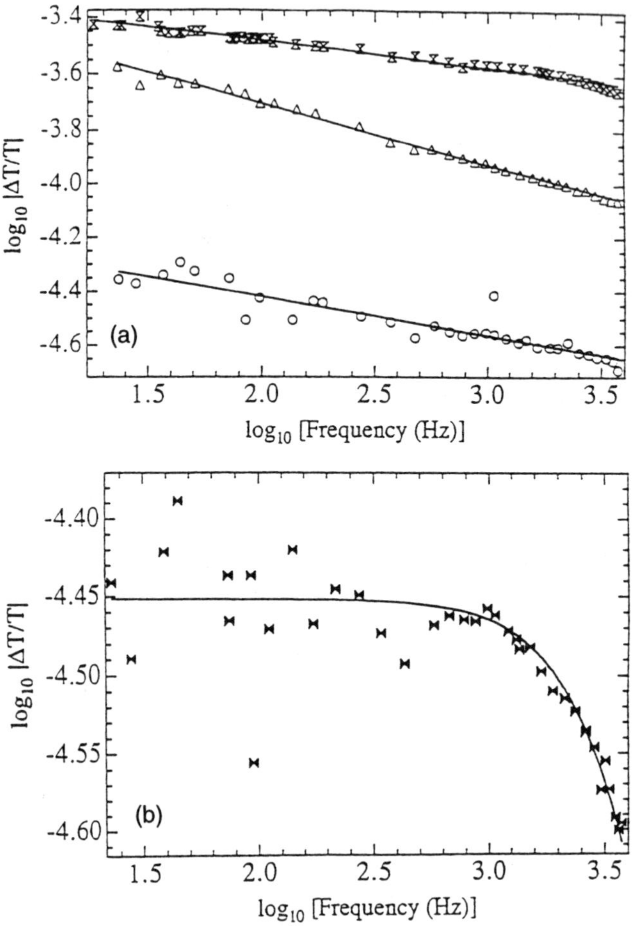

Fig. 3.11. Dependence of the photoinduced absorption features of P3OT on the chopping frequency. (**a**) The 1.25 eV feature is shown by the *double triangles* and fit to a power law with exponent 0.09. The 1.85 eV band is shown by *triangles* and fit to power law with exponent 0.22, and the 2.15 eV band is shown by *circles* and fit to power law with exponent 0.14. (**b**) The frequency dependence of the 1.05 eV feature with best fit monomolecular decay shown as *solid line*. Reproduced with permission of The American Physical Society from [28]

The following function was used to fit the relaxation kinetics [30]:

$$\Delta\alpha(\tau) \propto \int dt I_{pb}(t-\tau) \int dt' \theta(t-t') R(t-t') I_{pp}(t') . \tag{3.3}$$

In this expression t is the time delay between the pump and the probe pulses, and represents the pump pulse intensity profile, which was measured using second harmonic generation in a KDP crystal. The fit to a sech2 pulse shape yielded a 100 fs pulse width. represents the probe pulse intensity profile, which

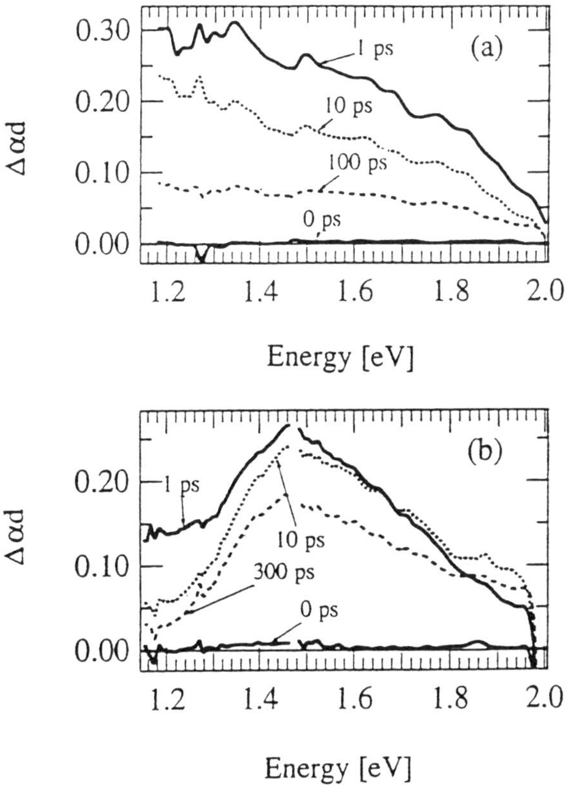

Fig. 3.12. Photoinduced absorption spectra at 300 K for various delay times after a 2.01 eV, 100 fs pump pulse for (**a**) BCHA-PPV and (**b**) BCHA-PPV/C_{60} (1%). Reproduced with permission of The American Physical Society from [30]

was obtained by deconvoluting the measured cross correlation of the probe and pump pulse. is a step function and is the response function which describes the dynamics of the system.

The decay of the PIA at 300 K in MEH-PPV is fit [30] using the response function with $t = 300$ fs, and a slow power-law tail, $t^{-0.33}$ (Fig. 3.14, filled circles). Upon cooling to 86 K, the fast component remains unchanged, but the longer lived component decays more slowly, $t^{-0.22}$ (solid diamonds). Upon addition of 1% C_{60}, the time constant of the fast exponential component decreases to 100 fs, and the power law decay becomes much slower. At 300 K (open circles) the latter follows $t^{-0.2}$, which at 86 K (open diamonds) becomes $t^{-0.15}$ [30]. This observation is consistent with photoconductivity measurements which showed an increase in the magnitude and lifetime of the photocurrent upon addition of 1% C_{60} [29]. The slight temperature dependence of the relaxation time ($t^{-0.15}$ at 86 K versus $t^{-0.2}$ at 300 K) is consistent with measurements of the temperature dependence of the LESR

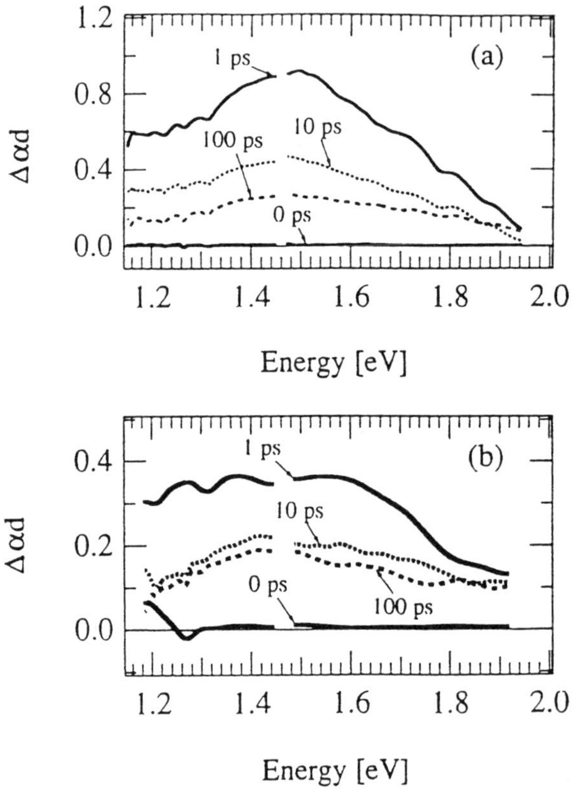

Fig. 3.13. Photoinduced absorption spectra at 300 K for various delay times after a 2.01 eV, 100 fs pump pulse for (**a**) MEH-PPV and (**b**) MEH-PPV/C_{60} (1%). Reproduced with permission of The American Physical Society from [30]

signal in MEH-PPV/C_{60} (50%) where an Arrhenius behavior was found for the back transfer process (electron on C_{60} transferring back to the polymer) with an activation energy of 15 meV [31]. This energy barrier inhibits the recombination process and may account for the slower kinetics observed at lower temperatures. From our measurements, however, it is not clear whether the activation barrier inhibits the back transfer rate (transfer of electron from C_{60}- to polymer) or whether it reduces the recombination rate of electrons and holes on the polymer chain (as is the case in pristine MEH-PPV) after the back transfer process is completed.

The relaxation kinetics of BCHA-PPV/C_{60} at 1.45 eV show a dramatic increase in lifetime over that of the pristine material [30]. However, in this PPV derivative, the decay of the PIA does not become metastable until approximately 7 ps after photoexcitation (Fig. 3.15). The observation that the dynamics of the PIA in BCHA-PPV/C_{60} are identical to that of the pristine polymer for the first 7 ps after photoexcitation is different from that

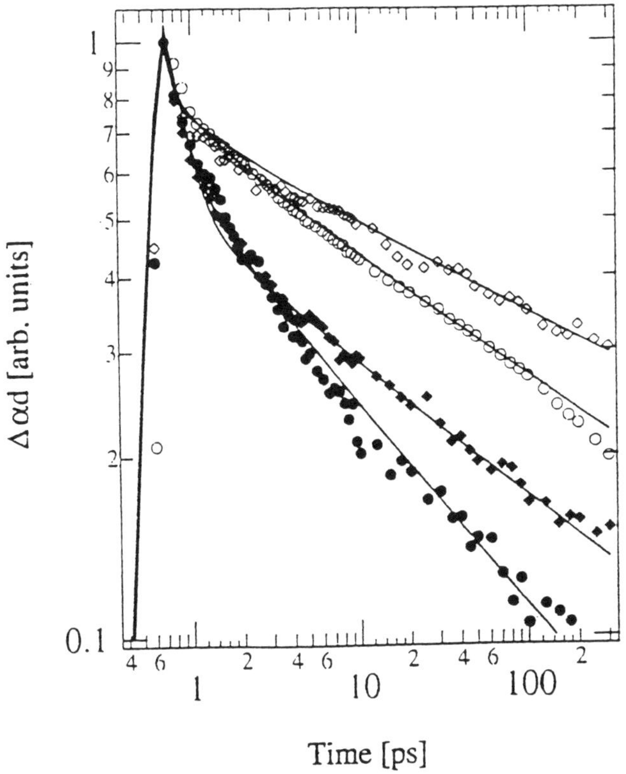

Fig. 3.14. The decay of the photoinduced absorption at 1.45 eV in MEH-PPV, at 300 K (*closed circles*), and at 86 K (*closed diamonds*), and at the same energy in MEH-PPV/C_{60} (1%) at 300 K (*open circles*) and at 86 K (*open diamonds*). The formula and parameters for the fits are decribed in the text. Reproduced with permission of The American Physical Society from [30]

observed in MEH-PPV/C_{60}, where the photoexcitations become metastable after only 300 fs.

These results imply that a large fraction of the initial photoexcitations undergo ultrafast charge transfer (within 1 picosecond) in MEH-PPV/C_{60} (1%) in contrast to in BCHA-PPV/C_{60} (1%) where charge transfer is somewhat slower [30].

Since 1% by weight corresponds to roughly 300 monomer units per C_{60} in MEH-PPV/C_{60} and 100 monomer units per C_{60} in BCHA-PPV/C_{60}, the diffusion time of the photoexcitations cannot account for the increased time neeed for all the photoexcitations to undergo charge transfer in BCHA-PPV/C_{60}, because the photoexcitations have to diffuse less far to reach a C_{60} molecule in this polymer than in MEH-PPV. In addition, the density of C_{60} in the polymer host exceeds the photoexcitation density used in this study,

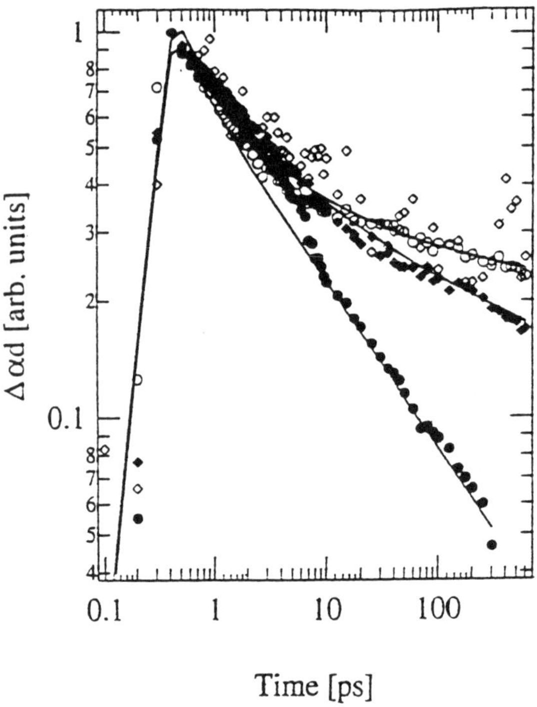

Fig. 3.15. The decay of the photoinduced absorption at 1.45 eV in BCHA-PPV, at 300 K (*closed circles*), and at 21 K (*closed diamonds*). Also shown are the dynamics in BCHA-PPV/C_{60} (1%) at the same energy, 300 K (*open circles*) and at 86 K (*open diamonds*). The fits are to $R(t) = At - a$ with $a = 0.43$ for pristine BCHA-PPV at 300 K and to $R(t) = At - a1 + Bt - a2$ with $a1 = 0.12$ and $a2 = 0.6$ at 21 K. At 300 K and 86 K in BCHA-PPV/C_{60} (1%) $a1 = 0.05$ and $a2 = 0.6$. Reproduced with permission of The American Physical Society from [30]

so that multiple electron transfer to a single C_{60} acceptor is not considered important for either PPV derivative [30].

Although currently no measurement of the ionization potential of BCHA-PPV has been reported, calculations indicate that the ionization potential of different PPV derivatives can differ by several hundred meV [49]. A difference in ionization potential between MEH-PPV and BCHA-PPV would influence the electron transfer rate. If the electron transfer occurred via an activated process, one would expect a temperature dependence on the electron transfer rate, contrary to observations. Another possibility is that tunneling though a potential barrier dominates the relaxation process. However, since the electron transfer rate in MEH-PPV/C_{60} and in P3OT/C_{60} (to be discussed below) are unaffected by the slightly different ionization potentials of these two polymers, we do not consider this mechanism to be the main reason for the difference in the charge transfer efficiency in the two PPV derivatives. We suggest that this delay in achieving charge transfer of all photoexcitations in BCHA-PPV/C_{60} is due to the influence of the bulky, symmetric side groups of this polymer, reducing the overlap of the excited state wave function with a C_{60} molecule.

Another possibility is the attenuation of the photoinduced electron transfer rate due to dielectric relaxation effects of the liquid crystalline (cholesteric) side chains of BCHA-PPV, as recently observed in donor–acceptor molecules in liquid crystalline matrices [50].

Another factor to consider is the more rigid conformation of BCHA-PPV as compared to MEH-PPV, as evidenced by the fact that the linear absorption spectrum in film or in solution for BCHA-PPV is virtually identical, whereas for MEH-PPV there is a blue shift in the absorption edge of approximately 200 meV in going from film to solution. This enhanced one-dimensionality would decrease the screening and increase the magnitude of electron–hole Coulomb correlations on the polymer backbone, thereby resisting the charge separation from BCHA-PPV onto C_{60}. A similar mechanism has been proposed to account for the lack of charge transfer in polydiacetylene/C_{60} composites [36], in which the excitons are bound by a Coulomb barrier of approximately 0.5 eV.

P3OT/C_{60} Composites. The time resolved photoinduced absorption (PIA) spectra of pristine P3OT and of P3OT with 1% C_{60} are shown in Fig. 3.16a and b, respectively [30]. The spectrum for the pristine material consists of two distinct features, one at 1.9 eV, and one at 1.2 eV, which we shall refer to as the high energy (HE) feature and the low energy (LE) feature, respectively. After 10 ps there is no change in the relative spectral weight of the HE and LE features; the spectrum at 250 ps can be scaled to precisely fit the spectrum at 10 ps [30]. The time evolution of the HE feature consists of two components, a fast exponential component with a lifetime of 800 fs, and a slow component which matches the dynamics of the LE feature [30]. The fast component of the HE feature matches that seen in poly(3-methylthiophene) [51] and polythiophene [52], and has been attributed to a self-trapped singlet exciton, or a singlet polaron-exciton. Upon adding C_{60} to P3OT, the PIA spectrum, decay kinetics, and intensity dependence all change dramatically [30].

At 1 ps after photoexcitation by a 100 fs pump pulse at 2.01 eV, the PIA spectrum for P3OT/C_{60} (1%) consists of a single broad peak at approximately 1.55 eV, Fig. 3.16 b, with virtually no evidence of the HE and LE features seen in pristine P3OT at 1.9 and 1.2 eV [30]. The risetime of this feature is resolution-limited, (250 fs). After photoexcitation, the peak shifts to the red, from 1.55 eV at 1 ps to 1.45 eV at 100 ps. The spectrum at 100 ps closely matches that seen in millisecond (ms) PIA experiments; the latter has been assigned to positively charged polaron band as discussed above [28, 31]. The ultrafast (< 1 ps) formation of the PIA band at 1.55 eV again

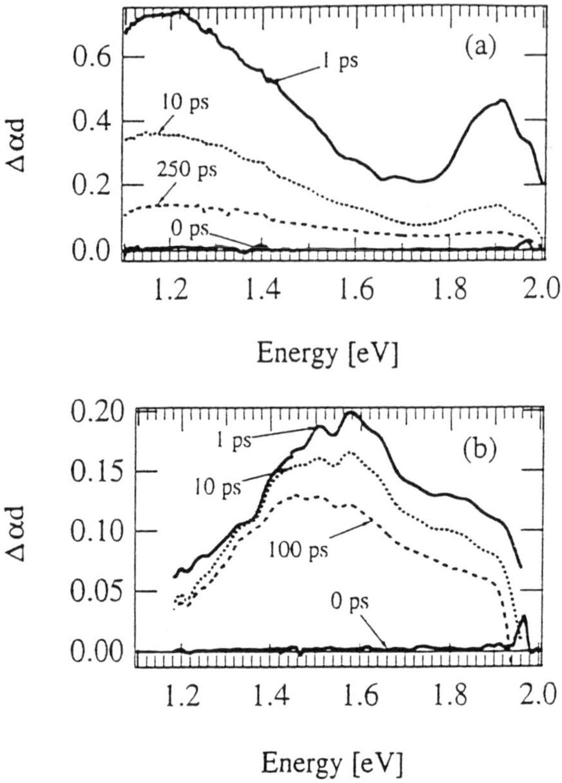

Fig. 3.16. Photoinduced absorption spectra at 300 K for various delay times after a 2.01 eV, 100 fs pump pulse for (**a**) P3OT and (**b**) P3OT/C_{60} (1%). Reproduced with permission of The American Physical Society from [30]

demonstrates that the electron transfer occurs on a subpicosecond time scale, confirming earlier estimates based on luminescence quenching.

In order to compare the time decay of the PIA in pristine P3OT to that of the peak of the PIA feature in P3OT/C_{60} (1%), we show the kinetics in P3OT at 1.45 eV (Fig. 3.17) [30]. The fit shown is to the response function with $t1 = 300$ fs and $t2 = 20$ ps. The decay of the photoexcitations in P3OT with 1% C_{60} fits the same response function used above for pristine P3OT. The lifetimes in P3OT/C_{60} (1%) are $t1 = 300$ fs and $t2 = 25$ ps. The ratio of the coefficients (B/A) used in the fitting increases by over an order of magnitude upon additon of 1% C_{60}, implying that a much larger fraction of the photoexcitations contribute to the long lived signal in the composite material than in pristine P3OT. In P3OT/C_{60} (10%), the lifetimes used to fit to the data are $t1 = 300$ fs and $t2 = 700$ ps. These results are consistent with those obtained earlier using P3OT/C_{60} (50%) [53] and with photoconductivity measurements made on P3OT/C_{60} (5%) composites (see below). The

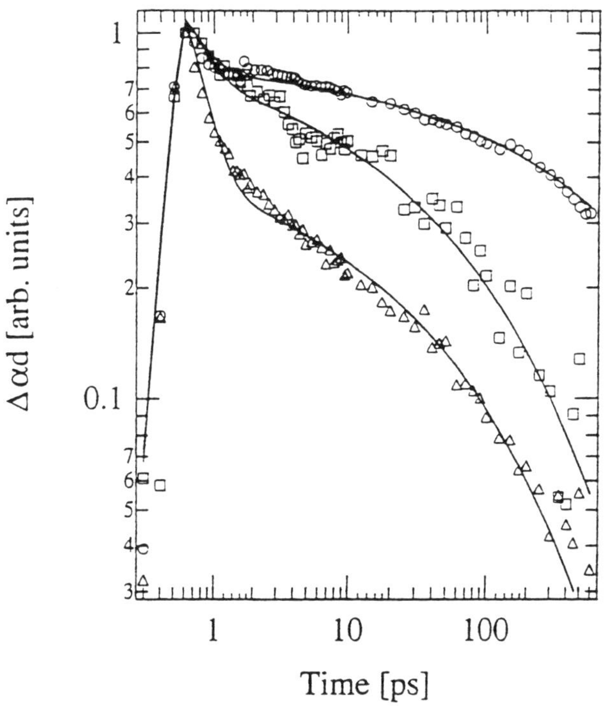

Fig. 3.17. Decay of the photoinduced absorption at 1.45 eV in pristine P3OT (*open triangles*), P3OT/C_{60} (1%) (*open squares*), and P3OT/C_{60} (10%) (*open circles*) at 300 K. The formula and fit parameters are described in the text. Reproduced with permission of The American Physical Society from [30]

increase in $t2$ observed in the 10% composite implies that not only does the charge transfer occur on an subpicosecond time scale, but the reverse transfer (i.e. the electron transferring from C_{60} back to P3OT) is inhibited, yielding metastable charge separated photoexcitations. Hence, in the 10% composite, both mechanisms discussed above are contributing to the increase in the yield of long-lived charged photoexcitations.

The addition of 1% C_{60} also affects the dependence of the PIA signal on the incident pump fluence [30]. In order to compare with the behavior of the peak of the PIA feature in P3OT/C_{60}, we consider the intensity dependence of the same spectral region (1.45 eV) in pristine P3OT. At this energy, pristine P3OT follows an I0.64 dependence at 1 ps delay time, which becomes I0.47 by 20 ps. An exciton is expected to decay geminately and therefore exhibit a linear intensity dependence. However, this superimposed HE band in P3OT has a long-lived species which follows a square-root dependence on the pump fluence. This results in an intensity dependence which is intermediate between linear and square root, which persists during the lifetime of the polaron-exciton. After the polaron-exciton has decayed ($t > 10$ ps) the intensity dependence of the entire PIA spectrum follows a square root behavior. In contrast, the P3OT/C_{60} (1%) system displays a linear dependence on pump fluence from the hundred femtosecond time scale to the hundred picosecond time scale [30].

An important observation displayed in Fig. 3.16 is that the nature of the primary photoexcitations in pristine P3OT is different than the ones in P3OT/C_{60} composites. Since the latter (Fig. 3.16b) has been well identified and characterized as due to positively charged polarons on P3OT main chains, it becomes difficult to argue that the primary photoexcitations in pristine P3OT (Fig. 3.16a) are also non-interacting charged polarons (the content of the C_{60} molecules within the matrix is low and should not effect the dielectric properties noticably). The exact assignment of the PIA bands in Fig. 3.16a has not been established yet.

3.2.4 Sub-Picosecond Photoinduced Dichrosim

PPV Derivatives and Their Composites with C_{60}. By measuring the dichroic ratio, the time resolution is improved since we detect only a 100 meV band of the probe pulse at a given energy [30]. The dichroic ratio is defined as the ratio of the photoinduced absorption with the pump and probe polarization vectors parallel, to that with the polarization vectors perpendicular [30, 54]:

$$\frac{\delta\alpha_{\parallel}}{\delta\alpha_{\perp}} = \frac{3(f_e + f_{p_{\parallel}})\alpha_{\parallel} + f_{p_{\perp}}\alpha_{\perp}}{(f_e + f_{p_{\parallel}})\alpha_{\parallel} + 3f_{p_{\perp}}\alpha_{\perp}} \tag{3.4}$$

In this expression the subscripts e, $p_{\parallel}$ and $p_{\perp}$ refer to neutral polaron-excitons, intrachain and interchain excitations, respectively, and f_i is defined by $f_i \sim \eta_i \Delta\alpha_i(\omega)$, where η_i is the quantum efficiency for generating excitations of type i, and is the change in absorption due to these excitations. Because of the strong delocalization of the π-electrons on an individual chain and the weak interchain coupling, the optical absorption matrix elements are strongest when the light is polarized along the chains. Highly oriented MEH-PPV, for example, shows $\alpha_{\parallel}/\alpha_{\perp} \sim 100$ [55]. Thus, since the initially excited electron-hole pairs should be predominantly on individual chains, the dichroic ratio should be 3 at the earliest times. In the pure material, the dichroic ratio will approach the isotropic value, $\delta\alpha_{\parallel}/\delta\alpha_{\perp} = 1$, after sufficient time has elapsed for interchain transfer or diffusion of the excitation over relatively long distances. If charge transfer from the polymer to C_{60} is rapid and if the absorption in the charge transferred state results from an interchain dipole transition, the charge transfer process will quickly destroy any excitation with a memory of the direction of the original transition dipole; hence, the dichroic ratio will approach the isotropic value more rapidly.

Measurements of the time decay of the dichroic ratio in MEH-PPV and BCHA-PPV show (Fig. 3.18) the initial value close to 3 in both polymers, confirming that intrachain excitations are photogenerated by the pump radiation

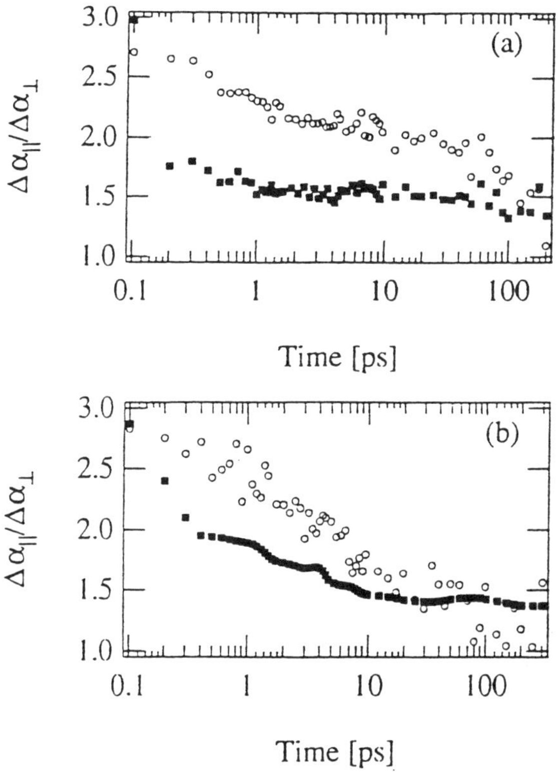

Fig. 3.18. (**a**) Dichroic ratio of MEH-PPV at 1.45 eV (*open circles*) and of MEH-PPV/C_{60} (1%) (*closed squares*) at 300 K; (**b**) Dichroic ratio of BCHA-PPV at 1.45 eV (*open circles*) and of BCHA-PPV/C_{60} (1%) (*closed squares*)

[30]. For the pure materials, the decay of the dichroism occurs over a relatively long time scale, with $\delta\alpha_{\|}/\delta\alpha_{\perp}$ dropping from 3 to 2 in approximately 10 ps. In both MEH-PPV/C_{60} and BCHA-PPV/C_{60} the initial dichroic ratio at 1.45 eV is again close to 3. The addition of 1% C_{60}, however, causes $\delta\alpha_{\|}/\delta\alpha_{\perp}$ to decay to less than 2 within a few hundred femtoseconds (Fig. 3.18).

Since charge transfer will quickly destroy any memory of the direction of the original transition dipole, the rapid decrease in the dichroism is attributed to the ultrafast charge transfer. This measurement sets an upper limit of 300 fs on the photoinduced electron transfer time from MEH-PPV to C_{60} [30].

P3OT and P3OT/C_{60} Composites. The dichroic ratio of the HE feature in pristine P3OT (Fig. 3.19, open circles) in the first picosecond after photoexcitation approaches the theoretical limit of 3, indicating a large contribution from intrachain excitations in this time regime [30]. By 10 ps after photoexcitation, the excitations decay and/or diffuse to polymer segments that are randomly oriented with respect to the initial segment, so that the dichroic ratio decreases to 1.6. The dichroism of the LE feature (Fig. 3.19, filled circles) begins at 2.25, then drops within several hundred femtoseconds to approximately 1.6, where it remains relatively constant for over 100 ps. The decay of the photoinduced dichroism discriminates the HE and LE PIA bands in pristine P3OT, even though their behavior is identical after 20 ps [30]. The definitive assignment of these two photoinduced absorption bands in P3OT can not be established with the available experimental data.

Upon adding 1% C_{60} to P3OT (Fig. 3.19, closed squares), the dichroic ratio at 1.5 eV decreases from 2.7 to 1.5 within 300 fs [30]. The high initial value is again attributed to the initial intrachain excitations on the polymer. The dichroism decreases rapidly to 1.5, and then remains relatively constant for over 100 ps. This behavior is attributed to the ultrafast charge transfer as described for the PPV derivatives above; i.e. once the electron transfers to a C_{60} molecule, the polarization memory is destroyed. This measurement sets an upper limit of 300 fs on the photoinduced electron transfer time in P3OT/C_{60} composites.

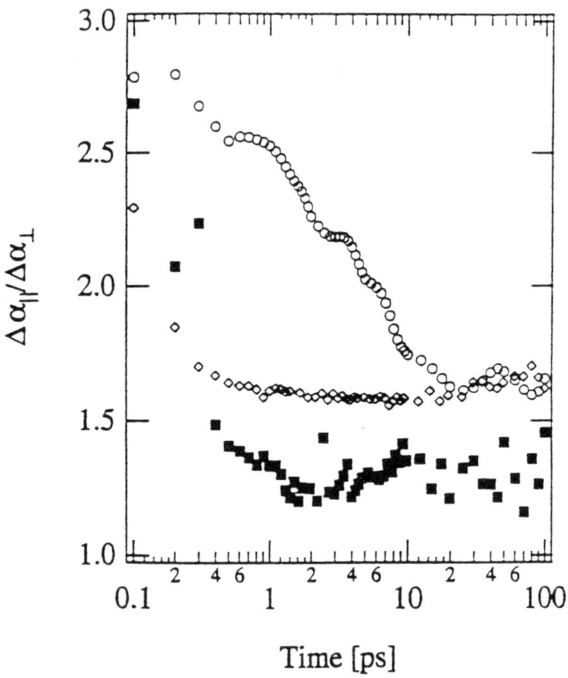

Fig. 3.19. Dichroic ratio for pristine P3OT at 1.87 eV (*open circles*), and at 1.21 eV (*open diamonds*), and for P3OT/C_{60} (1%) at 1.5 eV (*closed squares*), at 300 K

3.2.5 Steady State Infrared Photoinduced Absorption

In semiconducting, conjugated polymers, the quasi-one-dimensional electronic structure is strongly coupled to the chemical (geometrical) structure. As a result, the nonlinear excitations (solitons, polarons and bipolarons) are dressed with local structural distortions creating states with energies within the energy gap [8]. "New" infrared bands with large intensities (IRAV modes) are induced by doping and/or photoexcitation. Solitons, polarons and bipolarons are charged "defects" which break the local symmetry and therefore make Raman modes infrared active [56–58].

Figure 3.20 shows that there is no interaction between the conjugated polymer backbone and the C_{60} molecule in the ground state for the conjugated polymer/C_{60} composites [59]. The infrared spectrum, which is dominated by the vibrational absorptions of the conjugated polymer backbone, shows no change upon addition of 5% C_{60}. In the excited state, however, the photoinduced electron transfer and the long lived charge separation in conjugated polymer/C_{60} composites greatly enhances both the concentration and the lifetime of the photoinjected charges on the polymer backbone resulting in an enhancement of the strength of the IRAV modes (Fig. 3.21)

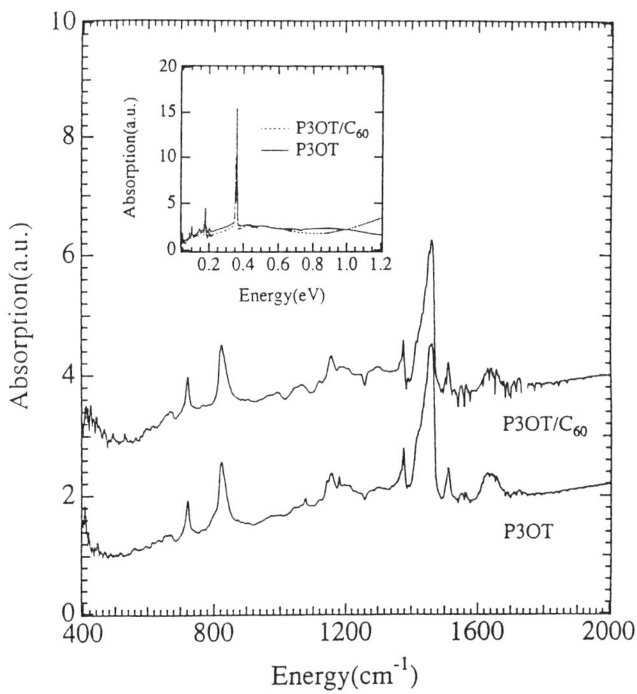

Fig. 3.20. Infrared absorption spectra of P3OT and P3OT/C_{60} (5%) films at 80 K. The inset shows the data over the whole IR range up to 1.0 eV. Reproduced with permission of The American Physical Society from [59]

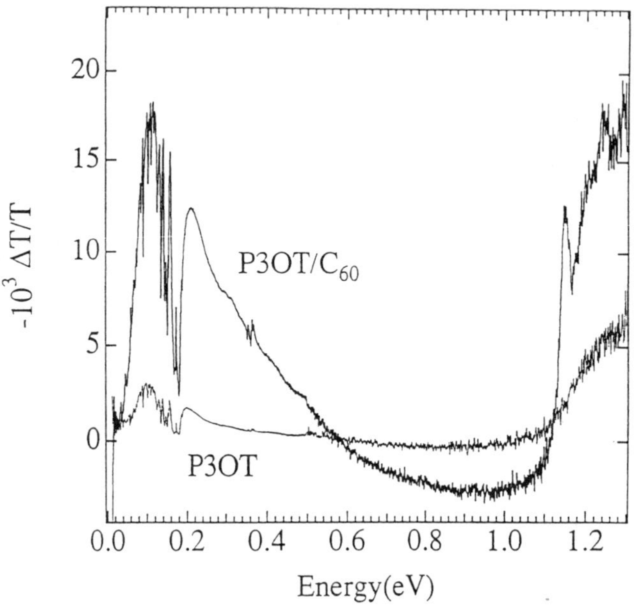

Fig. 3.21. Photoinduced IR absorption spectra of P3OT and P3OT/C_{60} (5%) at 80 K obtained by pumping with argon-ion laser at 2.41 eV with 50 mW cm^2. Reproduced with permission of The American Physical Society from [59]

[59]. In Fig. 3.21 the electronic absorption bands associated with the C_{60}- (at 1.15 and 1.25 eV) are also observable. Note that the more than one order of magnitude enhancement of the IRAV intensities does not alter the spectral positions of the IRAV modes. Thus, conjugated polymer/C_{60} composites may serve as valuable for the study of the IRAV modes of conjugated polymers.

3.2.6 Sensitization of Photoconductivity

Figure 3.22 shows the time-resolved transient photocurrent (PC) of a P3OT film containing 5% C_{60} and a pure P3OT film upon photoexcitation at 2.9 eV; the addition of 5% C_{60} to P3OT results in an enhancement of the photocurrent by nearly an order of magnitude [29]. The rise time is limited by the temporal resolution of the instrumental response (50 ps). The increase of the photocarrier generation efficiency results in the enhancement of the initial photocurrent. This enhancement for other non-dengenerate semiconducting polymers is demonstrated in Fig. 3.23 for the case of MEH-PPV/C_{60} composites with different C_{60} content [29]. The admixture of 1% of C_{60} results in an

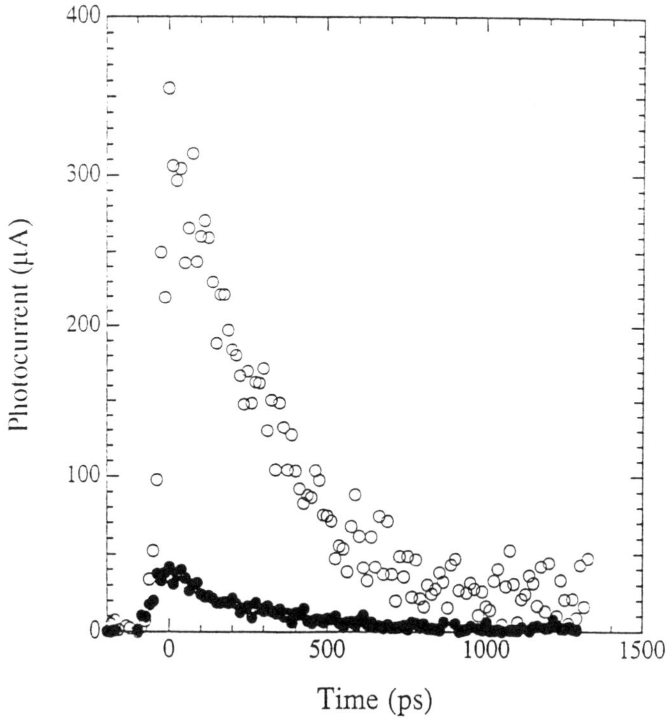

Fig. 3.22. Transient photoconductivity in P3OT alone (*full circles*) and in P3OT/C_{60} (5%) (*open circles*). Reproduced with permission of Elsevier Science from [32]

increase of initial photocurrent by an order of magnitude. This increase of the photocarrier generation efficiency is accompanied by the succesive increase in lifetime of the photocarriers upon adding more C_{60}. Thus, the ultrafast photoinduced electron transfer from the semiconducting polymer onto C_{60} not only enhances the charge carrier generation in the host polymer but also serves to prevent recombination by separating the charges and stabilizing them [29].

Figure 3.24 shows, on a semilog plot, the spectral response of the cw-PC of MEH-PPV alone and MEH-PPV/C_{60} composite for different concentrations of C_{60} [29]. These room temperature data are normalized to the constant incident photon flux of about 7.5×10^{14} photons/cm^{2}s. The photoconductive spectral response of MEH-PPV shows a sharp onset at about 2 eV which coincides with the optical absorption edge across the energy gap. However, the photoconductive response of the MEH-PPV/C_{60} composites increases

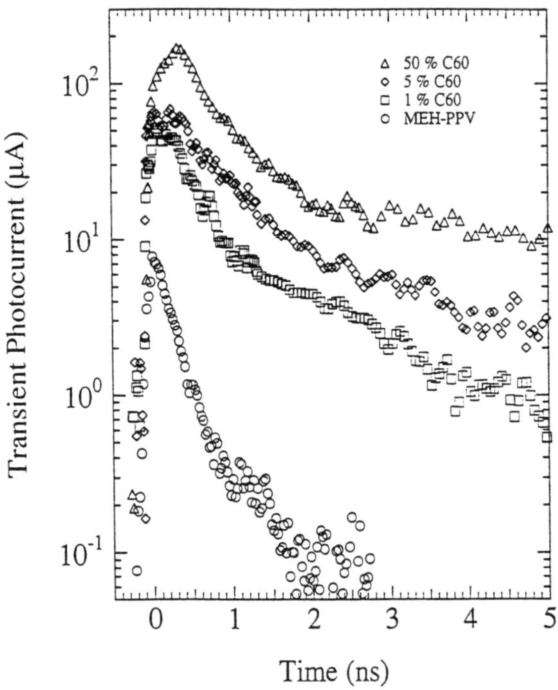

Fig. 3.23. The transient photoconductivity of MEH-PPV and MEH-PPV/C_{60} for various concentrations. Reproduced with permission of Elsevier Science from [32]

sharply at about 1.3 eV which is lower than the photoconductivity onset of the individual components alone [29]. This issue has been recently addressed by Zhang et al. [60] with calculations which show an enhancement of electron affinity of C_{60} upon photoexcitation, in agreement with the experimental results that photoexcited C_{60} is a better electron acceptor than C_{60} in the ground state [61]. Furthermore, the composite films exhibit a remarkably enhanced photoconductivity over the broad spectral range from the near infrared to the ultraviolet. At photon energies below 2 eV, the enhancement reaches several orders of magnitude [29]. Although the effect of the C_{60} on the mobility of the carrier should be considered (the mobility will probably decrease as a result of the deep trap sites associated with C_{60} clusters), the significant effect of C_{60} on the charge carrier generation efficiency is evident from the observation that even 1% C_{60} in the polymer matrix enhances the cw-PC by more than an order of magnitude [29]. This observation is in full agreement with the photoinduced electron transfer phenomenon which leaves metastable positive polarons on the polymer backbone after the electron transfer; i.e. "photodoping".

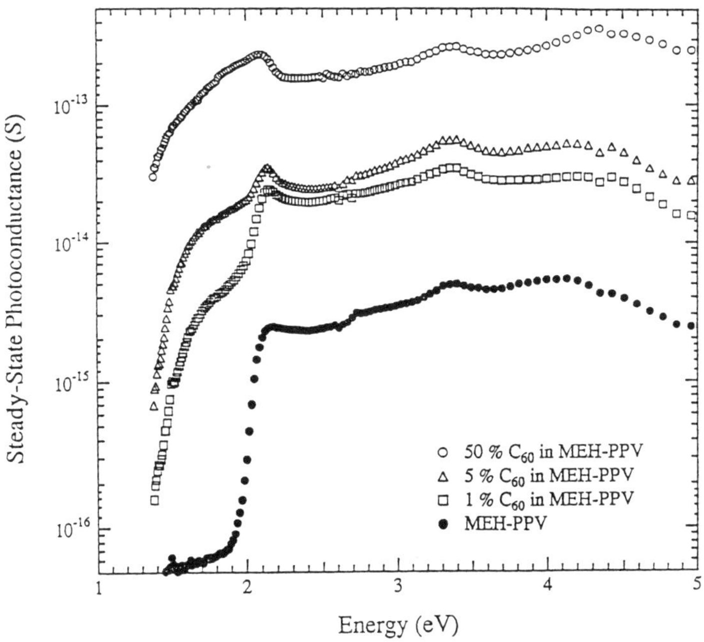

Fig. 3.24. Spectral response of the steady state photoconductivity of MEH-PPV alone and MEH-PPV/C_{60} for several concentrations at 300 K and biasing field of 104 V/cm. Reproduced with permission of Elsevier Science from [32]

3.2.7 Direct Experimental Evidence for the Metastable Charge Separation: Light Induced Electron Spin Resonance (LESR)

Definitive evidence of charge transfer and charge separation is obtained from LESR experiments [27]. Figure 3.25 shows the ESR signal upon illuminating the P3OT/C_{60} composites with light of $h\nu = E_{\pi-\pi^*}$ where $E_{\pi-\pi^*}$ is the energy gap of the conjugated polymer (donor). Two photoinduced ESR signals can be resolved; one at $g = 2.00$ and the other at $g = 1.99$ [27, 28, 31]. The higher g-value line is assigned to the conjugated polymer cation (polaron) and the lower g-value line to C_{60}^- anion. The assignment of the lower g-value line to C_{60}^- is unambiguous, for this low g-value was measured earlier [62]; the higher g-value is typical of conjugated polymers. The LESR signal vanishes above 200 K; this rules out permanent photochemical changes as the origin of the ESR signal and demonstrates the reversibility of the photoinduced electron transfer. The integrated LESR intensity shows two peaks with nearly equivalent intensities (Fig. 3.26) with the polymer cation radical signal showing a slightly higher spin resonance intensity due an additional contribution from the pristine polymer's response.

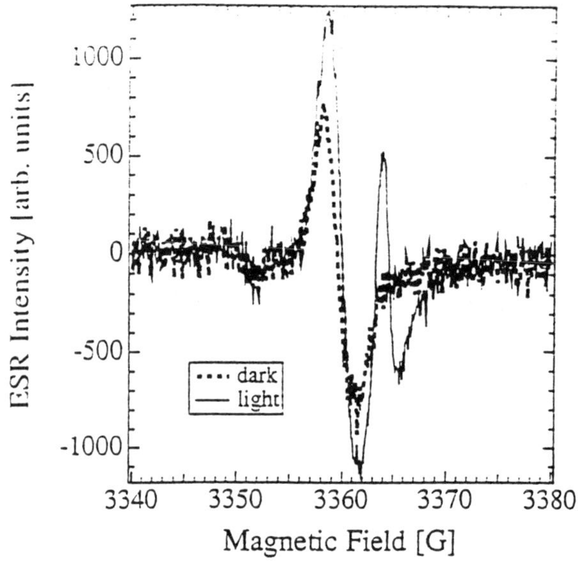

Fig. 3.25. Light induced electron spin resonance spectra of P3OT/C_{60} upon succesive illumination with 2.41 eV argonion laser with 100 mW. The dark ESR signal is also displayed as a dashed line. Reproduced with permission of The American Physical Society from [28]

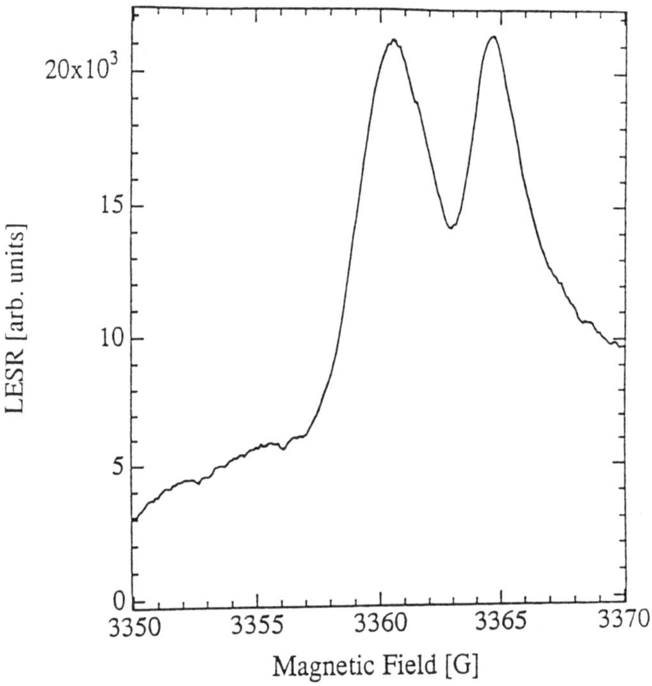

Fig. 3.26. Integrated LESR intensity [ESR (illuminated) – ESR (dark)] as displayed in Fig. 3.31 of P3OT/C_{60}. Reproduced with permission of World Scientific from [31]

The temperature dependence of the LESR signal intensity shows Arrhenius behaviour with activation energy of $\Delta\varepsilon = 15\,\text{meV}$. This result suggests a phonon assisted back relaxation mechanism of the photoinduced charge separated state.

3.3 Effect of the Surrounding Medium: Solution Studies

As observed in photoinduced charge transfer phenomena earlier and successfully described by the Marcus theory [63, 64], the rate and the lifetime of the charge transfer is a function of the polarity of the surrounding medium [2]. The photoinduced electron transfer in conjugated polymer/C_{60} composites has been investigated in two different media using 1,2-dichlorobenzene (ODCB, $\varepsilon = 9.93$) and p-xylene ($\varepsilon = 2.27$) [65]. The PIA spectrum of a 0.5 mg/ml solution of P3MBET in ODCB is displayed in Fig. 3.27. In this more polar solvent, a strong transition is observed centered at 1.47 eV with a second one at low energy with a maximum below 0.64 eV. The high energy transition at 1.47 eV is attributed to the superposition of a PIA signal due to a triplet–triplet absorption and a charged photoexcitation[65]. The low energy PIA band, however, is solely associated with charged excitations. Thus, in high polarity solvents the neutral photoexcitations (long lived triplet) and charged photoexcitations coexist.

Upon mixing C_{60} (5×10^{-4} M) into the ODCB solution of P3MBET, the high energy PIA band is quenched [65]. The resulting PIA spectrum exhibits the two peaks associated with charged excitations of the polymer only [65]. In addition to the features of the charged excitations on the P3MBET conjugated backbone, the PIA spectrum of the P3MBET/C_{60} solution in ODCB exhibits a small absorption at around 1.15 eV which is assigned to the optical absorption of C_{60}-.

The PIA spectrum of a 0.5 mg/ml solution of P3MBET in p-xylene, however, exhibits one strong, well defined band at 1.50 eV assigned to the triplet–triplet absorption band of the polymer backbone [65]. This clearly shows that the nature of the long lived photoexcitations in conjugated polymers can change from neutral triplets to charged photoexcitations by changing the polarity of the surrounding medium.

Upon mixing C_{60} into the p-xylene solution of P3MBET, the triplet–triplet absorption is quenched without an increase in the charged photoexcitation density. Since the observed PIA spectrum of P3MBET/ C_{60} solution in p-xylene is identified as the triplet–triplet absorption of C_{60} with significantly enhanced intensity, a triplet energy transfer from the conjugated polymer triplet state onto the C_{60} triplet state has been proposed [65]. The results account for the solvent dependent formation of the charged photoexcitations in conjugated polymers and for a shift between energy and/or electron transfer in co-solutions with C_{60} [65].

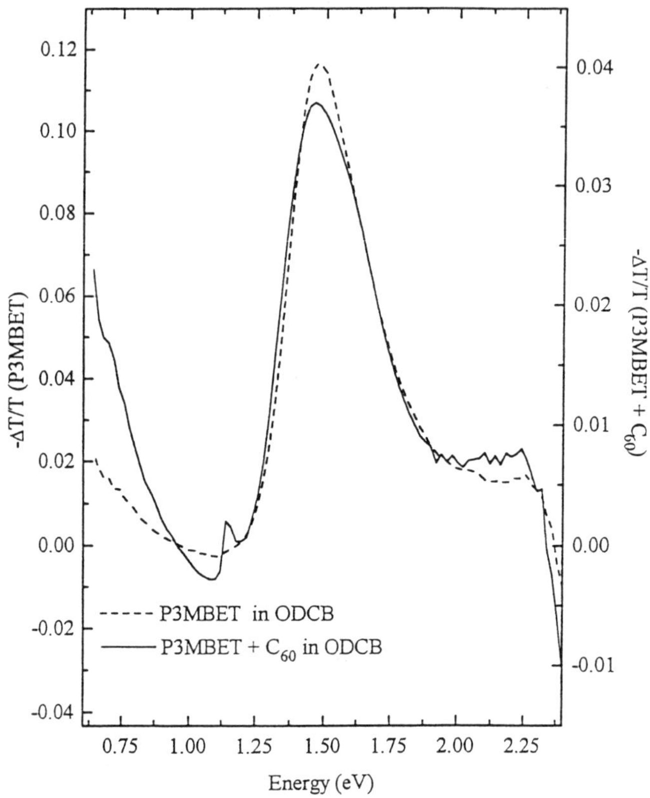

Fig. 3.27. Photoinduced absorption spectra of 0.5 mg/ml P3MBET and P3MBET/C_{60} (5×10^{-4} M) o-dichlorobenzene solutions

3.4 Photovoltaic and Photodetector Applications: A Molecular Approach to High Efficiency Photovoltaic Cells Using Photoinduced Electron Transfer

The use of organic bilayers with photoinduced electron transfer at the interface has been investigated heavily during the last couple of decades (for a summary of the early reports see for example [66, 67]). Tang [68] demonstrated photovoltaic activity in small molecular bilayers which were vacuum evaporated. Extensive literature exists on the fabrication of solar cells based on small molecule dyes as well as donor–acceptor systems (see for example [69] and references therein). Inorganic oxide semiconductors have also been used to facilitate electron transfer from organic dyes to achieve charge separation and photovoltaic conversion (see for example [70, 71] and references therein). Furthermore, conjugated polymer layers have been used in solar cells (see for example [72–74] and references therein). Yamashita et al. [75] reported a bilayer photodiode based on the organic donor tetratiafulvalene (TTF) and C_{60} [75]. The Au/TTF/C_{60}/Au devices do not exhibit rectification in the dark, indicating that there is no charge transfer in the ground state. Low charge carrier density results in negligible barriers, and the current–voltage (I–V) characteristic is nearly linear. Upon illumination, however, the I–V characteristic changes dramatically, exhibiting a large rectification ratio and photocurrent.

For photovoltaic cells made with pure conjugated polymers, energy conversion efficiencies were typically 10^{-3}–10^{-2}%, too low to be used in practical applications [72, 76–78]. The photoinduced electron transfer in composites of semiconducting polymers (as donors) and buckminsterfullerene, C_{60}, and its derivatives (as acceptors) has provided an alternative "molecular approach" to high efficiency photovoltaic conversion; the initial processes are analogous to the initial steps in natural photosynthesis. Since the time scale for photoinduced charge transfer is subpicosecond, more than 10^3 times faster than the radiative or nonradiative decay of photoexcitations, the quantum efficiency for charge transfer and charge separation from donor to acceptor is close to unity. Thus, photoinduced charge transfer across a donor/acceptor (D/A) interface provides an effective method to overcome early time carrier recombination in organic systems and thus to enhance their optoelectronic response. As noted above, for example, with the addition of only 1% C_{60}, the photoconductivity of MEH-PPV+C_{60} increases by an order of magnitude over that of pure MEH-PPV.

3.4.1 Conjugated Polymer/C_{60} Heterojunction Photodiodes

From the energy band diagram in Fig. 3.28 it is clear that the heterojunction formed at the interface between a semiconducting polymer and a C_{60} thin film should function as a diode with a rectifying current-voltage characteristic (analogous to a p-n junction, however, with a different mechanism based on molecular redox properties). In reverse bias, electron injection into the semiconducting polymer and electron removal from C_{60} are energetically unfavorable. This inherent polarity of the device results in very low reversed bias current densities. On the other hand, electron injection into C_{60} and electron removal from the semiconducting polymer are energetically favorable, thus resulting in relatively high current densities under forward bias. Analogous arguments for molecular diodes were first proposed by Aviram and Rattner years ago for Langmuir–Blodgett D-A structures [4, 79].

Figure 3.29 shows the current–voltage characteristics of a heterojunction device consisting of successive layers of ITO/MEH-PPV/C_{60}/Au; a schematic description of the device operation [34, 80] is shown in Fig. 3.30. Positive bias is defined as positive voltage applied to the ITO contact. Exponential turn-on up to 500 mV in forward bias is clearly observable; the rectification ratio is approximately 10^4.

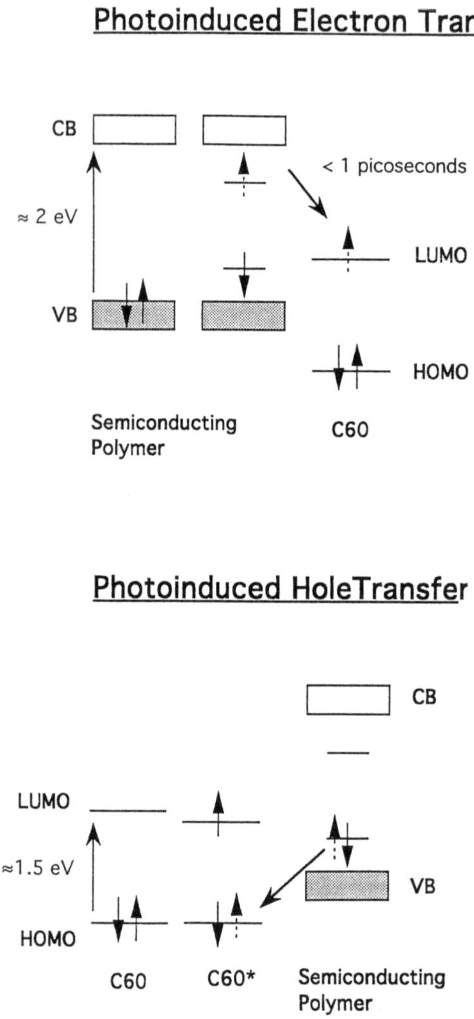

Fig. 3.28. Schematic energy band diagram for a photoinduced electron and hole transfer between semiconducting, conjugated polymers and buckminsterfullerene, C_{60}

In order to test if the ITO or the gold electrodes form blocking (or rectifying) contacts, the following three layer devices were prepared: gold/MEH-PPV/gold, gold/MEH-PPV/ITO, gold/C_{60}/gold, and gold/C_{60}/ITO. Since all such devices have symmetric and linear current–voltage characteristics, we conclude that gold and ITO form nonrectifying contacts both to MEH-PPV and to C_{60}, in agreement with the reported absence of rectification in Au/C_{60}/ITO devices [81]. We, therefore, attribute the rectifying behavior of the four layer device to the heterojunction between the semiconducting polymer and C_{60}.

The current–voltage characteristic of the device changes dramatically upon illumination by visible light. The open circuit voltage (V_{oc}) is 0.44 V

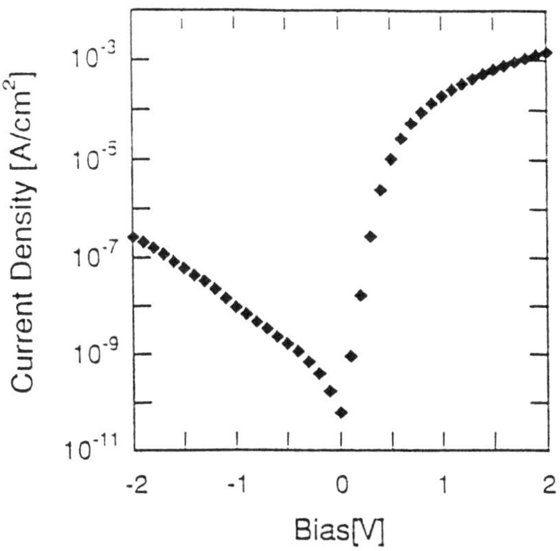

Fig. 3.29. Dark current vs voltage characteristics of the ITO/MEH-PPV/C_{60}/Au device at room temperature. Reproduced with permission of American Institute of Physics from [34]

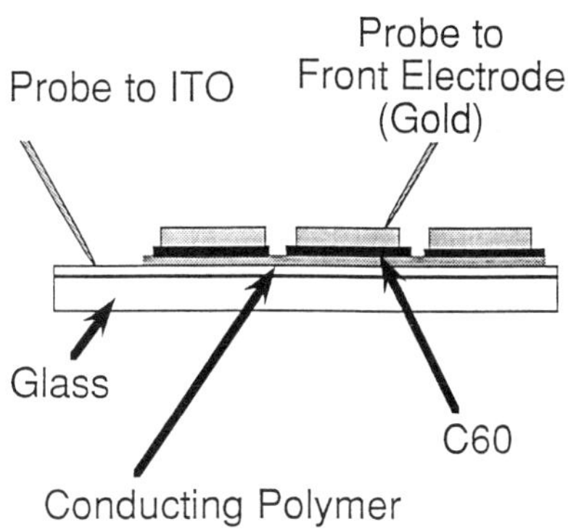

Fig. 3.30. Schematic cross-section of the heterojunction devices fabricated from MEH-PPV and C_{60}. Reproduced with permission of Elsevier from [102]

which saturates to around 0.53 V under stronger illumination. The short circuit current density (J_{sc}) is 2.08×10^{-6} A/cm^2, and the fill factor (FF) of 0.48 results in a power conversion efficiency of 0.04%. An open circuit voltage of about 0.5 V appears to be characteristic of the MEH-PPV/C_{60} interface. Similar values were obtained with ITO/MEH-PPV/C_{60}/Al devices and with Al/MEH-PPV/C_{60}/ITO devices.

As expected from the observation of fast photoinduced electron transfer from MEH-PPV to C_{60}, [27, 82] the major increase in both forward and reverse bias current results from photoinduced charge separation at the heterojunction interface. The increase in photo-current by nearly four orders of magnitude (from 1×10^{-9} A/cm^2 to 6×10^{-6} A/cm^2) upon illumination at -1 V (reverse) bias demonstrates that the heterojunction serves as a relatively sensitive photodiode.

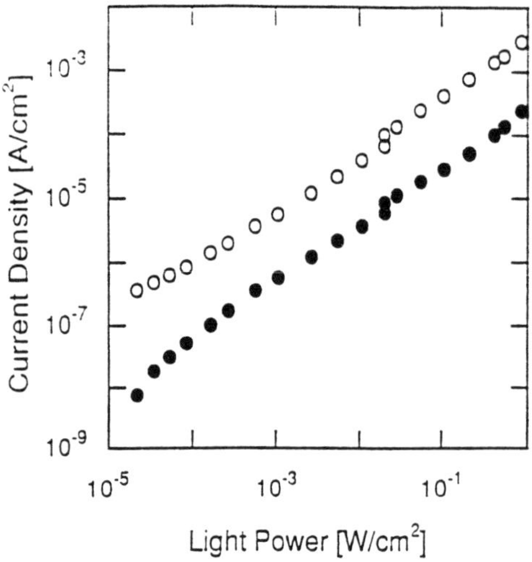

Fig. 3.31. Short circuit current (*closed circles*) and photocurrent at -1 V bias (*open circles*) as a function of light intensity for the ITO/MEH-PPV/C_{60}/Au device. Reproduced with permission of American Institute of Physics from [34]

Figure 3.31 shows the dependence of the short circuit current and the photocurrent at -1 V (reverse) bias as a function of the illumination intensity (514.5 nm); the data show no indication of saturation at light intensities up to approximately 1 W/cm^2; i.e. one order of magnitude greater than the terrestrial solar intensity.

The spectral dependence of the photoresponse of these bilayer heterojunction devices is displayed in Fig. 3.32. The onset of photocurrent at $h\nu = 1.7$ eV follows the absorption of MEH-PPV, which initiates the photoinduced electron transfer; [27, 82] note that illumination is from the ITO/MEH-PPV side

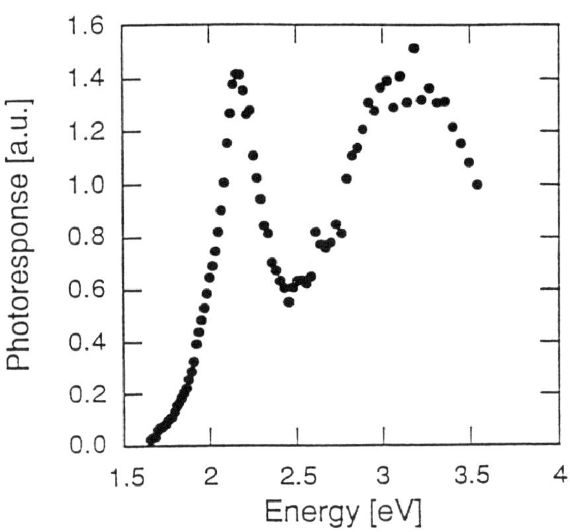

Fig. 3.32. Spectral response of the photocurrent in ITO/MEH-PPV/C_{60}/Au photodiode at (reverse) -1 V bias. Reproduced with permission of American Institute of Physics from [34]

of the device. The minimum in the photocurrent at $h\nu = 2.5\,\mathrm{eV}$ corresponds to the photon energy of maximum absorption of MEH-PPV. The MEH-PPV layer, therefore, acts as a filter which reduces the number of photons reaching the MEH-PPV/C_{60} interface. This implies that diffusion of charge carriers in these devices is limited; the photoactive region is restricted to a thin layer adjacent to the interface between the MEH-PPV and C_{60} layers. Assuming a charge carrier mobility of $10^{-4}\,\mathrm{cm^2/Vs}$, with an applied field of $10^5\,\mathrm{V/cm}$ and with a photoinduced carrier lifetime approximately 1 ns, the maximum distance a carrier could travel in the MEH-PPV is estimated to be only a few Ångstroms. Thus, the generation of photoexcitations which result in separated charge carriers occurs only at the heterojunction interface.

Yoshino et al. [42] also reported the optical response of a heterojunction device comprising a P3OT and C_{60} bilayer. The photoresponse of these devices shows a broad excitation profile ranging from 750 nm into the UV.

Tripathy et al. [83] reported the enhancement of the photocurrent upon mixing 10% C_{60} into poly(3-undecylthiophene) (P3UT) in Al/P3UT+C_{60}/ITO devices.

3.4.2 Diodes Made of Conjugated Polymer Composites with a Network of Internal Heterojunctions

A semiconducting polymer with asymmetric contacts (a low work function metal on one side and a high work function metal on the opposite side) functions as a "tunneling injection diode"; such devices have been described by Parker [84]. Conjugated polymer tunneling injection devices have been studied as electroluminescent light emitting diodes (LEDs) [77, 84–96]. A schematic cross-sectional view of such devices is displayed in Fig. 3.33.

In forward bias, tunneling injection diodes exhibit relatively high efficiency electroluminescence which is promising for flat panel and/or flexible, large area display applications. In reverse bias, on the other hand, the devices exhibit a strong photoresponse with a quantum yield >20% (el/ph at

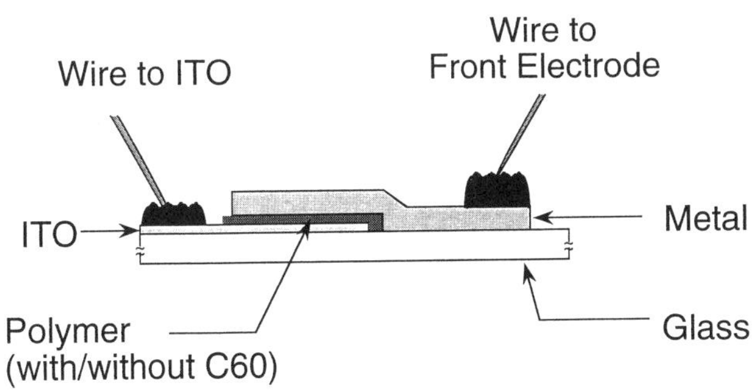

Fig. 3.33. Schematic cross-section of the tunnel diodes fabricated from conjugated polymer/C_{60} composites. Reproduced with permission of Elsevier Science from [32]

$-10\,V$ reverse bias) [77, 78]. Devices based on derivatives of polythiophene exhibit even better photoresponse (80% el/ph at $-15\,V$), competetive with UV sensitized Si photodiodes [77, 78]. A photovoltaic response was observed under zero bias conditions [76–78]. The energy band structure for such thin film devices can be approximated by the rigid band model. The charge carrier concentration in these undoped, pristine materials is sufficiently low that the semiconducting layer (a few thousand Ångstroms) is fully depleted; there is negligible band bending at the metal-polymer interface. Note, however, that the mobility of injected charge carriers is not symmetric; in most conjugated polymers, the electron mobility is lower than the hole mobility.

Although the quantum efficiency for photoinduced charge separation is near unity for a D/A pair, the conversion efficiency in a bilayer heterojunction device (see previous section) is limited [34]:

(i) Due to the molecular nature of the charge separation process, efficient charge separation occurs only at the D/A interface; thus, photoexcitations created far from the D/A junction recombine prior to diffusing to the heterojunction.

(ii) Even if charges are separated at the D/A interface, the photovoltaic conversion efficiency is limited by the carrier collection efficiency; i.e. the separated charges must be collected with minimum losses.

Consequently, interpenetrating phase separated D/A network composites would appear to be ideal photovoltaic materials [80, 97–99]. Through control of the morphology of the phase separation into an interpenetrating network, one can achieve a high interfacial area within a bulk material. Since any point in the composite is within a few nanometers of a D/A interface, such a composite is a "bulk D/A heterojunction" material. If the network is bicontinuous, as shown schematically in Fig. 3.34, the collection efficiency can be equally efficient.

Important progress has been made toward creating "bulk D/A heterojunction" materials [35, 78, 98, 99]. As shown in Fig. 3.35, the short circuit current, $I_{sc} = 0.5\,mA/cm^2$ under $20\,mW/cm^2$ illumination, corresponding to a collection efficieny of $\mu_c = 7.4\%$ electrons per incident photon [35] approximately two orders of magnitude higher than that of pure MEH-PPV tunneldiodes as well as of the MEH-PPV/C_{60} heterojunction device as described in the previous section. The electroluminescence quantum efficiency of this blend device was 10^3–10^4 times less than in pure MEH-PPV devices, consistent with the ultrafast photoinduced charge separation which quenches the emission of the donor [35].

Uniform films with even higher concentrations of soluble C_{60} derivatives were cast from 1,2-dichlorobenzene solutions containing up to 1:4 weight ratio MEH-PPV:C_{60} derivative. For devices made from these high concentration blends, charge collection efficiencies around $\mu_c = 45\%$ (electrons out /incident photon) and power conversion efficiencies around $\mu_e = 3\%$ (electrical power out/incident light power) have been realized [35]. The efficiencies are nearly

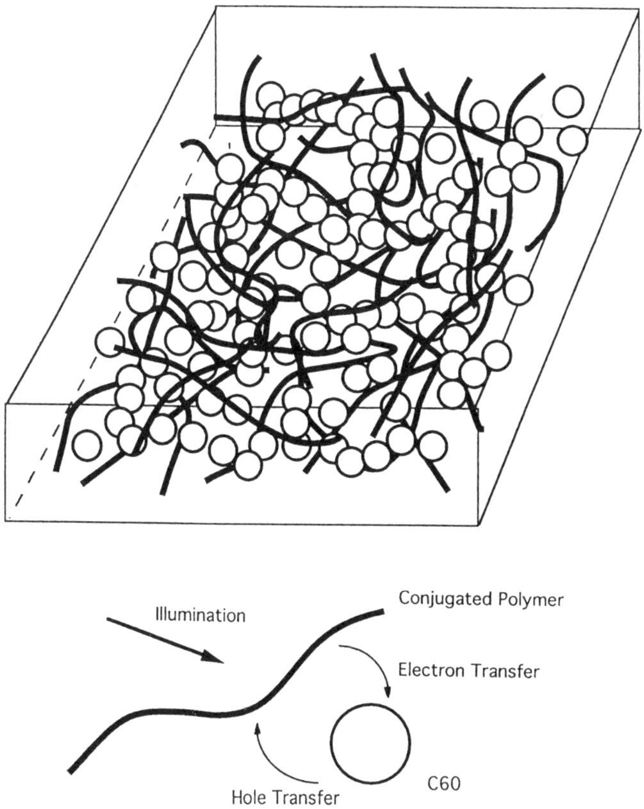

Fig. 3.34. Schematic illustration of the interpenetrating donor/acceptor (conjugated polymer/C_{60}) network of internal heterojunctions

independent of the incident illumination intensity as shown in Fig. 3.36. Furthermore, the internal efficiencies are even higher when corrected for the small thickness of the film which absorbs only 60% of the incident photons.

The excellent photosensitivity and relatively high energy conversion efficiencies obtained from the network heterojunction materials are promising. Further optimization of device performance can be achieved by opimization of the device physics:

(i) Optimize the choice of metallic electrodes to achieve good ohmic contacts on both sides for collection of the oppositely charged photocarriers,
(ii) Optimize the choice of the D/A pair (the energetics determine the open circuit potential), and
(iii) Optimize the network morphology of the phase separated composite material.

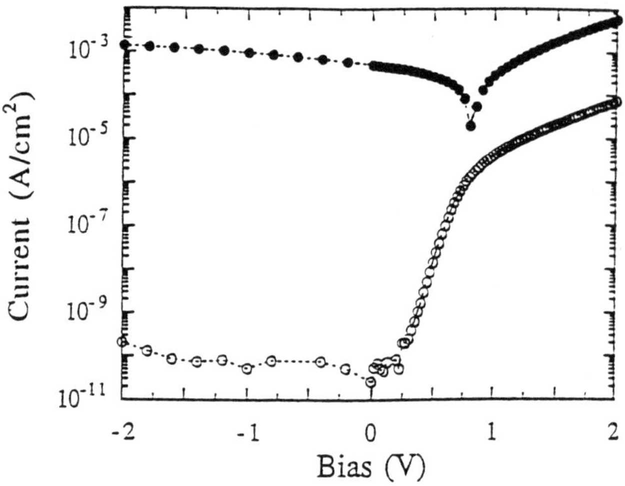

Fig. 3.35. Current-voltage characteristics of a Ca/MEH-PPV:[6,6]PCBM/ITO device in the dark (*open circles*) and under 20 mW/cm^2 illumination at 430 nm (*solid circles*). Reproduced with permission of AAAS from [35]

In addition, the band gap of the semiconducting polymer should be chosen for efficient harvesting of the solar spectrum.

The all-polymer approach to the interpenetrating D/A network has been demonstrated by using a conjugated polymer as photoelectron donor and another conjugated polymer as photoelectron acceptor [98, 99]. The carrier collection efficiency and energy conversion efficiency of conjugated polymer photovoltaic cells was improved by blending the two semiconducting polymers with different electron affinities. In films fabricated from a phase-separated polymer composite made with MEH-PPV, as donor, and cyano-PPV, CN-PPV, as acceptor, the photoluminescence and electroluminescence of both component polymers are quenched, indicative of rapid and efficient separation of photogenerated electron–hole pairs with electrons on the acceptor and holes on the donor. Diodes made with such a composite semiconducting polymer as the photosensitive medium show promising photovoltaic characteristics with carrier collection efficiency of 5% electrons/photon and energy conversion efficiency of 0.9%, 20 times larger than in diodes made with pure MEH-PPV and 100 times larger than in diodes made with CN-PPV [98]. The photosensitivity and the quantum yield increase with reverse bias voltage, to 0.3 A/Watt and 80% electrons/photon respectively at -10 V, comparable to results obtained from photodiodes made with inorganic semiconductors [98].

Because of the advantages which would be realized with polymer-based photovoltaics, such as low cost fabrication in large sizes and in desired shapes with mechanical flexibility, efficient "plastic" solar cells would have a major impact.

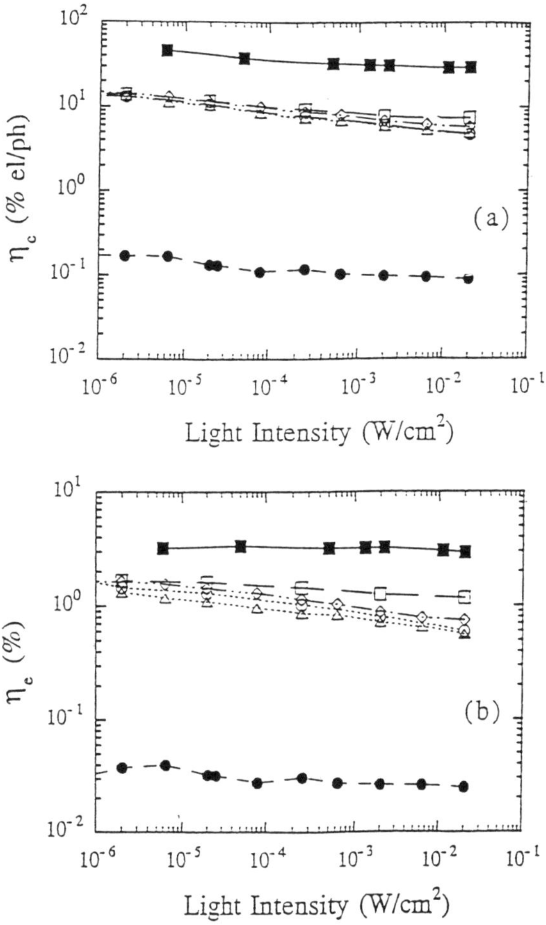

Fig. 3.36 a,b. Carrier collection efficiency, *hc*, and energy conversion efficiency, *he*, of a Ca/MEH-PPV: [6,6]PCBM (1:4)/ITO (*solid squares*), Ca/MEH-PPV: [6,6]PCBM (1:1)/ITO (*open squares*), Al/MEH-PPV: [6,6]PCBM (1:1)/ITO (*open diamonds*), Ca/MEH-PPV: [5,6]PCBM (1:1)/ITO (*open circles*), Ca/MEH-PPV: C_{60} (3:1)/ITO (*open triangles*) and Ca/MEH-PPV/ITO (*solid circles*). Reproduced with permission of AAAS from [35]

3.5 Future Areas

The discovery of an ultrafast, reversible, metastable photoinduced electron transfer in conjugated polymer/fullerene composites has opened a large area of scientific and technological interest. Photodetector and photovoltaic applications have reached the stage where development work in an industrial level is commencing. Nonlinear optical effects derived from the photoinduced electron transfer have potentials in areas such as all optical spatial light modulators, holographic memory units, photorefractive elements and optical limiters etc. Embedding the conjugated polymers and fullerenes in different matrices such as zeolites and liquid crystals, can reveal interesting effects due to decoupling of the individual D/A pairs in a dielectric environment which can be controlled. Yoshino and co-workers [100, 101] reported that doping the conjugated polymer/C_{60} composites with alkali atoms results in granular superconductivity of the C_{60} component embedded in the polymer matrix. Photoinduced electron transfer may increase the number of superconducting domains in these composites.

Acknowledgement. We gratefully acknowledge Alan Heeger and Fred Wudl for their support, encouragement and very valuble help. This work was supported by the Department of Energy (DOE No: DEFG0393ER12138), by the Office of Naval Research (ONR No: N00014-91-J-1235) and by the National Science Foundation (NSF No: DMR 90-12808). We gratefully acknowledge our co-workers who contributed to this body of work reported here while at the Institute for Polymers and Organic Solids at the University of California, Santa Barbara: Laura Smilowitz, Rene Janssen, Marwijn Christiaans, and Kwanghee Lee for photoinduced absorption studies, D. Braun, Zhang Chi, Gang Yu, and Jun Gao for preparation of devices, B. Kraabel, and D. Vacar for femtosecond photoinduced absorption studies, Changhee Lee for photoconductivity experiments, Myungsik Cha for nonlinear absorption studies, C. Gettinger, R. Wu for polymer synthesis, Jan Kees Hummelen and Gordana Srdanov for buckyball synthesis. We thank Vojislav Srdanov, from the Center for Quantized Electronic Structures at the University of California, Santa Barbara, for preparing C_{60} thin films; UNIAX Corporation Santa Barbara, for providing the MEH-PPV and BCHA-PPV; Valy Vardeny and X. Wei, from University of Utah, for PIAMDR studies.

References

1. M. A. Fox, J. Wayne E. Jones, D. M. Watkins: Chem. Eng. News, March 15, (1993).
2. *Photoinduced Electron Transfer*, M. A. Fox and M. Chanon (Eds.), Vol. 1-4, (Elsevier, Amsterdam, 1988).
3. *Photochemistry and Photobiology*, A. H. Zewail (Ed.), Vol. 1/2, (Harwood Academic Publ., Chur, 1983).
4. *Molecular Electronic Devices*, F. L. Carter (Ed.), Vol. 1/2, (Dekker, New York, 1982).
5. J. Guillet: *Polymer Photophysics*, (Cambridge University Press, New York, 1985).
6. S. E. Webber: Chem. Rev. **90**, 1469 (1990).
7. *Handbook of Conducting Polymers*, T. A. Skotheim (Ed.), Vol. 1/2, (Marcel Dekker, Inc, New York, 1986).
8. A. J. Heeger, S. Kivelson, J. R. Schrieffer, W.-P. Su: Rev. Mod. Phys. **60**, 781 (1988).
9. A. J. Heeger and P. Smith: in *Conjugated Polymers*, J. L. Bredas, R. Silbey (Eds.), Vol. pp. 141, (Kluwer Academic Publishers, Netherlands, 1991).
10. *Conjugated Polymeric Materials: Opportunities in Electronics, Optoelectronics and Molecular Electronics*, J. L. Bredas, R. R. Chance (Eds.), NATO-ARW Series E, Vol. 182, (Kluwer Academic Publishers, Dordrecht, 1990).

11. *Science and Application of Conducting Polymers*, W. R. Salaneck, D. T. Clark, E. J. Samuelsen (Eds.), Vol. Adam Hilger, Bristol, (1991).
12. Y. Lu: Solitons and Polarons in Conducting Polymers, (World Scientific, Singapore, 1988).
13. W.P. Su and J.R. Schrieffer: Proc. Natl. Acad. Sci. USA **77**, 5626 (1980).
14. H.W. Kroto, J.R. Health, S.C. O'Brien, R.F. Curl, R.E. Smalley: Nature **318**, 162 (1985).
15. W. Kratschmer, L.D. Lamb, K. Fostiropoulos, D.R. Huffman: Nature **347**, 354 (1990).
16. H.W. Kroto, A.W. Allaf, S.P. Balm: Chem. Rev. **91**, 1213 (1991).
17. P.M. Allemand, A. Koch, F. Wudl, Y. Rubin, F. Diederich, M.M. Alvarez, S.J. Anz, R.L. Whetten: J. Am. Chem. Soc. **113**, 1050 (1991).
18. R.C. Haddon, A.F. Hebard et al.: Nature **350**, 320 (1991).
19. A.F. Hebard, M.J. Rosseinsky, R.C. Haddon, D.W. Murphy, S.H. Glarum, T.T.M. Palstra, A.P. Ramirez, A.R. Kortan: Nature **350**, 600 (1991).
20. P.M. Allemand, K.C. Khemani, A. Koch, F. Wudl, K. Holczer, S. Donovan, G. Gruner, J.D. Thompson, Science **253**, 301 (1991).
21. F. Wudl and J.D. Thompson: J. Phys. Chem. Solids **53**, 1449 (1992).
22. Z.H. Kafafi, J.R. Lindle, R.G.S. Pong, F.J. Bartoli, L.J. Lingg, J. Milliken: Chem. Phys. Lett. **188**, 492 (1992).
23. L.W. Tutt and A. Kost: Nature **356**, 225 (1992).
24. J.S. Meth, H. Vanderzeele, Y. Wang: Chem. Phys. Lett. **197**, 26 (1992).
25. M. Cha, N.S. Sariciftci, A.J. Heeger, J.C. Hummelen, F. Wudl: Appl. Phys. Lett. in press. (1995).
26. M. Cha, N.S. Sariciftci, A.J. Heeger, J.C. Hummelen, F. Wudl: J. Opt. Soc. Am. B submitted (1995).
27. N.S. Sariciftci, L. Smilowitz, A.J. Heeger, F. Wudl: Science **258**, 1474 (1992).
28. L. Smilowitz, N.S. Sariciftci, R. Wu, C. Gettinger, A.J. Heeger, F. Wudl: Phys. Rev. B**47**, 13835 (1993).
29. C.H. Lee, G. Yu, N.S. Sariciftci, D. Moses, K. Pakbaz, C. Zhang, A.J. Heeger, F. Wudl: Phys. Rev. B**48**, 15425 (1993).
30. B. Kraabel, D. McBranch, N.S. Sariciftci, D. Moses, A.J. Heeger: Phys. Rev. B**50**, 18 (1994).
31. N.S. Sariciftci and A.J. Heeger: Int. J. Mod. Phys. B**8**, 237 (1994).
32. N.S. Sariciftci: Prog. Quantum Electron **19**, 131 (1995).
33. X. Wei, Z.V. Vardeny, N.S. Sariciftci, A.J. Heeger: Phys. Rev. B in press (1995).
34. N.S. Sariciftci, D. Braun, C. Zhang, V. Srdanov, A.J. Heeger, G. Stucky, F. Wudl: Appl. Phys. Lett. **62**, 585 (1993).
35. G. Yu, J. Gao, J.C. Hummelen, F. Wudl, A.J. Heeger: Science **270**, 1789 (1995).
36. N.S. Sariciftci, B. Kraabel, C.H. Lee, K. Pakbaz, A.J. Heeger, D.J. Sandman: Phys. Rev. B**50**, 12 (1994).
37. S. Morita, A.A. Zakhidov, K. Yoshino: Sol. State Commun. **82**, 249 (1992).
38. K. Yoshino, X.H. Yin, S. Morita, T. Kawai, A.A. Zakhidov: Sol. State Commun. **85**, 85 (1993).
39. K. Yoshino, X.H. Yin, K. Muro, S. Kiyomatsu, S. Morita, A.A. Zakhidov, T. Noguchi, T. Ohnishi: Jpn. J. Appl. Phys. **32**, L357 (1993).
40. S. Morita, S. Kiyomatsu, M. Fukuda, A.A. Zakhidov, K. Yoshino, K. Kikuchi, Y. Achiba: Jpn. J. Appl. Phys. **32**, L1173 (1993).
41. S. Morita, S. Kiyomatsu, X.H. Yin, A.A. Zakhidov, T. Noguchi, T. Ohnishi, K. Yoshino: J. Appl. Phys. **74**, 2860 (1993).
42. S. Morita, A.A. Zakhidov, K. Yoshino: Jpn. J. Appl. Phys. **32**, L873 (1993).

43. K. Yoshino, T. Akashi, K. Yoshimoto, S. Morita, R. Sugimoto, A. A. Zakhidov: Sol. State Commun. **90**, 41 (1994).
44. K. Yoshino, T. Akashi, K. Yoshimoto, M. Yoshida, S. Morita, A. A. Zakhidov: Mol. Cryst. Liq. Cryst. **256**, 343 (1994).
45. L. Smilowitz, A. Hays, A. J. Heeger, G. Wang, J. E. Bowers: J. Chem. Phys. **98**, 6504 (1993).
46. L. Smilowitz and A. J. Heeger: Synth. Met. **48**, 193 (1992).
47. Y. H. Kim, D. Spiegel, S. Hotta, A. J. Heeger: Phys. Rev. B**38**, 5490 (1988).
48. K. Pichler, D. A. Halliday, D. D. C. Bradley, R. H. Friend, P. L. Burn, A. B. Holmes: Synth. Met. **55–57**, 230 (1993).
49. J. L. Bredas and A. J. Heeger: Chem. Phys. Lett. **217**, 507 (1994).
50. K. Hasharoni and H. Levanon: J. Phys. Chem. **99**, 4875 (1995).
51. U. Stamm, M. Taiji, M. Yoshizawa, K. Yoshino, T. Kobayashi: Mol. Cryst. Liq. Cryst. A **182**, 147 (1990).
52. G. S. Kanner, X. Wei, B. C. Hess, L. R. Chen, Z. V. Vardeny: Phys. Rev. Lett. **69**, 538 (1992).
53. B. Kraabel, C. H. Lee, D. McBranch, D. Moses, N. S. Sariciftci, A. J. Heeger: Chem. Phys. Lett. **213**, 389 (1993).
54. M. Sinclair, D. Moses, A. J. Heeger: Phys. Rev. B**36**, 4296 (1987).
55. T. W. Hagler, K. Pakbaz, K. F. Voss, A. J. Heeger: Phys. Rev. B**44**, 8652 (1991).
56. B. Horowitz: Solid State Commun. **41**, 729 (1982).
57. E. Ehrenfreund, Z. V. Vardeny, O. Brafman, B. Horowitz: Phys. Rev. B**36**, 1535 (1987).
58. G. Zerbi, M. Gussoni, C. Castiglioni: in *Conjugated Polymers*, J. L. Bredas, R. Silbey (Eds.), Vol. pp. 435, (Kluwer Academic Publ., Dordrecht, 1991).
59. K. H. Lee, R. A. J. Janssen, N. S. Sariciftci, A. J. Heeger: Phys. Rev. B**49**, 5781 (1994).
60. G. P. Zhang, R. T. Fu, X. Sun, K. H. Lee, T. Y. Park: J. Phys. Chem. **99**, 12301 (1995).
61. J. W. Arbogast, A. O. Darmanyan, C. S. Foote, Y. Rubin, F. N. Diederich, M. M. Alvarez, S. J. Anz, R. L. Whetten: J. Phys. Chem. **95**, 11 (1991).
62. P. M. Allemand, G. Srdanov, A. Koch, K. Khemani, F. Wudl, Y. Rubin, F. Diederich, M. M. Alvarez, S. J. Anz, R. L. Whetton: J. Am. Chem. Soc. **113**, 2780 (1991).
63. R. A. Marcus: J. Chem. Phys. **24**, 966 (1956).
64. R. A. Marcus: Rev. Mod. Phys. **65**, 599 (1993).
65. R. A. J. Janssen, N. S. Sariciftci, A. J. Heeger: J. Chem. Phys. **100**, 8641 (1994).
66. H. Meier: *Organic Semiconductors*, (Verlag Chemie, Weinheim, 1974).
67. J. Simon and J. J. Andre: *Molecular Semiconductors*, (Springer, Berlin, 1985).
68. C. W. Tang: Appl. Phys. Lett. **48**, 183 (1986).
69. J. B. Whitlock, P. Panayotatos, G. D. Sharma, M. D. Cox, R. R. Sauers, G. R. Bird: Opt. Eng. **32**, 1921 (1993).
70. B. O'Regan and M. Gratzel: Nature **353**, 737 (1991).
71. W. A. Nevin and G. A. Chamberlain: IEEE Trans. Electron Devices **40**, 75 (1993).
72. J. Kanicki: in *Handbook of Conducting Polymers*, T. A. Skotheim (Ed.), Vol. 1, pp. 543, (Marcel Dekker, New York, 1986).
73. G. Horowitz: Adv. Mater. **2**, 287 (1990).
74. K. Uehara, A. Maekawa, A. Sugimoto, H. Inoue: Thin Solid Films **215**, 123 (1992).
75. Y. Yamashita, W. Takashima, K. Kaneto: Jpn. J. Appl. Phys. **32**, L1017 (1993).
76. H. Antoniadis, B. R. Hsieh, M. A. Abkowitz, M. Stolka, S. A. Jehneke: Polym. Preprints **34**, 490 (1993).

77. G. Yu, C. Zhang, A. J. Heeger: Appl. Phys. Lett. **64**, 1540 (1994).
78. G. Yu, K. Pakbaz, A. J. Heeger: Appl. Phys. Lett. **64**, 3422 (1994).
79. A. Aviram and M. A. Ratner: Chem. Phys. Lett. **29**, 277 (1974).
80. N. S. Sariciftci and A. J. Heeger: U.S. Patent, No: 5 331 183, (University of California, USA) (1994).
81. H. Yonehara and C. Pac: Appl. Phys. Lett. **61**, 575 (1992).
82. N. S. Sariciftci, L. Smilowitz, D. Braun, G. Srdanov, V. Srdanov, F. Wudl, A. J. Heeger: Proc. ICSM 92, Goteborg, Sweden, Vol. 55-57, pp. 3125, Synth. Met. (1992).
83. C. S. Kuo, F. G. Wakim, S. K. Sengupta, S. K. Tripathy: Sol. State Commun. **87**, 115 (1993).
84. I. D. Parker: J. Appl. Phys. **75**, 1656 (1994).
85. J. H. Burroughes, D. D. C. Bradley, A. R. Brown, R. N. Marks, K. Mackay, R. H. Friend, P. L. Burns, A. B. Holmes: Nature **347**, 539 (1990).
86. D. Braun and A. J. Heeger: Appl. Phys. Lett. **58**, 1982 (1991).
87. D. Braun, A. J. Heeger, H. Kroemer: J. Electron. Mat. **20**, 945 (1991).
88. D. Braun, G. Gustafsson, D. McBranch, A. J. Heeger: J. Appl. Phys. **72**, 564 (1992).
89. Y. Ohmori, Y. Manda, H. Takahashi, T. Kawai, K. Yoshino: Jpn. J. Appl. Phys. **29**, L837 (1990).
90. Y. Ohmori, M. Uchida, K. Muro, K. Yoshino: Sol. State Commun. **80**, 605 (1991).
91. Y. Ohmori, M. Uchida, K. Muro, K. Yoshino: Jpn. J. Appl. Phys. **30**, L1941 (1991).
92. H. Vestweber, A. Greiner, U. Lemmer, R. F. Mahrt, R. Richert, W. Heitz, H. Bassler: Adv. Mater. **4**, 661 (1992).
93. G. Gustafsson, Y. Cao, G. MTracy, F. Klavetter, N. Colaneri, A. J. Heeger: Nature **357**, 477 (1992).
94. G. Grem, G. Leditzky, B. Ullrich, G. Leising: Adv. Mater. **4**, 36 (1992).
95. C. Zhang, S. Hüger, K. Pakbaz, F. Wudl, A. J. Heeger: J. Electron. Mat. **22**, 413 (1993).
96. N. C. Greenham, S. C. Moratti, D. D. C. Bradley, R. H. Friend, A. B. Holmes: Nature **365**, 628 (1993).
97. N. S. Sariciftci and A. J. Heeger: U.S. Patent, No: 5 454 880, (1995).
98. G. Yu and A. J. Heeger: J. Appl. Phys. **78**, 4510 (1995).
99. J. J. M. Halls, C. A. Walsh, N. C. Greenham, E. A. Marseglia, R. H. Friend, S. C. Moratti, A. B. Holmes: Nature **376**, 498 (1995).
100. A. A. Zakhidov, H. Araki, K. Tada, K. Yakushi, E. Saiki, K. Yoshino: Phys. Lett. A**205**, 317 (1995).
101. H. Araki, A. A. Zakhidov, E. Saiki, N. Yamasaki, K. Yakushi, K. Yoshino: Jpn. J. Appl. Phys. Part 2, Letters **34**, L1041 (1995).
102. N. S. Sariciftci, L. Smilowitz, A. J. Heeger, F. Wudl: Synth. Met. **59** 333 (1993).

4. Optical Studies of Covalently Linked C_{60} Solids

A.M. Rao, P.C. Eklund

4.1 Introduction

In 1985, Smalley, Kroto and co-workers [1] discovered the all-carbon C_{2n} cage molecules, or "fullerenes" in carbon soot formed by laser vaporization of a rotating graphite target in an inert atmosphere (for a review see [2]). Subsequent discovery in 1990 to mass produce some of these molecules (e.g., C_{60} (Fig. 4.1a) and C_{70} (Fig. 4.1b)) in a carbon soot produced by striking an AC electric arc between two carbon electrodes in a He atmosphere [3, 4] has lead to considerable interdisciplinary research activity to understand the molecular and solid state properties of these beautifully symmetric molecules.

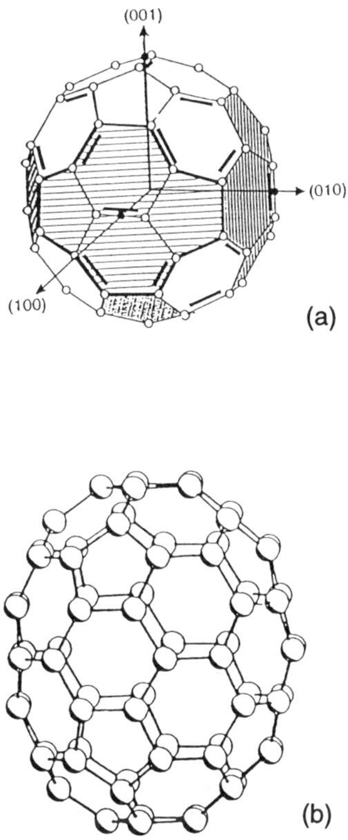

Fig. 4.1 a,b. Schematic view of a C_{60} molecule showing the octahedral arrangement of the pyracylene units. A pyracyclene unit contains two pentagons and two hexagons and a central reactive double bond. All rings associated with the pyracylene units are shaded in the figure

Consistent with the icosahedral symmetry, the C_{60} molecule has been shown by nuclear magnetic resonance (NMR) [5–8] to exhibit a single line, i.e. there are 60 equivalently bonded carbon atoms in the shell (Fig. 4.1a). Experimentally, the diameter for the nuclear shell is determined by NMR measurements to be 7.10 ± 0.07 Å [8]. The arrangement of C5 and C6 rings on the cage surface resemble the patchwork of pentagons and hexagons on the geodesic domes designed by architect Robert Buckminster Fuller, thus motivating the name "buckminsterfullerenes" or simply "fullerenes" for these C_{2n} molecules. The pentagons are required to achieve the curvature of the shell. The four chemical bonds required per C-atom can be realized if 30 double bonds are present in the structure of a C_{60} molecule; they are located at the fusion between hexagonal rings (Fig. 4.1a). The C_{70} molecule exhibits D_{5h} symmetry (Fig. 4.1b) and can be envisioned as an elongated C_{60} molecule obtained by adding a belt of 5 hexagons around the equatorial plane of the C_{60} molecule which is normal to one of the five-fold axes, and suitably rotating the two C_{60} hemispheres by 36 degrees so that they fit continuously onto the belt hexagons.

In this chapter we focus on a small corner of fullerene research devoted to understanding the assembly of C_{60} molecules into covalently bonded networks and covalently bonded crystalline materials. To do this, double bonds must first be broken on the C_{60} molecule and reconnected to neighboring molecules. Research has shown that this intermolecular bonding can be accomplished through several routes: (i) chemical [9–13], (ii) photochemical [14], (iii) alkali-metal doped C_{60}: MC_{60} (M=K,Rb) [15–18] or M_4C_{60} (M=Na) [19], (iv) high-pressure and temperature [20–25], (v) by electron impact [26], and (vi) in a plasma discharge [27]. Below, we discuss the formation, structure and vibrational properties of photoinduced, and high pressure and temperature-induced (HPHT) polymerized C_{60}. Polymerization in C_{60} via routes (i), (iii) and (iv), are discussed in Chaps. 10, 7 and 6, respectively, in this volume. Vibrational spectroscopy has played a key role in studying the covalent assembly of fullerenes, so we first discuss the vibrational modes of pristine C_{60} and review theory and experiment on the photopolymerization of solid C_{60} and related C_{60} polymers produced at high pressure and high temperature. Because of the high, icosahedral symmetry of C_{60}, out of a possible $174 = 180-6$ internal degrees of freedom, only 14 vibrational mode frequencies of C_{60} are allowed optically in first order. As we shall see, the covalent connection of these molecules into networks or crystalline solids strongly overrides the local intramolecular symmetry and render many new intra molecular vibrational modes optically active. The reader is referred is to Chaps. 5 and 8 for in-depth theoretical discussions of polymerization in C_{60} solids. Finally, recent experimental results on the identification of "charge transfer" excitons in solid C_{60} will be given, as they have been proposed as the "trigger mechanism" for the photopolymerization process [28, 29]. Detailed theoretical arguments concerning photopolymerization are presented in Chap. 8 of this volume.

4.2 Pristine C_{60}

Pristine C_{60} molecule can be thought of as a patchwork of pyracylene units. One such unit is shaded in Fig. 4.2a and isolates four rings: two pentagons and two hexagons, with a central, chemically reactive double bond. Addition reactions of C_{60} and C_{70} involve the scission of this double bond in the pyracylene unit[9-12]. For C_{60}, 30 such double bonds (C=C) provide numerous reactive sites for the attachment of a chemical side group. The reactivity of the double bonds in C_{60} indicated that chemical schemes to link together C_{60} into a polymer structure should also be possible, and several examples have been reported. This incorporation might be accomplished in one of two general ways [9]: "in-chain" addition, and "on chain" addition. In Fig. 4.2, these two types of C_{60} polymers are sometimes referred to as "pearl necklace" (or "in-chain") and "pendant" (or "on-chain") polymers. The one-dimensional, orthorhombic phase of alkali-metal (M) doped C_{60} (M_1C_{60}) is an example of the pearl necklace fullerene polymer and is shown in Fig. 4.3. The alkali metal atoms are singly charged (M^+), and for charge neutrality, the C_{60} polymer

(a)

(b)

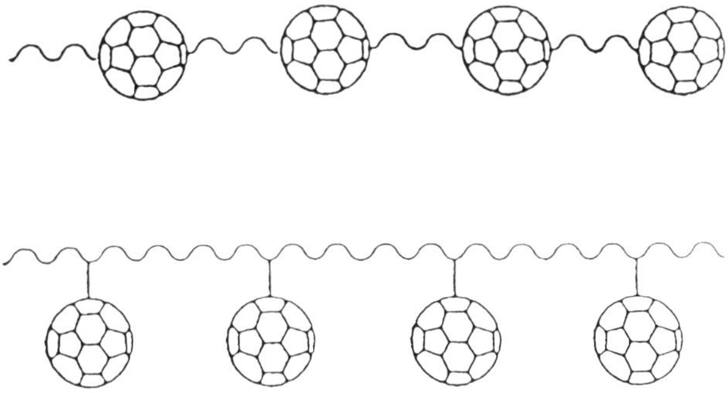

Fig. 4.2. Schematic representation of (**a**) "pearl necklace" and (**b**) "pendant chain" polymers that can be made with C_{60} fullerenes [9]

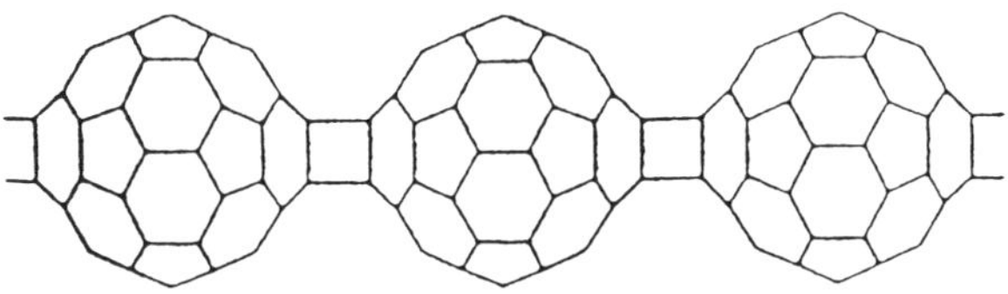

Fig. 4.3. A "pearl necklace" polymer in which the C_{60} molecules are covalently linked by a four-membered ring. This structure has been proposed for K_1C_{60} fulleride [16]

chains must be negatively charged. From detailed x-ray diffraction (XRD) [17] and electron spin resonance (ESR) studies [18] on M_1C_{60}, it has been concluded that the C_{60} anions are covalently linked to form a one-dimensional conducting $(C_{60})_n$ polyanion. We shall see later in this chapter that covalent bonding leads to polymerized C_{60} structures in two and three dimensions as well.

Solid C_{60} is a molecular crystal in which the molecules are coupled by weak van der Waals forces. In the pristine solid, the molecules are positioned in a close packed array on a face centered cubic (fcc) lattice ($a_0 = 14.7$ Å) in which the C_{60} nuclear shells of diameter (7.10 Å) are separated by a distance 2.92 Å, smaller than that found between the sheets of carbon in graphite (3.35 Å). The combination of this weak interaction and the spherical shape of the C_{60} molecule was also shown in several early NMR, differential thermal analysis (DTA) and diffraction studies to lead to the somewhat surprising result that the molecules were actually spinning rapidly about their fcc lattice positions above a temperature $T_{01} = 260$ K [8, 30–34]. As will be discussed in Sect. 4.1, we have proposed that this rotation is important for bringing reactive double bonds on adjacent molecules into the proper relative alignment to allow covalent bonding between fullerenes in the solid state. As the pristine solid is cooled below a temperature $T_{01} = 260$ K, an orientational ordering transition occurs in which the free molecular rotation about an arbitrary axis direction changes suddenly to a "ratchet-like" orientational hopping about equivalent (111) directions in the lattice [8, 30–33]. This orientational hopping has been investigated, for example, by x-ray [30, 31] and neutron [32, 34] studies. At the lower temperature $T_{02} = 80$ K, a second orientation ordering transition occurs, freezing out the motion and leaving the C_{60} system in a merohedrally (orientationally) disordered phase [31, 32]. However, two relative molecular orientations were found in the lattice: an electron rich double bond on one molecule faces either the electron deficient opening in a pentagon (83%) or hexagon (17%) of the adjacent molecule (Figs. 4.4a and 4.4b), where the % indicates the fraction of each relative orientation.

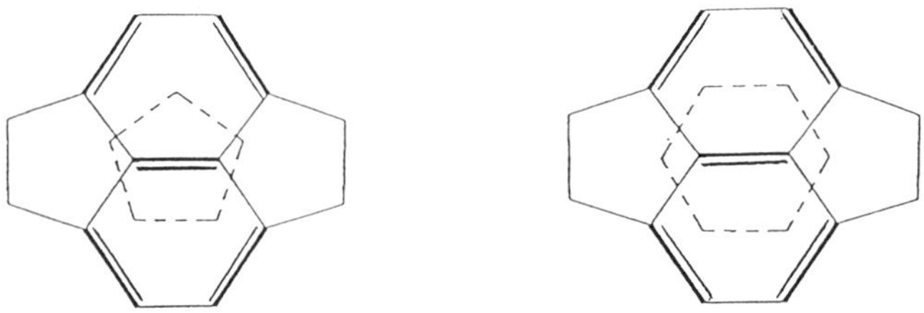

(a) (b)

Fig. 4.4 a,b. Two possible orientations of adjacent C_{60} molecules in the low temperature phase (see text)

4.3 Vibrational Modes of Solid C_{60}

We begin this section with an overview of the various vibrational modes that are possible in solid C_{60}. The vibrational modes of solid C_{60} can be divided into two classes: intermolecular vibrations (or lattice modes) and intramolecular vibrations (or "on-ball" modes). For a review of vibrational modes in alkali-metal (M) doped C_{60}, C_{70} and M_xC_{70}; see [2]. A useful, schematic view of the various vibrational modes in M_3C_{60} compounds is shown in Fig. 4.5 a–e which has been adapted from [35]. At low frequencies ($\omega \lesssim 200\,\mathrm{cm}^{-1}$), the compounds exhibit (a) molecular librational modes, (b) intermolecular (C_{60}–C_{60}) optic modes, and (c) IR-active C_{60}–M optic modes. At higher frequencies ($\omega > 200\,\mathrm{cm}^{-1}$), only intramolecular modes are found and they exhibit the schematic two-peak vibrational density of states (Fig. 4.5 a–e). The intramolecular modes below and above $\sim$700 cm^{-1} are, respectively, associated with radial and tangential character. Examples of radial mode displacements $H_g(1)$ (Fig. 4.5 a–ed) and tangential character $H_g(7)$ (Fig. 4.5 a–ee) mode are shown as examples.

Substantial information about the intramolecular modes for solid C_{60} have been obtained from Raman scattering, infrared spectroscopy, photoluminescence, and inelastic neutron scattering [2]. Raman and infrared spectroscopies

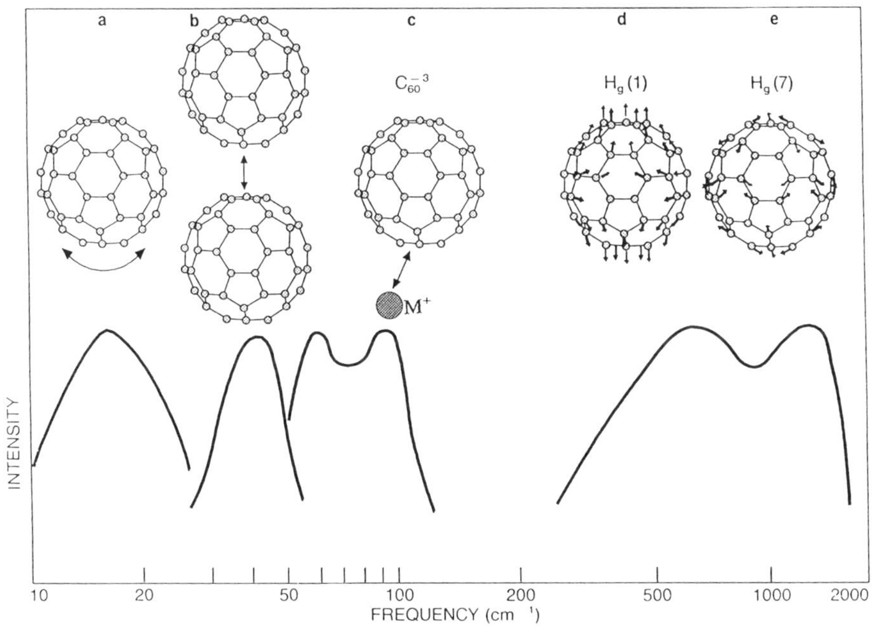

compounds (see text). M in the figure represents an alkali metal cation. After [35] provide the most quantitative methods for determining the vibrational mode frequencies and symmetries. The icosahedral (I_h) symmetry of an isolated molecule is very high, and this leads to a dramatic simplification of the vibrational and electronic states [36]. For an isolated C_{60} molecule, out of the $180 - 6 = 174$ vibrational degrees of freedom, only 46 distinct mode frequencies are consistent with the I_h symmetry of the molecule [36]. Ten of these 46 mode frequencies are Raman-active, i.e., 8 H_g + $2A_g$; four are IR-active, i.e., $4F_{1u}$ and the remainder are optically silent in first order process [36]. The H_g and F_{1u} modes are five-fold and three-fold degenerate, respectively, while the A_g modes are singly degenerate. Because solid C_{60} is a nearly ideal molecular solid, Raman and infrared studies of the solid state yield intramolecular mode frequencies very close to those of isolated molecules [37].

Depending on different experimental conditions, early Raman scattering studies of the intramolecular vibrations of thin solid C_{60} films produced significantly different results [38]. For example, different Raman spectra were obtained if the experiment was carried out in air, as opposed to vacuum or an inert atmosphere. The diversity in these early Raman scattering experiments were subsequently traced to two photochemical transformations of solid C_{60} which alter the structure of the pristine solid leaving the fullerene shells intact: (1) photopolymerization [14], and (2) photoenhanced diffusion of O_2 into interstitial voids in the fullerene lattice [39, 40]. To study these two photochemical transformations in solid C_{60}, we developed a procedure that allowed us to exclude oxygen from all stages of sample preparation to obtain oxygen-free, thin solid films of C_{60} [38].

In Fig. 4.6, we display the room temperature, first-order IR and Raman spectra of crystalline oxygen-free C_{60} films. The spectral lines are all identified with "on ball" or intramolecular modes. The IR transmission spectrum of a thin solid film of C_{60} on a KBr substrate (spectrum A in Fig. 4.6) shows four strong lines at 526, 576, 1182 and 1428 cm^{-1} that are assigned to the four anticipated F_{1u} mode frequencies. The Raman spectrum (spectrum B in Fig. 4.6), which was collected using low laser intensity (488 nm, 50 mW/mm^2) on a thin, oxygen-free C_{60} film vacuum deposited on Si (100) shows twelve lines as expected from group theoretical grounds. (Two features at 520 and 965 cm^{-1} are identified with the Si substrate.) Using polarization analysis, two of the 10 strong Raman lines of C_{60} are identified with the A_g modes [41]. The $A_g(1)$ mode at 493 cm^{-1} is the completely symmetric radial breathing mode. The tangential $A_g(2)$ mode at 1469 cm^{-1} is the "pentagonal pinch" in which the carbon atom displacements are correlated to cause the pentagons in C_{60} to alternately expand and contract. The remaining eight lines in the Raman spectrum are identified with the five-fold degenerate H_g symmetry modes. The experimental values for these mode frequencies obtained from solid films are in good agreement with theoretical calculations [42–46] on the isolated molecule, consistent with the view that solid C_{60} is nearly an ideal molecular solid.

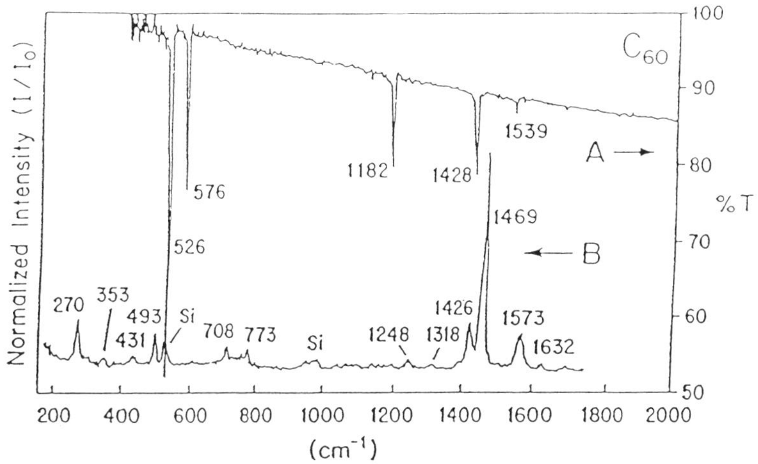

Fig. 4.6. Fourier transform infrared (FTIR) transmission (spectrum A; right axis) and Raman scattering (spectrum B; left axis) spectra of pristine C_{60}. The Raman spectrum was taken at low laser power density (~50 mW/mm^2) with the 488 nm Ar ion line on a C_{60} film ($d \sim 2000$ Å) deposited on a Si(100) substrate. A C_{60} film ($d \sim 5000$ Å) for the FTIR spectrum was deposited on a KBr substrate [14]

To obtain the intrinsic Raman spectrum of solid C_{60} a cylindrical lens was used in all our Raman scattering experiments to focus the laser beam to a 2 mm tall by 0.1 mm wide stripe, thereby reducing the power density on the sample by a factor of 20–100 over that obtained by a tightly focused spot produced by a spherical lens. This experimental point is very important, as solid C_{60} undergoes photopolymerization in an oxygen-free environment at relatively low laser power density (~100 mW/mm^2) [14]. Early Raman studies by several groups [47] incorrectly identified an unpolarized 1458 cm^{-1} line in photopolymerized C_{60} with the intrinsic pentagonal pinch mode. Subsequently, we showed that the 1458 cm^{-1} mode should be identified with a photochemically induced polymeric state of C_{60}[14]. Recently, changes in the Raman spectrum as a function of temperature and laser power of thin C_{60} films deposited on alumina substrates were reported by Pagnier et al [48]. They attributed the decrease in the A_g mode intensities to probable structural changes and not to a photochemical effect. It should be noted that: (i) the inherently large power densities (>20 000 mW/mm^2) in these later micro-Raman experiments were much greater than the densities 100 mW/mm^2 reported to produce photopolymerization. Furthermore, their data analysis did not account for the simultaneous polymerization induced by the incident laser and depolymerization caused by the laser heating of the film, and (ii) the laser power densities used in their macro-Raman study varied from 32 mW/mm^2 to 12 800 mW/mm^2. At low power densities (32–64 mW/mm^2) the Raman spectrum showed a single sharp line at 1469 cm^{-1} and a new line appeared at 1459 cm^{-1} for intermediate power densities (1360–2880 mW/mm^2). These observations are consistent with phototransformation of C_{60} films discussed in Sect. 4.1. However, at higher power densities (6400–12 800 mW/mm^2) a relatively broad peak was visible at 1463 cm^{-1} indicating that these power densities drive the system into a more disordered state [48].

Photopolymerization of solid C_{60} during Raman scattering studies was avoided. Detailed discussion of the photopolymerized state of C_{60} and the role of oxygen in several ways; e.g. by using a laser excitation energy less than the energy gap between the highest occupied molecular orbitals and the lowest unoccupied molecular orbitals (HOMO–LUMO gap); and by the sample temperature. Raman scattering studies using a visible laser (e.g., Ar ion laser) with an excitation energy greater than the HOMO–LUMO gap found that high incident powers could be used at temperatures $T < T_{01}$ [49], and for $T > 150\,°C$. For $T < T_{01}$, it was proposed that unfavorable orientations between molecules prevented the photochemical process and at high temperatures the thermal decomposition rate drives the system back to the pristine (monomeric) C_{60} [50].

In these low and high temperature regimes, higher laser intensity allows the observation of numerous ($\sim$ 100) sharp and weaker features (overtones and combination modes) in the second order infrared and Raman spectra of C_{60}. An analysis of such data has led to the determination of values for all of the 46 vibrational frequencies (Table 4.1). Finally, it should be mentioned that we also identified a few of the features in the Raman and infrared spectra with isotopic activation of a few otherwise optically silent modes [51]. Because of the 1.1% isotopic abundance of ^{13}C, approximately half the C_{60} molecules in a given film exhibit a reduced symmetry of a $^{13}C\,^{12}C_{59}$ fullerene. By virtue of the extra proton on the ^{13}C atom in the shell, the icosahedral symmetry of the molecule is broken, weakly activating otherwise silent intramolecular modes. Further work was carried out on ^{13}C enriched C_{60} to elucidate this point [52].

4.4 Polymerized C_{60} Structures

For polymerization to occur, it is clear that intramolecular bonds on adjacent fullerenes must be enticed to break and reform as covalent bonds between fullerene shells. As we shall see below, theoretical and experimental evidence indicates that the arrangement of pentagons and hexagons remains in tact. The weakest bonds are the π bonds, one of which is present in each of the 30 double bonds in C_{60}. The general criteria proposed for polymerization to occur are the same for neutral and monoanionic C_{60}, namely, a reduced intermolecular spacing and adequate rotational motion of the C_{60} molecules. In Sect. 4.1, we begin with the discussion of the photoinduced polymerization process in solid C_{60} and the topochemical requirements leading to intermolecular bonding. Next, we discuss briefly the polymerized fullerene structures

Table 4.1. The 46 mode frequencies for molecular C_{60}, determined by experiments

Even-parity		Odd-parity	
ν_i(R)	Frequency(cm^{-1})	ν_i(R)	Frequency(cm^{-1})
$\nu_1(A_g)$	497.5	$\nu_1(A_u)$	1143.0
$\nu_2(A_g)$	1470.0		
		$\nu_1(F_{1u})$	526.5
$\nu_1(F_{1g})$	502.0	$\nu_2(F_{1u})$	575.8
$\nu_2(F_{1g})$	975.5	$\nu_3(F_{1u})$	1182.9
$\nu_3(F_{1g})$	1357.5	$\nu_4(F_{1u})$	1429.2
$\nu_1(F_{2g})$	566.5	$\nu_1(F_{2u})$	355.5
$\nu_2(F_{2g})$	865.0	$\nu_2(F_{2u})$	680.0
$\nu_3(F_{2g})$	914.0	$\nu_3(F_{2u})$	1026.0
$\nu_4(F_{2g})$	1360.0	$\nu_4(F_{2u})$	1201.0
		$\nu_5(F_{2u})$	1576.5
$\nu_1(G_g)$	486.0		
$\nu_2(G_g)$	621.0	$\nu_1(G_u)$	399.5
$\nu_3(G_g)$	806.0	$\nu_2(G_u)$	760.0
$\nu_4(G_g)$	1075.5	$\nu_3(G_u)$	924.0
$\nu_5(G_g)$	1356.0	$\nu_4(G_u)$	970.0
$\nu_6(G_g)$	1524.5	$\nu_5(G_u)$	1310.0
		$\nu_6(G_u)$	1446.0
$\nu_1(H_g)$	273.0		
$\nu_2(H_g)$	432.5	$\nu_1(H_u)$	342.5
$\nu_3(H_g)$	711.0	$\nu_2(H_u)$	563.0
$\nu_4(H_g)$	775.0	$\nu_3(H_u)$	696.0
$\nu_5(H_g)$	1101.0	$\nu_4(H_u)$	801.0
$\nu_6(H_g)$	1251.0	$\nu_5(H_u)$	1117.0
$\nu_7(H_g)$	1426.5	$\nu_6(H_u)$	1385.0
$\nu_8(H_g)$	1577.5	$\nu_7(H_u)$	1559.0

that are obtained in the alkali-metal doped M_1C_{60} (M=K,Rb) and Na_4C_{60} compounds (Sect. 4.2), and the one-, two- and three-dimensional networks of interconnected fullerenes produced by the simultaneous application of high pressures and temperatures (Sect. 4.3). Vibrational spectroscopy is used to explore the results of the newly formed intermolecular bonds in all these polymer phases.

4.4.1 Photopolymerized C_{60}

The polymerization of solid C_{60} was proposed by Rao et al. [14] to explain the phototransformation of C_{60} in an oxygen-free environment. In a subsequent paper [49], we also showed that this process requires the molecules to be spinning about their lattice positions, to allow a favorable relative molecular alignment between adjacent C_{60} molecules. When this alignment involving double bonds on adjacent molecules is satisfied, we proposed that a "2 + 2" cycloaddition reaction leads to intermolecular bonding. This cycloaddition reaction is a well known photochemical or thermal mechanism leading to covalent bonding between a variety of molecules in close proximity in the solid state [53]. The "2+2" mechanism involves the reconfiguration of double bonds on adjacent molecules A and B. It has been shown in a variety of cases that if these double bonds (C=C), one on A and the other on B, are oriented parallel to one another, and separated by less than ~4.2 Å, then by photochemical or thermal assistance these double bonds can be broken and then reform as a four-sided ring of single bonds coupling the molecules together [53]. The simplest example of "2 + 2" cycloaddition is the dimerization of ethylene (C_2H_4) to form cyclobutane (C_4H_8) in which each carbon atom is found at a corner of a square, four-membered ring (Fig. 4.7a). In 1993, we proposed that two C_{60} molecules dimerize by this "2+2" cycloaddition mechanism, as shown schematically in Fig. 4.7b [14, 49]. Since a C_{60} molecule contains 30 chemically active double bonds, and furthermore, since adjacent molecules are separated by as little as ~3 Å in pristine solid C_{60} [31], we argued that solid C_{60} should satisfy the general topochemical requirements for "2+2" cycloaddition [49]. As mentioned above, neutron scattering studies have shown that the double bonds on adjacent C_{60} molecules are not likely to be parallel below $T_{01} \sim 260$ K [32] (Fig. 4.4 a,b). From a temperature dependent Raman study of photopolymerization in pristine C_{60} films we showed that photopolymerization was dramatically suppressed below T_{01}, consistent with the concept that parallel adjacent double bonds were unlikely in the solid state structure at low temperatures [49]. However, for $T > T_{01}$, two adjacent,

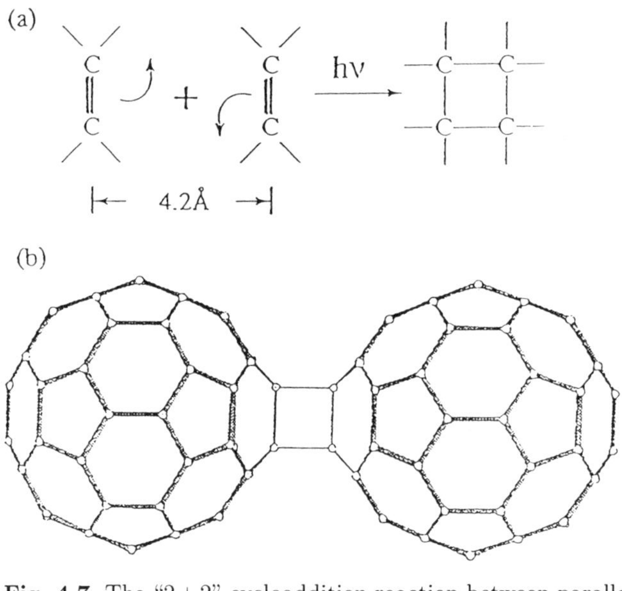

Fig. 4.7. The "2 + 2" cycloaddition reaction between parallel carbon double bonds on adjacent molecules is shown schematically in (**a**). A C_{60} dimer produced by the photochemical "2 + 2" cycloaddition reaction is shown in (**b**)

freely spinning molecules can adopt 30(30) = 900 favorable such alignments and the "2 + 2" cycloaddition reaction would then be expected to occur.

Based on ^{13}C NMR study of the C_{60}-benzyne adduct, Hoke et al. [54] reported direct experimental evidence for the attachment of the benzyne ring to C_{60} via "2 + 2" cycloaddition (Fig. 4.8) [54]. High-resolution ^{13}C NMR data on polymeric C_{60} phases produced under high pressure and temperature

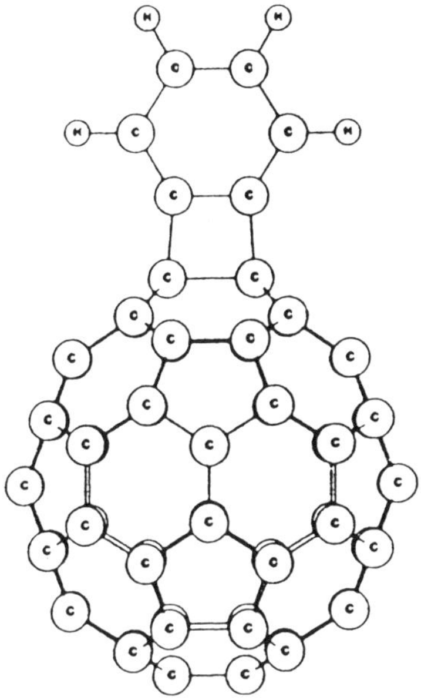

Fig. 4.8. Experimental evidence for the attachment of the benzyne ring to C_{60} via "2 + 2" cycloaddition. After [54]

[55] (discussed in Sect. 4.3) and on the $C_{60}{=}C_{60}$ dimer [13] also provide direct evidence for the "2 + 2" cycloaddition reaction involving C_{60}.

Direct evidence for photoinduced bonds between C_{60} molecules have also been obtained from laser desorption mass spectroscopy (LDMS) [14, 56] and atomic force microscopy studies [57] of photopolymerized C_{60} films. In the LDMS studies, identical experimental and desorption conditions for pristine and previously phototransformed C_{60} films were used and the resulting LDMS spectra compared. Using a pulsed nitrogen laser (337 nm, 5–10 ns, variable pulse energy of 0.5–10 mJ/cm^2) to ablate or vaporize clusters of C_{60} molecules from the film. Mass peaks were observed in the mass spectrum at the C_{60} mass value (720 atomic mass units (amu)) and at all integer multiples of this mass out to 12 × 720 amu (Fig. 4.9), consistent with the desorption of a distribution of covalently linked clusters of C_{60} molecules from the previously photopolymerized film surface [14, 56]. Under the same conditions, only a single peak at 720 amu was observed in the LDMS spectrum of the pristine (monomeric) C_{60} solid. Measurements at higher N_2 laser intensity also

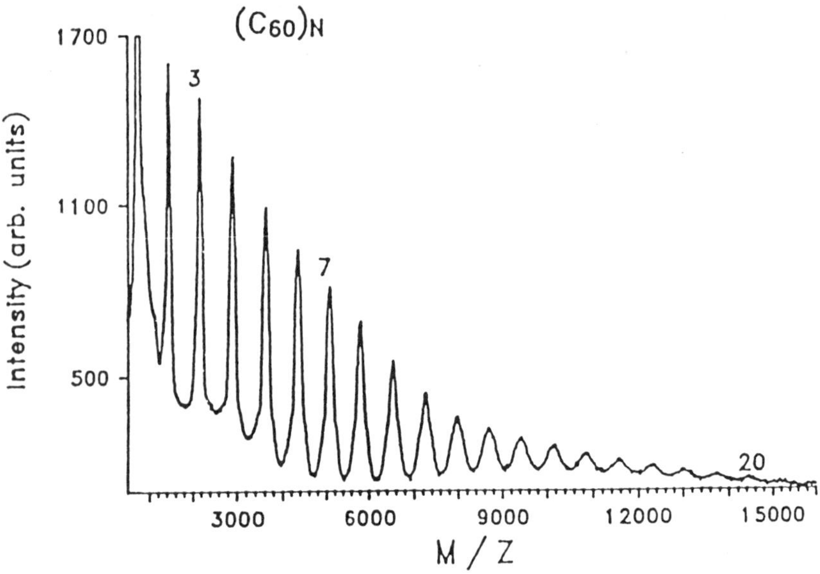

Fig. 4.9. Laser desorption mass spectrum (LDMS) of a phototransformed C_{60} film using high desorption laser fluence ($30\,mJ/cm^2$). A succession of 20 clear peaks are observed in the spectrum and are identified with large clusters of fullerene molecules $(C_{60})_N$ where $N = 2, \ldots, 20$ [14, 56]

showed that the desorption laser was itself capable of inducing higher levels of clustering and $(C_{60})_N$ clusters upto $N = 20$ were observed in previously photopolymerized films [56].

Hassanien et al. [57] have reported atomic force microscopy (AFM) measurements taken at 300 K on crystalline C_{60} films grown on mica and subsequently photopolymerized at 320 and 360 K using the 514.5 nm laser radiation from an Ar ion laser. Significant surface buckling due to the formation of covalent interfullerene bonds was observed in the AFM images of a C_{60} film photopolymerized at 300 K and a characteristic herringbone structure was reported (Fig. 4.10). Furthermore, polymer chains typically six–eight C_{60} units in length with a reduced intermolecular distance of 9.2 $\pm$0.1 Å were deduced from the AFM images such as shown in Fig. 4.10. For C_{60} films that were photopolymerized at 360 K, the AFM images revealed primarily C_{60} dimers and trimers with an unusually short interfullerene distance of 8.8 $\pm$0.1 Å [57].

Other strong evidence for covalent bonds between C_{60} molecules in the phototransformed solid phase come from Raman and IR spectroscopy. The overall effect of photopolymerization on the vibrational spectra of solid C_{60} is shown in Fig. 4.11. Since the molecules in the photopolymerized phase are now strongly coupled by covalent bonds, the I_h intramolecular symmetry is clearly broken and instead of only 14 optically-active modes ($2\,A_g + 8\,H_g + 4\,F_{1u}$), essentially all vibrational modes of the $(C_{60})_N$ oligomers are either

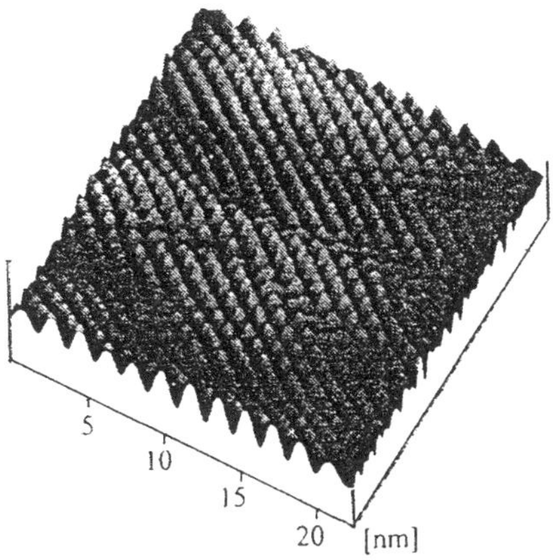

Fig. 4.10. Atomic force microscopy (AFM) image taken on a crystalline C_{60} film grown on mica and subsequently photopolymerized at 300 K using the 514.5 nm laser radiation from an Ar ion laser. The characteristic herringbone structure evident in the figure has been attributed to the surface buckling resulting from the formation of covalent inter-fullerene bonds

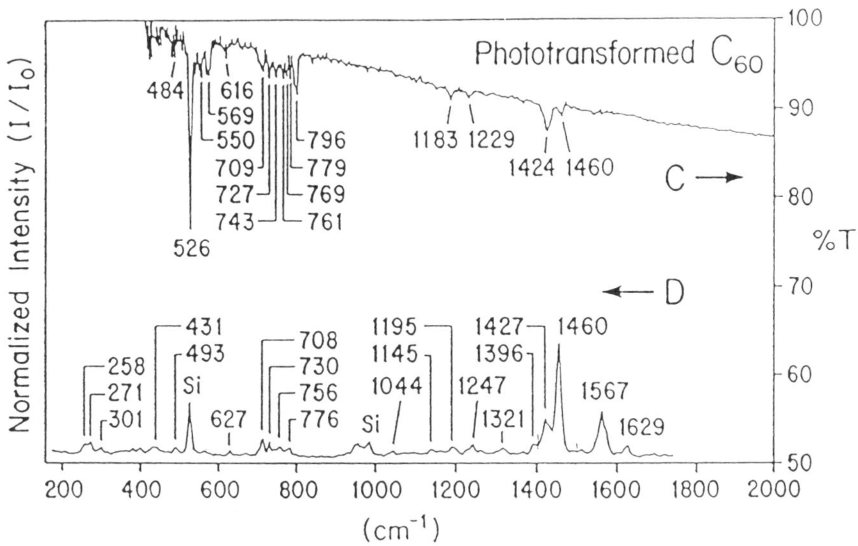

Fig. 4.11. Fourier transform infrared transmission (spectrum C, right axis) and Raman scattering (spectrum D, left axis) spectra of photopolymerized C_{60} films (see text)

Raman- or infrared-active. Thus a wonderfully rich spectrum was anticipated and, in fact observed, as shown in Fig. 4.11. These Raman and IR spectra should be compared to the much simpler spectra of the monomeric C_{60} solid (Fig. 4.6). The strongest and polarized A_{1g} line in the Raman spectrum (pentagonal pinch mode at 1469 cm^{-1}) in pristine C_{60} is observed downshifted to 1458 cm^{-1} in the photopolymerized material, loses intensity and becomes depolarized. The shift to lower frequency for the pentagonal pinch mode has been modeled using molecular dynamics for the C_{60} dimer [45, 46] and is in good agreement with the observed downshift of ~10 cm^{-1} induced photochemically [14].

Using the sensitivity of the pentagonal pinch mode ($A_g(2)$) to probe the nature of the intermolecular bonds, experiments were carried out on photopolymerized C_{60} films (~4500 Å on either suprasil or Si(100) substrates) to study the thermal stability of the intermolecular bonds [50]. In Fig. 4.12 a succession of high frequency Raman spectra taken at low excitation power for photopolymerized C_{60} are shown as a function of temperature $65°C < T < 260°C$. The evolution of the spectra with increasing temperature show the reversibility of the $2+2$ cycloaddition reaction. At the lowest temperature shown (65°C), the spectrum exhibits a broad Raman band at 1458 cm^{-1}, identified with the softened $A_g(2)$ mode of the photopolymer. As the temperature of the phototransformed film is increased, a narrow weak $A_g(2)$ line at 1469 cm^{-1} (characteristic of C_{60}) appears and then grows with increasing temperature at the expense of the diminishing 1458 cm^{-1} line. This was interpreted as clear evidence for a thermal transformation of the photopolymer to the monomeric solid. The competition between the photoinduced attachment of C_{60} monomers to form dimers and oligomers and the thermal scission of intermolecular bonds to generate monomers is shown in Fig. 4.13. There we plot the temperature dependence of the integrated intensity of the 1458 cm^{-1} band (dimers, trimers, ...) associated with the polymerized phase and the 1469 cm^{-1} band associated with isolated C_{60} monomers [58]. The solid curves in the figure are calculated according to a first order rate constant model [58]. The data analysis also showed that the polymerization rate depends linearly on the incident photon flux (Φ_0) at 488 nm. An Arrhenius plot of the Raman intensity ratio (I_{1459}/I_{1469}) was found to yield an activation energy of 1.25 eV for the thermal scission of intermolecular bonds [58]. A theoretical estimate of 2.4 eV for this chemical barrier was obtained by Adams et al. [45] (Chap. 5), while Porezag et al. [46] determined the value ~1.6 eV for this thermal energy barrier using the density functional tight binding (DF-TB) approach.

Onoe et al. [59] have used in-situ x-ray photoelectron (XPS) spectroscopy to determine the average number of four-membered rings formed on a C_{60} molecule as a function of irradiation time (Fig. 4.14). C_{60} films ($d \sim 1000$ Å) deposited on gold—overcoated stainless steel substrates were photopolymerized using the focused UV-visible radiation from a 500 W mercury lamp (power density ~1.5–2.0 mW/mm^2). The intensity of the C 1s shake-up satellite peaks which originate from the π–$\pi^\star$ transition of the π electrons in the valence region was used to monitor the degree of polymerization. With increasing irradiation time, the formation of the four-membered rings reduces the number of π electrons on a C_{60} molecule and thus leads to a decrease in the intensity of the C 1s shake-up satellite peaks. The results summarized in Fig. 4.14 were interpreted as follows: (i) after 100 h of

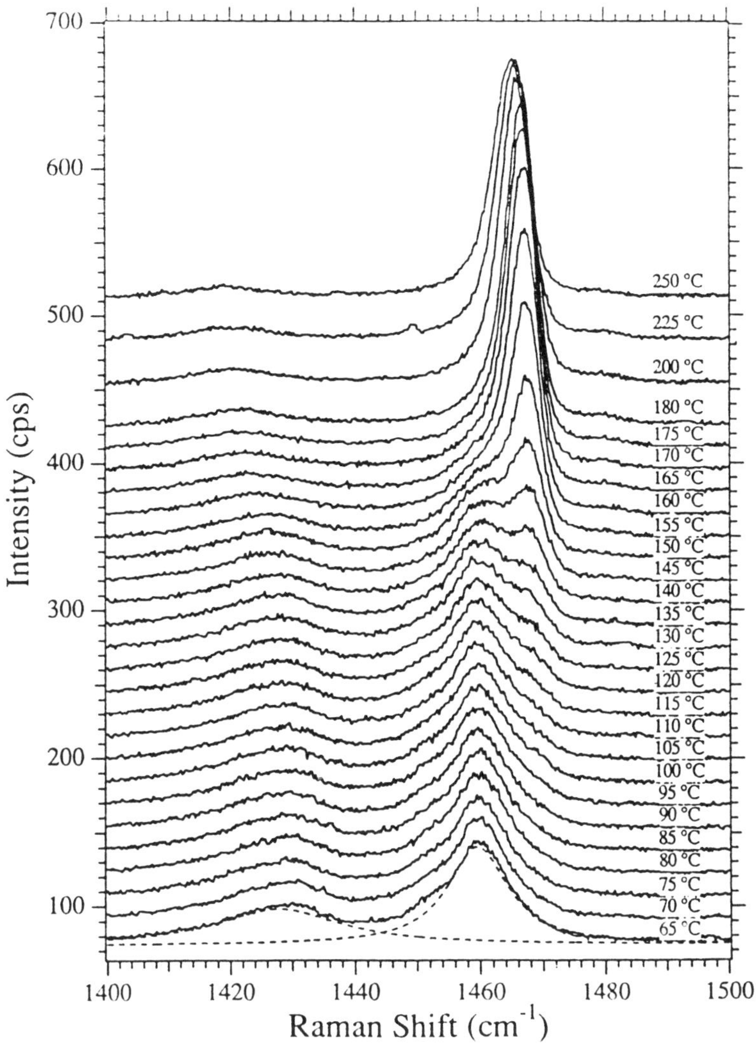

Fig. 4.12. Raman spectra obtained at various temperatures in the vicinity of the $A_g(2)$ "pentagonal pinch" mode for a photochemically polymerized C_{60} film ($d = 4500$ Å) on a suprasil substrate. The dashed curves in the bottom spectrum represent a Lorentzian lineshape fit to the data. With increasing temperature, the broad line at 1459 cm^{-1} identified with the photopolymer diminishes in intensity and the narrow line at 1469 cm^{-1} identified with pristine C_{60} grows at the expense of the broad line [50]

irradiation time, nearly all C_{60} molecules in the film are covalently bonded to neighboring C_{60} molecules to form dimers via a four membered ring [59]; (ii) after 150 h of exposure to radiation, a C_{60} polymer chain is obtained in which the C_{60} molecules possess at least two four-membered rings; and finally

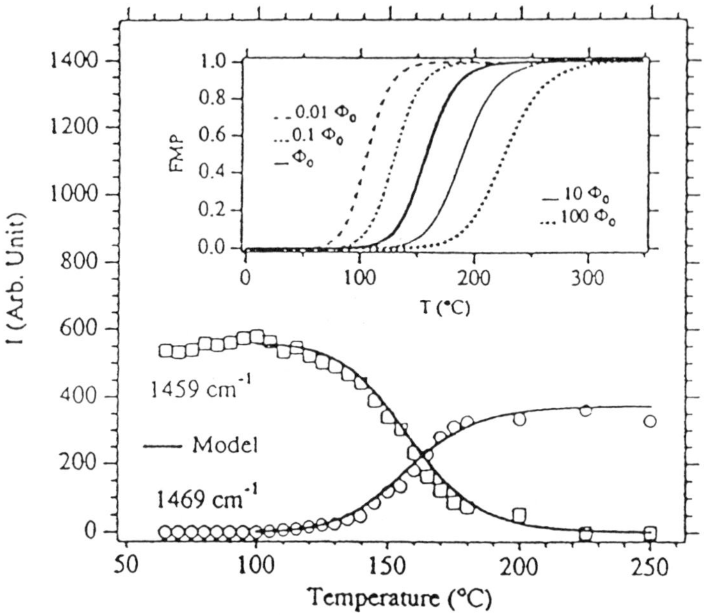

Fig. 4.13. Temperature dependence of the integrated intensity I obtained from analysis of the Raman spectra in Fig. 4.12 for the 1459 cm^{-1} (□) line (C_{60} polymer) and the 1469 cm^{-1} (○) line (C_{60} monomer) in comparison to rate equation model calculation (solid curves). Inset: the calculated fractional monomer population (FMP) in photopolymerized C_{60} as a function of temperature for several values of the incident light flux Φ_0 at 488 nm [58]

after 200 h of exposure time, a C_{60} polymer network is formed under the conditions used in their experiments [59]. An infrared spectrum of a C_{60} film polymerized under similar conditions indicated that the film was completely photopolymerized [59].

We have already discussed that the covalent bonding of two C_{60} monomers breaks the icosahedral symmetry of an isolated C_{60} molecule and therefore activates numerous new Raman- and infrared-allowed bands (Fig. 4.11). However, new interball or intermolecular modes are anticipated which we now discuss. They fall in the gap between the librational and intramolecular mode frequencies (see Fig. 4.5 a–e) and involve the stretching or twisting of the proposed four membered ring joining adjacent C_{60} molecules. In Fig. 4.15, we show for comparison the low frequency region (50–330 cm^{-1}) of the Raman spectrum for a photopolymerized (top curve) and pristine (bottom curve) film of solid C_{60}. As can be seen, a new Raman line is observed at 118 cm^{-1} in the photopolymerized film, in addition to several other new lines in the vicinity of the pristine $H_g(1)$ mode near ~270 cm^{-1}. The line at 118 cm^{-1} was identified with the interball stretching mode [14, 50]. This mode frequency in the C_{60} dimer (see the eigenvector in the inset to Fig. 4.15) was calculated

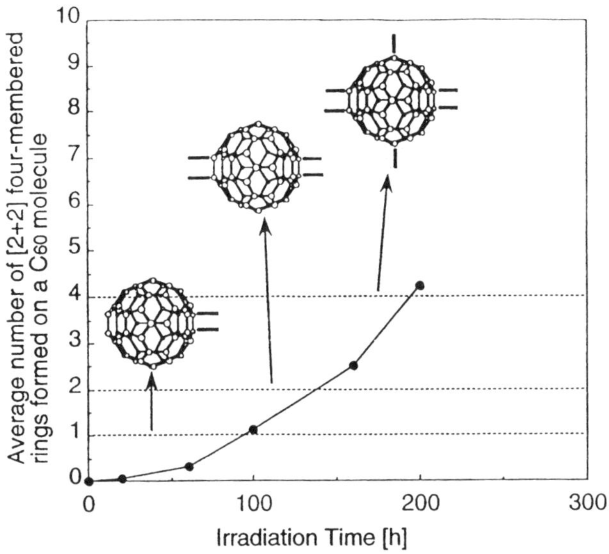

Fig. 4.14. The average number of four-membered rings formed on a C_{60} molecule as a function of irradiation time. In-situ x-ray photoelectron (XPS) spectroscopy was used in this study [59]

subsequently for the "2 + 2" dimer (Fig. 4.7b) by several groups: $104\,\mathrm{cm}^{-1}$ [60], $81\,\mathrm{cm}^{-1}$ [61], $69\,\mathrm{cm}^{-1}$ [45], and $89\,\mathrm{cm}^{-1}$ [46]. The Raman lines at higher frequency in Fig. 4.15 are identified with the photoinduced structural change which splits the five-fold degenerate $H_g(1)$ intramolecular mode of an isolated C_{60}. An infrared-active, torsional interball mode for the "2 + 2" dimer has also been predicted by Menon and Subbaswamy at [60] $227\,\mathrm{cm}^{-1}$ but has not yet been observed experimentally. It is also interesting to note that using the same calculational scheme Adams et al. [45] find the interball stretching mode in the C_{60} dimer to upshift from 69 to $119\,\mathrm{cm}^{-1}$ if the dimers are coupled together by similar four-membered rings to form the neutral linear C_{60} polymer chain [45]. This brings these theoretical results into closer agreement with experiment. However, there is no clear evidence for linear chains in photopolymerized C_{60}. In fact, XRD studies on a series of C_{60} films exposed to successively higher doses of UV light, showed a continual contraction of the fcc lattice constant from the initial value of 14.17 Å (pristine) to 13.80 Å [62]. The photopolymer XRD data to date suggest random intermolecular bonding with an average fcc structure.

In Fig. 4.16, the results of a density-functional based nonorthogonal tight-binding (DF-TB) method to investigate the vibrational properties near the intramolecular $A_g(2)$ mode frequency for five different $(C_{60})_N$ polymers is presented [46]. Geometries for three of the five structures are shown as insets in the figure. A downshift of $\sim 10\,\mathrm{cm}^{-1}$ for thc $A_g(2)$-derived mode is observed

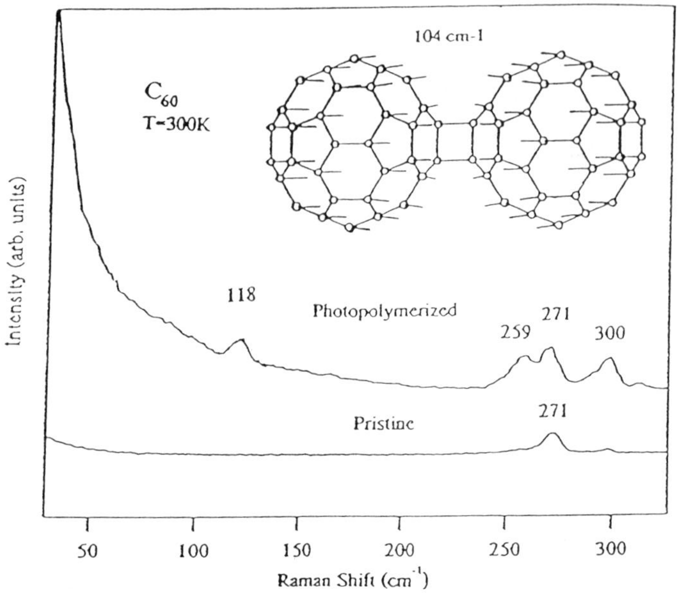

Fig. 4.15. Room temperature Raman spectra for pristine and photopolymerized C_{60} films in the region of the interball modes. The pristine C_{60} film shows only the $H_g(1)$ mode at 271 cm^1 in the figure, whereas the photopolymerized C_{60} film exhibits several other peaks in the vicinity of the $H_g(1)$ mode due to the reduced symmetry of the C_{60} polymer. The mode at 118 cm^1 is identified with the interball mode shown for the C_{60} dimer in the inset figure

in Fig. 4.16 for all polymerized structures, except for the square arrangement where the downshift is 16 cm^{-1}. For longer C_{60} polymer chains, the DF-TB method [46] finds another peak which is downshifted by ~20 cm^{-1}. Furthermore, these calculations obtained the following results for the covalent bonding induced changes in the vicinity of the $A_g(2)$ mode frequency: (i) C_{60} molecules with a single intermolecular bond per molecule (for example, the C_{60} dimer) exhibit a Raman peak which is downshift by ~10 cm^{-1}, (ii) C_{60} polymer structure with two polymer bonds (i.e. intermolecular bonds) per molecule formed at 90° angle (the central C_{60} molecule in the L-shaped C_{180} structure) exhibits a downshift of 16 cm^{-1}, and (iii) a C_{60} molecule with two polymeric bonds per molecule, but on opposite sides of the molecule (central C_{60} molecule in the linear C_{60} trimer) exhibits a peak downshifted by 20 cm^{-1} compared to the original $A_g(2)$ peak. As mentioned earlier, a downshift of ~10 cm^{-1} was experimentally determined for the $A_g(2)$ mode frequency when solid C_{60} photopolymerized. It has therefore been suggested [45, 46] that the photopolymerized C_{60} solid predominantly consists of C_{60} dimers, and to a small extent C_{60} trimers. This structural model has been used to interpret ultraviolet photoemission spectroscopic (UPS) data [63], but in our opinion is inconsistent with the LDMS (Fig. 4.9) and AFM (Fig. 4.10) discussed above.

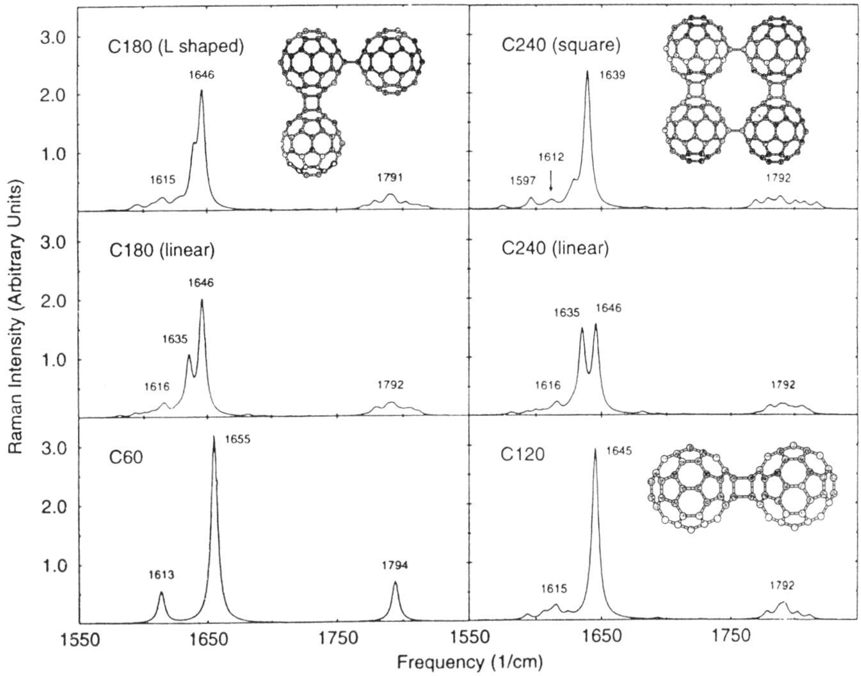

Fig. 4.16. Calculated Raman intensities near the intramolecular $A_g(2)$ mode frequency for five different $(C_{60})_N$ polymers. After [46]. Geometries for three of the five structures are shown as insets in the figure

Finally, it should be mentioned here that phototransformation of solid fullerenes is not specific to C_{60} alone. We have shown that it occurs in solid C_{70} as well [64]. However, the reaction rate for the phototransformation of C_{70} is much lower and therefore appears to have little practical value. The low phototransformation rate in C_{70} has been attributed to (i) the relatively reduced number (ten) of chemically active C=C bonds in C_{70} which are located only in the polar caps of a C_{70} molecule, and (ii) to the relative orientation of the C_{70} molecules in the solid state which seldom lead to the topological conditions required for the photochemical "2 + 2" cycloaddition reaction. Further details regarding photoinduced polymerization in solid C_{70} can be found in [64].

4.4.2 Charge Transfer Excitons: a Possible Trigger Mechanism for Photopolymerization

We conclude this section with the discussion of the proposed "trigger mechanisms" that initiate polymerization in fullerenes. The various routes to polymerization in C_{60} solids have sparked interest in a general mechanism. It seems that a contraction of the intermolecular separation is a reasonable first step for all of these processes. Of course, the details of how the intermolecular bonds then form is the key theoretical issue. A sufficient contraction can be induced by pressure, but high temperature also appears necessary to obtain certain C_{60} polymer phases [20, 21, 23, 24, 65]. In the case of M_1C_{60}, the Coulomb interaction between the alkali metal ions and the C_{60} anions has been proposed to drive the lattice distortion leading to the chain polymerization [66]. Recently, Tsubo and Nasu [28, 29] have calculated the optical absorption of solid C_{60} and have found "charge transfer" (CT) excitons between ~1.6 and 2.9 eV. These excitations are molecular in character but the weak intermolecular coupling allows the photo-excited electron to be delocalized on nearest neighbor molecules. The Coulomb attraction between the electron and hole is then postulated to lead to a reduction in the intermolecular separation and "trigger" the formation of intermolecular bonds. In this section, we present clear experimental evidence for these CT excitons by quantitative analysis and comparison of the optical absorption of C_{60} in the gas and solid phases. The molecular polarizibility of C_{60} in a truly isolated gas phase was determined experimentally and used to constructed the optical response of the "ideal" C_{60} molecular solid. The differences in the constructed optical response and the measured optical response of polycrystalline thin films is discussed in terms of charge transfer excitons. The experimental results are compared with theory.

The optical absorption of solid C_{60} has been calculated by several authors from one-electron calculations, such as the Huckel model, assuming that the π electrons move in the stationary, effective potential and neglecting electron–electron interactions. More sophisticated calculations are self-consistent, molecular orbital calculations which at least take into account some of the electron–electron interactions in their model Hamiltonians and use many-electron configurations as the basis wavefunctions. Since solid C_{60} is a nearly "ideal" molecular crystal, it is of interest to include the Coulomb interaction between electron and hole in the photoexcited (or excitonic) state and to allow for weak intermolecular wavefunction overlap. For the so called "charge-transfer" (CT) excitons, the hole is localized on the central molecule and the photoexcited electron is delocalized over neighboring C_{60} molecules. The degree of delocalization is variable, depending on the particular excitation, some of the excitons are found to be essentially localized on one molecule. Both "localized" (L) or Frenkel excitons and delocalized charge-transfer (CT) excitons are therefore possible and have been calculated [28, 29, 67–70]. Near the absorption edge, the calculations find that the largest oscillator strength in the spectrum lies primarily with localized (Frenkel) excitons [28, 29, 67–70]. However, the CT excitons become important a few hundred meV above the absorption threshold [28, 29, 69, 70].

4.4.3 Theoretical Predictions

Here, we briefly discuss the results of theoretical calculations for the Frenkel and charge transfer excitons. Details of these calculations can be found in Chap. 8 of this book. Shirley et al. [71] employed a model Hamiltonian H_{ex} for the exciton states in the solid derived from $h_u^9 t_{1u}^1$ molecular configurations given by

$$H_{ex} = H_{el} + H_{hole} + H_{eh}(1) + H_{eh}(2) \tag{4.1}$$

The first two terms in (4.1) represent the Hamiltonian for the noninteracting electron and hole and the second two terms describe the electron–hole interaction when the electron and hole reside on the same (Frenkel character) or different (CT character) molecules, respectively. When h_u and t_{1u} molecular orbitals are used, the resulting excitons can be singlet and triplet excitons with symmetries given by the direct product $F_{1u} \otimes H_u = H_g + G_g + F_{1g} + F_{2g}$. The lowest Frenkel type singlet exciton with $^1T_{2g}$ character and the triplet exciton with $^3T_{2g}$ character were predicted at 1.3 eV and 1.58 eV, respectively [71]. Higher energy singlet and triplet excitons have T_{1g}, G_g and H_g character. At energies 2.33 eV and above, excitons begin to take on charge transfer character with a high probability that electron and hole do not reside on the same molecule. Tsubo and Nasu [28, 29] have implemented an extended Hubbard model for the sixty π-electrons of the C_{60} molecule where the Hamiltonian contains terms for the intermolecular transfer and Columbic energy between orbitals on adjacent molecules. Intermolecular interactions were included only for the π-orbitals of neighboring molecules which come into closest contact in the fcc solid [28, 29]. Their calculated optical absorption spectra using excited exciton states of either type are shown in Fig. 4.17. Without the intermolecular Columbic repulsion terms, the calculated optical

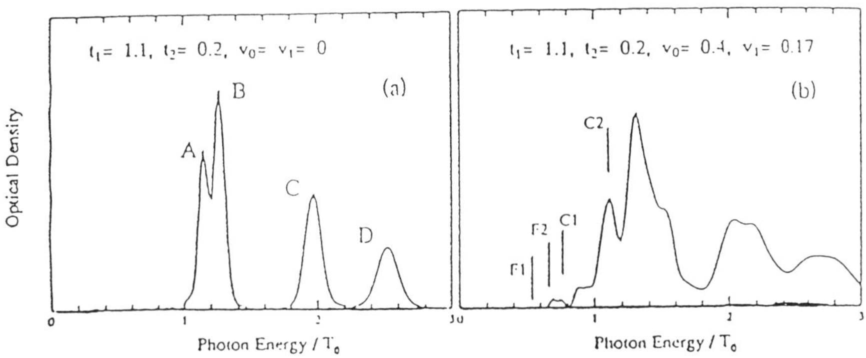

Fig. 4.17. Calculated absorption spectra of molecular (**a**) and solid C_{60} (**b**). Features labeled A, B, C, and D in (**a**) are completely Frenkel in character

absorption spectrum shows four prominent peaks which may be assigned, qualitatively, to Frenkel excitons of isolated C_{60} molecules. When the intermolecular Columbic repulsion is turned on, several new peaks appear in the absorption spectrum which are activated by the crystal field. The calculated probabilities that the electron and hole reside on the same molecule for two of the five new peaks were particularly low (F1 with 5.5% and F2 with 3.3% in Fig. 4.17) [28, 29]. Harigaya and Abe [69, 70] employed a similar Hamiltonian and used the Hartee–Fock approximation in the single excitation configuration interaction method. The authors calculated the density of excitations and the dipole-allowed optical absorption spectra for C_{60} and the results are discussed in Chap. 8.

4.4.4 Experimental Evidence for CT Excitons in C_{60}

To experimentally identify CT excitons Holden et al. [72] have carried out quantitative analysis of the gas phase and solid C_{60} optical absorption which we discuss below. Other experimental evidence reported in the literature include electro-absorption [73], photoconductivity [74], electromodulated luminescence quenching [74] and femtosecond pump and probe spectroscopy [75].

To measure the gas phase molecular polarizability, polycrystalline C_{60} films deposited on quartz substrates were transferred under inert atmosphere to quartz optical cells (path length 355 mm) and subsequently sealed at $< 10^{-8}$ torr vacuum. The C_{60} film (of measured thickness) was vaporized by placing the cell in a 1 m long furnace where gas phase absorption spectra in the range of 1.25 to 6 eV were obtained. For sufficiently small gas concentrations (N_g), the absorption coefficient (a_g) for gas phase C_{60}, can be related to the imaginary part of the molecular polarizibility (α_2) as $\alpha_2 \sim a_g \lambda_0 / 8\pi^2 N_g$, where λ_0 is the wavelength of light [72]. The real part of the molecular polarizibility, α_1, was obtained from a parametric Lorentzian oscillator fit to the absorption data,

$$\alpha_{\mathrm{M}}^{\mathrm{Lor}}(\omega) = \alpha_0 + (\mathrm{e}^2/m)\Sigma_j(f_j/(\Omega_j^2 - \omega^2 - \mathrm{i}\omega\Gamma_j)), \tag{4.2}$$

where the f_j, ω_j and Γ_j are the oscillator strengths, positions and widths, respectively, and α_0 accounts for oscillator strength from transitions above the data range (< 6 eV). In the fitting procedure to absorption data, α_0 was adjusted to match the dielectric constant obtained previously by ellipsometry on thin, solid films [76]. The dielectric function, $\epsilon(\omega) = \epsilon_1 + \mathrm{i}\epsilon_2$, of the "ideal molecular solid" (where there is no intermolecular interaction except the dipole screening from neighboring molecules) was constructed using $\alpha_{\mathrm{M}} \sim \alpha_1^{\mathrm{Lor}} + \mathrm{i}\alpha_2$ in the Claussius-Mossotti relationship [77],

$$(\epsilon(\omega) - 1)/(\epsilon(\omega) + 2) = (4\pi/3)N_{\mathrm{s}}\alpha_{\mathrm{M}}(\omega) \tag{4.3}$$

where N_{s} is the molecule density for solid C_{60}. The experimental dielectric function for fcc C_{60} has been obtained previously by ellipsometry [76] and

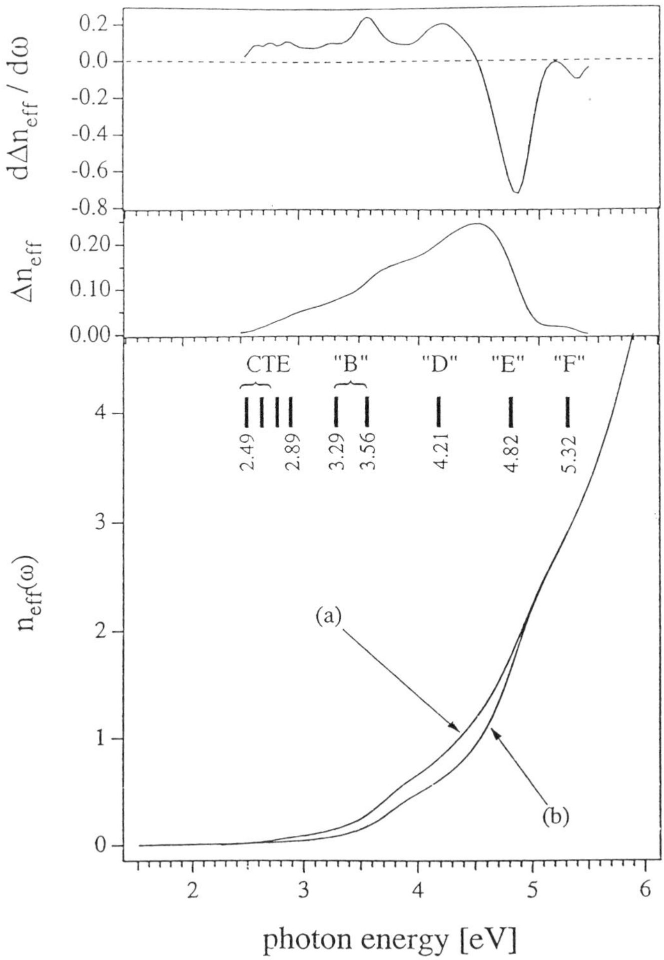

Fig. 4.18. Integrated oscillator strength comparison of ideal to real solid. *Lower panel*: Integrated oscillator strength for (**a**) the real solid film and (**b**) the ideal molecular solid. *Center panel*: Difference of (**a**) and (**b**), $\Delta n_{\text{eff}} = n_{\text{eff}}^{\text{real solid}} - n_{\text{eff}}^{\text{ideal solid}}$ is plotted as a function of photon energy. *Upper panel*: Derivative of Δn_{eff} is plotted as a function of photon energy. See text for discussion of other labels in the figure

R, T spectroscopy [78]. Figure 4.18 shows $\epsilon_2^{\text{real solid}}$, $\epsilon_2^{\text{ideal solid}}$ and ϵ_2^{gas}. Comparing $\epsilon_2^{\text{ideal solid}}$ to ϵ_2^{gas}, it can be seen that the effect of dipole screening is a strong redshift and redistribution of apparent oscillator strengths. For simple cases, the degree of "renormalization" can be shown to be proportional to the oscillator strength of the transition and so the effect is most prominent above 3 eV in C_{60}, where there are intense dipole-allowed transitions [72].

Comparing $\epsilon_2^{\text{ideal solid}}$ to $\epsilon_2^{\text{real solid}}$ Holden and Eklund [72] concluded that: (i) major features of the real solid above 3.5 eV can be very nearly generated by assuming the effective polarizibility of a C_{60} molecule in the solid is the same as the true molecular polarizibility of C_{60}. Above 3 eV, C_{60} may be characterized as an ideal "molecular solid"; (ii) the broad resonance observed between 2.5 eV and 3.1 eV in the real solid is not duplicated in the ideal molecular solid. This new solid-state resonance was assigned to CT excitons.

A comparison of $\epsilon^{\text{ideal solid}}$ and $\epsilon^{\text{real solid}}$ is not expected to be perfect since in the real solid intermolecular wavefunction overlap occurs and excitations can, in some cases, be significantly delocalized. As a result, a redistribution of oscillator strength will occur for C_{60} molecules residing in a interacting solid fcc lattice. The nature of the redistribution is illustrated in Fig. 4.18 by plotting the integrated oscillator strength which is essentially the "effective number of electrons per molecule" (n_{eff}) contributing to the optical properties over a finite frequency range [79]. Both $n_{\text{eff}}^{\text{ideal solid}}(\omega)$ and $n_{\text{eff}}^{\text{real solid}}(\omega)$ (plotted in the lower panel) begin to diverge slightly at 2.5 eV and rejoin completely ~5.4 eV [72]. This is more apparent in the middle panel where the difference $\Delta n_{\text{eff}} = n_{\text{eff}}^{\text{real solid}} - n_{\text{eff}}^{\text{ideal solid}}$ is plotted. A redistribution of the oscillator strength has occurred completely within this energy range and it is reasonable to expect that no further redistribution will occur at higher energies.

The derivative of Δn_{eff} with respect to energy is plotted in the top panel of the figure. A peak or dip in this curve corresponds to a region of gain or loss of oscillator strength as the interaction in the solid are turned on, i.e. when the C_{60} molecules can "feel" their neighbors in the fcc solid. Features in the d $\Delta n_{\text{eff}}/\text{d}\omega$ spectrum have been indicated by corresponding vertical bars and frequencies in the lower panel of the figure. The bands labeled "CTE" in the figure correspond to the proposed charge transfer excitons predicted by Tsubo and Nasu [28, 29].

If the charge transfer excitons are the "trigger" for photopolymerization, the one would expect irradiation of pristine C_{60} in the region 2.5–3.1 eV to be particularly effective in producing the photopolymerized material. No experimental results on this prediction have been reported.

4.5 Photo-induced Hardening of C_{60} Crystals

A variety of hard carbons, such as, diamond-like carbons synthesized by chemical vapor deposition or similar methods exhibit hardness similar to that of diamond. Kozlov et al. [80] have reported that fullerenes subjected to the right pressure and temperature conditions can yield hard carbons with useful properties. Tachibana et al. [81] have investigated the effect of light illumination (from a halogen lamp: 2 mW/mm^2) on the hardness of C_{60} crystals in air and inert atmospheres. Their study found that (i) the room temperature Vickers microhardness of the (111) faces of C_{60} crystals increases from an initial value 18 kgf/mm^2 with increasing exposure time and saturates to ~35 kgf/mm^2; (2) the microhardness does not return to the intrinsic hardness value after the illumination is turned off suggesting that the photoinduced microhardening of C_{60} is an irreversible process at room temperature and is different from the "photoplastic effect" in which the crystals slowly recover their intrinsic microhardness when the illumination is turned off; (3) the photo induced hardening is effective in the temperature range of 298–380 K. Moreover, the intrinsic hardness was almost completely recovered when the illuminated crystals with higher hardness were annealed at temperatures higher than 400 K for ~10 min in air or vacuum. The origin of the increased photoinduced hardness was attributed to the presence of a photopolymerized C_{60} skin on the (111) faces of the crystal [81]. Using the intensity ratio of Raman lines at ~1469 and ~1458 cm^{-1}, they were able to correlate microhardness with photopolymerization. Photoinduced polymerization in C_{60} is normally inhibited in the presence of air (oxygen) [82]. Thus, it appears reasonable to assume that the rate of increase of the microhardness with irradiation time would be higher if the sample was irradiated in vacuum than in air.

4.6 Pressure-induced C_{60} Polymers

Initial high pressure (P) experiments on polycrystalline C_{60} powders at room temperature were carried out to investigate the resilience of the C_{60} molecules. Rapid application of pressure of GPa at room temperature on C_{60} powders was found to stop the rotational motion of the C_{60} molecules and establish an orientationally disorder state. Slow application of quasi-hydrostatic pressures ($P \sim$ 6–7 GPa) at high temperatures (~1000°C) and nonhydrostatic pressures (15 GPa $< P <$ 25 GPa) at room temperature were found to destroy the C_{60} cages, yielding amorphous sp^2 bonded [83, 84] or sp^3 bonded carbons [84, 85], polycrystalline cubic diamond [86, 87], microcrystalline graphite [84, 85] and other unidentified solid carbon structures [85, 88]. However, using XRD, NMR, Raman scattering and infrared spectroscopy, Nunez-Regueiro et al. [21] and Iwasa et al. [20] showed that the simultaneous application of high temperature (~400–900°C) and high pressure (~2 GPa $< P <$ 8 GPa), inhibited the formation of an orientationally disordered phase and, polymerized C_{60} crystal lattices were proposed to form.

Three new structural phases of polymerized C_{60} were discovered using a high pressure–high temperature "belt apparatus" [21]: (i) a tetragonal (T) phase obtained at low pressure ~3 GPa and ~600°C, (ii) a rhombohedral (R) phase obtained at pressure ~5 GPa and ~700°C, and (iii) an orthorhombic phase obtained at ~8 GPa and ~300°C. Figure 4.19 suggests the origin of the R, T and O polymerized C_{60} phases is due to particular contraction of the intermolecular separation in the parent C_{60} fcc structure. The solid circles in the figure schematically represent the C_{60} lattice positions for the parent

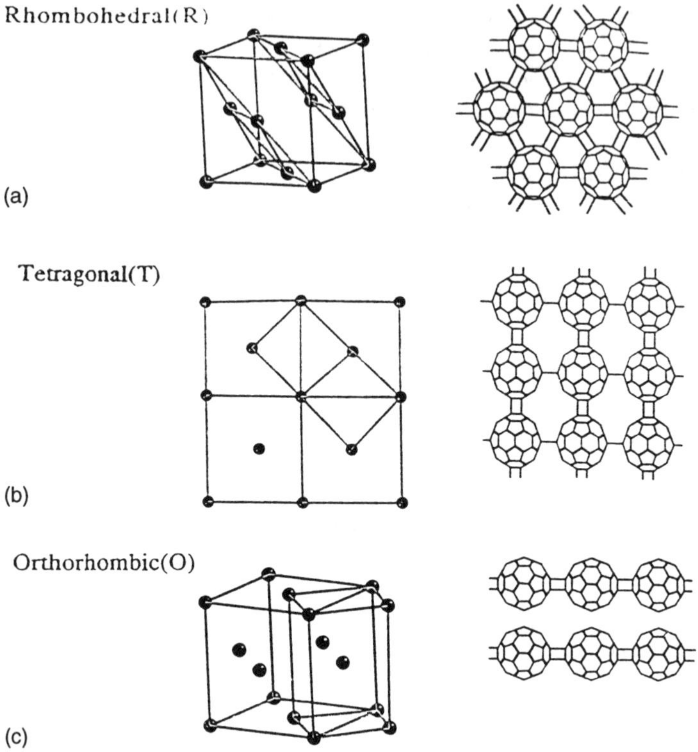

Fig. 4.19 a–c. Origin of the R, T and O polymerized C_{60} phases due to particular contraction of the intermolecular separation in the parent C_{60} fcc structure. The solid circles in the figure schematically represent the C_{60} lattice positions for the parent fcc structure. In the orthorhombic cell, and the polymer chains are parallel to the a-axis of the orthorhombic cell

fcc structure ($a_o = 14.17$ Å). The R phase is obtained via an intermolecular contraction in the (111) planes (Fig. 4.19a), while the T phase is obtained via a contraction of the intermolecular spacing in the (001) planes (Fig. 4.19b). The O phase is obtained by forming long, one-dimensional chains of neutral C_{60} molecules (Fig. 4.19c), similar to the structure originally proposed for the polyanion chains in KC_{60} or RbC_{60} [17, 18] (see Fig. 4.3). Also, similar to M_1C_{60}, the chains form via a contraction of the intermolecular separation along the $\langle 110 \rangle$ directions. This one-dimensional polymerization leads to an orthorhombic cell ($a = b \neq c$) with the chains aligned parallel to the a axis (Fig. 4.19c). The coupling between adjacent planes in the R and T phases is the much weaker van der Waals interaction. Structural calculations using a tight binding potential model [89] for the R phase found a shortened intermolecular distance of 9.17 Å, in good agreement with the experimental value of 9.2 Å. Furthermore, as in the M_1C_{60} polymer and the C_{60} dimer, the calculated molecular shape of the C_{60} molecules in the R and T phases exhibits significant elongation along the direction of the intermolecular bonds [21]. From a detailed structural analysis for the R phase, Oszlanyi and Forro [23] estimated a 5% elongation of the C_{60} molecules in the R network.

The R and T phases represent two dimensional (2D) fullerene polymer structures. In the original work, Nunez-Regueiro et al. [21] proposed that the covalent inter-fullerene coupling involved the four-membered ring, i.e. the "2 + 2" cycloadditive bond. Later, this proposal was confirmed for the O and R phases in their ^{13}C NMR experiments depicted in Fig. 4.20 [55]. Two (six) resonances were identified for the othorhombic (rhombhohedral) fullerene polymers [55]. The origin of these lines has been attributed to the inequivalent carbon sites induced by the four membered rings joining C_{60} molecules. Also, ^{13}C NMR confirmation of the four-membered ring and the D_{2h} symmetry for the C_{60} dimer was later reported by Wang et al. [13]. The C_{60} dimer was synthesized in a vibrating mill by a solid state reaction of C_{60} with potassium cyanide [13].

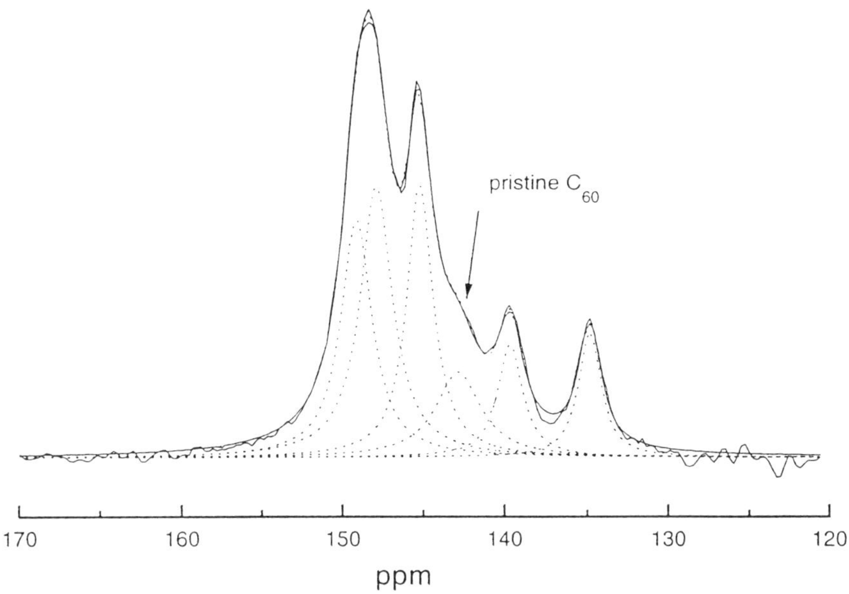

Fig. 4.20. ^{13}C MAS NMR spectrum of the rhombhohedral fullerene polymers [55]

Similar to the C_{60} and C_{70} photopolymers, the R, T and O polymer phases were found to be insoluble in toluene. Also, similar to the photopolymer, annealing in an inert atmosphere at $\sim$200°C and ambient pressure, is sufficient to drive a back transformation to pristine fcc C_{60} [20, 21]. These new phases are collectively referred to below as high-pressure, high-temperature (HPHT) phases. A structural model for these HPHT phases was obtained by a Rietveld analysis of the powder XRD data [21]. In addition to the R, T, and O phases described above, Iwasa et al. [20] reported another polymerized structure exhibiting a fcc structure obtained at 300–400°C and at a pressure of ~5 GPa. This fcc structure (F) was found to exhibit a contracted lattice parameter $a_0 = 13.6$ Å [20]. The structure of the photopolymer is most similar to this F phase [62]. However, the photopolymer undoubtedly exhibits considerably more disorder stemming from a light-induced, random formation of intermolecular bonds. At higher temperatures between 500–800°C and at 5 GPa, the R phase with a hexagonal lattice parameter of $a = 9.22$ Å and $c = 24.6$ Å was also obtained by Iwasa et al. [20]. Table 4.2 summarizes all the polymerized fullerite structures obtained by Nunez-Regueiro [21] and Iwasa et al. [20].

Table 4.2. Structural parameters for pristine C_{60} and pressure polymerized C_{60} solids

Phase (Symmetry label)	Cell Parameters (Å)	Distance between centers of molecules (Å)	Volume per C_{60} molecules (Å)^3
C_{60}(Fm3m)	$a_0 = 14.17$	10.02	711
F phase (Fm3m)	$a_0 = 13.6$	9.62	629
	14.253	14.436	
R phase (R3m)	$a_0 = 9.22$; $c_0 = 24.6$	9.22	603
	$a_0 = 9.19$; $c_0 = 24.5$	9.20	597
T phase (Immmm)	$a_0 = 9.09$; $c_0 = 14.95$	9.09	618
O phase (Immmm)	$a_0 = 9.26$; $b_0 = 9.99$	9.26	650
	$c_0 = 14.22$		

In Fig. 4.21, we compare the laser desorption mass spectroscopy (LDMS) spectrum for a HPHT synthesized polymeric C_{60} powder to that for pristine C_{60} and phototransformed C_{60} films. Analysis of the XRD powder pattern for this sample indicated that the sample was highly textured and contained approximately 50% O, 45% T and 5% R phases [90]. This sample was prepared by simultaneously exposing a pressed C_{60} pellet to 2 GPa hydrostatic pressure and 350°C for 30 min [24]. All three mass spectra in Fig. 4.21 are plotted with the same vertical scale and were measured under exactly identical laser desorption conditions: ~10^{-7} Torr vacuum, pulsed Nd:YAG laser at 355 nm (third harmonic) focused to a spot size of 1 mm^2, ~5 ns pulse width, pulse energy ~2 mJ/mm^2. The mass spectra of both the photopolymer and the HPHT polymer exhibited ~12 additional peaks at integer multiples of the C_{60} mass whereas pristine C_{60} was found to exhibit very weak dimer and trimer peaks due to photopolymerization induced by the desorption laser in the mass spectrometer. It should be noted that the laser desorption conditions used to obtain the spectra in Fig. 4.21 are slightly different from those used for the spectra described in Fig. 4.9. However, similar to the LDMS spectra discussed in Fig. 4.9, a succession of ~12 clear peaks in Fig. 4.21

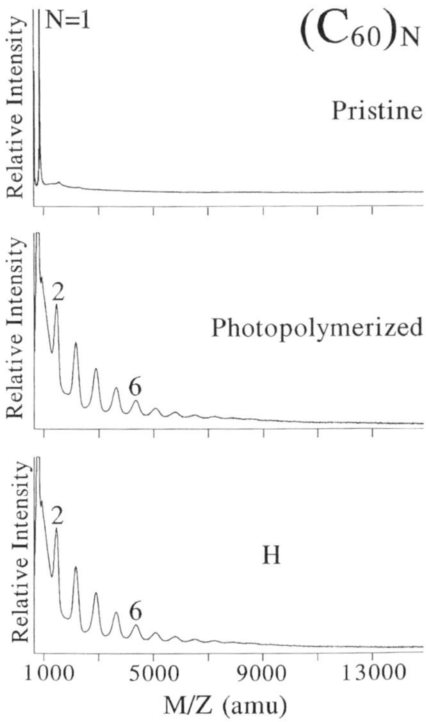

Fig. 4.21. LDMS spectra of pristine C_{60} film, phototransformed C_{60} film and HPHT synthesized polymeric C_{60} powder (see text)

is observed and interpreted as direct evidence for clusters of cross-linked fullerene molecules $(C_{60})_N$.

Due to the reduced symmetry of the C_{60} molecules in the HPHT phases, these phases were expected to exhibit a richer vibrational spectra relative to that of pristine C_{60} and vibrational spectroscopic studies were carried out [91]. Polycrystalline powders of the HPHT polymers and pristine C_{60} were dispersed in KBr pellets and their infrared transmission spectra are shown in Fig. 4.22 together with that for the photopolymerized C_{60} film. Unfortunately, the high pressure and temperature conditions necessary to obtain a pure T phase sample were not achieved in the belt apparatus and a mixed O and T phase sample was produced. We also show in Fig. 4.22 the infrared transmission spectrum for this mixed-phase "O + T" sample synthesized at 11 GPa and 600°C. A Rietveld analysis of this sample indicated the sample composition as ~50% O + ~50% T. The infrared spectrum at the bottom of Fig. 4.22 was found consistent with a mixture of the 2D R phase and a second phase which was identified with a 3D R phase whose structures may be realized by compressing a cross-linked C_{60} layer, such as found in the R phase, against a neighboring R layer leading to interlayer bonding and a 3D fullerene polymer. Rietveld analysis of the XRD data for

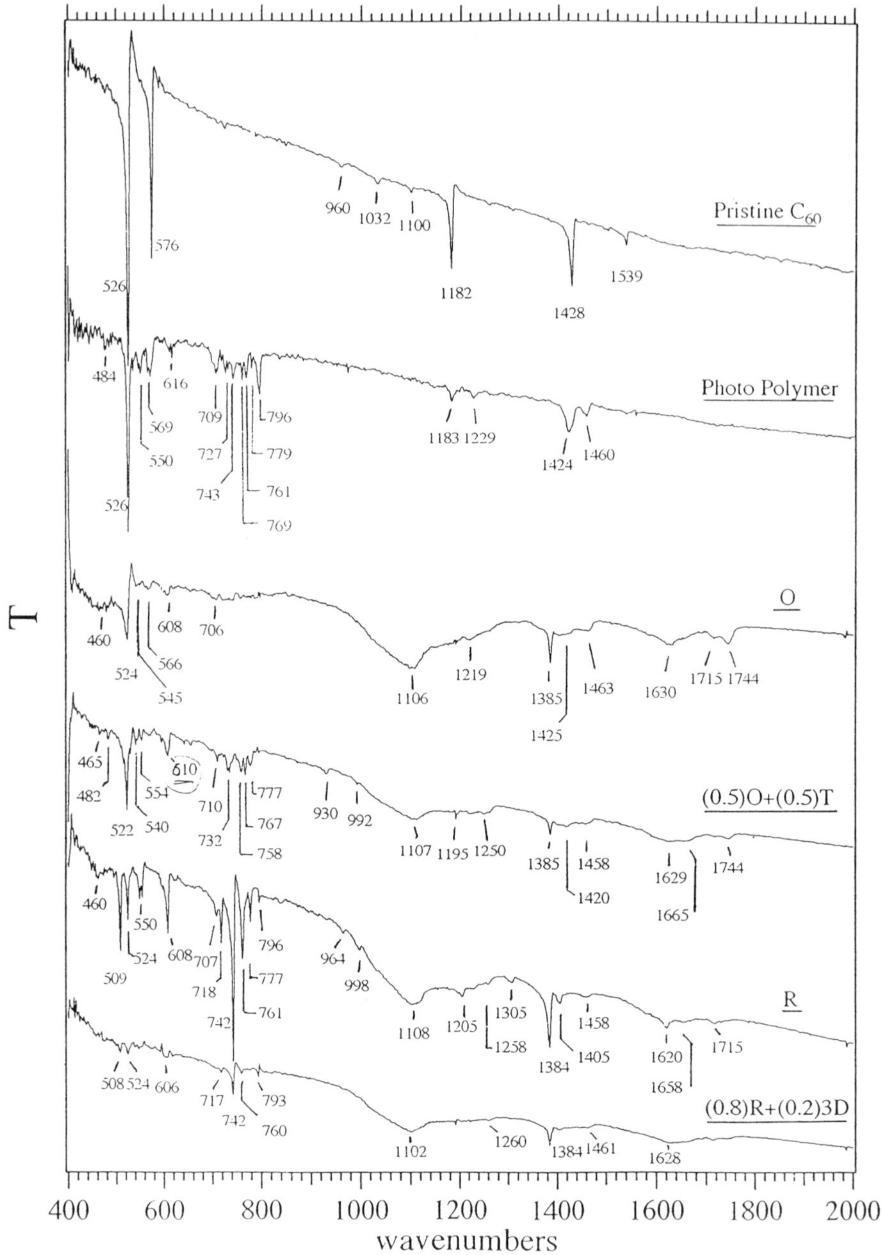

Fig. 4.22. Room temperature FTIR transmission spectra for pristine C_{60} powder, photopolymerized C_{60} film and the HPHT polymers (see text)

the 2D R + 3D R fullerene polymer (whose infrared and Raman spectra appear in Figs. 4.22 and 4.23, respectively), indicated that the sample was predominantly the 2D R phase (80%) and only ~20% of the sample exhibited interplanar contraction required for the 3D R phase. As shown in Figs. 4.22 and 4.23, respectively, the infrared and Raman spectra for the 2D R + 3D R mixed phase sample exhibit broader and weaker lines which were identified with the 3D phase in the sample.

In Fig. 4.22, the intensity of all the four F_{1u}-derived modes of the fullerene polymers is reduced relative to those of pristine C_{60}, as expected. The covalent bonding between neighboring C_{60} molecules not only lifts the degeneracy of the F_{1u} modes but also (i) soften these intramolecular modes relative to those in pristine solid C_{60}. The observed mode softening was attributed to the loss of C=C bonds in the molecular cages, which are broken to reform as intermolecular four-membered rings. (ii) Activates many new infrared-active modes (for example, the new infrared-active modes in the 400–500 cm^{-1} and 700–800 cm^{-1} regions in Fig. 4.22). These new intramolecular bands are expected to be associated with predominantly radial character modes. Their intensities were found sensitive to the particular fullerene polymer phase, exhibiting weak intensity for the 1D O phase and relatively stronger intensity for the 2D R phase. The sharpness of these new lines were interpreted to indicate that the fullerene molecules do remain intact in the polymerized structures, consistent with similar conclusions drawn from the Rietveld analysis of the XRD data for the HPHT polymers. Interestingly, a quite strong mode softening of ~45 cm^{-1} is also observed for the $F_{1u}(4)$ derived modes in the HPHT fullerene polymers. Implications of this mode softening are discussed below.

The room temperature Raman spectra for the fullerene polymers are shown in Fig. 4.23 and approximate center frequencies for selected Raman peaks are indicated in the figure. The strongest line at 1469 cm^{-1} in pristine C_{60} (pentagonal pinch mode) is observed downshifted to 1460 cm^{-1} in the spectrum for photopolymerized C_{60}. As noted above, this band loses intensity and becomes depolarized. It is clear that the Raman spectra of the HPHT C_{60} polymers are quite rich in structure, similar to that observed for the photopolymer. However, clear differences exist which still need to be elucidated theoretically. As was the case for the F_{1u}-derived modes, a splitting of the H_g-derived modes is observed and several new Raman lines are also activated by the loss of icosahedral molecular symmetry. This is particularly evident in the vicinity of the pentagonal pinch mode frequency which is shown on an expanded frequency scale in Fig. 4.24. The Raman spectrum for the T phase in Fig. 4.24 was derived by subtraction of the O phase spectrum from the ~ 50% O + ~ 50% T phase spectrum (Fig. 4.24) [91]. The data in the figure are represented by dots and the solid curves through the data represent the results of Lorentz oscillator fits to the data. Individual Lorentz oscillators are shown below the data superimposed on a flat background and are intentionally displaced downward for clarity. The center frequencies for

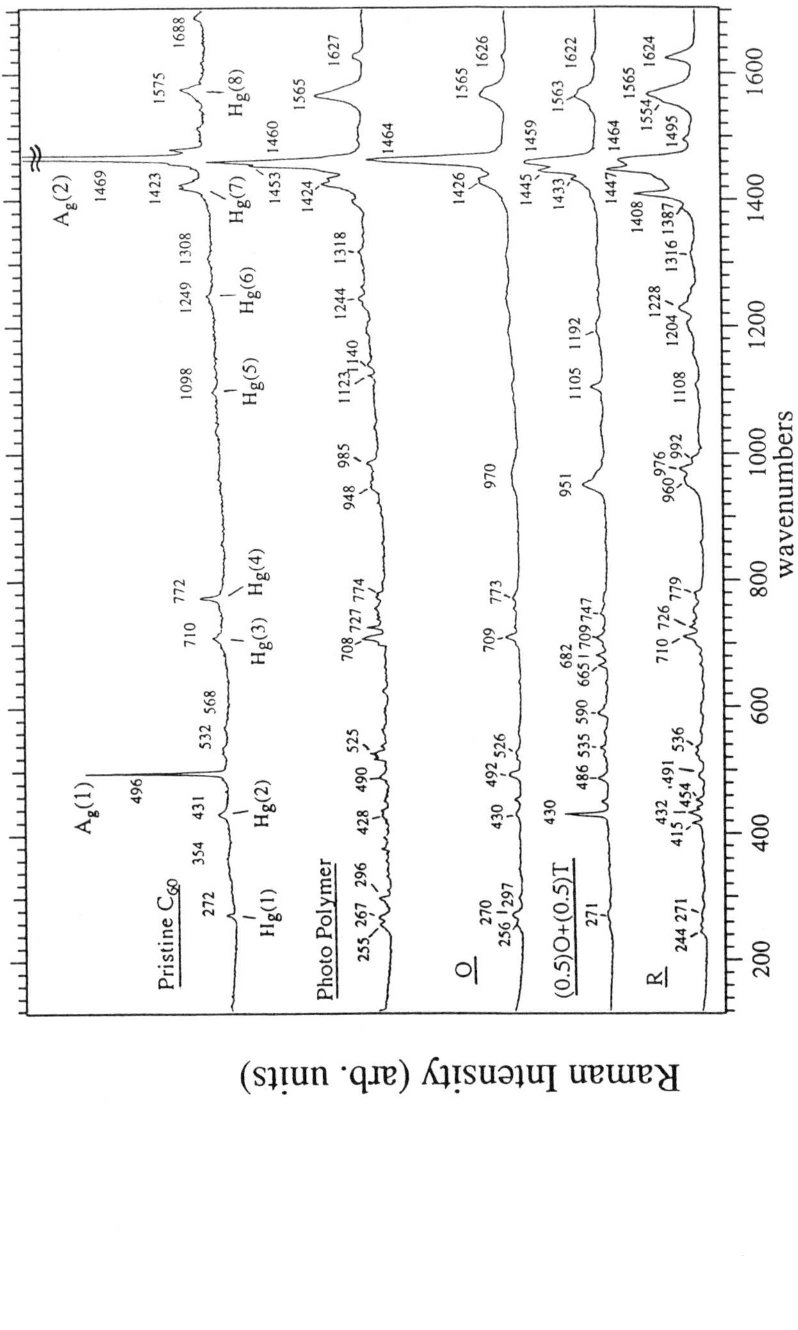

Fig. 4.23. Room temperature Raman spectra for pristine C_{60} film, photopolymerized C_{60} film and the HPHT polymers (see text)

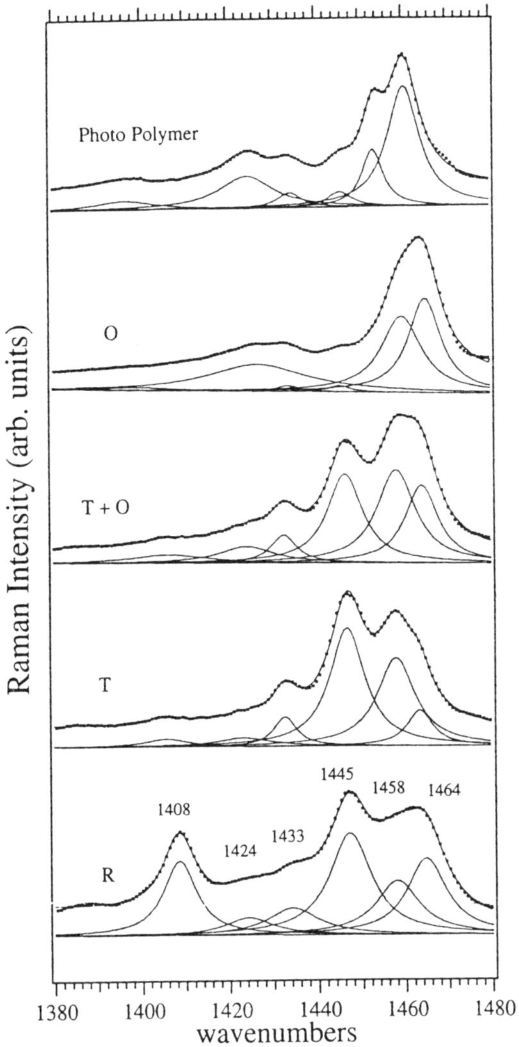

Fig. 4.24. Raman spectra for a photopolymerized C_{60} and the HPHT polymers in the vicinity of the pentagonal pinch mode frequency. Individual Lorentz oscillators for each spectrum are shown below the data superimposed on a flat background

the Lorentz oscillators in the spectra of all the HPHT polymers in Fig. 4.24 were observed very nearly at the same frequency; only the relative intensities change. The frequencies are: $1408\,cm^{-1}$, $1424\,cm^{-1}$, $1433\,cm^{-1}$, $1446\,cm^{-1}$, 1458 and $1464\,cm^{-1}$.

If it is assumed that the strongest peak in the Raman spectra for the HPHT phases is always the $A_g(2)$-derived mode, then the degree of frequency softening as a function of the intermolecular separation (a_I) can be obtained. This is shown in Fig. 4.25, where the upper sloping line is a least squares fit to the frequency of the strongest Raman line versus the a_I for the particular HPHT phase. The values for the frequency position and the intermolecular

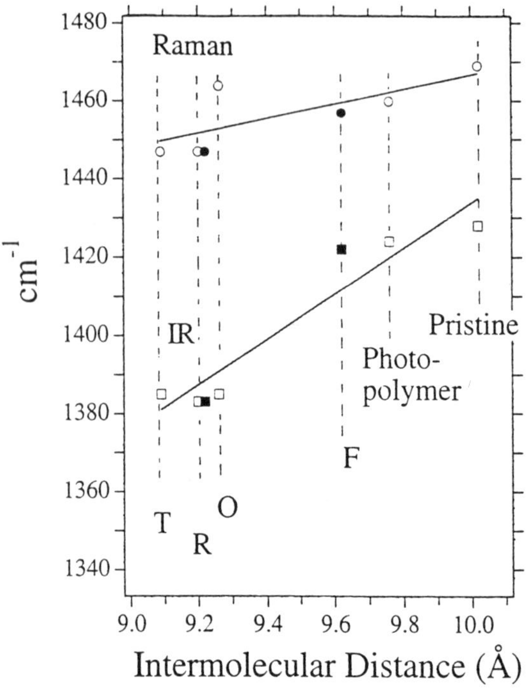

Fig. 4.25. The degree of frequency softening as a function of the intermolecular separation (a_I) for the $A_g(2)$-derived and the $F_{1u}(4)$-derived modes of the photopolymer and HPHT polymers

distance for the F phase were obtained from [20]. The least squares fit yields $\Delta\omega/\Delta a_I \sim 19\,cm^{-1}/Å$ for this $A_g(2)$-derived mode. Similarly, the infrared-active $F_{1u}(4)$-derived mode frequency versus a_I is plotted in Fig. 4.25 and a value of $58\,cm^{-1}/Å$ for $\Delta\omega/\Delta a_I$ is obtained. For comparison, the $A_g(2)$-derived mode and the $F_{1u}(4)$-derived mode frequencies in the alkali metal (M) doped C_{60} (M_xC_{60}) exhibited, respectively, a softening of only 6 and $14\,cm^{-1}$ per electron transferred from the alkali metal to the C_{60} molecule [92].

4.7 M_1C_{60} Polymers

Before we discuss the proposed trigger mechanisms for photopolymerization in Sect. 5, we review briefly the vibrational properties of alkali metal (M) doped C_{60} solids. Detailed discussion of polymerization in these systems can be found in Chap. 7 [15]. M_1C_{60} (M = K, Rb) has been found to undergo a number of structural and electronic phase transitions [16–18, 25, 93, 94]. At elevated temperature (410–460 K), the M_1C_{60} phase has been observed with M = K. Rb, Cs, where the M ion is in an octahedral site, forming a rocksalt (NaCl) crystal structure [93–98]. The reported lattice constants for $K_{1.4}C_{60}$, $Rb_{0.9}C_{60}$, Cs_1C_{60} are 14.07 Å, 14.08 Å, and 14.12 Å, respectively, somewhat smaller than that for undoped C_{60} (14.17 Å) [94]. Upon slow cooling to room temperature, a first-order phase transition to an orthorhombic phase is formed in which linear chains of C_{60} molecular anions were proposed to form along the $< 110 >$ direction. This is an example of a "charged" pearl necklace C_{60} polymer (see Fig. 4.2a), and this structure retains inversion symmetry [16, 99]. Detailed Rietveld analysis of the x-ray powder diffraction data for the orthorhombic K_1C_{60} and Rb_1C_{60} has been reported by Stephens et al. [17]. They proposed the formation of covalent bonds between the C_{60} molecular anions along the chain direction via "2 + 2" cycloaddition. On the other hand, Raman scattering data has been interpreted to indicate that rapid quenching of the high temperature fcc structure to temperatures below room temperature leads to $Rb_2(C_{60})_2$ dimers with no inversion symmetry present in the lattice [99]. It has also been proposed that mixed alkali metal doped fullerene based solids, such as Na_2RbC_{60} and Na_2CsC_{60} exhibit polymerized structures with fullerenes connected via the four membered rings [25].

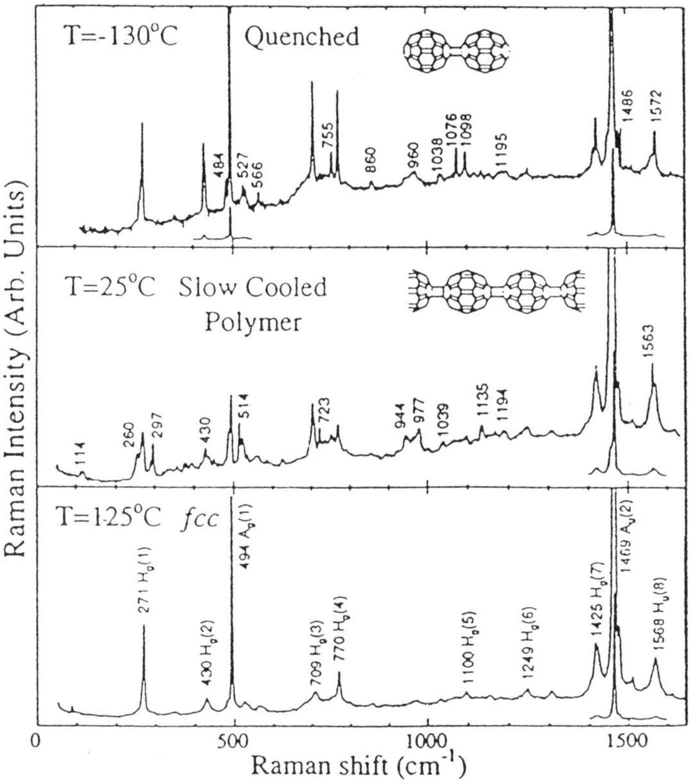

Fig. 4.26. Raman spectra of Rb_1C_{60} in its fcc rocksalt structure (*bottom panel*), its slow cooled polymeric structure (*middle*) and the rapidly quenched dimer structure (*top*). The inset sketches represent the dimer and polymer structures of the C_{60} molecules in the quenched and slow-cooled polymeric states of Rb_1C_{60}, respectively (see text)

Figure 4.26 shows the Raman spectra obtained for Rb_1C_{60} in the high temperature fcc ionic solid (bottom), the slow cooled polymeric structure (middle) and the rapidly quenched dimer structure (top). In the high temperature fcc phase, the 10 Raman-active modes of C_{60} exhibit reasonably narrow linewidth. The intermolecular bonding in the polymerized phase breaks the I_h symmetry of the C_{60}-anions and many more modes are found Raman-active, similar to the effect in the photopolymerized C_{60} phase. The rapidly quenched Rb_1C_{60} dimer phase exhibits additional lines which have been attributed to the breakdown of inversion symmetry in the dimer lattice [99]. As shown in the inset for the quenched Rb_1C_{60} in Fig. 4.26, the four membered ring was also the proposed mechanism for the intermolecular bonding in the $Rb_2(C_{60})_2$ dimers [99]. However, recent Rietveld analysis of the x-ray diffraction data for quenched Rb_1C_{60} have resulted in the proposed structure where two ${C_{60}}^{1-}$ anions are covalent bonded via a single bond to form the $(C_{60})_2^{2-}$ dimer [100] (Fig. 4.27).

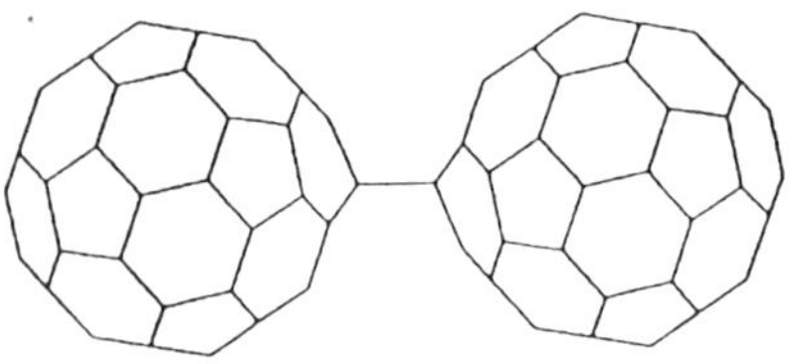

Fig. 4.27. Two ${C_{60}}^{1-}$ anions are covalent bonded via a single bond to form the $(C_{60})_2^{2-}$ dimer [100]

Unlike all other M-doped C_{60} phases, the M_1C_{60} phase is surprisingly stable in air under ambient conditions. The infrared transmission spectra for a Rb_1C_{60} film (deposited on Si substrate) before and after air-exposure are depicted in Fig. 4.28. Infrared lines which correspond to those for pristine C_{60} and Rb_6C_{60} were also observed in the spectrum obtained for the Rb_1C_{60} film before air-exposure (Fig. 4.28), indicating that pristine C_{60} and Rb_6C_{60} co-exist as impurity phases in the sample. After air exposure for $\sim$10 min, the infrared spectrum taken on the same sample showed the infrared lines corresponding to Rb_1C_{60} while those corresponding to Rb_6C_{60} were absent (Fig. 4.28) [101]. In addition, the intensity and the width of the ESR signal measured at room temperature on polycrystalline Rb_1C_{60} sample did not change even after exposing the sample to air for several days [101]. Based on these observations, it was concluded that Rb_6C_{60} phase was destroyed while the Rb_1C_{60} polymeric phase is relatively much more air-stable.

4.7.1 Na_4C_{60} Polymer

In contrast to other alkali cations, the relatively small size of the sodium cation (Na^+) enables multiple occupancy of the octahedral interstitial sites in the fcc C_{60} sub-lattice to obtain Na_xC_{60} compounds with $x = 6$ and 10. Compositions with $x = 2$, 3, 6 and 10 have been reported to exhibit a face

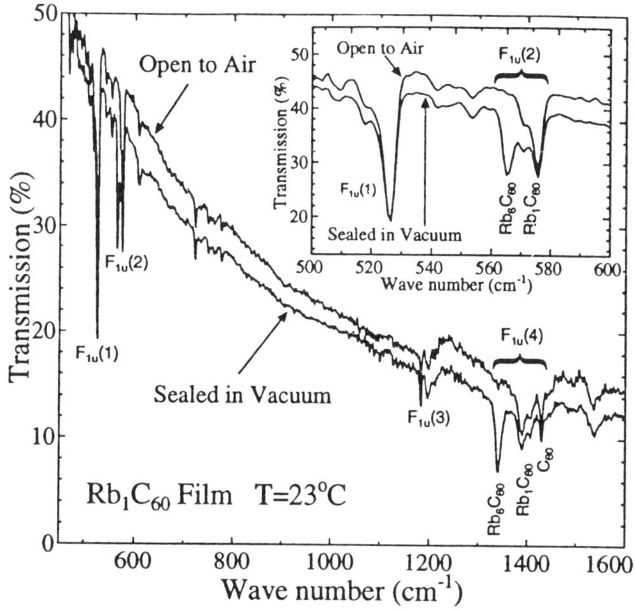

Fig. 4.28. Infrared transmission spectra for a Rb_1C_{60} film (deposited on Si substrate) before and after air-exposure [101]

centered structure [102–104]. Recent systematic search conducted by Oszlanyi et al. [19] for other possible phases of Na_xC_{60} uncovered a new 2D polymerized Na_4C_{60} phase in which it was proposed that each C_{60} molecule forms four "single" bonds within the plane spanned by the vectors $(\mathbf{a}+\mathbf{c}+\mathbf{b})$ and $(\mathbf{a}+\mathbf{c}-\mathbf{b})$ (bold lines in Fig. 4.29a). The samples were prepared by a prolonged

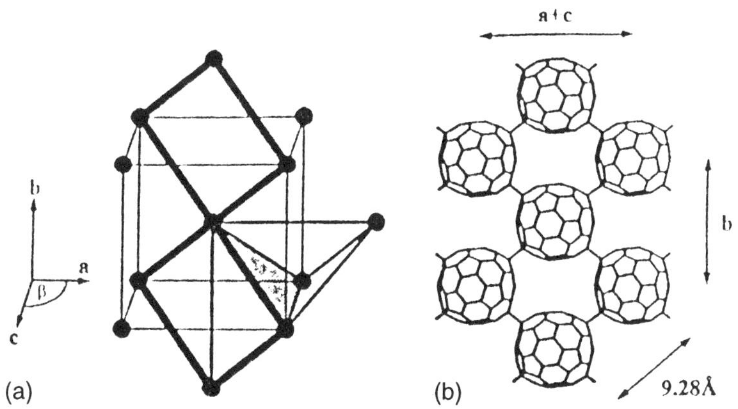

Fig. 4.29 a,b. A 2D polymerized Na_4C_{60} phase in which each C_{60} molecule forms four "single" bonds within the plane spanned by the vectors $(\mathbf{a}+\mathbf{c}+\mathbf{b})$ and $(\mathbf{a}+\mathbf{c}-\mathbf{b})$ (bold lines in the figure) [19]

(two weeks) reaction between Na with C_{60} at 200°C in sealed quartz capillary tubes. All reflections in the x-ray diffraction pattern obtained on $Na_4.2C_{60}$ were indexed with a body-centered monoclinic unit cell with $a = 11.24$ Å, $b = 11.72$ Å, $c = 10.28$ Å and $\beta = 96.2$ degrees (Fig. 4.29). The nearest interfullerene distance was 9.28 Å, identical to the inter fullerene distance in $(C_{60})_2^{2-}$ dimer anion. Assuming this structure is correct, the Na_4C_{60} compound represents the only alkali metal doped fullerene polymer in which the C_{60} anions are bonded by single bonds.

References

1. H. W. Kroto, et al.: Nature **318**, 162 (1985).
2. M. S. Dresselhaus, G. Dresselhaus, P. C. Eklund: *Science of Fullerenes and Carbon Nanotubes*, (Academic press, New York, 1996).
3. R. E. Haufler, et al.: J. Phys. Chem. **94** 8634 (1990).
4. W. Kratschmer, et al.: Nature **347**, 354 (1990).
5. R. Taylor, et al.: J. Chem. Soc. Chem. Commun. **20**, 1423 (1990).
6. R. D. Johnson, G. Meijer, D. S. Bethune: J. Am. Chem. Soc. **112**, 8983 (1990).
7. R. Tycko, et al.: J. Phys. Chem. (1991).
8. R. D. Johnson, D. S. Bethune, C. S. Yannoni: Acc. Chem. Res. **25**,169 (1992).
9. R. Taylor, D. R. M. Walton: Nature **363**, 685 (1993).
10. P. J. Fagan, J. C. Calabrese, B. Malone: Acc. Chem. Res. **25**(3), 134 (1992).
11. J. M. Hawkins: Acc. Chem. Res. **25**, 150 (1992).
12. F. Wudl: Acc. Chem. Res. **25**, 157 (1992).
13. G. Wang, et al.: Nature **387**, 583 (1997).
14. A. M. Rao, et al.: Science **259**, 955 (1993).
15. H. Kuzmany and J. Winter: *Fullerene Polymers and Fullerene Polymer Composites* (Springer Verlag).
16. S. Pekker, et al.: Science **265**, 1077 (1994).
17. P. W. Stephens, et al.: Nature **370** 636 (1994).
18. O. Chauvet, et al.: Phys. Rev. Lett. **72**, 2721 (1994).
19. G. Oszlanyi, et al.: Phys. Rev. Lett **78**, 4438 (1997).
20. Y. Iwasa, et al.: Science **264**, 1570 (1994).
21. M. Nunez-Regueiro: Phys. Rev. Lett. **74**, 278 (1995).
22. M. E. Kozlov and K. Yakushi: J. Phys. Condens. Matter **7**, L209 (1995).
23. G. Oszlanyi and L. Forro: Solid State Commun. **93**, 265 (1995).
24. I. O. Bashkin: J. Phys. Condens. Matter **6**, 7491 (1994).
25. Q. Zhu, et al.: Phys. Rev. B**52**, R723 (1995).
26. Y. B. Zhao, et al.: Appl. Phys. Lett. **64**, 577 (1994).
27. N. Takahashi, et al.: J. Appl. Phys. **74**, 5790 (1993).
28. T. Tsubo and K. Nasu: Solid State Commun. **91**, 907 (1994).
29. T. Tsubo and K. Nasu: Jpn. J. Appl. Phys. **34**, 261 (1995).
30. P. A. Heiney, et al.: Phys. Rev. Lett. **66**, 2911 (1991).
31. P. Heiney, et al.: J. Phys. Chem. Solids **53**, 1333 (1992).
32. W. I. F. David, et al.: Europhys. Lett. **18**, 219 (1992).
33. R. Tycko, et al.: Phys. Rev. Lett. **67**, 1886 (1991).
34. W. I. F. David, et al.: Nature **353** (1991).
35. A. F. Hebard, Physics Today **45**, 26 (1992).
36. G. Dresselhaus, M. S. Dresselhaus, P.C. Eklund: Phys. Rev. B**45**, 6923 (1992).

37. C.I. Frum: Chem. Phys. Lett. **176**, 504 (1991).
38. P.C. Eklund, et al.: Thin Film Solids **257** 185 (1995).
39. A.M. Rao, et al.: J. Mat. Res. **8**, 2277 (1993).
40. C.C. Eloi, et al.: J. Mat. Res. **8**, 3085 (1994).
41. P.C. Eklund, et al.: J. Phys. Chem. Solids **53**, 1391 (1992).
42. F. Negri, G. Orlandi, F. Zerbetto: Chem. Phys. Lett. **144**, 31 (1988).
43. R.E. Stanton and M.D. Newton: J. Phys. Chem. **92**, 2141 (1988).
44. R.A. Jishi, R.M. Mirie, M.S. Dresselhaus: Phys. Rev. B**45**, 13685 (1992).
45. G.B. Adams, et al.: Phys. Rev. B**50**, 17471 (1994).
46. D. Porezag, et al.: Phys. Rev. B**52**, 14963 (1995).
47. S.J. Duclos, et al.: Solid State Comm. **80**, 481 (1991).
48. T. Pagnier, M. Barral, G. Lucazeau: J. Phys. Chem. Solids **58**, 185 (1997).
49. P. Zhou, et al.: Chem. Phys. Lett. **211**, 337 (1993).
50. Y. Wang, et al.: Chem. Phys. Lett. **217**, 413 (1994).
51. K.-A. Wang, et al.: Phys. Rev. B**48**, 11375 (1993).
52. S. Guha, et al.: Phys. Rev. Lett. **72**, 3359 (1994).
53. K. Venkatesan and V. Ramamurthy: *Photochemistry in organized and Constrained media.* Vol. Ch. 4. (VCH, New York, 1991), 133.
54. S.J. Hoke, et al.: J. Org. Chem. **57**, 5069 (1992).
55. Goze, et al.: Phys. Rev. B**54**, R3676 (1996).
56. D.S. Cornett, et al.: J. Phys. Chem. **97**, 5036 (1993).
57. A. Hassanien, et al.: Appl. Phys. Lett. **70**, 417 (1997).
58. Y. Wang, et al.: Chem. Phys. Lett. **211**, 341 (1993).
59. J. Onoe and K. Takeuchi: Phys. Rev. Lett. **79**, 2987 (1997).
60. M. Menon, K.R. Subbaswamy, M. Sawtarie: Phys. Rev. B**49**, 13966 (1994).
61. D.L. Strout, et al.: Chem Phys. Lett. **214**, 576 (1993).
62. P.W. Stephens, private communication.
63. G.P. Lopinski, J.R. Fox, J.S. Lannin: Chem. Phys. Lett. **239**, 107 (1995).
64. A.M. Rao, et al.: Chem. Phys. Lett. **224**, 106 (1994).
65. O. Bethoux: Proceedings of the Materials Research Society, Boston, 1993, Abstract No. G2.9: p. 202.
66. Y. Iwasa: *Optical Properties of Low Dimensional Materials*, T. Ogawa and Y. Kanemitsu (Eds.), (World Scientific, Singapore, 1995).
67. S.G. Louie and E.L. Shirley: J. Phys. Chem. Solids **54**, 1767 (1993).
68. E.L. Shirley and S.G. Louie: Phys. Rev. Lett. **71**, 133 (1993).
69. K. Harigaya and S. Abe, (unpublished).
70. K. Harigaya: *Fullerene Polymers and Fullerene Polymer Composites*, Chapter 9 in this volume.
71. E.L. Shirley, L.X. Benedict, S.G. Louie: Phys. Rev. B**54**, 10970 (1996).
72. J.M. Holden: Ph.D thesis, University of Kemtucky, USA (1996).
73. S. Jeglinski, et al.: Synth. Met. **557**, 49 (1992).
74. S. Kazaoui, R. Ross, N. Minami: Phys. Rev. B**52**, 665 (1995).
75. A. Nakamura, et al.: J. Lumin. **66–67**, 383 (1996).
76. S.L. Ren, et al.: Appl. Phys. Lett. **59**, 2678 (1991).
77. C. Kittel: *Solid State Physics*, (John Wiley and Sons).
78. Y. Wang, et al.: Phys. Rev. B**45**, 14396 (1992).
79. F. Wooten: *Optical Properties of Solids*, (Academic Press, New York, 1972).
80. M.E. Kozlov, et al.: Appl. Phys. Lett. **66**, 1199 (1995).
81. M. Tachibana, H. Sakuma, K. Kojima: J. Appl. Phys. **82**, 1 (1997).
82. P. Zhou, et al.: Appl. Phys. Lett. **60**, 2871 (1992).
83. C.S. Yoo and W.J. Nellis: Science **254**, 1489 (1991).
84. J.-J. Hodeau, Phys. Rev. B**50**, 10311 (1994).
85. C.S. Yoo, Appl. Phys. Lett. **61**, 273 (1992).

86. H. Hirai: Carbon **31**, 1095 (1993).
87. M. Nunez-Regueiro: Nature **355**, 237 (1992).
88. F. Moshary: Phys. Rev. Lett. **69**, 466 (1992).
89. C.H. Xu and G.E. Scuseria: Phys. Rev. Lett. **74**, 274 (1995).
90. A.M. Rao: Appl. Phys. A, (1997).
91. A.M. Rao, et al.: Phys. Rev. B**55**, 4766 (1997).
92. P.C. Eklund, et al.: Thin Solid Films **257**, 211 (1995).
93. J. Winter and H. Kuzmany: Solid State Comm. **83**, 1321 (1992).
94. Q. Zhu, et al.: Phys. Rev. B**47**, 13948 (1993).
95. T. Yildirim, Phys. Rev. Lett. **71**, 1383 (1993).
96. J.E. Fischer and P.A. Heiney: J. Phys. Chem. Solids **54**, 1725 (1993).
97. D.M. Poirier and J.H. Weaver: Phys. Rev. B**47**, 10959 (1993).
98. T. Tycko, et al.: Phys. Rev. B**48**, 9097 (1993).
99. M.C. Martin, et al.: Phys. Rev. B**51**, 3210 (1995).
100. G. Oszlanyi, et al.: Phys. Rev. B**51**, 12228 (1995).
101. D. Koller, et al.: Appl. Phys. Lett, (1995).
102. M.J. Rossiensky, et al.: Nature **356**, 416 (1992).
103. T. Yildirim, et al.: Nature **360**, 568 (1992).
104. W. Andreoni, et al.: Phys. Rev. Lett. **72**, 848 (1994).

5. First-Principles Molecular Dynamical Studies of Polymerized C_{60}

G.B. Adams, J.B. Page

This chapter is primarily a summary of first-principles quantum molecular dynamics simulations of fullerene polymerization carried out by our group at Arizona State University. The majority of the chapter deals with simulations of C_{60} polymerization as related to photopolymerization (also known as laser-modification) of C_{60} solid. This theoretical work was motivated by the experimental work of Rao et al. [1], which is discussed in Chap. 4 of this book. The final section of the chapter deals with simulations of C_{119}, which is believed to be a polymeric linking of two C_{60}'s with one carbon atom removed. These simulations were motivated by the experimental work of McElvany et al. [2]. This second simulation study, in particular, illustrates the strength of using first-principles quantum molecular dynamics to simulate novel materials.

5.1 Method: First-Principles Quantum Molecular Dynamics

A first-principles, electronic-structure-based, total-energy calculation uses as input only the types and positions of the atoms in the system to be simulated. From this minimum of information, the output of the calculation is the electronic energy levels, or band structure, the total energy of the system, and if required, the forces on the atoms. In a molecular dynamics simulation, known forces are used with Newton's laws to move atoms over short time steps. First-principles quantum molecular dynamics (QMD) combines a parameter-free, electronic-structure-based, calculation of energy and forces with molecular dynamical atomic movement. As such, QMD consists of three primary steps: (1) solve $H\Psi = E\Psi$ for some input atomic configuration; (2) calculate the forces on the atoms from the Hellmann–Feynman Theorem; and (3) move the atoms over a short time interval using $F = ma$. The electrons, being of small mass, are assumed to respond instantly to the motion of the heavy nuclei, and thus are always assumed to be in their instantaneous ground state. This is the well known Born-Oppenheimer, or adiabatic, approximation.

The first calculation to include the electronic structure of a covalent system in a molecular dynamics simulation was published in 1986 by Sankey and Allen [3]. In that work, the electronic structure calculation was considerably simplified by using an empirical tight-binding Hamiltonian of the Harrison et al. form [4]. In an empirical tight-binding method, the matrix elements of the Hamiltonian are adjusted to fit various experimental or theoretical observations. At least two groups have developed empirical tight-binding methods for use with carbon [5, 6], and each of these methods has been applied to the study of fullerene polymerization [7, 8].

Quantum molecular dynamics goes beyond empirical tight-binding to calculate, from first principles, the matrix elements of the Hamiltonian. Car and Parrinello [9] and others [10] have developed first-principles molecular dynamics methods which follow the traditional approach of expanding the wavefunction in planewaves of a periodic system. These methods make no fundamental approximations not also made in static total energy calculations, and thus should enjoy the same level of accuracy; however, they require large computer resources, and large systems with hundreds of atoms require the use of parallel computing machines and/or extended computing times.

Our first-principles quantum molecular dynamics (QMD) method was developed by Sankey and Niklewski [11] and introduced in 1989. It combines the simplicity of empirical tight-binding techniques with the rigor of the local density approximation (LDA) and the use of pseudopotentials. The wavefunction is expanded as a linear combination of a minimal basis of pseudo-atomic-orbitals (PAOs) at each atomic site. For the systems studied here, we have used one s- and three p-orbitals per atom. The choice of local orbitals over planewaves means that the method's calculations are performed entirely in real space, so that molecules or clusters of atoms can be handled as easily as a periodic solid. The local orbitals are obtained from a self-consistent LDA pseudopotential calculation of the free atom, but with the boundary condition that the wavefunction vanish at a confinement radius r_c, instead of the usual free-atom boundary condition that the wavefunction vanish at infinity. The use of r_c slightly excites the atom from its ground state, and crudely represents confinement effects of the electron in the solid or molecular system being simulated; moreover, a key practical effect of r_c is to introduce a cutoff for the range of wavefunction overlap between an atom and its neighbors, thus reducing the number of overlap interactions that must be calculated.

Our QMD method was originally applied to silicon, and in those simulations the selected value of r_c was 5.0 Bohr radii [11]. This value was chosen to give a third-neighbor model in bulk diamond silicon and was essentially the smallest value which seemed to be a reasonable choice for realistic simulations. It has since been used in reliable and realistic simulations for silicon solids [11, 12], including amorphous silicon [13, 14], silicon molecules [15], and silicon surfaces [16–19]. For bulk diamond carbon, a geometric scaling of this third-neighbor model yields an r_c of 3.3 Bohr radii. This small carbon r_c was used successfully in simulating three-fold-coordinated carbon [20, 21] as well as four-fold-coordinated carbon [12]. However, the smaller the selected value of r_c, the more the atoms are excited above their ground state. For a carbon

r_c of 3.3 Bohr radii, the excitation is sufficiently large that four-fold coordinations become energetically preferred over three-fold coordinations to an unphysical extent. Thus, 3.3 Bohr radii is an unreliable r_c for simulating carbon structures having both three-fold- and four-fold-coordinated atoms. Accordingly, as we began to simulate novel covalently-connected fullerene solids such as solid C_{28} and solid C_{20} [22], which contain both three-connected and four-connected atoms, we adopted an improved convention for selecting r_c.

An improvement over the third-neighbor-model convention is to relate the confinement radius to the confinement-induced increase in energy for the s- and p-valence state free-atom energy levels. As discussed above, the value $R_c = 5.0$ Bohr works well for Si. For this value, we find that the free-atom silicon $3p$ energy level is raised by 0.191 Ryd, and the free-atom silicon $3s$ energy level is raised by 0.156 Ryd. The improved convention is to choose a confinement radius for which the s- and p-valence state free-atom energy levels are raised by about these same amounts. For example, at an r_c of 4.1 Bohr radii, we find that the carbon 2p energy level is raised by 0.193 Ryd and the carbon 2s energy level is raised by 0.165 Ryd. This larger r_c for carbon yields the correct energetic relationship between the diamond and graphitic forms of carbon and so provides for accurate simulations of carbon systems having both three-fold- and four-fold-coordinated atoms. In addition to our applications of this constant-energy-increase convention in the case of carbon [23, 24], it has also been used in successful simulations of materials with more than one atomic species, such as SiC [25] and SiO [26]. Therefore, in general, even for different atomic species occurring in the same material, we have found that maintaining a constant confinement-induced orbital-energy increase provides the right energetic relationship for modeling the bonding between the different types of atoms.

Nevertheless, there is a disadvantage in using the larger confinement radius for carbon. The earlier r_c of 3.3 Bohr produced the correct bond lengths for graphite, diamond, and C_{60}, all within two percent. The r_c of 4.1 Bohr radii results in carbon bond lengths which are about four percent too long. However, we are able to recover the experimental bond lengths for carbon by calculating the Coulomb integrals in the electron double-counting correction, U_{ee} (see [11]), using, in the charge density only, local orbitals which have an r_c equal to 96% of 4.1 Bohr. This technique produces experimental fullerene bond lengths and graphite and diamond bond lengths within two percent of experiment, with no significant effects on the relative energies of the various carbon structures. In addition, as discussed below, fullerene vibrational frequencies calculated with this technique are markedly improved over frequencies calculated using the previous, smaller, r_c [27].

We have described above several approximations used in our first-principles QMD method. The local density approximation and the use of norm-conserving pseudopotentials are standard approximations. The expansion of the wavefunction in terms of confined local orbitals was an innovation. We make one other major approximation. We replace the Kohn–Sham LDA self-consistent energy functional by the considerably simplified Harris [28] energy functional. In this approximation, the input charge density is a sum of neutral atom charge densities, and the output charge density is included in the energy functional only to first order.

Porezag et al. [29] have developed another approximate first-principles method in which the wavefunction is expanded in terms of confined PAOs. They refer to this method as the density-functional tight-binding (DF-TB) method. They use an orbital confinement scheme based on the work of Eschrig and Bergert [30]; and, instead of making the Harris energy-functional approximation, they neglect the three-center and crystal-field integrals, leading to Hamiltonian matrix elements dependent only on internuclear distances. For carbon, the short range repulsive part of the tight-binding potential is fit to self-consistent LDA-based calculations of graphite and diamond. Geometry optimization is accomplished by means of a conjugate gradient algorithm. Porezag et al. [31] have applied this method to fullerene polymerization, and as a part of the discussion below we will compare their work with our own.

5.2 Simulation One: Polymerized C_{60}

5.2.1 Structure and Energy

The first question to answer about polymerized C_{60} is, what is its structure; i.e, what is the lowest-energy configuration for a C_{60} dimer? We have sought to answer this by performing QMD relaxation simulations for many likely configurations of paired C_{60}'s. In a QMD relaxation, we begin with a starting configuration which is usually far from relaxed, and at a temperature of 0 K. The atoms then respond to the QMD forces and move towards equilibrium, transferring the initial high atomic potential energy into kinetic energy of atomic motion. In order to arrive at a relaxed final configuration, this kinetic energy must be removed. We do this by testing the sign of the quantity force × velocity for each atom. When this quantity is negative, indicating that force and velocity for an atom are in opposite directions, we set the velocity for that atom to zero. In this manner all the excess kinetic and potential energy is eventually removed, and the system finally achieves a relaxed configuration.

The theoretical structure of dimerized C_{60} has been investigated by several other groups [7, 8, 32], as well as our own [23]. Of these, only ours has used a first-principles technique for the relaxation of various trial structures. The other groups have employed either empirical tight-binding [7, 8] or semiempirical MNDO (modified neglect of differential overlap) methods [7, 32] for geometry optimization. Two of these groups [7, 32] have then used first-principles techniques to calculate the energies of their empirically-optimized geometries. The important result is that, despite using somewhat different sets of starting configurations, all of the groups have come to the

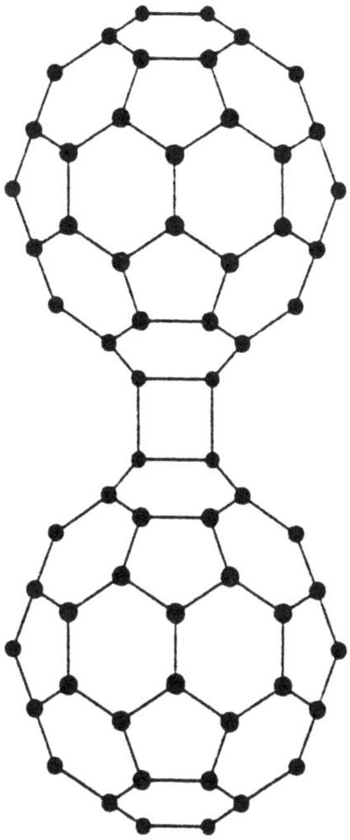

Fig. 5.1. The relaxed C_{60} dimer configuration which is found to have the lowest energy of any C_{60} dimer yet simulated. It is the only C_{60} dimer configuration found to have an energy lower than the energy of two isolated I_h C_{60}'s. The configuration is characterized by having parallel 6-6 bonds (the "double" bonds in I_h C_{60}) on the adjacent C_{60} molecules. The point symmetry is D_{2h}. After [23]

same conclusion as to the lowest-energy C_{60} dimer configuration, namely that shown in Fig. 5.1. In this chapter, we will describe in detail our own search for the lowest-energy configuration of a C_{60} dimer, compare with the results of other searches where applicable, and describe briefly those trial configurations tested by the other groups but not by ourselves.

We have run QMD relaxation simulations for seven different possible C_{60} dimer geometries. The six simulated geometries which failed to produce the low-energy structure are shown in Fig. 5.2. Before giving a detailed description of the low-energy geometry of Fig. 5.1, we will first describe our six simulations which resulted in higher-energy final geometries, and then give a brief description of higher-energy geometries simulated by other groups.

Figure 5.2a shows the final configuration which results from simulating a polymeric connection between single atoms on neighboring C_{60} molecules. This simulation began with 1.5 Å separating the connecting atoms on two relaxed C_{60} molecules. The orientation of these two C_{60}'s was such that the upper and lower balls were related by a rigid translation along the line connecting their centers-of-mass and passing through the connecting atoms. As a result of the QMD simulation the two C_{60} molecules were pushed apart; that is, we find that the atom-to-atom dimer, for this particular initial orientation, is unbound. We did not try other atom-to-atom dimer configurations which could be obtained by rotating either of the C_{60}'s about the axis connecting the

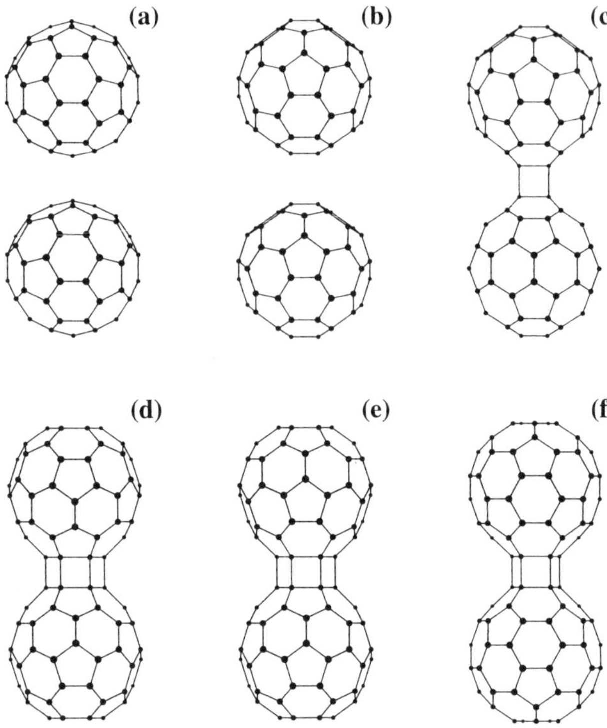

Fig. 5.2. Six relaxed C_{60} dimer configurations which are found to be higher in energy than two isolated I_h C_{60}'s. (**a**) The two C_{60}'s are connected by single atoms. This configuration is found to be unbound. (**b**) The two C_{60}'s are connected across parallel single (or 5-6) bonds. This configuration is also found to be unbound. (**c**) A single (5-6) bond on the upper C_{60} is parallel to a double (6-6) bond on the lower C_{60}. (**d**) Six-membered rings on the neighboring C_{60}'s are brought together double-bond-to-double-bond and single-bond-to-single-bond. (**e**) Same as (**d**), except that the upper C_{60} has been rotated by 60° about the line connecting the two centers-of-mass. (**f**) The two C_{60}'s are brought together five-membered ring to five-membered ring. After [23]

two centers-of-mass. Menon et al. [8] have simulated the atom-to-atom dimer configuration obtained by rotating the upper C_{60} of Fig. 5.2a by 180° about the axis connecting the two centers-of-mass. In contrast to our simulation for Fig. 5.2a, this rotated orientation aligns like bonds of the neighboring C_{60} molecules and results in a repetition of the lowest-energy dimer of Fig. 5.1.

Figure 5.2b shows the result of simulating a polymeric connection between two parallel single bonds (also known as long bonds or 5–6 bonds) of the neighboring C_{60} molecules. The simulation began with the two single bonds on neighboring C_{60}'s separated by 1.5 Å. Here too we find that the initial configuration is unbound. We did not try the alternative single-bond-to-single-bond dimer, which can be obtained by rotating either of the C_{60} molecules by 180° about the axis connecting the two centers-of-mass. Both of these single-bond-to-single-bond dimers have been simulated by Strout et al. [7] using their empirical tight-binding method. This group found that the two dimers were equal in energy, bound, and higher in energy than two isolated I_h C_{60}'s by 1.8 eV. Menon et al. [8], also with an empirical tight-binding method, found the single-bond-to-single-bond dimer to have an energy of 1.56 eV, again with respect to the energy of two isolated I_h C_{60}'s.

Figure 5.2c shows the result of simulating a polymeric connection between a single bond on one C_{60} molecule and a parallel double bond (also known as a short bond or a 6–6 bond) on the other. We find this configuration to be bound. (By "bound," we mean that the QMD relaxation finds the configuration to be trapped within at least a shallow potential well.) However, the energy of this configuration is found to be higher than the energy of two isolated I_h C_{60}'s by 0.36 eV. Empirical tight-binding calculations [7] have found the energy of this configuration to be 1.0 eV higher than the energy of two isolated I_h C_{60}'s.

Figures 5.2d and 5.2e show the results of simulating a polymeric connection between two neighboring C_{60}'s in which six-membered rings on the neighboring balls are brought face-to-face. In Fig. 5.2d the six-membered rings on the neighboring C_{60}'s are brought together double-bond-to-double-bond and single-bond-to-single bond. This configuration is found to have an energy of 0.83 eV with respect to two isolated I_h C_{60}'s. In Fig. 5.2e the upper C_{60} is rotated by 60° about the line between the centers-of-mass of the two C_{60}'s so that the six-membered rings are brought together double-bond-to-single-bond and vice versa. This configuration is found to have an energy of 0.78 eV with respect to the energy of two isolated I_h C_{60}'s. Menon et al. [8], in an empirical tight-binding simulation of the configuration of Fig. 5.2d, found an energy of 2.76 eV. Matsuzawa et al. [32], in an MNDO simulation of the configuration of Fig. 5.2d, found a different relaxed structure, having only two covalent connections between the balls rather than six. The two covalent connections are at opposite ends of the connecting six-membered rings; the other four atoms of each connecting six-membered ring retreat away from the connection, leaving the six-membered ring distorted in a saddle-like shape. Their ab initio density-functional-theory energy calculation of this MNDO relaxed configuration gave an energy 3.74 eV higher than the energy of two isolated I_h C_{60}'s, much higher than our first-principles energy of 0.83 eV. This difference in energy is not too surprising given the great difference in the final geometries. The difference in final geometries is interesting, but may not be significant, as it would not be surprising to find several local energy minima for such a high-energy configuration. Matsuzawa et al. [32] also performed an MNDO relaxation for the configuration of Fig. 5.2e, again finding only two covalent connections between saddle-shaped six-membered rings. They did not compute an ab initio energy for this latter configuration.

Finally, Fig. 5.2f shows the result of simulating a polymeric connection between two neighboring C_{60}'s in which five-membered rings of the neighboring balls are brought face-to-face. This configuration was also found to be bound but quite high in energy, 3.24 eV higher than for two isolated I_h C_{60}'s. Menon et al. [8] have simulated this dimer with an empirical tight-binding method and find an energy of 6.36 eV.

We now describe briefly the trial configurations simulated by other groups but not by ourselves. All energies reported for these other configurations are relative to the energy of two isolated I_h C_{60}'s. Matsuzawa et al. [32] have simulated a dimer in which a double bond on one C_{60} molecule is placed over a six-membered ring on the other. For this double-bond-to-six-membered-ring dimer, they have found a final geometry having two covalent connections between the two C_{60} balls, and having a first-principles energy of 3.39 eV. Strout et al. [7] have simulated this same dimer, finding a different final geometry having four covalent connections between the two C_{60}'s, and having an MNDO energy of 4.4 eV. Menon et al. [8] have simulated a dimer in which a single atom on one C_{60} molecule is placed over the center of a six-membered ring on the other. For this single-atom-to-six-membered-ring dimer they find an energy of 6.36 eV. This same group has also simulated a dimer similar to that of Fig. 5.2f, but with one of the two C_{60} balls rotated by 180° about the axis connecting the two centers-of-mass, so that the single bonds of the facing five-membered rings are staggered rather than eclipsed. For this staggered five-membered-ring-to-five-membered-ring dimer they have found an energy of 2.76 eV. In addition, Menon et al. [8] have simulated a dimer similar to Fig. 5.1 but with one of the two C_{60} balls rotated by 90° about the axis connecting the two centers-of-mass. This dimer was found to be unbound. They have also simulated a dimer with a starting geometry having a single atom on one C_{60} ball placed over a double bond of the other. This simulation reproduced the dimer of Fig. 5.1. Finally, both Menon et al. [8] and Strout et al. [7] have simulated more exotic dimers which require breaking of no more than two bonds on one or both of the C_{60}'s. Such dimers, since they require broken bonds, have higher barriers to formation than any of the dimers discussed above, and the lowest reported empirical tight-binding energy for any such dimer is 3.9 eV.

Before going on to describe in detail the lowest-energy dimer of Fig. 5.1, we should point out that it is by no means the lowest-energy configuration for 120 carbon atoms. By breaking three bonds on each of two I_h C_{60}'s and then combining the two, Strout et al. [7] have created a "bucky peanut" with a lower first-principles energy than two isolated I_h C_{60}'s. In addition, a 120-atom "capsule" formed by a 60-atom buckytube capped with two hemispheres of I_h C_{60} has a calculated energy 15.6 eV less than two isolated I_h C_{60}'s, and the lowest-energy 120-atom fullerene yet simulated, T_d C_{120}, has a calculated energy 18.8 eV less than two isolated I_h C_{60}'s [20]. These last two fullerenes are shown in Fig. 5.3. Such fullerenes cannot be called C_{60} dimers, and their

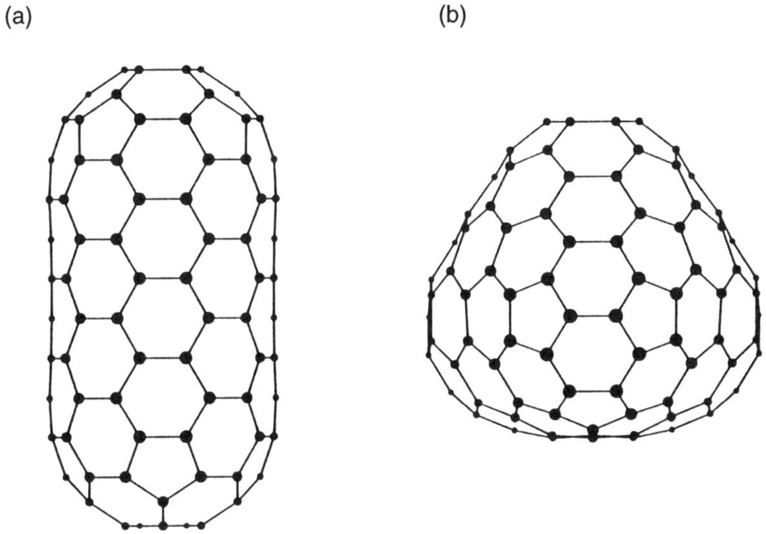

Fig. 5.3. Two 120-atom fullerene cages. (**a**) A "buckycapsule" formed by a 60-atom buckytube capped with two hemispheres of I_h C_{60}. (**b**) T_d C_{120}, the lowest-energy 120-atom fullerene known. Each is much lower in energy than the dimer of Fig. 5.1

formation from I_h C_{60}'s would require many broken bonds and involve large energy barriers. Hence, fullerenes such as those shown in Fig. 5.3 are not expected to be present under C_{60} polymerization conditions. On the other hand, ball-like 120-atom fullerenes, such as T_d C_{120}, are expected to be present in samples of large fullerenes produced by the Krätschmer–Huffman technique [33]. However, here we are concerned with C_{60} dimers, and among those dimer configurations which require no broken bonds, all of the groups ([7, 8, 23, 32]) which have simulated various trial dimer configurations have found the same lowest-energy configuration, namely the double-bond-to-double-bond dimer shown in Fig. 5.1, to which we now return.

Figure 5.1 shows the result of simulating a polymeric connection between two parallel double bonds of the neighboring C_{60} molecules. We find the energy of this configuration to be lower than the energy of two isolated I_h C_{60}'s by 0.48 eV (0.0040 eV/atom) [34]. Empirical tight-binding calculations find the energy of this dimer configuration to be from 0.4 eV [7] to 0.6 eV [8] higher than the energy of two isolated C_{60}'s. First-principles energy calculations using MNDO-optimized geometries of this dimer yield energies varying from 0.6 eV higher ([7], based on the Becke–Lee–Yang–Parr (BLYP) method [35, 36]), to 0.4 eV higher ([32], based on nonlocal density functional theory with the exchange-correlation potential of [37]), to 0.7 eV lower ([7] based on LDA-DFT) than the energy of two isolated I_h C_{60}'s. Porezag et al. [31], using the first-principles DF-TB (density-functional tight-binding) method [29], have relaxed this dimer and found an energy 0.30 eV lower than that of two isolated I_h C_{60}'s. In addition, Porezag et al. [31] have calculated the energy of their DF-TB relaxed geometry using all-electron, self-consistent methods based on the LDA and on the generalized-gradient approximation (GGA), finding energies of -1.2 eV (LDA) and -0.32 eV (GGA) with respect to two isolated C_{60}'s. By virtue of the energetics alone, we must conclude that the dimer of Fig. 5.1 is a more likely candidate for the structure of polymerized C_{60} than any other dimer structure yet simulated.

We now discuss the calculated structural properties of the dimer of Fig. 5.1. We find the length of the intermolecular bonds connecting the two C_{60}'s to be 1.59 Å, and the length of the other bonds in the four-membered ring (the intramolecular four-ring bonds) to be 1.58 Å. Porezag et al. [31], using the first-principles DF-TB method, find (to this accuracy) an identical result. (Empirical tight-binding calculations [7] give 1.56 Å and 1.62 Å, respectively.) For the bonds between the four-coordinated atoms and their three-coordinated neighbors (there are eight of these), we find a length of 1.51 Å, as do Porezag et al. [31]. Finally, we find that the centers-of-mass of the two C_{60}'s are, after polymerization, separated by 9.02 Å, while Porezag et al. [31] find 9.03 Å. (Empirical tight-binding calculations [7] yield 9.2 Å.) In C_{60} solid, the individual C_{60} molecules are separated by about 10 Å. It has been reported [1] that the altered x-ray diffraction pattern of the photopolymerized film is consistent with an average contraction of ~0.1 Å in the lattice constant. However, since the polymerizing connection between C_{60}'s may be expected to occur in random directions, this finding cannot be converted into a quantitative photopolymerized dimer length.

We have also simulated a linear trimer of C_{60}'s connected in the manner of Fig. 5.1, as is shown in Fig. 5.4a. We find the energy of this configuration to be lower than the energy of three isolated I_h C_{60}'s by 0.92 eV (0.0051 eV/atom). Porezag et al. [31] find, for this same trimer structure, an energy of -0.58 eV, again using the DF-TB method. Note that, according to both of these linear trimer calculations, the additional pair of polymer bonds does not provide as much energetic advantage as did the first pair. Consider creating the trimer of Fig. 5.4a by adding a third C_{60} ball to the dimer of Fig. 5.1. As reported above, we find an energy lowering of 0.48 eV for the pair of polymer bonds in the dimer. However, creating a trimer from the dimer, by adding a second pair of polymer bonds to the opposite side of one of the two C_{60}'s, only provides an additional energy lowering of 0.44 eV. Similarly, Porezag et al. [31] find an energy lowering of 0.30 eV for the first pair of polymer bonds and a slightly smaller energy lowering of 0.28 eV for the second pair. These authors [31] have also simulated a linear quadrimer of C_{60}'s connected in the same manner, finding an energy 0.86 eV lower than the energy of four isolated I_h C_{60}'s. So for the third pair of polymer bonds, like the second pair, they find an energy lowering of 0.28 eV. From these calculations we conclude that the energetic cost of distorting both sides of a C_{60} molecule, so as to polymerize on both sides, is 0.02–0.04 eV.

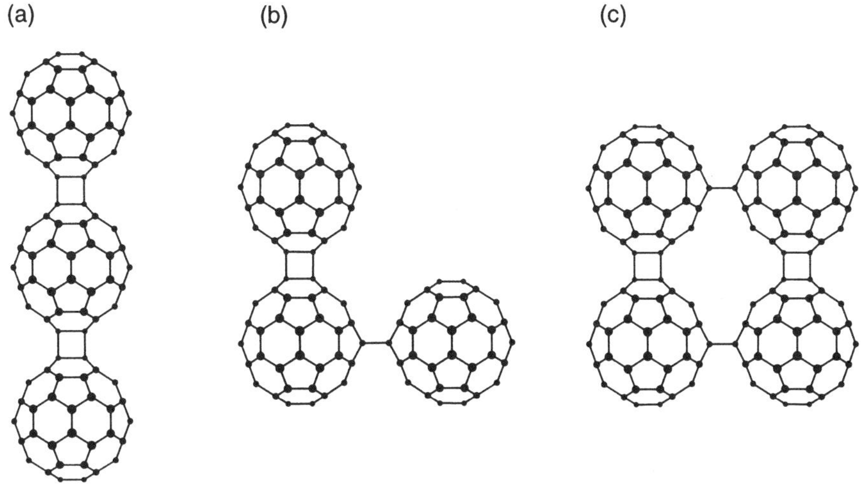

Fig. 5.4. (**a**) A relaxed C_{60} linear trimer. (**b**) An L-shaped C_{60} trimer. (**c**) A square C_{60} quadrimer. The linear trimer configuration is a result of our first-principles quantum molecular dynamics simulation. The molecules in (**b**) and (**c**) have been simulated by Porezag et al. [31]

Porezag et al. [31] have also simulated an L-shaped C_{60} trimer and a square C_{60} quadrimer (see Fig. 5.4), for which they find energies of -0.60 eV (trimer) and -1.20 eV (square), each with respect to the energy of a corresponding number of isolated I_h C_{60}'s. Thus, they find that, for the L-shaped trimer, the second pair of polymer bonds provides almost the same energetic advantage as did the first. Unlike the linear trimer case, there is nearly no extra cost for the second distortion of the central C_{60}. Similarly, for the C_{60} square, each of the four pairs of polymer bonds provides an energetic advantage of 0.30 eV. It appears that if the second distortion of the doubly-polymerized C_{60} ball is 90° from the first, rather than opposite as in the linear case, then there is essentially no extra energy cost. This implies that an infinite zigzag chain of polymerized C_{60}'s will be lower in energy than an infinite linear chain.

We have simulated both the linear and the zigzag infinite chains of polymerized C_{60}'s (Fig. 5.5). Every polymerizing connection between the C_{60} balls is made in the manner of the dimer of Fig. 5.1. In each case we have used a unit cell containing two C_{60} balls, and we have performed each calculation using two special k-points (see [38]) in the one-dimensional electronic Brillouin zone. The linear chain is found to be 0.0074 eV/atom lower in energy than isolated I_h C_{60} (0.44 eV lower per C_{60} ball or per polymer bond), while the zigzag chain is found to be 0.0080 eV/atom lower in energy (0.48 eV per polymer bond). Thus, in agreement with Porezag et al. [31], we find that a second pair of polymer bonds which is 90°, rather than 180°, from the first

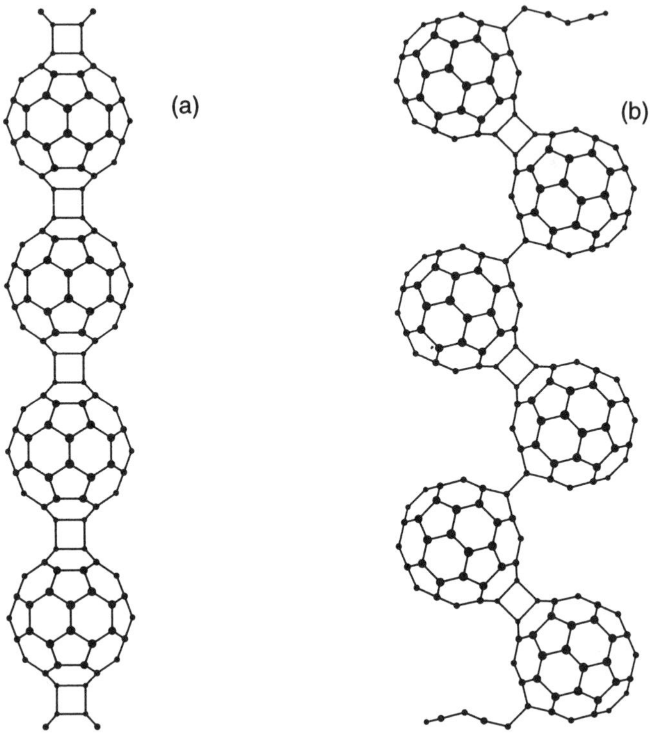

Fig. 5.5. Sections of simulated infinite C_{60} polymer chains. (**a**) Linear chain. (**b**) Zigzag chain. The zigzag chain has a lower calculated energy than the linear chain, as discussed in the text

provides the same energy lowering as did the first pair of polymer bonds. We find that the linear chain has an intermolecular bond length of 1.60 Å, an intramolecular four-ring bond length of 1.59 Å, a bond length between four-connected and three-connected atoms of 1.52 Å, and a center-of-mass separation distance of 9.13 Å. Each of these is longer than in the dimer case. The authors of [39] reported x-ray diffraction evidence for linear C_{60} polymer chains in RbC_{60}; this evidence indicates a center-of-mass separation of 9.12 Å. Since one electron per C_{60} is transferred from the Rb atoms to the linked C_{60} balls, we have repeated our calculation for the linear infinite chain using C_{60}'s which are singly-charged instead of neutral [40]. We find no significant changes in the lowest-energy bond lengths due to the addition of a single electron per C_{60}, and our calculated coordinates are found to give good agreement with structural information about RbC_{60} obtained by means of inelastic neutron scattering [40].

5.2.2 Vibrational Properties

We now report the calculated vibrational properties of the dimer of Fig. 5.1, the linear trimer of Fig. 5.4a, and the linear and zigzag infinite chains of Fig. 5.5. We have calculated these vibrational properties via a direct, first-principles computation of the dynamical matrix; that is, we displace each symmetrically unique atom by a small amount (0.0125 Å for these calculations) in each of six directions ($\pm x, \pm y, \pm z$), and for each displacement compute the first-principles force (per unit displacement) on each of the atoms in the system. Averaging the results of the plus and minus displacements eliminates cubic and higher odd-order anharmonicity. The remaining entries in the dynamical matrix arise from symmetrically equivalent atoms and are obtained by applying point symmetry operations. Our results for the first-order IR- and Raman-active vibrations of I_h C_{60} are compared with experiment [41] in Table 5.1. The average discrepancy between our calculation and experiment is 3.7%. We note that this average discrepancy is significantly improved over our older calculation [27], which used an r_c for carbon of 3.3 Bohr radii, and in which the average discrepancy is 10.3%. For the DF-TB method, the average error is about 10% [29]. Our method has also been applied with similar success to C_{70}, C_{76}, and C_{84} [42].

For the infinite chains, we have calculated the zero-wavevector vibrations in a unit cell containing two C_{60} balls. For the zigzag chain, this is the smallest

Table 5.1. Measured and calculated frequencies for the spectroscopically-active normal modes of C_{60}

Experimental frequency (cm^{-1})	Assignment	Theoretical frequency (cm^{-1})	Discrepancy
495[a]	$A_g(1)$	494	−0.2%
1468[a]	$A_g(2)$	1607	9.5%
271[a]	$H_g(1)$	259	−4.4%
431[a]	$H_g(2)$	427	−0.9%
710[a]	$H_g(3)$	694	−2.3%
772[a]	$H_g(4)$	760	−1.6%
1100[a]	$H_g(5)$	1103	0.3%
1250[a]	$H_g(6)$	1328	6.2%
1425[a]	$H_g(7)$	1535	7.7%
1574[a]	$H_g(8)$	1628	3.4%
526[b]	$T_{1u}(1)$	522	−0.8%
576[b]	$T_{1u}(2)$	570	−1.0%
1183[b]	$T_{1u}(3)$	1227	3.7%
1429[b]	$T_{1u}(4)$	1560	9.2%

[a] Reference [41], C_{60} powder at 300 K
[b] Reference [41], C_{60} film at 300 K

possible unit cell. For the linear chain, all C_{60} balls are identical, but we have chosen a two-ball unit cell in order to compare with the results for the dimer. Since we have thus doubled the unit cell and halved the Brillouin zone, we actually obtain the linear chain modes for wavevectors zero and π/a in a one-ball-per-cell description, where a is the lattice constant. The point symmetry for each of the dimer, the trimer, and the two-ball-per-cell linear chain is D_{2h} ($D_2 \times i$). There are eight irreducible representations [43], all one-dimensional, so that all of the vibrational modes are nondegenerate. The IR-active vibrations belong to the odd-parity irreducible representations B_{1u}, B_{2u}, and B_{3u}, whereas the first-order Raman-active modes belong to the even-parity irreducible representations A_g, B_{1g}, B_{2g}, and B_{3g}. For the linear chain, we add the constraint that the mode displacement patterns are identical on each ball, i.e. that the mode is of zero wavevector in a one-ball-per-cell description. We note that in the linear chain, the point symmetry origin can be taken either midway between the balls or at the ball centers, with the same point symmetry description in each case. For the zigzag chain, the point symmetry is C_{2h}. Again, all of the vibrational modes are nondegenerate, with IR-active vibrations belonging to the odd-parity irreducible representations A_u and B_u, and the first-order Raman-active modes belonging to the even-parity irreducible representations A_g and B_g.

Figure 5.6a shows our calculated frequencies for the IR-active modes of C_{60} and of the dimer, trimer, and linear and zigzag infinite chains. Experiment [1] shows an abundance of new lines appearing between 700 and $800\,\mathrm{cm}^{-1}$ in the IR spectrum of phototransformed C_{60} solid. We note a similar abundance of new lines in the calculated dimer frequencies of Fig. 5.6a. In order to complete the comparison with experiment we need to know the strength of these new lines. We have now expanded our method to include first-principles computation of IR strengths [42]. The results of applying this method to I_h C_{60} and the C_{60} dimer are shown in Fig. 5.7. While for C_{60} our calculated strength for the $T_{1u}(3)$ IR-active mode (at $\sim 1200\,\mathrm{cm}^{-1}$) is much stronger than observed experimentally, we note that the dimer frequencies between 700 and $800\,\mathrm{cm}^{-1}$ are calculated to have significant strength, in good agreement with the experimental result for photopolymerized C_{60} solid.

Also evident in Fig. 5.6a, for the dimer, the trimer, and the infinite chains, are a number of new IR-active modes at frequencies less than $200\,\mathrm{cm}^{-1}$. For the dimer, there are two of these new IR-active modes, at ~ 0 and $14\,\mathrm{cm}^{-1}$, and their mode patterns are shown in Fig. 5.8. Note that for the mode of Fig. 5.8a the restoring forces are so small that we calculate a frequency which is essentially zero (in fact, it is small and imaginary). That our method should be inaccurate in the calculation of these near-zero frequencies is not surprising. As can be seen in Table 5.1, our calculated frequencies for I_h C_{60} are too low for the low-frequency modes and too high for the high-frequency modes (possibly indicating that our method calculates bond-bending forces which are too small, and bond-stretching forces which are too large). For the lowest-

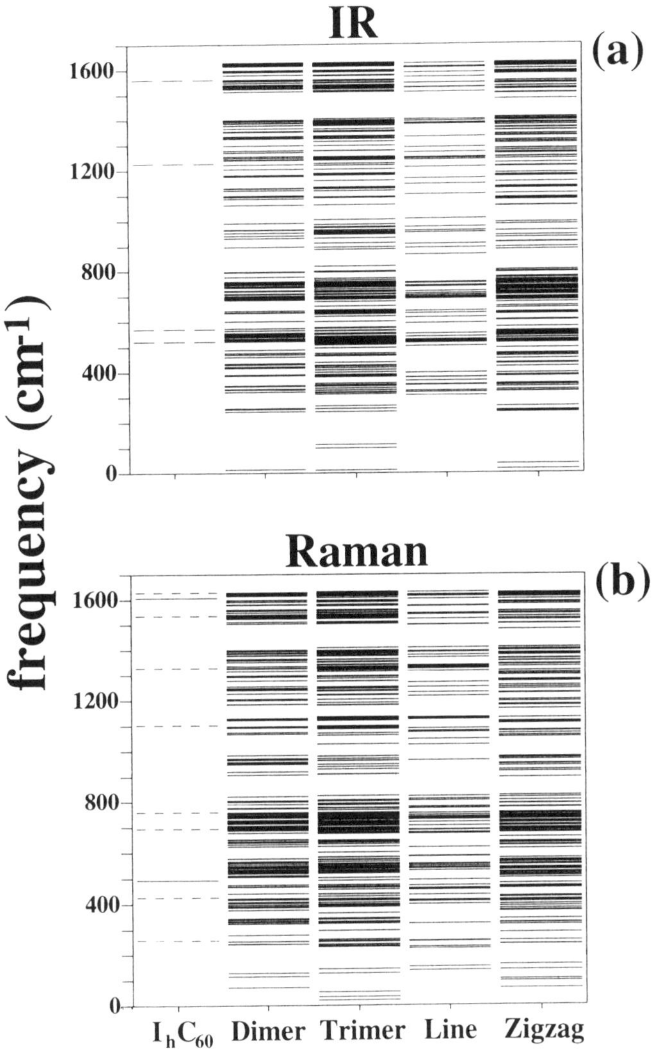

Fig. 5.6. Calculated IR and Raman frequencies of I_h C_{60}, the C_{60} dimer of Fig. 5.1, the C_{60} trimer of Fig. 5.4(a), and the C_{60} polymer chains of Fig. 5.5. (**a**) Infrared-active mode frequencies. For I_h C_{60} these are the threefold degenerate T_{1u} modes. (**b**) Raman-active mode frequencies. For I_h C_{60} these are the nondegenerate A_g modes and the fivefold-degenerate H_g modes. All of the modes in the dimer, trimer, and infinite chains are nondegenerate

frequency I_h C_{60} mode, $H_g(1)$, our calculation is $12\,\mathrm{cm}^{-1}$ low (-4.4%), and we expect errors of this or greater magnitude as frequencies become lower. As a result, for frequencies of less than $100\,\mathrm{cm}^{-1}$, our percentage errors can be large. For the DF-TB method of Porezag et al. [29], the situation is some-

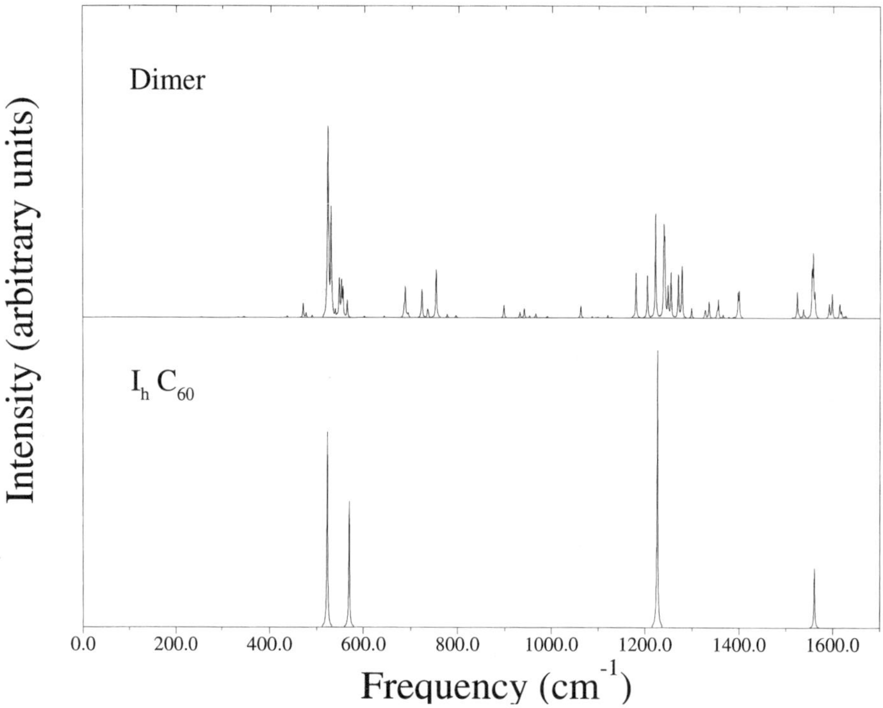

Fig. 5.7. First-principles calculated IR strengths of the vibrational modes of I_h C_{60} and of the C_{60} dimer of Fig. 5.1. The IR peaks have been broadened into Lorentzians of width one wavenumber (FWHM). For the dimer, the strong new peaks in the region 700–800 cm^{-1} are in good agreement with experimental IR spectra of photopolymerized C_{60}

what different. For high-frequency modes of I_h C_{60}, these authors' calculated frequencies are higher than ours – for spectroscopically active modes above $800\,cm^{-1}$ the average discrepancy of DF-TB is 14.2%, whereas ours is 5.7%. On the other hand, for the lowest-frequency modes the DF-TB frequencies of Porezag et al. [29] are more accurate than ours. Indeed, for $H_g(1)$ they obtain zero percent error. Hence, for the near-zero-frequency polymer modes, we generally expect DF-TB frequencies to be higher and more accurate than ours. For the ~ 0 and $14\,cm^{-1}$ IR-active modes of Fig. 5.8, the DF-TB calculations of Porezag et al. [31] yield frequencies of 22 and $35\,cm^{-1}$. (Note that long-range forces, such as the interball forces in pristine solid C_{60} which are responsible for the low-frequency librational modes in that system, will also likely affect the measured values of near-zero frequency modes of covalently connected balls. Because of orbital confinement, both our QMD method and the DF-TB method are ill-suited to modeling these kinds of long-range forces.) Fig. 5.7 shows that our calculated IR strength for each

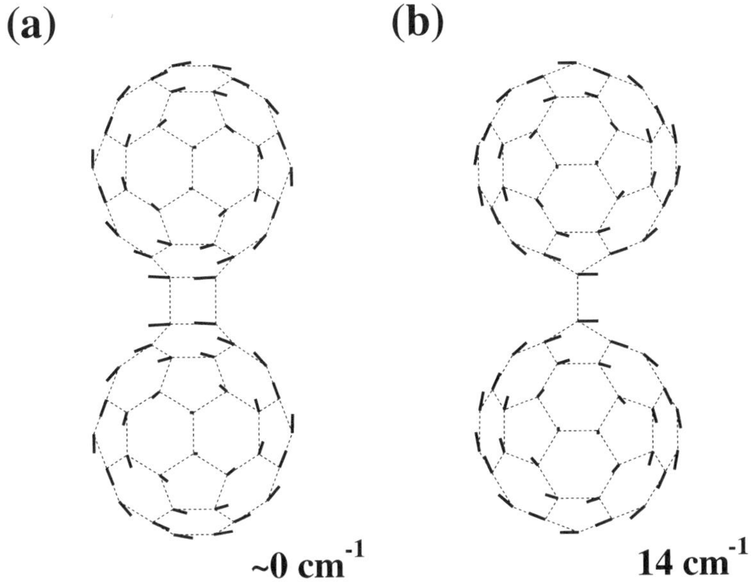

Fig. 5.8. Calculated mode patterns and frequencies for the IR-active interball modes of the C_{60} dimer of Fig. 5.1. As discussed in the text, for near-zero frequencies our calculated values are not expected to be accurate

of these two modes is insignificant. For the dimer there is one other odd-parity mode in this frequency range, belonging to the irreducible representation A_u, for which we calculate a frequency of $18\,cm^{-1}$ and Porezag et al. [31] find a frequency of $23\,cm^{-1}$. This mode is torsional, with the two C_{60} balls twisting in opposite directions about the axis connecting the two centers-of-mass.

For the trimer, there are five low-frequency IR-active modes, with calculated frequencies from ~0 to $114\,cm^{-1}$, as shown in Fig. 5.9. There is again a torsional mode which is not spectroscopically active, with a calculated frequency of $10\,cm^{-1}$. For the linear chain, there are no IR-active low frequency modes; however, shown in Figs. 5.10a and b are the two low-frequency linear chain modes which are related to the IR-active dimer modes of Fig. 5.8. Since adjacent balls move π out of phase in these modes, they are not IR-active in the linear chain. There is also a spectroscopically-inactive torsional mode in the linear chain, for which we find a frequency of $18\,cm^{-1}$. Finally, for the zigzag chain there are three low-frequency IR-active modes, two of which are shown in Fig. 5.10. The mode not shown is torsional in the two-ball unit cell, with the two full balls shown in Fig. 5.10d twisting against each other about their common axis. This mode has a calculated frequency of $36\,cm^{-1}$, like the mode of Fig. 5.10d.

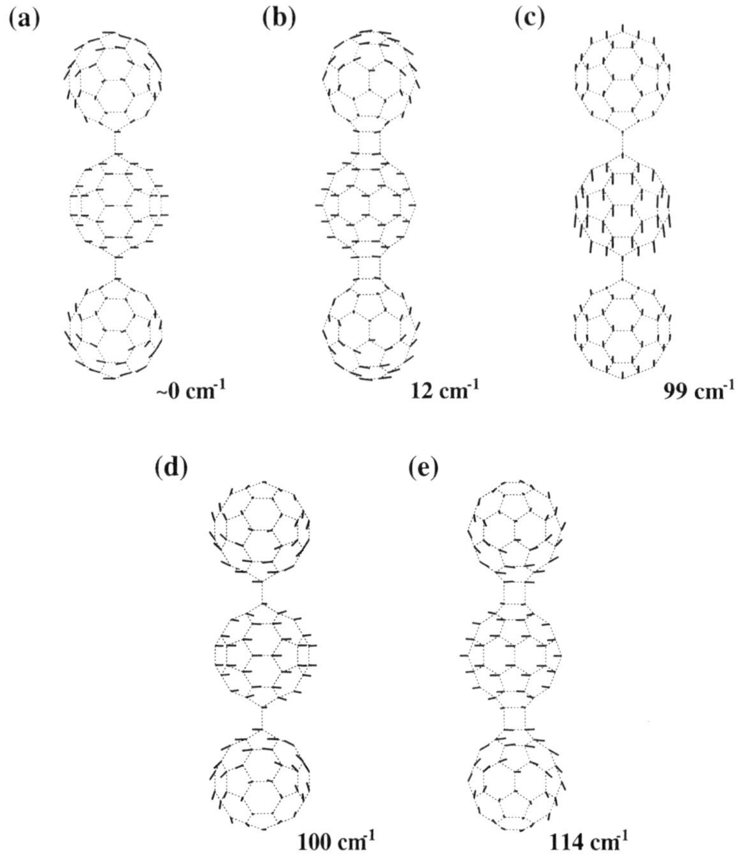

Fig. 5.9. Calculated mode patterns and frequencies for the IR-active interball modes of the C_{60} trimer of Fig. 5.4(a). Below about 100 cm^{-1}, our calculated frequencies are not expected to be accurate, as discussed in the text

Figure 5.6b shows our calculated frequencies for the Raman-active modes of C_{60}, the dimer, the trimer, and the linear and zigzag infinite chains. In this case, we calculated the Raman scattering strengths using our first-principles force constants together with the bond-charge polarizability model and I_h C_{60} polarizability parameters derived by Guha et al. [44] by means of fitting the experimental I_h C_{60} Raman spectra of Chase et al. [41]. These experimental spectra were taken with incident laser light in the near infrared (10640 Å) to avoid resonance effects which occur at the upper frequency end of the I_h C_{60} Raman spectrum when using incident laser light in the green (5145 Å). In our previous publication of a similar calculation [23] we used the polarizability parameters of Snoke and Cardona [45], which had been derived from spectra taken with incident laser light in the green. Our theoretical unpolarized

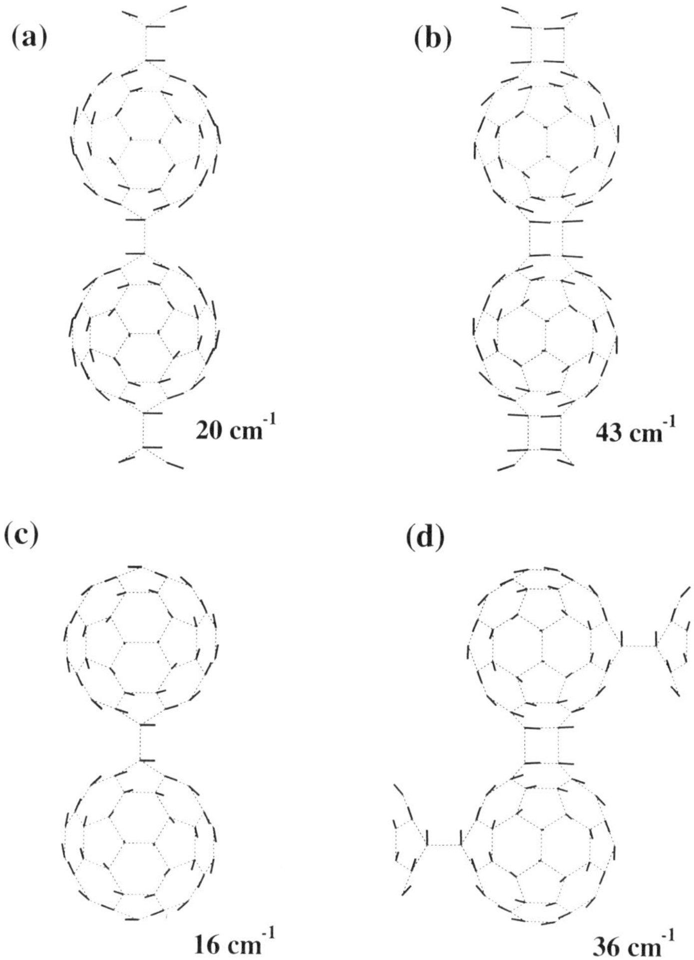

Fig. 5.10. (**a**) and (**b**) Calculated mode patterns and frequencies for those modes of the C_{60} linear chain (see Fig. 5.5(a)) which correspond to the IR-active dimer modes of Fig. 5.8. These are not IR-active in the linear chain. (**c**) and (**d**) Calculated mode patterns and frequencies for the IR-active modes of the zigzag chain. In (**c**) the connections to the neigboring unit cells cannot be seen

Raman spectra for C_{60}, the dimer, the trimer, and the linear and zigzag chains are shown in Fig. 5.11.

Of particular interest in Fig. 5.11b is the calculated shift for the peak which, in I_h C_{60}, is the $A_g(2)$ peak, i.e. the peak corresponding to the "pentagonal-pinch" mode. This calculated shift is from $1607\,cm^{-1}$ in I_h C_{60} to $1598\,cm^{-1}$ for the dimer, a downshift of $9\,cm^{-1}$, and to $1577\,cm^{-1}$ in the linear chain, a downshift of $30\,cm^{-1}$. The calculated $9\,cm^{-1}$ downshift in going

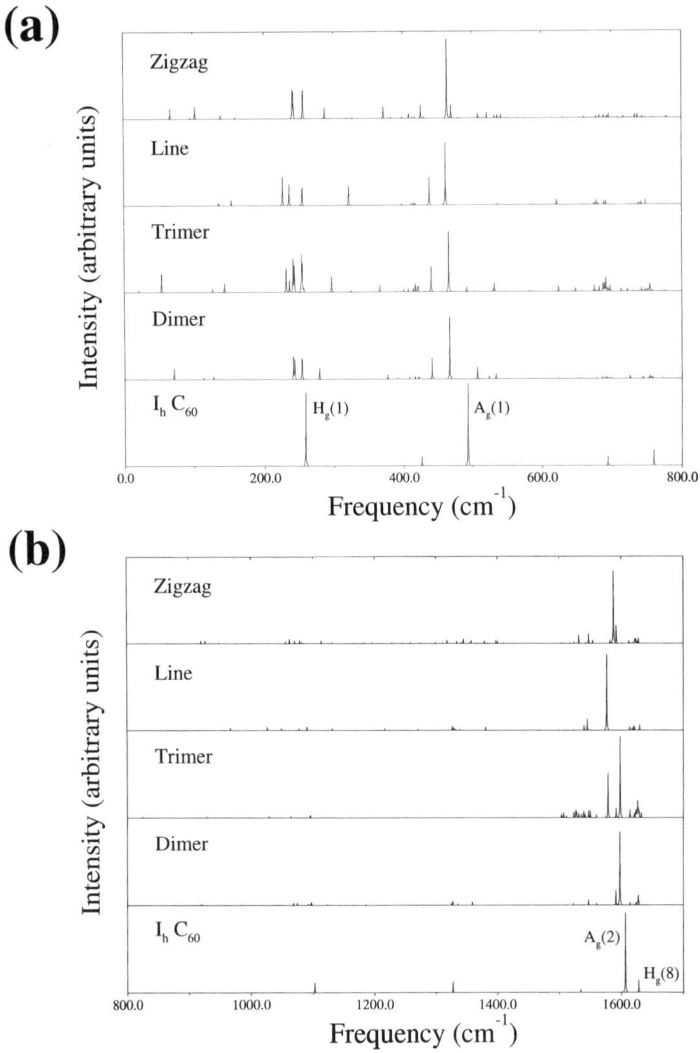

Fig. 5.11. Calculated unpolarized Raman spectra for I_h C_{60}, the C_{60} dimer of Fig. 5.1, the C_{60} trimer of Fig. 5.4(a), and the infinite C_{60} polymer chains of Fig. 5.5. Part (**a**) is for the region 0–800 cm^{-1} and part (**b**) is for the region 800–1700 cm^{-1}. The Raman strengths were computed from a bond-charge model as discussed in the text, and the peaks have been broadened into Lorentzians of width one wavenumber (FWHM). There is no correlation of vertical scales between panels, but the vertical range for each panel in (**a**) is identical to the vertical range in the corresponding panel of (**b**). In (**a**) note the new lines which appear in the region 0–200 cm^{-1} in the polymer spectra. In (**b**) note the shift in the strongest line, i.e., the I_h C_{60} $A_g(2)$ peak. For the dimer, this calculated shift is in excellent agreement with experimental Raman spectra for photopolymerized C_{60}

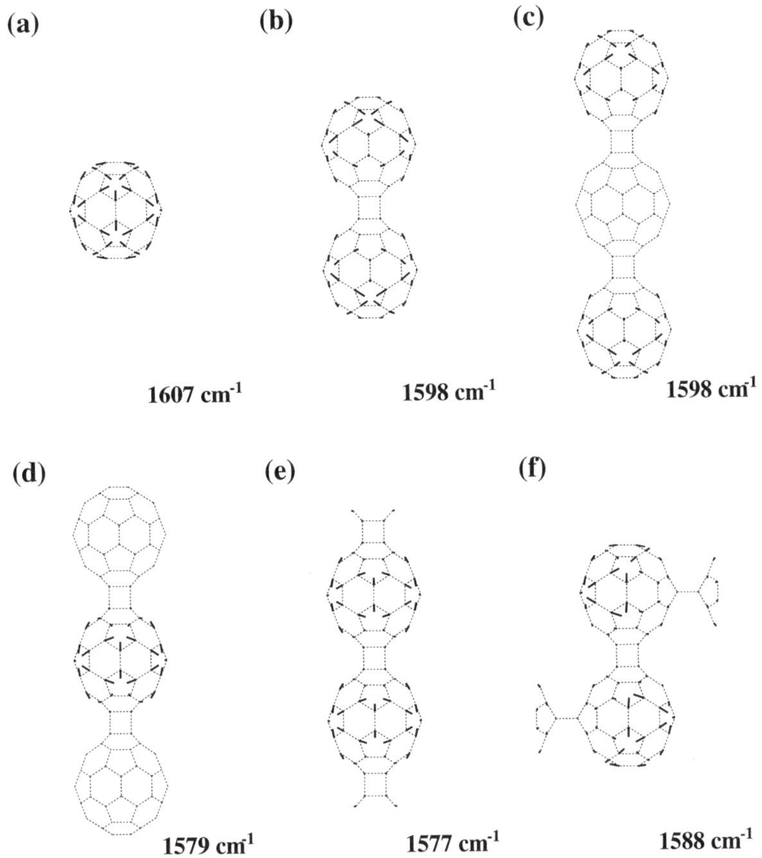

Fig. 5.12. (**a**) Calculated mode pattern and frequency for the I_h C_{60} $A_g(2)$ mode. (**b**)–(**f**) Calculated mode patterns and frequencies for the strongest Raman-active modes in the (**b**) dimer, (**c**)–(**d**) trimer, (**e**) linear chain, and (**f**) zigzag chain spectra of Fig. 5.11(b). All of the mode patterns in (**b**)–(**f**) are variations on the pentagonal-pinch pattern of (**a**), with suppressed atomic motion near the polymer connections

from I_h C_{60} to the dimer agrees well with the observed shift of $10\,\mathrm{cm}^{-1}$ in going from pristine solid C_{60} to the photopolymerized solid [1]. Porezag et al. [31] have used the DF-TB method to calculate the vibrational frequencies of I_h C_{60} and of this same C_{60} dimer, and for the $A_g(2)$ peak they obtain a downshift of $10\,\mathrm{cm}^{-1}$. In Fig. 5.12 we compare our calculated vibrational mode patterns for this shifting peak. In I_h C_{60} the pattern is purely pentagonal pinch, but in the dimer and the infinite chains, while the patterns are similar, the amplitudes of atoms near the interball connections are small. For the trimer, we find two peaks: one at $1598\,\mathrm{cm}^{-1}$, downshifted by $9\,\mathrm{cm}^{-1}$ as in the dimer, and a peak at $1579\,\mathrm{cm}^{-1}$, a downshift of $28\,\mathrm{cm}^{-1}$, in close agreement with our 30 cm^{-1} calculated downshift for the linear infinite chain. For comparison, Porezag et al. [31] calculate these linear trimer downshifts as 9 cm^{-1} and 20 cm^{-1}, respectively. As is shown in Fig. 5.12, the peak at 1598 cm^{-1} corresponds to a vibrational mode with displacements occuring primarily on the outer balls of the trimer, each of which has a single polymerizing connection, while the peak at 1579 cm^{-1} is for a vibrational mode that is largely confined to the central ball of the trimer, which has two linearly-opposite polymerizing connections. On the other hand, the L-shaped trimer, the C_{60} square, and the infinite zigzag chain have C_{60} balls with two polymeric connections that are at right angles to one another. Porezag et al. [31] have found that, for the L-shaped trimer and the C_{60} square, pentagonal-pinch-like modes on C_{60} balls with two polymeric connections are downshifted from I_h C_{60} $A_g(2)$ by 16 cm^{-1}. Similarly, in our calculation for the infinite zigzag chain, we find a pentagonal-pinch-like mode downshifted by 19 cm^{-1} with respect to the I_h C_{60} $A_g(2)$ (see Fig. 5.12f). Compare this calculated downshift of 19 cm^{-1} with our 30 cm^{-1} calculated downshift of this same mode for the linear infinite chain. This smaller downshift is another indication of the smaller net distortion created when the second polymeric connection occurs at 90° to, rather than linearly opposite, the first.

Figure 5.11a compares the calculated unpolarized Raman spectra of I_h C_{60}, the C_{60} dimer, the trimer, and the linear and zigzag infinite chains in the region 0–800 cm^{-1}. Note the predicted splitting of the $H_g(1)$ peak in going from I_h C_{60} to the dimer. In I_h C_{60} we find a single peak at 259 cm^{-1}; in the C_{60} dimer this peak is split into peaks at 279, ~253, and ~242 cm^{-1}. Experiment [1] shows a peak at 270 cm^{-1} for pure C_{60}, and peaks at 301, 271, and 258 cm^{-1} for phototransformed C_{60} solid. Our calculated shifts of +20 and −17 cm^{-1} agree reasonably well with the measured shifts of +31 and −12 cm^{-1}. Porezag et al. [31] calculate these shifts as +30 cm^{-1} and −13 cm^{-1}, respectively. For the $A_g(1)$ peak, we calculate a shift of −26 cm^{-1}, whereas experiment [1] shows no shift. Porezag et al. [31] calculate a shift of −2 cm^{-1}, in much better agreement with experiment. We must conclude that, while our approximations (Harris energy functional and confined orbitals) give better absolute vibrational frequencies for I_h C_{60}, the approximations of Porezag et al. [29] (neglect of three-center and crystal field contributions to the matrix elements, with a different orbital confinement scheme), as used in [31], give a better accounting of the shifts in frequency upon dimerization, at least for the lower-frequency modes. The underlying cause of this difference in the two methods is unknown.

We turn now to the new low-frequency modes (in the range 0–200 cm^{-1}) which appear in the Raman spectra of the C_{60} dimer, trimer, and linear and zigzag chains. In the dimer there are three of these Raman-active low-frequency modes (see Fig. 5.11a), for which the calculated mode patterns and frequencies, at 71, 114, and 128 cm^{-1}, are shown in Fig. 5.13. Porezag et al. [31] calculate these frequencies at 89, 130, and 146 cm^{-1}. The most

Raman-active of the three modes, according to the strengths which result from our bond-charge polarizability parameters, is the ball-ball vibrational mode (see Fig. 5.13a) at 71 cm^{-1}. We note that the bond-polarizability parameters of Snoke and Cardona [45] also select this ball-ball vibrational mode as the most Raman-active of these three modes, both in our earlier calculation [23] and in the calculation of [31]. However, the strongest low-frequency mode appearing in experimental Raman spectra of photopolymerized C_{60} is reported in the range 115 cm^{-1} [46] to 118 cm^{-1} [47]. Accordingly, Porezag et al. [31] have investigated the frequency of this ball-ball mode with an all-electron SCF-LDA method [48], determining the LDA-calculated upper limit for the frequency of this mode to be 92 cm^{-1}. Menon et al. [8], in an empirical tight-binding calculation in which the interactions have exponential tails, have calculated this same mode to have a frequency of 101 cm^{-1}. It may be the case that neglected long-range interactions, such as van der Waals interactions between neighboring dimers, contribute in such a way as to increase the frequency of the ball-ball mode; or, it may be that the bond-charge polarizability parameter models, fit from I_h C_{60} Raman data, are inadequate to determine realistically the relative strengths of these three low-frequency Raman-active modes. Given that these modes involve strongly the bonds of the four-coordinated carbon atoms (see Fig. 5.13, of which there are none in I_h C_{60}, a failing of this sort would not be unreasonable. (For example, the mode of Fig. 5.13a requires considerable bending of the bonds between the four-coordinated atoms and their three-coordinated neighbors.) This possible explanation of the experiment-theory discrepancy is particularly attractive, since Lopinski et al. [46] have found three low-frequency Raman active peaks, at 85, 115, and 140 cm^{-1}, with the mode at 115 cm^{-1} having the greatest

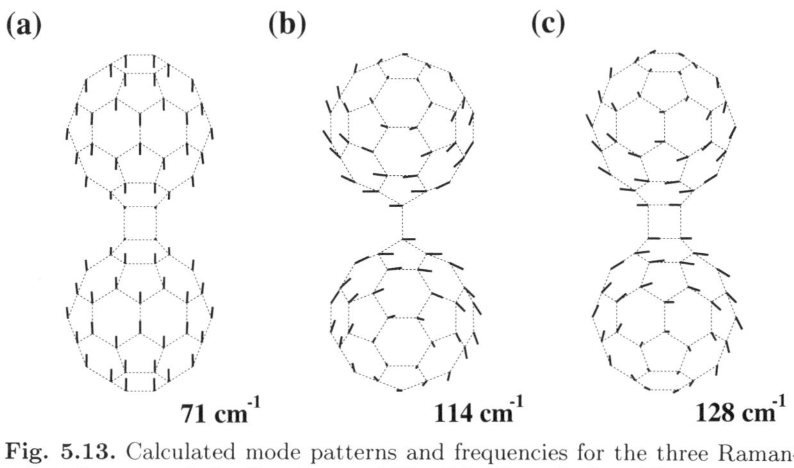

Fig. 5.13. Calculated mode patterns and frequencies for the three Raman-active interball modes of the C_{60} dimer of Fig. 5.1. The calculated Raman strengths of these modes can be seen in Fig. 5.11(a). After [23]

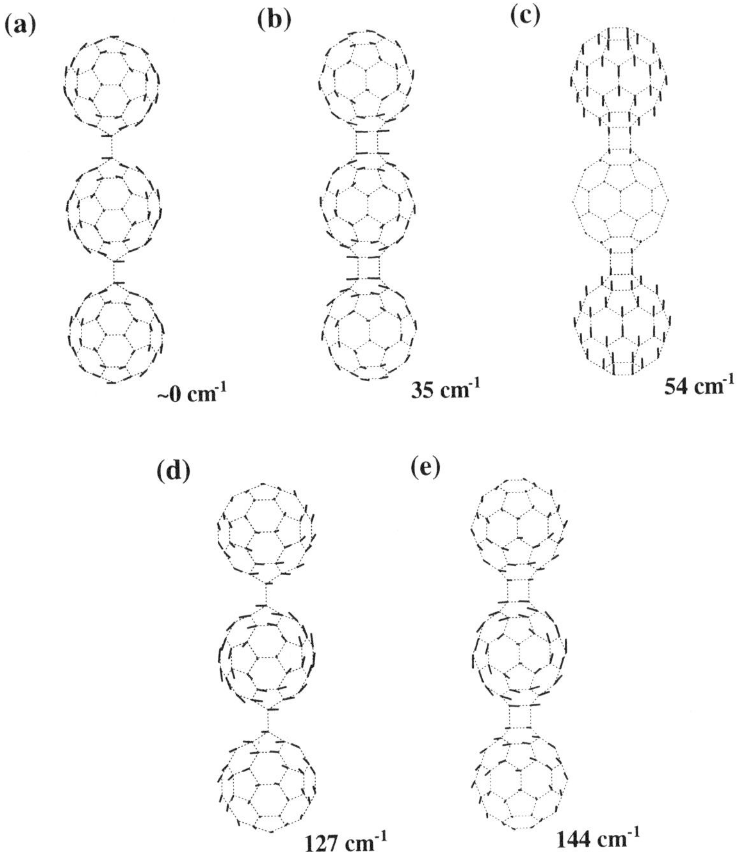

Fig. 5.14. Calculated mode patterns and frequencies for the five Raman-active interball modes of the C_{60} trimer of Fig. 5.4(a). The calculated Raman strengths of these modes can be seen in Fig. 5.11(a)

strength. A first-principles calculation of the relative Raman strengths of these three modes is needed.

It is also possible that the 115–118 cm^{-1} mode seen in the Raman spectra of photopolymerized C_{60} is due to a vibrational mode of some molecule other than the dimer. For example, Porezag et al. [31] find that the L-shaped trimer has a Raman-active mode at 118 cm^{-1}. In our calculation for the linear trimer there are six Raman-active modes in the range 0–200 cm^{-1}, for which we calculate frequencies ranging from zero to 144 cm^{-1}. The mode patterns and calculated frequencies for five of these linear trimer modes are shown in Fig. 5.14. Left out of this figure is a Raman-active torsional mode which has a calculated frequency of 21 cm^{-1}. For the linear chain there are two Raman-active modes in this range, for which the patterns and frequencies are shown in Figs. 5.15d and e. Also shown in Fig. 5.15 are three other low-frequency linear chain

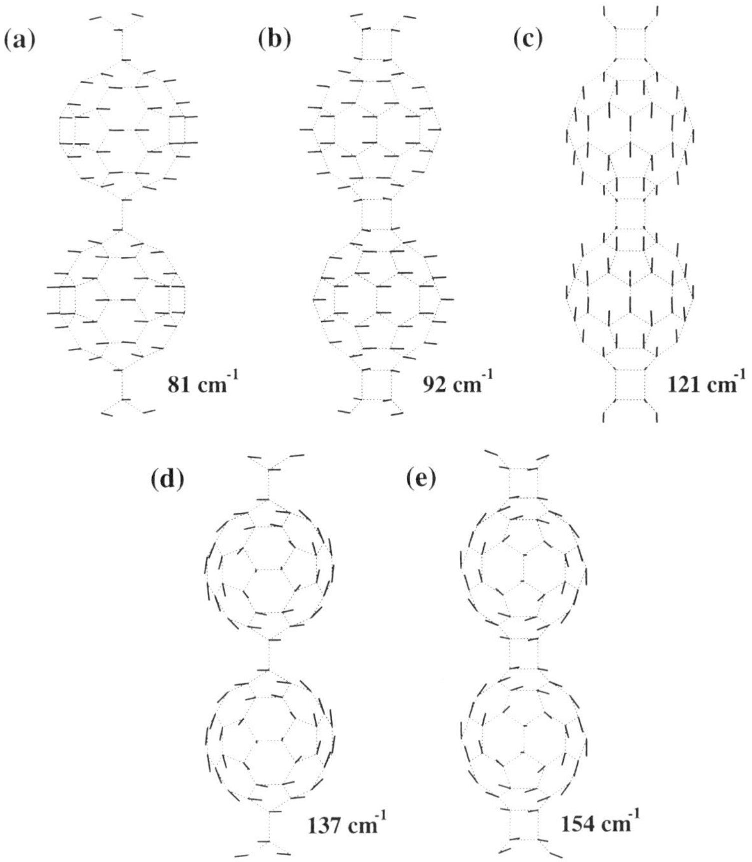

Fig. 5.15. (**a**) and (**b**) Calculated mode patterns and frequencies for the "transverse" ball-ball modes of the C_{60} linear chain of Fig. 5.5(a). These modes are spectroscopically inactive. (**c**) Calculated mode pattern and frequency for the "longitudinal" ball-ball linear chain mode, which corresponds to the Raman-active dimer mode shown in Fig. 5.13(a). This mode is not Raman-active in the chain. (**d**) and (**e**) Calculated mode patterns and frequencies for the two Raman-active interball modes of the linear chain. These correspond to the Raman-active dimer modes shown in Figs. 5.13(b) and (c). After [23]

modes which are not spectroscopically active, but which are interesting to compare with the Raman-active dimer mode of Fig. 5.13a. Fig. 5.15c shows the computed "longitudinal" ball-ball stretching mode which corresponds to the dimer mode of Fig. 5.13a, and Figs. 5.15a and b show the companion "transverse" ball-ball modes. As mentioned earlier, the linear-chain torsional mode, with a calculated frequency of $18\,\mathrm{cm}^{-1}$, is not spectroscopically active. For the zigzag chain, there are five low-frequency Raman-active modes,

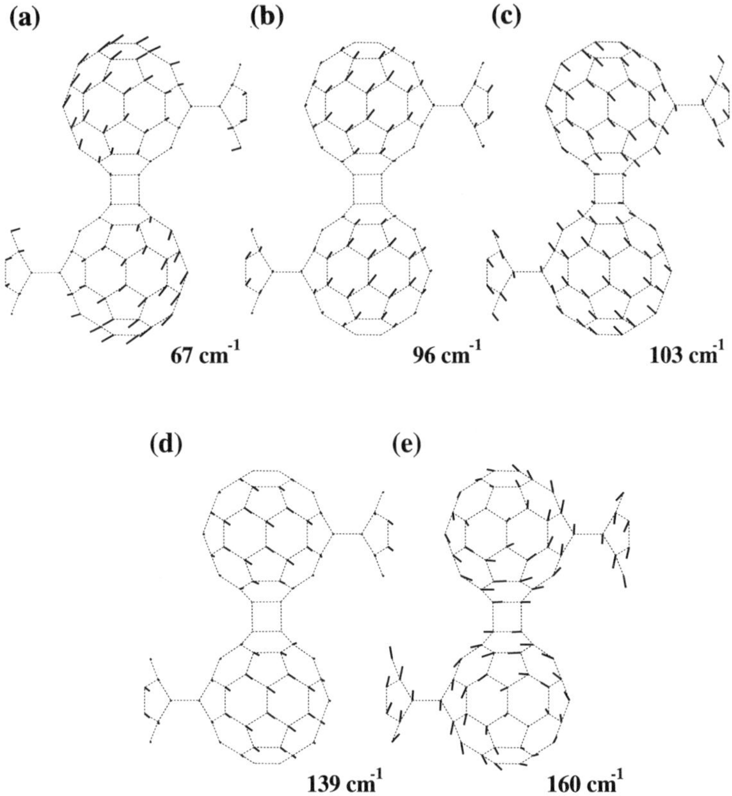

Fig. 5.16. Calculated mode patterns and frequencies for the five Raman-active interball modes of the C_{60} zigzag chain of Fig. 5.5(b). Modes (**b**) and (**d**) have torsional character and, as such, are difficult to portray. In (**b**) each C_{60} ball is rotating about a line which passes through its center-of-mass and which is roughly perpendicular to the zigzig chain direction. In (**d**) each C_{60} ball is rotating about a line which passes through its center-of-mass and is roughly parallel to the direction of the zigzig chain. The calculated Raman strengths of these modes can be seen in Fig. 5.11(a)

shown in Fig. 5.16. The strongest of these modes, at a calculated frequency of $10^3\,\mathrm{cm}^{-1}$, is shown in Fig. 5.16c.

5.2.3 Electronic Eigenvalues

In Fig. 5.17, we compare, in the near-gap region, the calculated electronic eigenvalues of C_{60} and the C_{60} dimer and trimer, with the band structure of the linear and zigzag infinite chains. Lopinski et al. [46] have measured ultra-violet HeI photoemmision (UPS) spectra for both pristine and photopolymer-

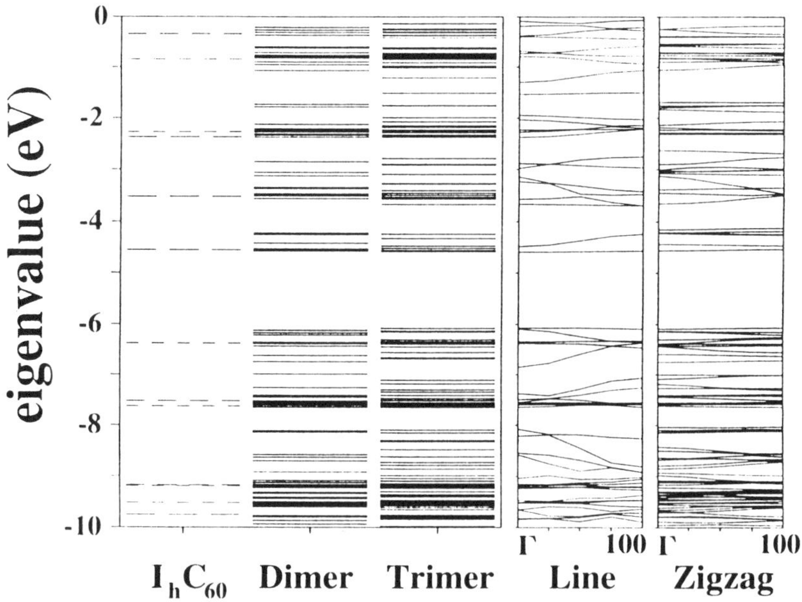

Fig. 5.17. The calculated electronic eigenvalues for I_h C_{60}, the C_{60} dimer of Fig. 5.1, and the C_{60} trimer of Fig. 5.4(a), compared with calculated band structures for the C_{60} polymer chains of Fig. 5.5, all in the near gap region. For I_h C_{60}, the degeneracies of the levels are shown by the number of subdivisions of the horizontal lines. For the linear and zigzag chains, the eigenvalues are plotted for five wavevectors in the 100 direction, from the origin (Γ) to the Brillouin zone edge. For all five configurations, the Fermi level is just below -6.0 eV, but for each of the latter four, the Fermi level is about 0.3 eV higher than for I_h C_{60}

ized C_{60} solid. In their spectrum for the photopolymerized material, shown in Fig. 5.18a, there is a shoulder just above the pure C_{60} highest occupied molecular orbital (HOMO) edge, and a new peak at ~2 eV below the pure C_{60} HOMO edge. As Fig. 5.17 shows, our calculation of the eigenvalues of the C_{60} dimer is in agreement with these experimentally observed features. In addition, Lopinski et al. [46] have used our calculated electronic eigenvalues for I_h C_{60} and the C_{60} dimer to calculate theoretical densities of electronic states (Fig. 5.18b), and have then taken the difference between these densities of states and compared with difference spectra from their UPS experiments. The agreement between theory and experiment, as shown in Fig. 5.18, is excellent, and provides further evidence supporting the dimer model for the structure of photopolymerized C_{60}. In addition, our calculated eigenvalues are in qualitative agreement with optical absorption data reported by Wang et al. [49]. An extensive discussion of the electronic structure of C_{60} polymers is given in Chap. Eight of this book.

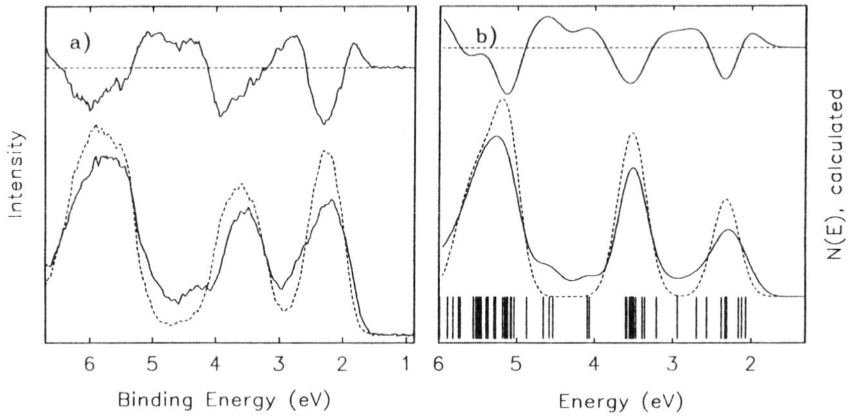

Fig. 5.18. (**a**) Background-subtracted ultraviolet photoemission spectra for C_{60} (dashed) and laser-modified C_{60} (solid). (**b**) Calculated electronic density of states for I_h C_{60} (dashed) and the C_{60} dimer of Fig. 5.1. The corresponding difference spectra are shown at the top of each of (**a**) and (**b**). The eigenvalue stick spectrum shown in (**b**) is for the same dimer valence levels as shown in Fig. 5.17, shifted in energy by 4.3 eV to facilitate comparison with experiment. After [46]

5.2.4 Energy Barrier to Dimer Separation

Finally, we report our calculation of the energy barrier to C_{60} dimer separation. We began with the relaxed C_{60} dimer, and in 100 steps changed the coordinates to those of two separated, relaxed I_h C_{60}'s. This change was made by linear interpolation of coordinates, with no atomic relaxation at the intermediate steps. At each step, we calculated the energy with our first-principles method. The results of this simulation are presented in Fig. 5.19, where the energy per carbon atom is plotted versus step number. The energy barrier to dimer separation, defined to be the difference in energy between the relaxed dimer and the highest energy configuration in the linear interpolated separation, is found to be 2.4 eV. Since the coordinates were not relaxed at each step in the linear interpolated separation, this 2.4 eV should be regarded as a maximum theoretical value for the energy barrier to separation. Compare this to the experimental result, in which, from an analyis of the thermal decomposition of the phototransformed material, the activation energy for thermal dissociation of the proposed polymer is estimated to be 1.25 eV [50]. Porezag et al. [31] have done a simulation similar to ours, but in which they have allowed all but the four-connected atoms to relax at each step. In this simulation, they find a barrier energy of 1.9 eV. In addition, they have performed another simulation in which only one of the two intermolecular-binding pairs of four-connected atoms was not allowed to relax at each step. As a result, the polymeric connection was broken one bond at a time, and the resulting energy barrier was a smaller 1.6 eV, closer yet to the experimental estimate.

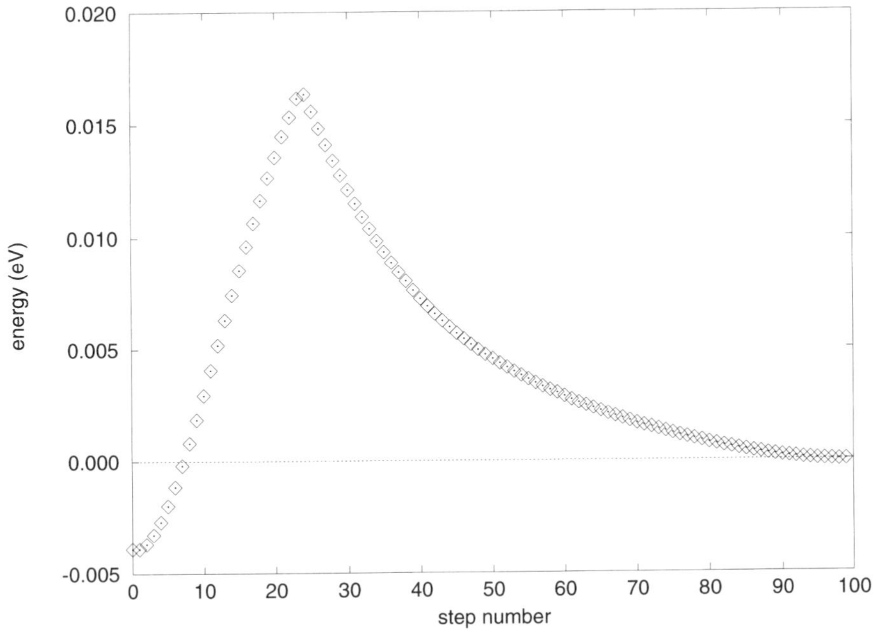

Fig. 5.19. Energy barrier for separation of the C_{60} dimer, computed via linear interpolation of coordinates. The simulation begins at step 0 with the bound C_{60} dimer of Fig. 5.1. At step 24, the separating C_{60}'s attain their maximum energy, which is 0.020 eV/atom, or 2.4 eV, higher in energy than the bound C_{60} dimer. Finally, at step 100, the two C_{60}'s are completely separated and have recovered their individual, relaxed I_h configurations. As discussed in the text, this value of 2.4 eV should be regarded as an upper bound. After [23]

5.2.5 Summary for Photopolymerized C_{60}

We have presented here a summary of current first-principles theoretical simulations supporting the double-bond-to-double-bond dimer model, as shown in Fig. 5.1, for the structure of photopolymerized C_{60} solid. We have simulated seven different dimer models, and several other dimer models have been simulated by other groups. All of these simulations have found that the double-bond-to-double-bond dimer has the lowest energy of any simulated dimer model which does not allow for bond breaking. We find that the energy of this dimer is 0.48 eV lower than the energy of two isolated (and relaxed) I_h C_{60} molecules. This finding is in good agreement with an SCF-GGA total energy calculation [31], which uses coordinates relaxed by means of a DF-TB approach and yields a dimer binding energy of 0.32 eV. This is the only dimer model, excluding those which require bond breaking, that has been found to be lower in energy than two isolated I_h C_{60}'s. Longer polymer chains of C_{60}'s, connected in the same manner as the dimer of Fig. 5.1, are found to be still lower in energy [23, 31], but first-principles calculations of vibrations [23, 31] support the dimer model. In particular, phototransformed C_{60} is identified experimentally by a 9–10 cm^{-1} downshift in that high frequency Raman mode which is $A_g(2)$ in pristine C_{60} [1, 46]; for the C_{60} dimer of Fig. 5.1, two different first-principles calculations [23, 31] find this mode to be downshifted by 9–10 cm^{-1}, in excellent agreement with experiment. Longer polymer chains are calculated to have larger shifts [23, 31]. The calculated electronic eigenvalues of the double-bond-to-double-bond dimer, in comparison with the calculated electronic eigenvalues of I_h C_{60}, are found to be in excellent agreement with difference UPS spectra between pristine and photopolymerized C_{60} solid. Finally, the energy barrier to separation for the dimer has been calculated to be at most 1.6 eV [31], in reasonable agreement with an experimental estimate of 1.25 eV [50] for the energy of polymer dissociation.

An important disagreement between theory and experiment occurs for the three low-frequency Raman-active modes of the dimer. Two different calculations [23, 31] have found the strongest of these three modes to be the one of the lowest frequency, at less than 92 cm^{-1}; while experiment [46] has seen three peaks with the strongest being the second of the three, at 115 cm^{-1}. However, both of these theoretical works relied on a parameterized model [45] for the calculation of Raman strengths, and the transferability of that model is questionable at low frequencies. It is also possible that photopolymerized C_{60} is not composed primarily of dimers and that the 115 cm^{-1} mode is due to a vibrational mode of some other C_{60} cluster. A more recent experiment [51], described below, has partially resolved this question.

Burger, Winter, and Kuzmany (BWK) [51] have phototransformed solid C_{60} at a number of temperatures, ranging from 260 to 400 K. The material phototransformed at high temperature (380 K) was found to have the strongest Raman-active interball mode at 96 cm^{-1}, with only a "weak shoulder" found at 117 cm^{-1}. On the basis of good agreement in the low-frequency region (70–400 cm^{-1}) between the experimental Raman spectrum and the calculated Raman spectrum of Porezag et al. [31], BWK were able to show that solid C_{60} photopolymerized at 380 K is most likely to consist primarily of the dimer shown in Fig. 5.1, although they were not able to rule out the L-shaped trimer or the square quadrimer. The shift of the pentagonal-pinch mode for this 380 K sample was found to be 7 cm^{-1} [51], still in reasonable agreement with the calculated shifts of 9 cm^{-1} [23] and 10 cm^{-1} [31] for the double-bond-to-double-bond dimer.

On the other hand, for solid C_{60} phototransformed at room temperature (300 K), BWK found the strongest Raman-active interball mode at 117 cm^{-1} and a pentagonal-pinch-mode shift of 10 cm^{-1}, in agreement with earlier experimental work [1, 46, 47]. In addition, BWK identified a splitting of the pentagonal-pinch mode with a second peak having a downshift of 17 cm^{-1} with respect to the I_h C_{60} $A_g(2)$ peak. Based on these differences between the Raman spectra of solid C_{60} phototransformed at different temperatures, BWK were able to show that the material phototransformed at room temperature does not consist primarily of dimers. However, BWK found unsatisfactory agreement in the low-frequency region (70–400 cm^{-1}) between the experimental Raman spectrum of the 300 K sample and the calculated Raman spectrum for any one of the C_{60} polymers simulated by Porezag et al. [31], including dimer, linear and L-shaped trimers, and linear and square quadrimers. In particular, the worst agreement was found for the linear trimer and quadrimer, indicating that these species are unlikely to be present in the 300 K sample. So, while the primary component of solid C_{60} phototransformed at room temperature has yet to be identified, it is likely to be a nonlinear cluster of more than two C_{60}'s. Note added in proof: Consistent with this idea, a recent experiment has demonstrated that the room temperature photopolymer is composed of small closed oligomers. (T. Pusztai, G. Oszlányi, G. Faigel, K. Kamarás, L. Gránásy, S. Pekker: Solid State Commun. III, 595 (1999).)

5.2.6 Other Forms of Polymerized C_{60}

Linear C_{60} polymer chains which were first found in RbC_{60}, as mentioned above, have now been identified in KC_{60} [52] and CsC_{60} [53]. In addition, another type of charged C_{60} polymer, a dimer with a charge of minus two and connected by a single carbon–carbon bond (see Fig. 5.20a), has been identified in A_1C_{60} solids (where A is K, Rb, or Cs) which have been subjected to rapid cooling from high temperature [54, 55]. Semiempirical MNDO cal-

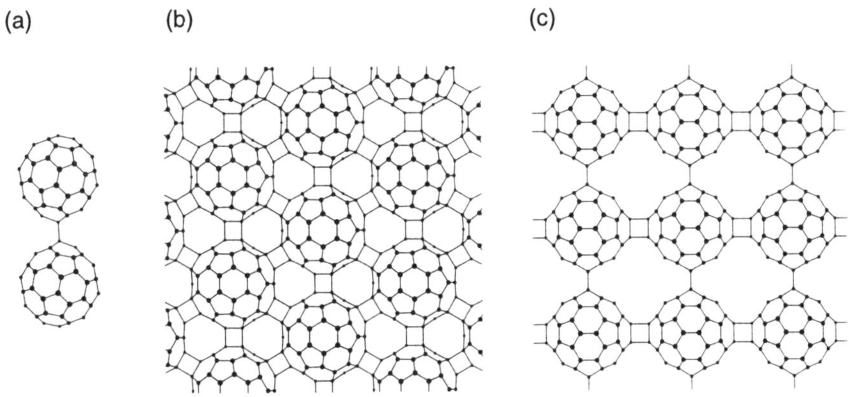

Fig. 5.20. Other forms of polymerized C_{60}. (**a**) The atom-to-atom C_{60} dimer is bound when it has a charge of minus two. This dimer has been identified in A_1C_{60} compounds which have been rapidly cooled from high temperature (A is K, Rb, or Cs). (**b**) and (**c**) Proposed structures of the planar phases of polymerized neutral C_{60} which have been induced by the application of high pressure and high temperature to *fcc* solid C_{60}. (**b**) is the proposed structure of the rhombohedral phase and (**c**) is the proposed structure of the tetrahedral phase

culations [56] have found that the atom-to-atom dimer of Fig. 5.2a, though unbound as a neutral molecule, is bound when the molecule has a charge of minus two, thus creating the dimer shown in Fig. 5.20a. QMD simulations of these and other polymers of charged C_{60}, including calculations of vibrational frequencies and theoretical Raman spectra, have recently been completed [57].

Returning to neutral molecules, planes of polymerized C_{60} have been created under conditions of high pressure and high temperature [58, 59]. This discovery has motivated both empirical tight-binding [60] and first-principles [61] simulations of planes of polymerized C_{60} in which the C_{60} balls are joined by four or six parallel-double-bond connections (see Fig. 5.20b and c). We are currently carrying out QMD simulations of these structures, including calculations of vibrational frequencies and theoretical Raman spectra.

5.3 Simulation Two: C_{119}

The discovery of large odd-numbered carbon clusters, specifically C_{119}, C_{129}, and C_{139}, in the mass spectra of commercial fullerene soot, and the subsequent production of these clusters as the products of ozone/fullerene reactions [2], led us to apply our first-principles QMD method [11] in an investigation of their possible structure. In this investigation, we attempted to predict, based on particular assumptions, a structure for C_{119} as a joining of two C_{60}'s with one atom missing. The results of this investigation, first published in [24], along with subsequent work, are summarized here.

5.3.1 Predicting the Structure

McElvany et al. [2] were the first to report detection of these odd-numbered clusters. Their first detection was in a "commercially available, unchromatographed mixture of fullerenes, as well as in toluene extracts of various fullerene soots." Subsequently, they were able to produce these clusters by bubbling ozone through toluene solutions of pure C_{60}, pure C_{70}, and a 1:1 mixture (by weight) of pure C_{60} and pure C_{70}. Significantly, the pure C_{60}/ozone reaction produced only the C_{119} cluster; the pure C_{70}/ozone reaction produced only the C_{139} cluster; and the mixture/ozone reaction produced all three clusters, C_{119}, C_{129}, and C_{139}. Their conclusion was that the ozone exposure of C_{60} or C_{70} "causes an opening of the fullerene cage and a subsequent reaction with another C_{60} or C_{70} to yield a larger carbon cluster". Thus C_{119}, C_{129}, and C_{139} are assumed to correspond to dimers of C_{60}, C_{60}/C_{70}, and C_{70}, less one carbon atom, respectively.

We have made the above assumption plus two others. First, we have assumed that the fullerene attacked by ozone combines with an intact fullerene before healing itself, i.e. that the ozone removes one carbon atom, leaving three dangling bonds which then attach to another fullerene. Second, we

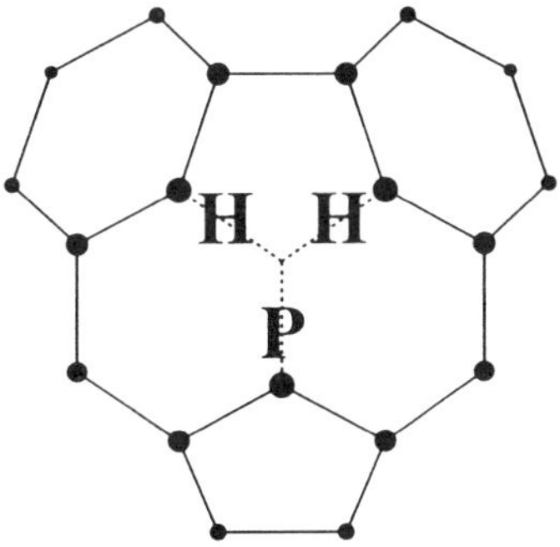

Fig. 5.21. Schematic diagram of a portion of a C_{60} molecule with a single atom removed. The dotted lines represent the bonds which were originally connected to the atom that has been removed. The letters H, H, and P represent the dangling bonds that remain after the removal. Each H dangling bond emanates from an atom which is a part of an intact hexagon. The P dangling bond emanates from an atom which is a part of an intact pentagon

have assumed that there is little energy available for annealing the resulting dimer-like structure. Specifically, in our simulation we have attached a C_{59}, being an I_h C_{60} with one atom removed, to an intact I_h C_{60}, at 0 K. We then subjected this initial structure to a QMD relaxation.

We have tried seven different ways of connecting the C_{59} (an I_h C_{60} with one atom missing) to an intact C_{60}. To dicuss these seven ways, we first note that there are three dangling bonds on the C_{59}, as indicated by the letters H, H, and P in Fig. 5.21. One of these dangling bonds (which was a double bond before the carbon atom was removed) emanates from a carbon atom at a vertex of a pentagon and is denoted by the letter P. Each of the other two dangling bonds (which were single bonds before the carbon atom was removed) emanates from a carbon atom at a vertex of a hexagon and is denoted by the letter H. Fig. 5.22 shows diagrammatically the seven initial configurations we tried for attaching the C_{59} to the intact C_{60}. The lines in this figure represent, for each of the seven configurations, that part of the intact C_{60} to which the dangling bonds P, H, and H attach. More precisely, the letters P, H, and H represent the positions of the atoms of the C_{59}, with respect to the structure of the C_{60}, at the beginning of each of the seven QMD simulations. In each case the C_{59} was placed close enough to the C_{60} so that bonds of 1.6 Å or less were formed between the atoms P, H, and H of the C_{59} and atoms of the C_{60}.

Also in Fig. 5.22, adjacent to each of the seven diagrams, we list an energy. In each case, this is the energy of the final relaxed structure that resulted from the QMD relaxation of the initial configuration described by the corresponding diagram. Each of these energies is with respect to the energy/atom of a relaxed I_h C_{60} molecule. Note that each of these energies is positive, meaning that an input of energy is required to form C_{119}'s from C_{60}'s. The simulation which results in our predicted structure for C_{119}, namely the lowest-energy structure we have found within our assumed constraints, begins with the initial configuration described by the diagram labeled 3 meV.

It is important to note that the final structures are usually quite different from the initial configurations described in Fig. 5.22. We illustrate this point in Fig. 5.23, in which we show four frames from the QMD relaxation which

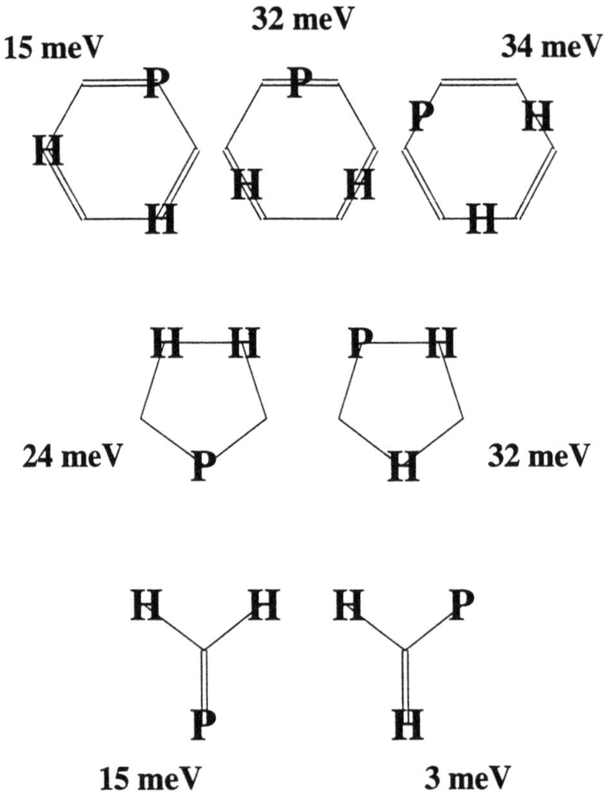

Fig. 5.22. Diagrams illustrating the seven different starting configurations for our simulations of C_{119}. The P's and H's represent the atoms with dangling bonds on a C_{59}, i.e., a C_{60} with one atom removed (see Fig. 5.21). The lines represent that part of an intact I_h C_{60} to which the P and H atoms are attached. The energies listed are energies/atom with respect to I_h C_{60} for the final structures which result from the starting configurations described by these diagrams. See the text for further explanation. After [24]

results in our predicted structure. In frame (a), we show a side view of the initial configuration. This is the initial configuration described in Fig. 5.22 by the diagram labeled 3 meV. The C_{59} is on the bottom, with the intact I_h C_{60} on the top. The dangling bonds of the C_{59} are attached to three atoms of the intact I_h C_{60} which surround a common neighbor atom, which we call the central atom. In frame (b), the central atom has come down into the area between the two molecules. In this frame that atom is momentarily six-fold coordinated. In frame (c), two of the six bonds have broken, one to the top half of the new molecule, and one to the bottom half. Finally, in frame (d) we show the final relaxed form of the molecule. We stress here that the final coordination of the molecule is an output of our first-principles QMD method, a result which demonstrates vividly the utility of first-principles molecular dynamics.

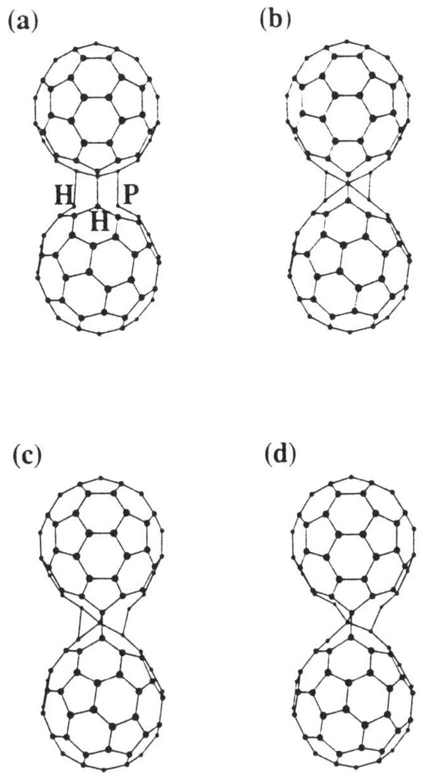

Fig. 5.23. Four frames from the first-principles molecular dynamics simulation which resulted in our predicted structure for C_{119}. Frame (**a**) shows the initial configuration represented by the diagram labeled 3 meV in Fig. 5.22. Frame (**d**) shows the final relaxed structure. See the text for additional details. After [24]

5.3.2 Properties of the Predicted Structure

In Fig. 5.24, we show the relaxed molecule again, this time rotated so as to make apparent the C_2 symmetry of the final structure. We henceforth refer to this molecule as C_2 C_{119}. We note that the symmetry of the initial configuration was C_1. Also evident in Fig. 5.24a are the two seven-membered rings which are formed at the center of the molecule. For I_h C_{60}, our method gives bond lengths of 1.399 and 1.447 Å. In C_2 C_{119}, there are 60 symmetrically distinct atoms (59 on each half plus the unique central atom) and 180 bonds which occur in 90 symmetrically distinct pairs. In our calculated structure, the shortest bond length is 1.386 Å and the longest is 1.568 Å. A bond length distribution for C_2 C_{119}, with the bond lengths separated into bins of width 0.001 Å, is presented in Fig. 5.25. The ten longest bonds are the ten bonds (five are symmetrically distinct) on the three four-coordinated atoms. These ten bonds, which might be called the "intermolecular bonds", are all found to be longer than 1.5 Å. Since the two halves of C_2 C_{119} are quite similar

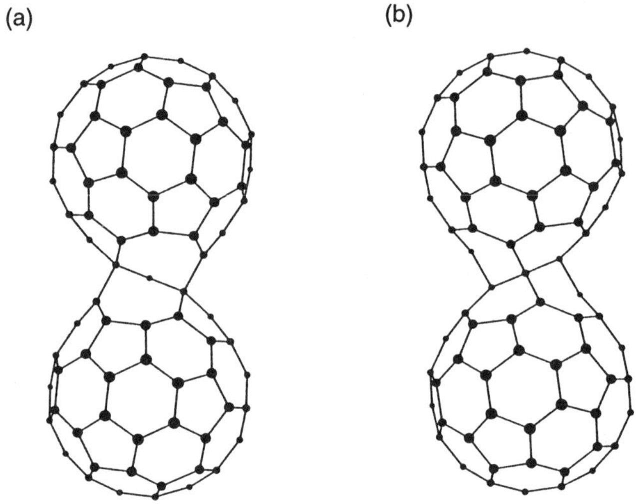

Fig. 5.24. Our predicted structure for C_{119}, rotated so as to show the C_2 symmetry, with the C_2 axis normal to the page. This is the same structure shown in Fig. 23(d). In (**a**) the two seven-membered rings can be seen. The configuration in (**b**) is the same as that in (**a**) except for a rotation by 180° about the long molecular axis. After [24]

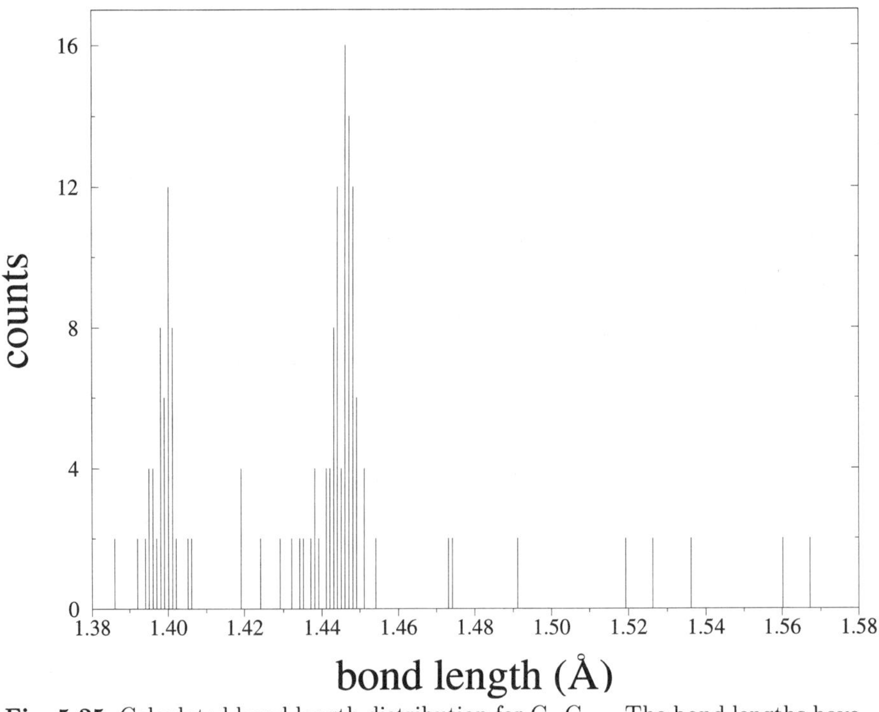

Fig. 5.25. Calculated bond length distribution for C_2 C_{119}. The bond lengths have been sorted into bins of width 0.001 Å. The shortest bond length is 1.386 Å, and the longest is 1.568 Å

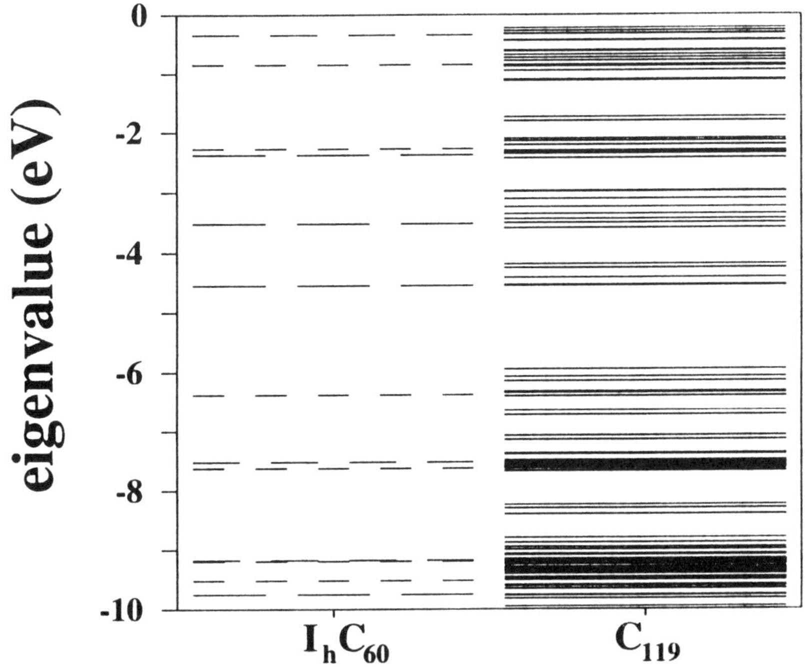

Fig. 5.26. Calculated electronic eigenvalues in the near-gap region for C_2 C_{119}, compared with those for I_h C_{60}. Degeneracies, where they exist, are shown by multiple horizontal lines at those levels. The highest occupied molecular orbital (HOMO) is just below −6.0 eV for I_h C_{60} and just above −6.0 eV for C_2 C_{119}. After [24]

to I_h C_{60}, it is not surprising that for C_2 C_{119} a distribution of calculated bond lengths is bunched around the bond lengths of I_h C_{60} (21 pairs within 0.005 Å of 1.399 Å, and 29 pairs within 0.005 Å of 1.447 Å). In fact, in order to resolve all 90 symmetrically distinct bonds, we must sort the calculated bond lengths into bins of width less than 2×10^{-5} Å.

At the time of writing, we have seen only one other proposed model for the structure of C_{119} [62]. This is a model proposed on the basis of general chemical principles and which has not yet been theoretically tested for stability. In this model, there is a three-membered ring where the C_{60} and the C_{59} are joined, and the C_{59} has a nine-membered ring and a four-membered ring; therefore, it is unlikely to be energetically competitive with ours. However, it is certain that the actual structure of C_{119} will be determined by growth kinetics, as well as energetic considerations. In order to facilitate comparison with eventual experiment, we present below the calculated electronic and vibrational properties for our predicted structure.

In Fig. 5.26 we present the calculated electronic eigenvalues of C_2 C_{119}, in the near-gap region, in comparison with those of I_h C_{60}. The calculated

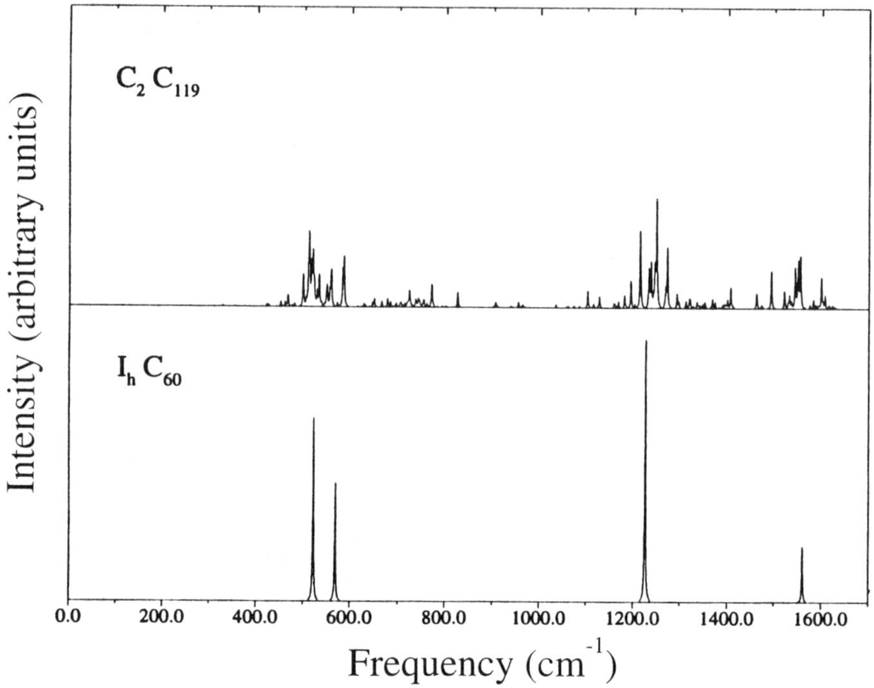

Fig. 5.27. The first-principles calculated IR strengths for the vibrational modes of I_h C_{60} and of C_2 C_{119}. The IR peaks have been broadened into Lorentzians of width one wavenumber (FWHM)

HOMO-LUMO gaps are: for I_h C_{60}, 1.83 eV; and for C_2 C_{119}, 1.40 eV. In Figs. 5.27 and 5.28 we present the calculated vibrational spectra. Because of the low symmetry, all 351 nonzero normal vibrational modes are nondegenerate. Furthermore, for C_2 symmetry, all of the normal modes are both IR- and Raman-active. In Fig. 5.27 we present the results of our first-principles calculation of the IR activity [42], and in Fig. 5.28 we present our calculated unpolarized Raman spectrum. In each case the calculated vibrational spectrum for C_2 C_{119} is compared with a similar calculation for I_h C_{60}. As was the case for polymerized C_{60}, we have calculated the Raman spectra using our first-principles force constants, together with the bond-charge polarizability model and I_h C_{60} polarizability parameters fit by Guha et al. [44] to the spectra of [41].

5.3.3 Summary for C_{119}

Our predicted structure for C_{119}, shown in Figs. 5.23d and 5.24, is based upon specific assumptions about the conditions under which C_{119} is made, and is the result of first-principles molecular dynamics simulations of seven

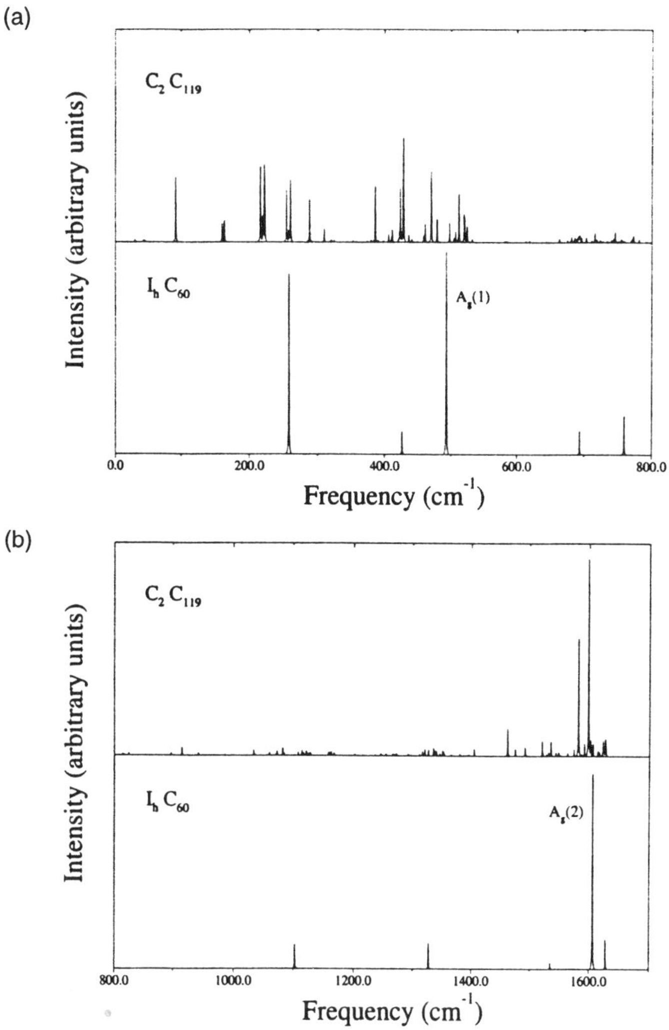

Fig. 5.28. Calculated unpolarized Raman spectrum for C_2 C_{119} compared with a similar spectrum for I_h C_{60}. The Raman strengths were computed from a bond-charge model as discussed in the text, and the peaks have been broadened into Lorentzians of width one wavenumber (FWHM). (**a**) Range 0–800 cm^{-1}. (**b**) Range 800–1700 cm^{-1}. The vertical scales in the two panels are uncorrelated; however, the vertical range for each panel of (**b**) is identical to the vertical range for the corresponding panel of (**a**)

different trial configurations, the seven which seemed to us most likely to give the lowest-energy structure under the assumed conditions. The predicted structure is quite different from the trial configuration from which it was produced, illustrating powerfully the utility of first-principles quantum molecular dynamics in the simulation of novel materials. This predicted C_{119} structure may be thought of as a joining of two C_{59} balls, with one carbon atom "in between." It has C_2 symmetry, with three four-coordinated atoms and two seven-membered rings. The energy/atom of this C_2 C_{119} molecule is found to be only 0.003 eV/atom higher than the energy of I_h C_{60}.

A sample of C_{119} has now been produced by a two-step high-performance-liquid-chromatography (HPLC) separation of a mixed sample created by thermal decompostion of $C_{120}O$ [63]. NMR spectroscopy of this sample indicates that C_{119} has twofold symmetry with two sets of 58 three-coordinated atoms and three four-coordinated atoms, two of which are alike, all in agreement with our predicted model. The authors of [63] have, in addition, used the semiempirical MNDO method (AM1 parameterization) to calculate the energies of several other C_2 C_{119} isomers having three four-coordinated atoms, and find our suggested model to be the lowest in energy. Raman spectroscopy of the same sample is now underway [64].

Recently a comprehensive experimental/theoretical Raman study of C_{119} has confirmed our suggested model [65].

Acknowledgement. We wish to thank O.F. Sankey for many helpful discussions. In addition, we wish to thank G. Lopinski and D. Porezag for providing preprints, figures, and other useful information. JBP acknowledges support from NSF grants DMR 9014729, DMR 9510182, and DMR 9624102.

References

1. A.M. Rao, P. Zhou, K.-A. Wang, G.T. Hager, J.M. Holden, Y. Wang, W.-T. Lee, X.-X. Bi, P.C. Eklund, D.S. Cornett, M.A. Duncan, I.J. Amster: Science **259**, 955 (1993).
2. S.W. McElvany, J.H. Callahan, M.M. Ross, L.D. Lamb, D.R. Huffman: Science **260**, 1632 (1993).
3. O.F. Sankey, R.E. Allen: Phys. Rev. B **33**, 7164 (1986).
4. W.A. Harrison: *Electronic Structure and the Properties of Solids* (Freeman, San Francisco, 1980); R. Enderlein, W.A. Harrison: Phys. Rev. B **30**, 1867 (1984); E.A. Kraut, W.A. Harrison: J. Vac. Sci. Techol. B **2**, 409 (1984).
5. M. Menon, K.R. Subbaswamy: Phys. Rev. Lett. **67**, 3487 (1991); M. Menon, K.R. Subbaswamy, Phys. Rev. B **47**, 12754 (1993).
6. C.H. Xu, C.Z. Wang, C.T. Chan, K.M. Ho: J. Phys. Condens. Matter **4**, 6047 (1992); C.Z. Wang, C.H. Xu, C.T. Chan, K.M. Ho: J. Phys. Chem. **96**, 3563 (1992).
7. D.L. Strout, R.L. Murry, C. Xu, W.C. Eckhoff, G.K. Odom, G.E. Scuseria: Chem. Phys. Lett. **214**, 576 (1993).

8. M. Menon, K. R. Subbaswamy, M. Sawtarie: Phys. Rev. B **49**, 13966 (1994).
9. R. Car, M. Parrinello: Phys. Rev. Lett. **55**, 2471 (1985).
10. M. Needles, M. C. Payne, J. D. Joannopolous: Phys. Rev. Lett. **58**, 1765 (1987); D. C. Allen, M. P. Teter: Phys. Rev. Lett. **59**, 1136 (1987); G. Galli, R. M. Martin, R. Car, M. Parrinello: Phys. Rev. Lett. **62**, 555 (1989); D. Hohl, R. O. Jones, R. Car, M. Parrinello: Chem. Phys. Lett. **139**, 540 (1987).
11. O. F. Sankey, D. J. Niklewski: Phys. Rev. B **40**, 3979 (1989).
12. G. B. Adams, M. O'Keeffe, A. A. Demkov, O. F. Sankey, Y.-M. Huang: Phys. Rev. B **49**, 8048 (1994).
13. D. A. Drabold, P. A. Fedders, O. F. Sankey, J. D. Dow: Phys. Rev. B **42**, 5135 (1990).
14. D. A. Drabold, J. D. Dow, P. A. Fedders, A. E. Carlsson, O. F. Sankey: Phys. Rev. B **42**, (1990).
15. O. F. Sankey, D. J. Niklewski, D. A. Drabold, J. D. Dow: Phys. Rev. B **41**, 12750 (1990).
16. O. F. Sankey, G. B. Adams, X. Weng, J. D. Dow, Y.-M. Huang, J. C. H. Spence, D. A. Drabold, W.-M. Hu, R. P. Wang, S. Klemm, P. A. Fedders: Superlattices Microstruct. **10**, 407 (1991).
17. G. B. Adams, O. F. Sankey: Phys. Rev. Lett. **67**, 867 (1991).
18. G. B. Adams, O. F. Sankey: J. Vac. Sci. Techol. A **10**, 2046 (1992).
19. Y.-M. Huang, J. C. H. Spence, O. F. Sankey, G. B. Adams: Surface Science, **256**, 344 (1991).
20. G. B. Adams, O. F. Sankey, J. B. Page, M. O'Keeffe, D. A. Drabold: Science **256**, 1792 (1992).
21. M. O'Keeffe, G. B. Adams, O. F. Sankey: Phys. Rev. Lett. **68**, 2325 (1992).
22. G. B. Adams, O. F. Sankey, J. B. Page, M. O'Keeffe: Chem. Phys. **176**, 61 (1993).
23. G. B. Adams, J. B. Page, O. F. Sankey, M. O'Keeffe: Phys. Rev. B **50**, 17471 (1994).
24. G. B. Adams, J. B. Page, M. O'Keeffe, O. F. Sankey: Chem. Phys. Lett. **228**, 485 (1994).
25. A. A. Demkov, O. F. Sankey: Phys. Rev. B **48**, 2207 (1993).
26. A. A. Demkov, J. Ortega, O. F. Sankey, M. Grumbach: Phys. Rev. B **52**, 1618 (1995).
27. G. B. Adams, J. B. Page, O. F. Sankey, K. Sinha, J. Menéndez, D. R. Huffman: Phys. Rev. B (RC) **44**, 4052 (1991).
28. J. Harris: Phys. Rev. B **31**, 1770 (1985).
29. D. Porezag, Th. Frauenheim, Th. Köhler, G. Seifert, R. Kaschner: Phys. Rev. B **51**, 12947 (1995).
30. H. Eschrig, I. Bergert: Phys. Status. Solidi B **90**, 621 (1978); H. Eschrig, *Optimized LCAO Method and the Electronic Structure of Extended Systems*, (Akademie-Verlag, Berlin, 1988).
31. D. Porezag, M. Pederson, T. Frauenheim, T. Köhler: Phys. Rev. B **52**, 14963 (1995).
32. N. Matsuzawa, M. Ata, D. A. Dixon, G. Fitzgerald: J. Phys. Chem. **98**, 2555 (1994).
33. W. Krätschmer, L. D. Lamb, K. Fostiropoulos, D. R. Huffman: Nature **374**, 354 (1990); L. D. Lamb, D. R. Huffman, R. K. Workman, S. Howells, T. Chen, D. Sarid, R. F. Ziolo: Science **255**, 1413 (1992).
34. Small differences between numbers appearing in this chapter and numbers published in our previous works [23, 24] are due to two factors. First, we have made our criterion for full relaxation more stringent by an order of magnitude. That criterion is now that there be no energy change greater than 10^{-6} eV/atom over

20 time steps (about 20 fs). Secondly, we have corrected a minor error in the programing of that part of the Hellmann-Feynman forces due to the exchange-correlation term in the Hamiltonian. The net effect of these two factors is to change energies by less than 0.0002 eV/atom, and to change vibrational frequencies, for frequencies of more than 200 cm^{-1}, by one percent or less.
35. A. D. Becke: Phys. Rev. A **38**, 3098 (1988).
36. C. Lee, W. Yang, R. G. Parr: Phys. Rev. B **37**, 785 (1988); B. Mielich, A. Savin, H. Stoll, H. Preuss: Chem. Phys. Lett. **157**, 200 (1989).
37. U. von Barth, L. Hedin: J. Phys. C **5**, 1629 (1972).
38. H. J. Monkhorst, J. D. Pack: Phys. Rev. B **13**, 5188 (1976).
39. O. Chauvet, G. Oszlanyi, L. Forró, P. W. Stephens, M. Tegze, G. Faigel, A. Jánossy: Phys. Rev. Lett. **72**, 2721 (1994).
40. J. R. Fox, G. P. Lopinski, J. S. Lannin, G. B. Adams, J. B. Page, J. E. Fischer: Chem. Phys. Lett. **249**, 195 (1996).
41. B. Chase, N. Herron, E. Holler: J. Phys. Chem. **96**, 4262 (1992).
42. G. B. Adams, J. B. Page, O. F. Sankey: unpublished.
43. J. A. Salthouse, M. J. Ware: *Point Group Character Tables and Related Data* (Cambridge University Press, Cambridge, 1972).
44. S. Guha, J. Menéndez, J. B. Page, G. B. Adams: Phys. Rev. B **53**, 13106 (1996); S. Guha, J. Menéndez, J. B. Page, G. B. Adams, Phys. Rev. B **56**, 15431 (1997).
45. D. W. Snoke, M. Cardona: Solid State Commun. **87**, 121 (1993).
46. G. P. Lopinski, J. R. Fox, J. S. Lannin: Chem. Phys. Lett. **239**, 107 (1995).
47. Y. Wang, J. M. Holden, Z.-H. Dong, X.-X. Bi, P. C. Eklund: Chem. Phys. Lett. **211**, 341 (1993).
48. M. R. Pederson, K. A. Jackson: Phys. Rev. B **41**, 7453 (1990); M. R. Pederson, K. A. Jackson in *Density Functional Methods in Chemistry*, edited by J. K. Labanowski, J. W. Andzelm (Springer-Verlag, New York, 1991); K. A. Jackson, M. R. Pederson: Phys. Rev. B **42**, 3276 (1990); M. R. Pederson, K. A. Jackson: Phys. Rev. B **43**, 7312 (1991).
49. Y. Wang, J. M. Holden, A. M. Rao, P. C. Eklund, U. D. Venkateswaran, D. Eastwood, R. L. Lidberg, G. Dresselhaus, M. S. Dresselhaus: Phys. Rev. B **51**, 4547 (1995).
50. Y. Wang, J. M. Holden, X.-X. Bi, P. C. Eklund: Chem. Phys. Lett. **217**, 3 (1994).
51. B. Burger, J. Winter, H. Kuzmany: Z. Phys. **101**, 227 (1996).
52. P. W. Stephens, G. Bortel, G. Faigel, M. Tegze, A. Jánossy, S. Pekker, G. Oszlányi, L. Forró: Nature **370**, 636 (1994).
53. F. Bommeli, L. Degiorgi, P. Wachter, Ö. Legeza, A. Jánossy, G. Oszlányi, O. Chauvet, L. Forró: Phys. Rev. B **51**, 14794 (1995).
54. G. Oszlányi, G. Bortel, G. Faigel, L. Gránásy, G. M. Bendele, P. W. Stephens, L. Forró: Phys. Rev. B **54**, 11849 (1996).
55. J.-L. Sauvajol, E. Anglaret, R. Aznar, D. Bormann, B. Hennion: Solid State Commun. **104**, 387 (1997).
56. J. Kürti, K. Németh: Chem. Phys. Lett. **256**, 119 (1996).
57. G. B. Adams, J. B. Page: in preparation.
58. O. Béthoux, M. Núñez-Regueiro, L. Marques, J.-L. Hodeau, M. Perroux in *Proceedings of the Materials Research Society, Boston, 1993, Abstracts of Contributed Papers* (MRS, Pittsburgh, 1993) Abstract No. G2.9, p. 202.; M. Núñez-Regueiro, L. Marques, J.-L. Hodeau, O. Béthoux, M. Perroux: Phys. Rev. Lett. **74**, 278 (1995).
59. Y. Iwasa, T. Arima, R. M. Fleming, T. Siegrist, O. Zhou, R. C. Haddon, L. J. Rothberg, K. B. Lyons, H. L. Carter Jr., A. F. Hebard, R. Tycko, G. Dabbagh, J. J. Krajewski, G. A. Thomas, T. Yagi: Science **264**, 1570 (1994).

60. C.H. Xu, G.E. Scuseria: Phys. Rev. Lett. **74**, 274 (1995).
61. S. Okada, S. Saito: Phys. Rev. B **55**, 4039 (1997).
62. R. Taylor: J. Chem. Soc., Chem. Commun. 1629 (1994).
63. A. Gromov, S. Ballenweg, S. Giesa, S. Lebedkin, W.E. Hull, W. Krätschmer: Chem. Phys. Lett. **267**, 460 (1997).
64. S. Lebedkin: private communication.
65. S. Lebedkin, H. Rietschel, G.B. Adams, J.B. Page, W.E. Hull, F.H. Hennrich, H.-J. Eisler, M.M. Kappes, W. Krätschmer: J. Chem. Phys. **110**, 11768 (1999).

6. Covalent Bonding Between Fullerenes

M. Menon, K.R. Subbaswamy

6.1 Introduction

The recent discovery of phototransformation of solid C_{60} [1] into a covalently bonded solid has generated considerable interest. When exposed to visible or ultraviolet light, C_{60} molecules arranged in an fcc lattice by a weak van der Waals type inter-molecular bonding appear to undergo a phase transformation resulting in a photopolymerized state [1]. The well known photochemical 2+2 cycloaddition reaction has been proposed as the mechanism for this transformation [1, 2]. In this reaction, parallel double bonds on adjacent molecules (or monomers) separated by less than 4.2 Å [2] are proposed to break and reform into a four-membered ring cross-linking the molecules. This model for the covalent bonding between C_{60} molecules was strongly supported by the theoretical investigations of Menon et al. [3] who studied various possible covalent C_{60} dimer bonding configurations to determine the most stable bonding geometry using the generalized tight-binding molecular dynamics (GTBMD) technique. Subsequently other theoretical [4–8] and experimental [9–14] studies have supported this conclusion.

This leads to the next natural question of whether other carbon molecules with less than the perfect icosahedral symmetry of C_{60} (I_h) can also cross-link via covalent bond formation. The theoretical study of covalent bonding between C_{70} molecules indicated the bonding to be highly directional, strongly favoring certain relative intermolecular orientations [15, 16] The most stable bonding occurring for the relaxed structure which involved the formation of a four-membered ring between the polar caps on adjacent monomers. Since the number of reactive bonds per C_{70} is smaller than that for C_{60}, the theoretical results lead to the prediction of smaller cross-section for C_{70} dimer formation. This was confirmed by the experimental work in which the phototransformation of solid C_{70} film was found to have a much lower cross-section than that observed for C_{60} [15, 16].

In this review we present a summary of our results for the investigations of covalent bonding between C_{60} and C_{70} dimers and present investigations of other lower symmetry fullerene dimers.

6.2 Theoretical Methods

The details of the technique can be found elsewhere [17–19]. The method is a generalization of the conventional tight-binding molecular dynamics formulation so as to apply to nontetrahedral, and multicoordinated covalent systems. The explicit inclusion of nonorthogonality of the orbitals in the scheme allows us to treat interactions between atoms without any artificial cutoff. The Hellmann–Feynman theorem is used to perform molecular dynamics optimization with no symmetry restrictions. The present scheme is an improvement over the conventional orthogonal tight-binding schemes where the overlap interactions are neglected. This incorporation of nonorthogonality has yielded much improved agreement with available *ab initio* results for small silicon [17,19] and carbon [18] clusters when compared with the conventional orthogonal schemes. The most noteworthy improvement is in the area of vibrational frequencies where the orthogonal schemes have been found to be inadequate. The present scheme has also been found to be very reliable in obtaining good agreement with experimental and local density approximation (LDA) results for the structural and vibrational properties of fullerenes and nanotubes [20].

The method is orders of magnitude faster than *ab initio* or LDA techniques and, therefore, permits a much larger sampling of the configuration space to determine minimum energy structures. The tight-binding method can also be used to obtain valuable electronic structure information. The electronic tight-binding parameters (e.g. $V_{ss\sigma}, V_{pp\pi}$, etc.) used here for carbon are generated using Harrison's [21] universal parameter scheme. This allows us to keep the number of adjustable parameters to a minimum. These parameters can be fitted simply either to a dimer or the crystalline solid for experimental bond length and frequencies, although fitting at either end does not substantially alter results at the other end owing to the transferability. In our case we fixed the parameters to minimize the errors at both ends.

6.3 Covalent Bonding Between Fullerenes

6.3.1 C_{60} Dimer

Motivated by the work of Rao et al. [1] we proceeded to investigate the various possible covalent C_{60} dimer bonding configurations to determine the most stable bonding geometry relevant to polymer formation using our MD scheme.

The C_{60} molecule has perfect icosahedral (I_h) symmetry with two distinct bond lengths between carbon atoms. In order to perform a detailed computational analysis of covalent bonding between C_{60} dimers, we considered the following possible relative orientations (see Fig. 6.1) as the initial configurations in our molecular dynamics relaxations: C_{60} units facing each other with

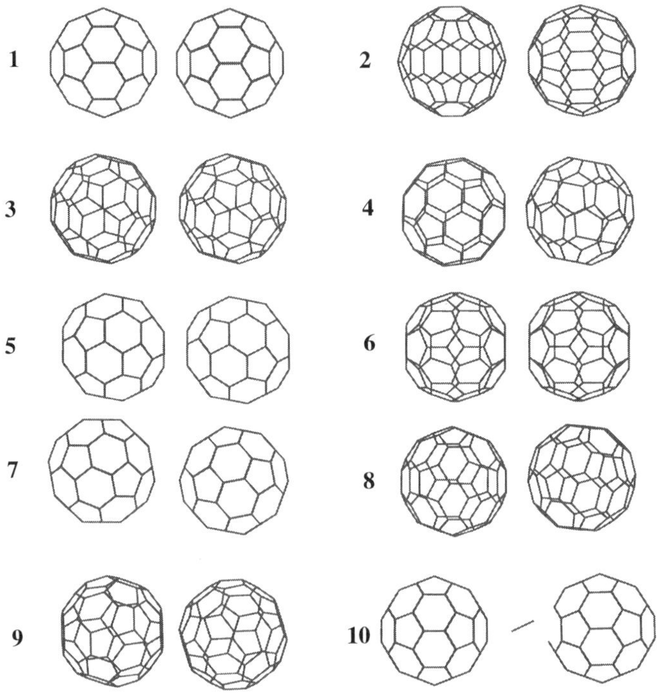

Fig. 6.1. Initial geometries for various starting configuration

the hexagonal edges closest to each other in parallel (**1**) or perpendicular (**2**) orientations of the double bonds; the pentagonal edges parallel to each other (**3**); the pentagonal faces of the dimers facing each other in an eclipsed (**4**) or staggered (**5**) geometry; hexagonal faces parallel (**6**); the corners of the units as the closest approach (**7**); the corner of one C_{60} near a bridge site over the double bond of the other molecule (**8**); the corner of one C_{60} over the center of a hexagonal face of the other (**9**), and the C_{58}-C_2-C_{60} geometry that was discussed as a possible alternative linkage in [1] and later rejected by the same authors (**10**). In all our simulations the initial configurations were chosen in such a way that the C_{60} dimers were within covalent bonding range of each other (≈2 Å). Damped molecular dynamics relaxation was then performed by removing 1% of each velocity at each time step. No symmetry restriction is imposed while the optimization is being carried out. The process is stopped when the largest velocity component of any atom falls below 54 m/s (corresponding to a temperature of about 10 K).

In Figs. 6.2 and 6.3 we show the initial and final configurations from our molecular dynamics relaxation for two different starting geometries (**7** and **8**, respectively). In both cases, the final geometry is the same, which is, in fact, the most stable relevant relaxed geometry obtained. The initial configurations were taken so as not to bias them in favor of symmetric bonding. The final

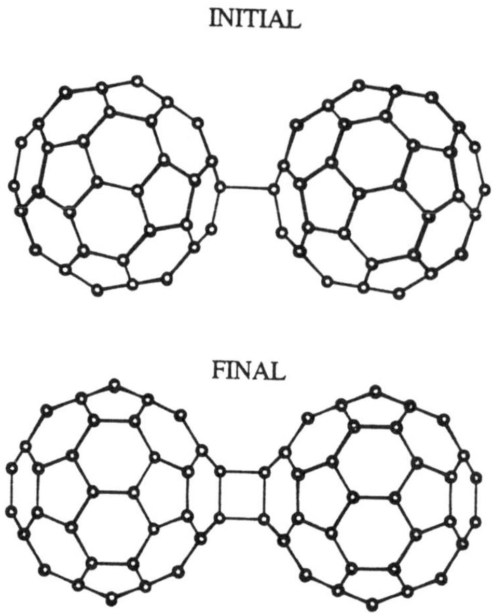

Fig. 6.2. Initial and final geometries for starting configuration **7** in Sect. 6.3. The final structure shown is the most stable found in our optimization

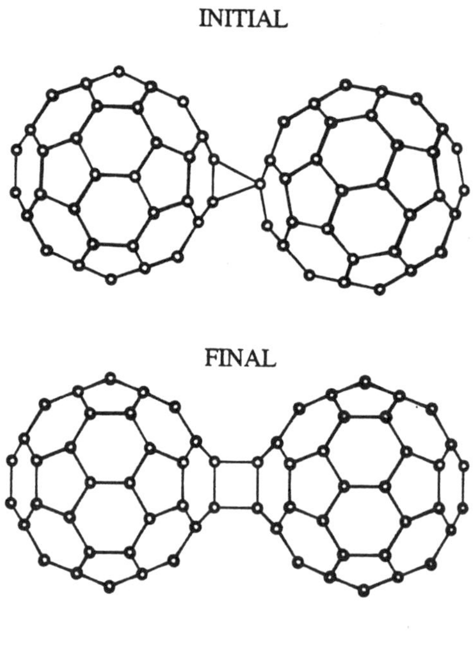

Fig. 6.3. Initial and final geometries for starting configuration **8**. in Sect. 6.3. (Same as Fig. 6.1, but for a different starting geometry.)

geometry shows a "bucky dumb-bell" structure, with almost all the deviations from the perfect icosahedral symmetry confined to the nearest and next-nearest carbon atoms surrounding the four atoms participating in the inter-ball bonding. The dimer exhibits D_{2h} symmetry. The four atoms involved in the inter-ball linkage form the vertices of a square of side 1.51 Å. In Table 6.1 the total energies are listed for all the relaxed structures considered here.

Table 6.1. Relative energies for various C_{60} dimer configurations with respect to the "bucky dumb-bell"

Structure	Relative Total Energies ΔE (eV)
1	0.0
2	no bonding
3	0.96
4	5.76
5	2.16
6	2.16
7	unstable
8	unstable
9	5.76
10	4.80
Two isolated C_{60}	-0.40

The starting geometry **2** gave no bonding and, in fact, led to a repulsion between the balls. The geometry **3** which is similar to **1** was found to be about 1 eV higher in energy than the relaxed structure shown in Fig. 6.2. Several starting geometries (e.g. **4**, **5**, **6**, and **9**) led to relaxed structures with substantially higher energies (See Table 6.1). The relaxed C_{58}-C_2-C_{60} geometry (**10**), which involves the breaking of the cage framework, is found to be significantly higher in energy than the bucky dumb-bell (by 4.8 eV). Subsequent experimental analysis by the authors of [1] has ruled out this structure. Considering the extensive sampling of initial orientations, it is reasonable to conclude that the 4-fold cyclic linkage shown in Figs. 6.2 and 6.3 is the minimum energy configuration relevant to polymerization.

In Fig. 6.4 we show the energy levels in the vicinity of the Fermi level for an isolated C_{60} and the bucky dumb-bell (our lowest energy configuration for the dimer). An isolated C_{60} has a five-fold degenerate highest occupied molecular orbital (HOMO) and a three-fold degenerate lowest unoccupied molecular orbital (LUMO) with a gap of 1.10 eV. As seen in this figure, the formation of the dimer has removed the degeneracy of energy levels near the Fermi level as a consequence of the lowered symmetry. The narrowing of the optical gap is consistent with the observed darkening of the photoinduced film [1].

We have also determined the vibrational mode frequencies for the bucky dumb-bell by explicitly constructing the dynamical matrix. As expected, most

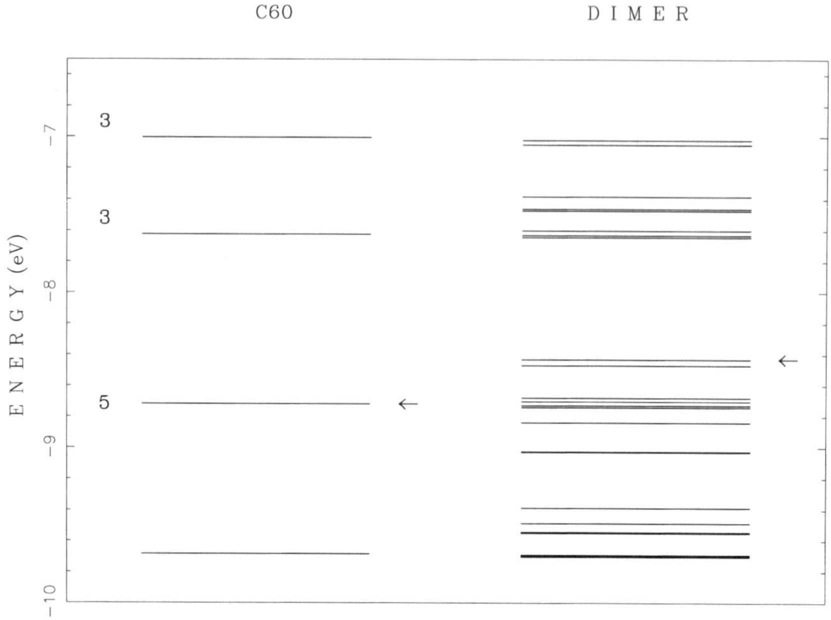

Fig. 6.4. Energy levels in the vicinity of the Fermi level for an isolated C_{60} and the bucky dumb-bell (our lowest energy configuration for the dimer)

of the modes are easily recognized as monomer-derived, with all degeneracies removed. There are six additional modes associated with intermolecular vibrations. In Fig. 6.5 we show the atomic displacement patterns for a Raman active mode at $101\,\mathrm{cm}^{-1}$ (a symmetric distortion of the dumb-bell). Experimentally, a new mode at $116\,\mathrm{cm}^{-1}$ has been observed in the Raman spectrum of the photopolymerized C_{60} film, [1] which is in good agreement with the calculated frequency of the distortion mode displayed in Fig. 6.5.

In Fig. 6.6 we show the low frequency vibrational spectra for the bucky dumb-bell (top) and the isolated C_{60} (bottom). The lowering of the perfect icosahedral (I_h) symmetry for the dimer gives rise to a removal of degeneracies of vibrational modes present in the isolated C_{60} resulting in many additional

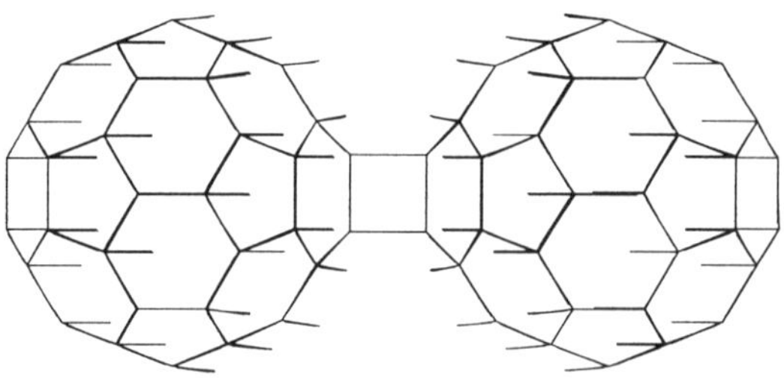

Fig. 6.5. Vibrational displacement patterns for the bucky dumb-bell for a symmetric distortion mode at $101\,\mathrm{cm}^{-1}$

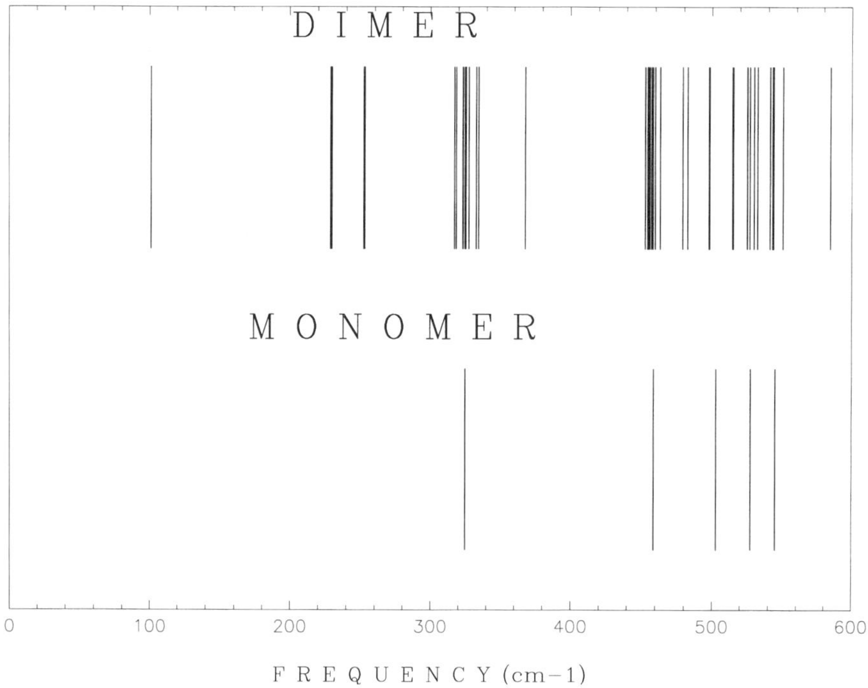

Fig. 6.6. Low frequency vibrational spectra for the bucky dumb-bell (top)and the isolated C_{60} (bottom). The inter molecular stretch and librational modes dominate the lower end of the spectrum for the dimer

modes shown in the figure. The intermolecular stretch and librational modes dominate the lower end of the spectrum for the dimer.

We find the dimer configuration to be slightly less stable than two isolated C_{60} monomers ($\approx$0.4 eV). We have estimated an upper bound to the dissociation barrier for the dimers, starting with the minimum geometry shown in Fig. 6.2, by calculating the total energy when the inter-ball separation is increased while keeping all other bonds and angles frozen. From this and other supplementary calculations we estimate the barrier to be in the range of 2–4 eV.

6.3.2 C_{70} Dimer

The isolated C_{70} molecule has D_{5h} symmetry where a cylindrical belt of 10 atoms has been added to C_{60} to yield a prolate spheroidal structure (Fig. 6.7).

It is reasonable to assume that covalent intermolecular bonding between large carbon clusters can proceed most easily via the rearrangement of adjacent reactive, intramolecular double bonds. The perfect icosahedral symmetry (I_h) in C_{60} allows for the presence of 30 reactive double bonds in the

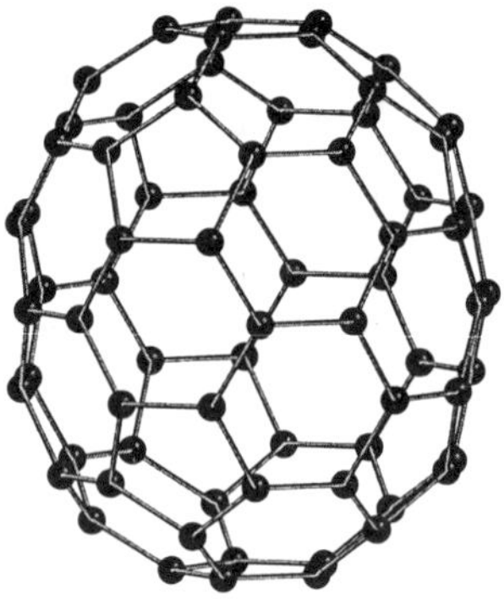

Fig. 6.7. C_{70} molecule

molecule, distributed isotropically. The lower symmetry of C_{70} (D_{5h}), on the other hand, leads to four discrete shorter bond lengths in C_{70} [22] and is expected to exhibit interesting consequences in the bonding mechanism, making it curvature dependent [24]. Balch et al. [23] found that [$Ir(CO)Cl(PPh_3)_2$] complex prefers to bond at one of the five C=C double bonds in a polar cap of the C_{70} molecule, rather than to add across the bonds in the hexagons of the equatorial belt, suggesting that the polar caps are "C_{60}-like" and the belt bonds are sufficiently different, and relatively unreactive [24].

It is, therefore, clear that a detailed understanding of the intermolecular bonding mechanism between fullerene molecules requires that all inequivalent configurations must be considered. With this in mind, we carried out a detailed theoretical analysis of covalent bonding between C_{70} monomers considering several important initial relative orientations between monomers. In all our simulations the four initial configurations were chosen in such a way that the C=C bonds on adjacent C_{70} monomers were well within covalent bonding range of each other ($\approx$2 Å). Damped molecular dynamics relaxation was then performed with no symmetry constrained to determine stable bonding configurations.

The four different starting configurations considered in our molecular dynamics simulations are: (**1**) equatorial bonds parallel, (**2**) polar pentagons parallel, (**3**) one equatorial and one polar pentagonal bond parallel, and finally, (**4**) polar double bonds parallel. All these configurations with the exception of **2** satisfy the topochemical requirements [25] for 2+2 cycloaddition.

Configuration **1** is found to be the least bound on relaxation, with an intermolecular bond length of 1.90 Å. This can be viewed as due to the absence of electron rich C=C double bonds around the equatorial belt. All the deviations in the bond lengths from a pristine C_{70} molecule were found to be confined to nearest, and next-nearest neighbor carbon atoms surrounding the four atoms participating in the intermolecular bonding. The final geometry for configuration **1** exhibits D_{2h} symmetry. The symmetries, relative energies and bond lengths of the cross-linking bonds for this and other stable dimers

Table 6.2. Symmetries, relative energies and bond lengths of the cross-linking bonds obtained from the GTBMD calculations for the four different starting configurations for C_{70} dimers in Sect. 6.3.2

Geometry	Symmetry	Relative Energy (eV)	Length of Cross-linking bond (Å)
I	D_{2h}	9.70	1.90
II	C_1	5.00	1.57
III	D_{5h}	9.66	1.57
IV	C_{2h}	0	1.51

found in this work are presented in Table 6.2. The final, relaxed configuration of the most stable dimer (**4**) is shown in Fig. 6.8. The relaxed configuration **2** is found to be more stable (Table 6.2) with an intermolecular bond length of 1.57 Å. Once again, all the deviations from pristine C_{70} are found to be confined to the hemispheres facing each other. The relaxed configuration **2** has C_1 symmetry. The relaxed configuration **3** is found to be relatively higher in energy, showing maximum strain on the bonds involved in the intermolecular linking. The optimized geometry has D_{5h} symmetry. The relaxed configuration **4**, which involves the formation of a four-membered ring between the polar caps on adjacent monomers, is found to be the most stable. This configuration **4** is geometrically possible in either fcc or hcp solid C_{70}. The square, four membered ring, and the intermolecular C-C bond length of 1.51 Å, are both found identical to that obtained earlier for the lowest energy configuration for the C_{60} dimer [3]. Here again, deviations from pristine C_{70} are found to be confined to the nearest and next-nearest neighbor carbon atoms which surround the atoms participating in the intermolecular bonding.

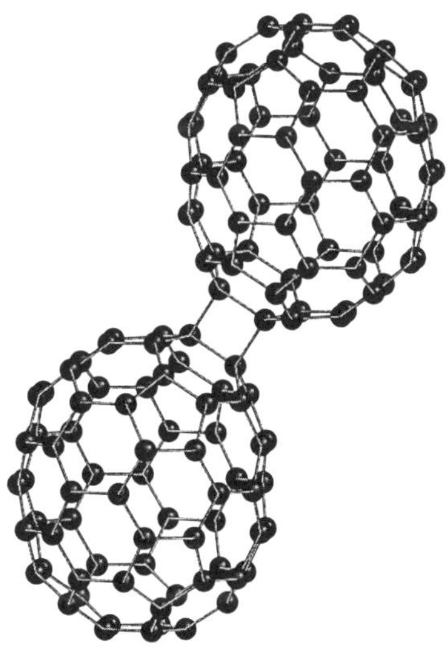

Fig. 6.8. Most stable configuration (**4**) for C_{70} dimer

6.3.3 Curvature Dependence of Covalent Bonding

In view of the substantially lower cross-section observed for the photopolymerization of solid C_{70} film when compared with the solid C_{60} film, it is reasonable to expect the bonding energy of the fullerene dimers to be curvature dependent. While the dimer formation for both C_{60} and C_{70} are found to be endothermic reactions (i.e., dimers are less stable than the separated monomers), we find that exothermic reactions accompany the formation of smaller fullerene dimers. In Fig. 6.9 we show the most stable dimer bonding configurations for C_{20}, C_{28}, C_{36} and C_{50} relevant to polymerization. Individually, these monomers relax to considerably lower symmetry geometries when subjected to symmetry unrestricted molecular dynamics optimization. The symmetries are further lowered on dimer formation.

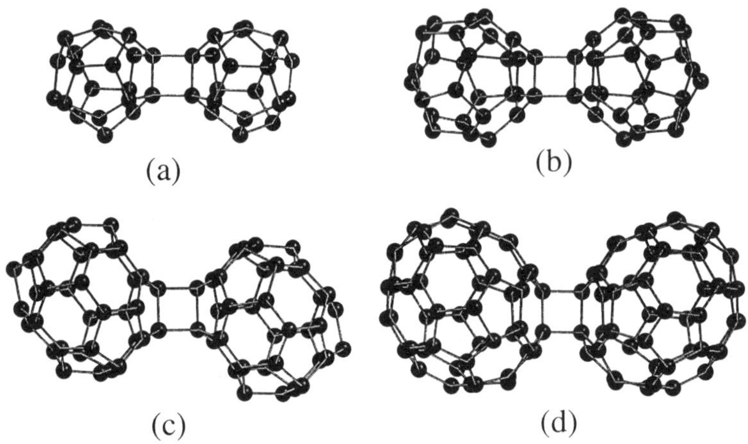

Fig. 6.9. Dimer bonding configurations for (**a**) C_{20}, (**b**) C_{28}, (**c**) C_{36} and (**d**) C_{50}

In Fig. 6.10 we plot the difference in the binding energy between the dimer configuration and the two isolated monomers for all the cages studied in this work. We see that C_{60} is the least endothermic of 6 fullerenes whose dimer formation is endothermic.

6.3.4 Summary

In conclusion, we have investigated the covalent bonding configurations of fullerene dimers using the generalized tight-binding molecular dynamics scheme. We find the stable dimer configurations to consist of monomer units linked via four-sided rings consisting of single bonds giving rise to so called "dumb-bell" structures. These single bonds have their origin in parallel double bonds of the original monomers. Consideration of the energetics of the dimer formation for several fullerenes suggests that this reaction may be thermodynamically most favorable for C_{60}.

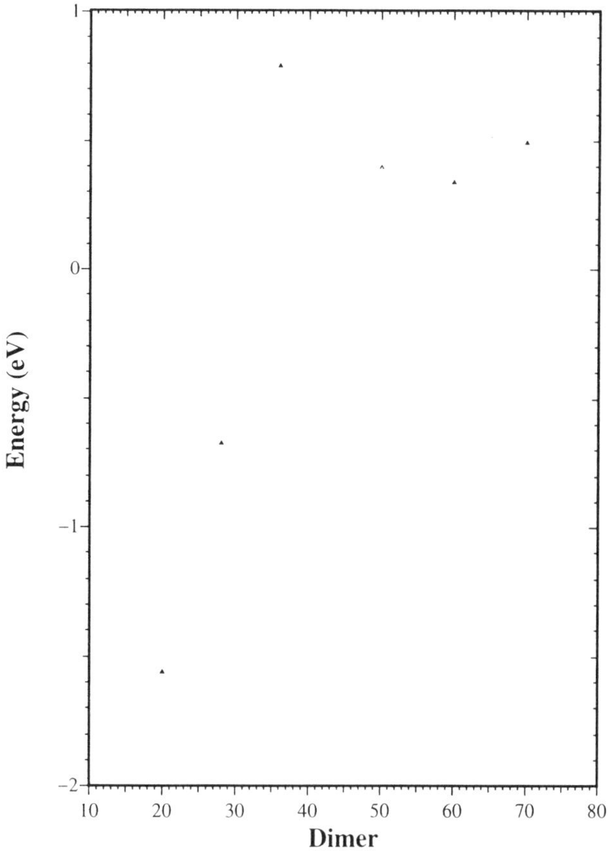

Fig. 6.10. Binding energies for dimer formation

References

1. A. M. Rao et al.: Science **259**, 955 (1993).
2. P. Zhou et al.: Appl. Phys. Lett. **60**, 2871 (1992).
3. M. Menon, K. R. Subbaswamy, M. Sawtarie: Phys. Rev. B **49**, 13966 (1994).
4. D. L. Strout, et al.: Chem. Phys. Lett. **214**, 576 (1993).
5. M. Pederson and A. A. Quong: Phys. Rev. Lett. **74**, 2319 (1995).
6. G. Adams et al.: Phys. Rev. B **50**, 17471 (1994).
7. D. Porezag, M. Pederson, Th. Frauenheim, Th. Köhler: Phys. Rev. B (in press).
8. S. Stafström, M. Boman, J. Fagerstöm, Europhys. Lett. **30**, 295 (1995).
9. S. Pekker, A. Janossy, L. Mihaly, O. Chauvet, M. Carrard, L. Forro: Science **265**, 1077 (1994).
10. S. R. Flom, R. G. S. Pong, F. J. Bartoli, Z. H. Kafafi: Mol. Crys. Liq. Cryst. **256**, 289 (1994).
11. R. D. Beck, P. Weis, J. Rockenberger, M. M. Kappes: J. Phys. Chem. **99**, 3990 (1995).

12. R. Winkler, T. Pichler, H. Kuzmany: Z. Phy. B Condens Matter **96**, 39 (1994).
13. K. L. Akers, C. Douketis, T. L. Haslett, M. Moskovits: J. Phys. Chem. **98**, 10824 (1994).
14. P. W. Stephens, G. Bortel, G. Faigel, M. Tegze, A. Janossy, S. Pekker, G. Oszlanyi, L. Forro: Nature, **N6491**, 636 (1994).
15. A. M. Rao, M. Menon, P. C. Eklund, K. R. Subbaswamy: Chem. Phys. Lett. **224**, 106 (1994).
16. M. Menon, A. M. Rao, K. R. Subbaswamy, P. C. Eklund: Phys. Rev. B **51**, 800 (1995).
17. M. Menon and K. R. Subbaswamy: Phys. Rev. B **47**, 12754 (1993).
18. M. Menon, K. R. Subbaswamy, M. Sawtarie: Phys. Rev. B **48**, 8398 (1993).
19. M. Menon and K. R. Subbaswamy, Phys. Rev. B **55**, 9231 (1997).
20. M. Menon, E. Richter, K. R. Subbaswamy: J. Chem. Phys. **104**, 5875 (1996).
21. W. A. Harrison: *Electronic Structure and the Properties of Solids* (Freeman, San Francisco, 1980), and references therein.
22. K. Harigaya: Chem. Phys. Lett. **189**, 79 (1992).
23. A. L. Balch et al.: J. Am. Chem. Soc. **113**, 8953 (1991).
24. R. C. Haddon, Science, **261**, 1545 (1993).
25. K. Venkatesan and V. Ramamurthy: *Biomolecular Photoreactions in Crystals* Ch. 4, (Ed.) V. Ramamurthy.
26. S. N. Khanna and P. Jena: Phys. Rev. Lett. **69**, 1664 (1992).
27. M. Menon and K. R. Subbaswamy: Phys. Rev. Lett. **67**, 3487 (1991).
28. M. Menon and K. R. Subbaswamy: Phys. Rev. B **47** 12754 (1993).
29. M. van Schilfgaarde and W. A. Harrison, Phys. Rev. B **33**, 2653 (1986); M. van Schilfgaarde and W. A. Harrison, J. Phys. Chem. Solids, **46**, 1093 (1985).
30. A. M. Rao, et al.: Science **259**, 955 (1993).
31. K. M. Creegan, J. L. Robbins, W. K. Robbins, J. M. Millar, R. D. Sherwood, P. J. Tindall, D. M. Cox, A. B. Smith, III, J. P. McCauley, Jr., D. R. Jones, R. T. Gallagher: J. Am. Chem. Soc. **114**, 1103 (1992).
32. Y. Elemes, S. K. Silverman, C. Sheu, M. Kao, C. S. Foote, M. M. Alvarez, R. L. Whetten: Angew. Chem. Int. Ed. Engl. **31**, 351 (1992).
33. P. Zhou et al.: Appl. Phys. Lett. **60**, 2871 (1992).
34. J. D. Coyle: *Introduction to Organic Photochemistry* (Wiley, New York, 1986) p. 61.
35. See the special issue Acc. Chem. Res. **25**, 98 (1992) for a review of fullerene-based solids.

7. Polymerized Fullerite Structures: Experiment and Theory

M. Núñez-Regueiro, L. Marques, J.L. Hodeau, C.H. Xu, G.E. Scuseria

7.1 Introduction

The C_{60} molecule discovered by Kroto et al. [1] is such an extremely stable closed-cage cluster that it can be viewed in some respects as a super-atom or new chemical species [2]. Considerable research involving chemical and physical properties has been done since the discovery of a method allowing its production on a sufficiently large scale [3].

Much of the actual interest in buckminsterfullerene comes from the fact that it is more reactive than was originally expected from considering its aromatic resonances [4]. The poor delocalization of the electrons that results from the strained pentagonal loops adjacent to the benzenoid rings induces a chemical reactivity similar to that of an electron-deficient polyalkene. Having no attached atoms or groups, fullerenes cannot undergo substitution reactions. However, different type of groups can be easily added to the molecule through opening of the double bonds, the total number of them being limited, in principle, only by spatial hindrance. Nonetheless, the increasing number of sp^3 hybridized atoms undermines the cohesion of the molecule as is shown by progressive fluorine addition, where the stability of the cluster is strongly reduced beyond $C_{60}F_{40}$ [5].

As the C_{60} molecules gather in a solid they behave as very weakly interacting spheres and form the compact face-centered cubic structure, held together by feeble van der Waals forces. Its symmetry is $Fm\bar{3}$, and above room temperature the molecules rotate almost freely; there are only local orientational ordering interactions due to the non-perfect sphericity of C_{60} and the weak crystalline field. However, as the temperature is lowered, Coulomb coupling between the molecules becomes important, and below 260 K an orientational ordering transition develops, changing the crystal structure to a new cubic one of $Pa\bar{3}$ symmetry [6]. The molecules dispose themselves with the electron-rich double bonds facing electron depleted pentagons of their nearest neighbours [7]. A non-negligible residual disorder is always present, though, because there is an energetically competing interaction caused when double bonds face hexagons [8], as explained in detail by several theoretical models [9]. At very low temperatures, we end up with a solid that has some properties of an orientational glass, but at ambient pressure the pristine C_{60} solid does not develop chemical bonding between its individual molecules. In this chapter we review the experimental and theoretical work done on the carbon phases [10, 11] obtained from fullerite under compression where the C_{60} molecules are now chemically bonded into a polymerized lattice.

In the first part of this chapter we present some cases of polymerization under pressure such as those of organic molecules. We then present an overview of known, and some hypothetical, carbon phases, as well as describe briefly their transformations under compression. Separately, we survey the experiments done on fullerenes and try to classify the different types of carbon phases obtained when C_{60} is compressed until the collapse of the cage. We conclude the first part with a detailed inspection of all the work that has been reported on polymerized C_{60} structures. In the second part we present first-principles calculations of the energy and stability of different configuration of C_{60} dimers. In the final section, we build up periodic polymerized structures derived from the original fcc lattice by connecting the C_{60} molecules using several elemental bonding configurations of C_{60} dimers. Their calculated energy and other properties are described, and we try to assemble for a certain number of the conceivable structures the evidence that they may indeed exist.

7.2 Bonding Under Pressure

7.2.1 Organic Molecules Under Pressure

The volume decrease induced by pressure is an obvious way of increasing the interaction among clusters or molecules leading to chemical reaction. Molecules weakly bonded by van der Waals forces approach each other rapidly on compression. As the distance between carbon atoms on different molecules approaches a threshold value near that of the usual C–C bond, unsaturated molecules should, in principle, be able to develop chemical bonding. Thus, larger molecules are obtained, including chain and network polymers. Unsaturated organic materials are particularly prone to be unstable at high pressures with respect to associative, crosslinking reactions which form denser, more saturated species [12]. Examples are abundant including alkenes [13], alkynes [14], nitriles [15], carbonyls [16], carbon sulfide [17], cyanogen [18] and even benzene [19]. The materials obtained are generally stable, after the release of the applied pressure, with new physical properties, such as colour, Raman spectra, infrared absorption or texture. Most of these materials are in a gaseous or liquid state before application of pressure. It is often difficult to determine a precise crystallographic structure of the new pressure-induced phases that have a glassy or amorphous arrangement. There are cases where the molecules are allowed to distribute on a regular lattice during solidification, but then an orientational disorder is always present. The predictions of possible polymerized phases taking into account the pre-existing lattice are thus uncertain. As an example, we can cite the pressure-induced polymerization of acetylene, where the early theoretical calculations [20] favoured a very crosslinked product rather than the observed linear polymerization [21] into transpolyacetylene.

7.2.2 Carbon Under Pressure and New Possible Carbon Structures

It is useful at this point to review briefly the different allotropes of carbon and its modification under pressure, to show the rich variety of behaviour that can be expected (and are actually found) in C_{60}.

Carbon atoms posses four valence orbitals that can adopt three different geometries according to the suitable hybridization. Its allotropes with ordered structures are: cubic and hexagonal diamond, hexagonal and rhombohedral graphite and fullerenes and its derivatives. The number of disordered forms is even larger, ranging from the diamond-like carbons to amorphous carbon fibers, passing through such materials as carbon glass, all of them with different but interesting properties. It is important to note that the reproducibility of the obtained materials in passing from one method of preparation to another is far from ideal, as evidenced by the innumerable types of diamond-like materials that have been reported to date.

Since cubic and hexagonal diamond are denser phases, they are more stable above pressures of about 2 GPa at room temperature. Indeed, application of pressure on hexagonal graphite can transform it into hexagonal diamond [22] (with the same ABAB stacking). This transition is, however, reversible unless the sample is heated above 1000 K under pressure, and may undergo an intermediate stage where a transparent graphite is stabilized [23]. The transformation of hexagonal graphite to cubic diamond has, due to the different stackings, such slow kinetics that it is unobservable in normal time spans. Only when the graphite lattice is dissolved in a molten metal under 5 GPa pressure can cubic diamond crystals precipitate from the supersaturated metal-carbon solution [24].

One special class of hypothetical solids actively studied by theoreticians contains the three-dimensional carbon networks with coordination three, usually known as schwarzites. Initially proposed by Mackay and Terrones [25], a schwarzite is characterized by regions of negative curvature involving polygons larger than hexagons. Special schwarzites containing seven- and eight-membered rings have been considered and their energies minimized by using various potential models. Using hexagons and heptagons, Vanderbilt and Tersoff have built a schwarzite of cubic diamond lattice (D168) [26]. Lenosky et al. have also studied schwarzites of diamond and simple cubic lattices (D216 and P216 respectively) [27]. On the other hand, the schwarzite of the bcc lattice (P192) initially proposed by Mackay and Terrones contains hexagons and octagons, and so do the two solid forms (P24 and D24) proposed by O'Keeffe et al. [28]. Other schwarzites with different geometries and unit cell sizes have been generated by Townsend et al. using a computational algorithm [29]. Many properties of these structures have been calculated using empirical potentials, atomic-orbital-based methods, and ab initio local density approximation (LDA) approaches. The energies of these structures, have been predicted to be between that of graphite and icosahedral C_{60} fullerene. The conducting properties of these schwarzites were found to range from semimetals to insulators [30, 31, 32]. The reported bulk moduli and elastic constants of the schwarzites reveal that these structures could be strong materials of rather low density. Other negative curvature structures related to schwartzites have also been proposed [33]. It has been speculated that schwarzites could be present in carbon soot, a hypothesis that has not been corroborated experimentally. The question of new carbon phases formed from bound C_{60} molecules has, however, been almost ignored.

7.2.3 C_{60} Under Pressure

As C_{60} behaves as an unsaturated organic molecule it is plausible to envision bonding between fullerene molecules under sufficiently high pressures. The highly symmetric C_{60} molecule can be expected to give birth to a large number of phases, both ordered and disordered. It may even be a precursor to new unknown forms of plain carbon. C_{60} has thirty reactive double bonds; a double bond can be broken and then participate in an intramolecular link to one of the twelve nearest-neighbour molecules in the fcc lattice. We immediately foresee a large number of possible polymerization configurations, hindering the obtention of a single structure, as is the case in the primitive $Fm\bar{3}$ cubic phase which never attains a perfect orientational ordering. As noted for the other organic molecules, the freezing in place of the orientational disorder will prevent the formation of ordered polymerized solids. We also know that pressure increases the interaction between neighbouring molecules and increases the ordering temperature, as shown experimentally [34] and explained in different theoretical models [35]. Gradual application of pressure favours the hexagon–double-bond interaction with respect to the pentagon–double-bond interaction [36], until it predominates at sufficiently high-pressures [37]. However, inevitable rapid compressions in high pressure apparatus (>2 GPa) causes the molecules to freeze in a rotational glass phase [38]. The key factor in optimizing the long range order of the polymerized solid will then be to keep the spheroids rotating as freely as possible by increasing the synthesis temperature. The system will, in this way, more completely homogenize the number of bonds per molecule and ordered phases will thus be stabilized.

Evidence of pressure-induced bonding between molecules came from the first experiments performed at room temperature. The study of the evolution [39] of the structure of C_{60} in a diamond anvil cell showed that, under hydrostatic pressure, the molecule was stable up to 20 GPa and that the sample conserved its ambient pressure crystal structure. On the other hand, it became apparent that non-hydrostatic loading caused a phase transformation to a lower symmetry crystal structure at a critical pressure of about 16 GPa. Electrical resistivity measurements [40] using sintered diamond anvils exhibited similar behaviour. The resistivity of the fullerene first decreased but at higher pressures it became an insulator, which was interpreted in terms of the formation of sp^3 bonding between the molecules and related to the previous experiment. This type of behaviour of the resistivity was later seen also in C_{70} and $C_{60}I_4$ [41].

Further experiments lead to the first evidence that C_{60} could collapse under stringent pressure conditions. Rapid non-hydrostatic compressions above 20 GPa transformed the samples instantaneously into polycrystalline diamond [42], identified both by X-ray and Raman spectra [43]. Shock wave experiments [44] have as well yielded cubic diamond in the same range of pressure and temperatures. Once more, however, sample preparation conditions are critical in determining the degree of crystallinity of the samples. Amorphous sp^3 carbon, that is crystallographically analogous to amorphous silicon, is also obtained by non-hydrostatic compressions [45], shock waves [46] and shear deformation high-pressures experiments [47]. Cubic diamond crystallites of 1000 Å are obtained from pristine C_{60} within well-defined [48] experimental conditions such as sample proportions, pressure of the transformation or reaction kinetics.

A large number of experiments have been performed in natural diamond cells under hydrostatic conditions [49, 50, 51]. In all this cases transparent and amorphous materials have been recovered. They exhibit Raman spectra typical of a diamond-like material implying mixed sp^2-sp^3 bonding. Some research groups have been able to follow the pressure-dependence of intramolecular fullerene phonon Raman lines and have come to the conclusion that the final material has linked clusters that are possibly partially collapsed and certainly with a high degree of disorder. The progressive and disordered polymerization up to very high pressures (around 30 GPa) has been proposed to gradually weaken the individual clusters and those with too much bonding disintegrate into amorphous carbon [52].

In contrast, the sudden compression of C_{60} under strong anisotropic stress, causes it to break down suddenly, probably due to its softness towards anisotropic stresses [53]. Measurements [54] of the temperature of the sample at the moment of the transformation, which is violent with a small popping noise, show an increase of several hundred degrees. The sudden collapse of the molecule frees the heat of formation of fullerenes, which is larger than those of graphite and diamond. In this way the carbon atoms are given the energy necessary to rearrange into the structure stable at high pressure, i.e. diamond, possibly via an intermediate unstable defective graphite phase [55].

Experiments at temperatures higher than room conditions were also performed by several groups. The early shock wave measurements [56] found that at high temperatures and pressures C_{60} was transformed into graphite or sp^2

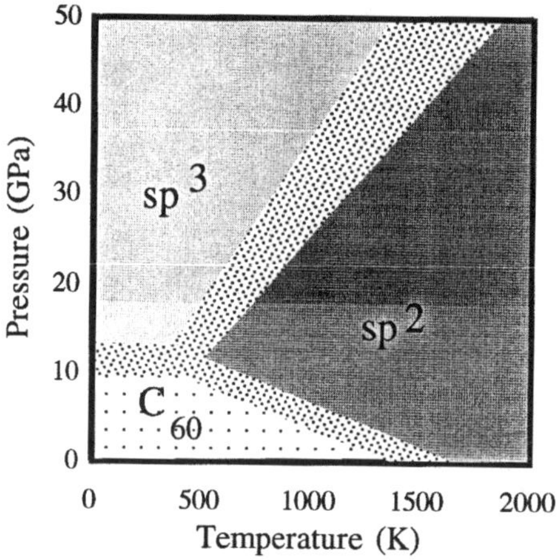

Fig. 7.1. Figure 7.1 Schematic *P-T* diagram showing the region of stability of fullerite. In the high-pressure region it is converted to materials with different degree of sp^3 hybridization, while at higher temperatures it is converted to graphitic or amorphous sp^2 carbon

amorphous materials. This was confirmed by systematic, detailed studies using a belt apparatus [45], allowing the preparation of a rough phase diagram (see Fig. 7.1) showing the (P, T) region where samples with intact C_{60} cage were recovered after the pressure-temperature treatment [57].

7.2.4 Polymerizing C_{60}

In parallel to the pressure experiments we have already described, several other groups have obtained evidence of polymerization of fullerite under different conditions. The most exciting for its simplicity and potential application is photochemical [58] polymerized C_{60}, which is extensively discussed in the Chap. 4 of this volume. The alkali-metal-doped (A-doped) C_{60} compound A_1C_{60} (A = K, Rb) has been proposed to polymerize spontaneously into a chain structure [59] with fascinating electronic properties as described in Chap. 7. Other methods of polymerization have been reported through plasma [60] or electron-induced proccesses [61].

The pressure work reviewed previously reported irreversible linking of C_{60} molecules, with total or partial disintegration of their cage structure. However, samples produced at the high temperature side of the stability region of fullerite (see Fig. 7.1), exhibited small but distinct distortions from the fcc structure [62]. On the other hand, infrared measurements under pressure [63] have shown that samples retrieved from about 1 GPa revealed evidence of polymerization that was reversible if the recovered samples were heated above 473 K at room pressure.

The first report of a new pressure-induced fullerite structure with ordered polymer bonds and stabilized at room temperature was the rhombohedral lattice obtained at 5 GPa and 1000 K [64]. As in photochemical polymerization, this new phase was metastable at normal pressure and temperature conditions, transforming back to fcc C_{60} only by heating above 500 K. Important differences were noted with respect to the normal fcc fullerite. Although the distance between high density planes remained the same as in the original lattice, the separation between molecules in this plane had decreased fom 10 Å to 9.17 Å, a distance that corrresponds to dimerized [65] C_{60}. In this new rhombohedral phase, which will be described in detail later, each molecule in the high density plane connects via "2 + 2 cycloaddition" to its six nearest neigbours in the plane. Compression close to the limit of molecular integrity probably causes a deformation of the cage that alters the symmetry and allows this reaction without violating the usual Woodward–Hoffmann rules.

The same rhombohedral phase and another compressed fcc phase were then announced [66]. Although a detailed structure for these two phases was not proposed in this report, very clear evidence of polymerization was presented. For example, the infrared spectra of the new phases showed a much richer structure than the pristine fcc fullerite, indicating a significant reduction of the I_h molecular symmetry. Most of the peaks observed in C_{60} were either split or shifted in the compressed phases, with several intense peaks in spectral regions which would be otherwise featureless in the pristine material. These spectral characteristics were reminiscent of the infrared spectra of photopolymerized C_{60}. Similar evidence was found from Raman scattering spectra and absorption spectra in the ultraviolet region. Nuclear magnetic resonance (NMR) spectra moreover showed evidence of non-equivalent sites, as expected from molecules deformed by bonding.

Detailed crystallographic evidence on three well-defined phases originated from "2 + 2 cycloaddition" of C_{60} molecules synthesized under high pressure and temperature were then published [10]. Even though the X-ray diffraction patterns did not allow an unambiguous structural determination due to sample strain, domains and molecular disorder, the agreement between the data and X-ray pattern simulations using the atomic positions generated from the results of tight-binding calculations unequivocally indicated the intermolecular bonding arrangement [11]. As a result of this analysis it came clear that in addition to the rhombohedral lattice, a tetragonal one where each molecule in the (001) fcc plane is bonded to four nearest neighbours and also an orthorhombic phase in which chains of polymerized C_{60} molecules are formed could be obtained. On the other hand, measurements of compressibility [67, 68] under pressure have shown phase transitions near the conditions which allow the formation of orthorhombic C_{60}. Finally there are several other reports of phases obtained from fullerite with partial characterization that could correspond to other polymerized structures, including a 3-D polymerized phase with expected exciting properties of hardness [69].

7.2.5 C_{60} Dimers

The reaction between two C_{60} molecules to form a C_{120} adduct is the simplest process invoked in C_{60} polymerization. Several experimental reports have documented this primary process. Yeretzian et al. [70] have reported that laser desorption of a C_{60} film yielded coalescence products with mass distributions peaked at $(C_{60})n$ for $n = 2$–5. Campbell et al. [71] have observed fusion processes in $C_{60} + C_{60}^{+}$ collisions with energies from 200 to 500 eV in the center-of-mass reference frame. The polymerization of a C_{60} solid film using photolysis has been reported by Rao et al. [58] and Eklund and coworkers [72–75]. These authors detected, using X-ray diffraction, a slight contraction of the fcc C_{60} lattice consistent with a 116 cm^{-1} intermolecular mode observed by Raman scattering.

Theoretical work on reactions of C_{60} with itself has encompassed several different processes, including simulations of cluster collisions and potential energy surface studies of C_{60} dimers. Collision studies [76–79] under high-energy conditions, yielded fusion products consistent with the experiments of Campbell et al. [70] and Yeretzian et al. [71]. Theoretical studies of C_{60} dimers [65,80–84] covering many possible bonding mechanisms and geometries have provided plausible candidates for rationalizing experimental information on polymerized C_{60} structures. Strout et al. [65] have proposed different classes of C_{60} dimers: "2 + 2" and "2+4" cycloaddition products, dimers where one C_{60} molecule reaches out to bridge the other C_{60} molecule, and a series of coalescence products reminiscent of a peanut, which require multiple bond rearrangement, named "buckypeanuts". Semiempirical tight-binding (TB) and MNDO calculations predicted that the energies of the $2+2$ isomers and peanuts are comparable, and lower than the energy of the $2+4$ and bridged isomers. Of the four $2+2$ isomers studied (covering all the possible combinations between reactions of single and double bonds in C_{60}), the $2+2$ cycloaddition of double bonds (denoted as 66/66) was consistently predicted to be the most stable structure based on semiempirical and ab initio calculations at the STO-3G self-consistent field Hartree–Fock level of theory. Since formation of the peanuts requires the breaking of six bonds, these authors [65] suggested that peanuts are likely to be formed only under high-energy conditions [70, 71] such as in laser desorption experiments. Thus, theoretical calculations by Strout et al. [65] support the view that the most likely products of polymerization experiments are the C_{60} dimers of 66/66 scheme.

Other authors have also studied products of C_{60} dimerization. Menon et al. [80] have reported TB molecular dynamics calculations where the initial C_{60} dimer configurations were selected from a variety of different possibilities. The relative orientations of the two C_{60} molecules at the starting point of the molecular dynamics simulations included combinations of bond-to-bond (single and double), face-to-face (pentagon and hexagon), corner-to-corner, corner-to-bond, and corner-to-face arrangements. Additionally, these authors also considered a C_{58}-C_2-C_{60} initial geometry. All these structures were found to relax to a few different geometries, including a 'bucky-dumbell' which is geometrically identical to the "2 + 2" cycloadduct of double bonds and was found to have the lowest energy. Based on the reaction patterns of experimental fullerene derivatives, Matsuzawa et al. [81] proposed that C_{60} dimers could also be formed by 1,2 and 1,4 addition. Four dimers of C_{60} were proposed, namely, 1,2, C_{2v}-1,4, C_{2h}-1,4, and 1,2 + 1,4. It should be pointed out that 1,2 addition is anologous to 2 + 2 cycloaddition of double bonds. Theoretical calculations at the MNDO/AM-1, PM-3, and LDA of density functional theory, all predict that the product of 1,2 addition is more stable than the others by 70–90 kcal/mol. The dimerization reaction to form the 1,2 adduct was found to be slightly endothermic, whereas all others were predicted to be strongly endothermic. In agreement with the above studies [65,80], Matsuzawa et al [81] suggested that "2 + 2" cycloaddition is the most favorable scheme for C_{60} dimerization. Many authors have pointed out that such dimerization is symmetry-allowed photochemically but is symmetry-forbidden thermally.

Pederson et al [82] have performed all-electron, full potential LDA calculations on a C_{60} dimer formed by "2 + 2" cycloaddition. Their results suggest that the "2 + 2" dimer is more exothermic than the isolated C_{60} molecules by approximately 1.0–2.2 eV. G. B. Adams et al. [83] have recently carried out local-orbitals-based first principles calculations on seven different configurations of C_{60} dimers and one infinite C_{60} chain [83]. Again, "2 + 2" cycloaddition of double bonds was found to yield the lowest energy. The infinite C_{60} chain of similar dimerization, was found to be 0.41 eV (per 120 atoms) lower in energy than the dimer. Using Tersoff's empirical potential, Bacsa and Lannin [84] have investigated the intermolecular vibrational modes observed in their photon-induced Raman spectra of ultrathin C_{60} films. With this potential, "2 + 2" cycloaddition of double bonds was found to be stable with respect to C_{60} monomer separation.

All these calculations consistently predict that "2 + 2" cycloaddition of double bonds is the most favourable scheme for the polymerization of C_{60}. The binding energy of the "2 + 2" C_{60} dimer with respect to C_{60} + C_{60} calculated by various methods is listed in Table 7.1. In the following, we will implicitly refer to a C_{60} dimer as the "2+2" cycloadduct from the interaction of two double bonds on neighbouring molecules, unless otherwise specified.

Table 7.1. Binding energy (eV) of "2 + 2" C_{60} dimer with respect to C_{60} + C_{60}

Method	Ref.	Energy
TB	[65]	−0.4 (eV)
MNDO	[65]	2.1 (eV)
LDA	[65]	0.7 (eV)
BLYP	[65]	−0.6 (eV)
TB	[80]	−0.6 (eV)
AM1/LDFT	[81]	0.67 (kcal/mol)
NLDFT	[81]	−0.37(kcal/mol)
LDA	[82]	1.2 (eV)
FP	[83]	0.47 (eV)

The 2 + 2 C_{60} dimer has D_{2h} symmetry. In most studies, its geometry has been characterized by the center-center distance between the two C_{60} molecules, and the bond-lengths in the 2 + 2 four-membered ring. The values obtained for these properties in the different theoretical calculations are presented in Table 7.2. Some general trends are observed in this data. The bond-lengths in the 2 + 2 four-membered ring are fairly long whereas a contraction (compared to fcc C_{60}) in the center to center to distance is observed consistent with the experimental X-ray values (9.1–9.2 Å) in the polymerized fullerites.

Table 7.2. Calculated properties of the "2 + 2" C_{60} dimer: center–center distance (CCD) in Å, bondlengths (Å) in the "2+2" 4-membered ring ($R1$ is the intermolecular bond, and $R2$ is the substituted bond), and HOMO-LUMO gap E_g (eV). The E_g for an isolated C_{60} is presented for comparison purposes

Method	Ref.	CCD	$R1$	$R2$	E_g (dimer)	E_g (C_{60})
MNDO	[65]	9.2	1.56	1.62		
TB	[80]		1.51	1.51	0.8	1.1
MNDO	[81]		1.55	1.60		
LDA	[82]	9.1	1.60	1.57	1.2	
FP	[83]	9.022	1.588	1.578	1.5	1.7
Tersoff	[84]		1.616			
TB	[11]				1.3	1.6

Also listed in Table 7.2 is the HOMO-LUMO or electronic band gaps predicted in several calculations. The HOMO-LUMO gap of the dimer was found to be narrower than that of an individual C_{60} molecule due to its reduced symmetry and the resultant reduced degeneracy of HOMO- and LUMO-derived bands. Vibrational properties of the C_{60} dimer have been calculated by various methods. In particular, these studies have confirmed that the Raman line observed at $116\,cm^{-1}$ is an intermolecular mode. Strout et al. [65] obtained an intermolecular stretching mode of $81\,cm^{-1}$ in their TB calculations. Using a different TB model, Menon et al. [80] found a $101\,cm^{-1}$ ode to be Raman active. Adams et al. [83] first principles study revealed a similar mode of $69\,cm^{-1}$ having strong intensity. A similar intermolecular mode at $91\,cm^{-1}$ was found with Tersoff potential [84]. Pederson et al. [82] have also carried out vibrational density of states (DOS) calculations which emphasized on the symmetry reduction effect on the structure of the DOS.

7.3 The Polymerized Fullerite Structures

The number of possible structures is large and we shall here concentrate on only a restrained number that we believe are the most probable ones. We have already described several local arrangements of dimers, and we shall extend the discussion first to one-dimensional polymerized chains, then to two-dimensional planes and finally to three-dimensional polymerized structures.

7.3.1 1-D Polymers

The simplest case is obtained by the extension to chains of the dimers formed from "2 + 2" cycloaddition. In this case each molecule is connected to two neighbours via the opening of opposing double bonds. The chain structure shown in Fig. 7.2a is the one that has been shown [10] to correspond to the orthorhombic phase consistently obtained at 600–700K between 1 to 8 GPa by different groups [67, 68] and is also observed in the A_1C_{60} compounds [85]. Recent NMR experiments [86] indeed find remarkably similar spectra for the pressure-induced phases obtained and for the A_1C_{60} materials, indicating an analogous distortion of the C_{60} molecule in both cases.

The X-ray powder pattern of C_{60} compressed at 8 GPa and heated at 600 K is shown in the Fig. 7.3. It can be indexed in a orthorhombic body centered lattice with cell parameters shown in Table 7.3. The structure can be viewed as a distortion of the C_{60} fcc cell by compressing the molecules along ⟨110⟩ fcc direction [87]. In this direction, the interball distance is much smaller (9.26 Å) than in other directions, and this was found to correspond to the polymerization direction [11]. The dotted line is the best simulation of the experimental pattern and is obtained by using rings of charge, indicating that a libration disorder exists around the chain direction. It must be emphasized that the X-ray diffraction pattern is poor, with only a few peaks, so the lattice parameters are highly correlated. However, we find experimentally very good agreement with the theoretical values for intermolecular distances along the polymerization direction.

Table 7.3. Structural parameters of the compressed fullerite phases that have been identified up to date

phases	space group	cell parameters/Å	distance between the centers of the molecules/Å	volume/ molecule Å^3
pristine	Fm$\bar{3}$	$a = 14.17$	10.02	711
rhomohedral	R$\bar{3}$m	$a = 9.19$	$\underline{9.20}$	597
		$c = 24.50$	9.74	
tetragonal	Immm	$a = b = 9.09$	$\underline{9.09}$	618
		$c = 14.95$	9.86	
orthorhombic	Immm	$a = 9.26$	$\underline{9.26}$	650
		$b = 9.88$	9.88	
		$c = 14.22$	9.82	

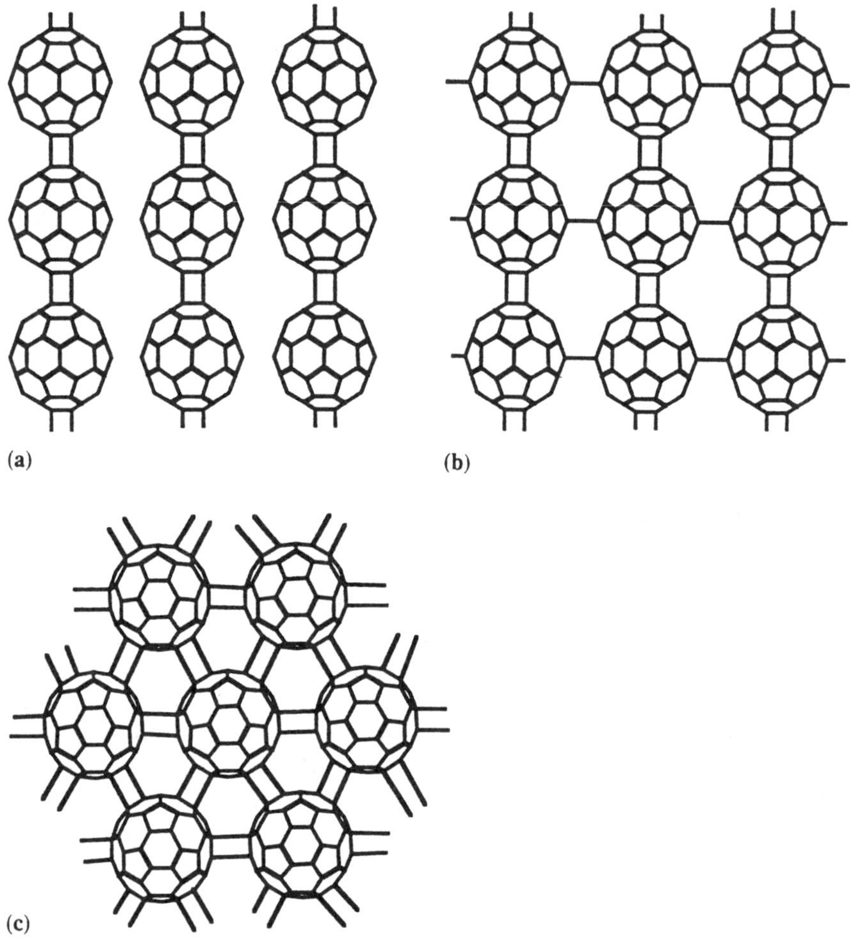

Fig. 7.2. The three experimentally observed polymerized fullerite structures all issued from 2 + 2 cycloaddition bonding: (**a**) orthorhombic structure where each molecule is bonded to the two nearest neighbours along a ⟨110⟩ fcc direction forming parallel chains; (**b**) "tetragonal" lattice where each molecule is bonded to its four nearest neighbours in a (100) fcc plane; (**c**) rhombohedral structure where each molecule is bonded to its six neighbours in a high-density (111) fcc plane

It is clear that the polymerized chains need not be completely ordered, i.e., there can be an orientational disorder due to possible rotations around the chain axes and/or static disorder. On the other hand, at room temperature there certainly exist different types of twisting (bending) excitations that could also explain the diffuseness observed in our X-ray spectra. As in the case of fcc fullerite, the study of the behaviour of these twisting and bending modes should introduce us to a rich physics, though for the time being the

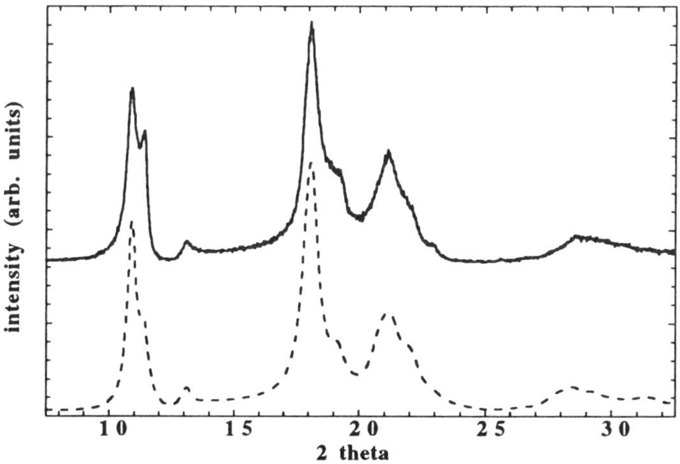

Fig. 7.3. X-ray spectrum from an orthorhombic-phase-compressed sample (*full line*) together with the simulation using the theoretically derived results (*dashed line*)

poor crystallization of the samples will hinder complete and quantitative experimental investigations.

The orthorhombic phase has also been obtained using lower pressure ($P < 2\,\mathrm{GPa}$) apparatus [88]. Another related phase that has been encountered in the same pressure/temperature (5 GPa; 600 K) domain [66] presents larger peaks, broader that those of any other polymerized phase. It is difficult to establish clearly the real structure; its peaks can be indexed in a fcc lattice of parameter $a = 13.6\,\text{Å}$, suggesting that it can be the result of partial polymerization, inducing an overall (albeit disordered) lattice.

As we have seen previously, other organic reactions may occur between C_{60} molecules. So it is possible to think about additional chain structures formed through those polymerization reactions, though they should be less stable than the $2+2$ products. If we restrict ourselves to chain structures possible candidates are the $1+4$ polymerizations, involved for instance in the dimerization normal to the molecular plane of anthracene [89] where the C_{60} molecules are placed opposing hexagons in phase (Fig. 7.4a) or at antiphase (Fig. 7.4b) [81]. In the first case we have one molecule per unit cell of the chain to compare with two molecules for the second case with an intermolecular distance of 8.3 Å in both cases.

Another C_{60} chain polymer can be build up by molecules polymerized by 1,2+1,4 cycloadditions (Diels–Alder reaction), where a double bond of one molecule must be in front of an hexagon of the adjacent molecule, once more with two molecules per unit cell (Fig. 7.4c). Finally, we cannot ignore the very dense case of $3(2+2)$ cycloadditions where the confronted hexagons

Fig. 7.4. Hypothetical polymerized phases: (**a**) to (**e**) linear chains formed as described in the text; (**f**) possible 2-D structure derived by connecting the (**a**) chains

are totally bound also with two structural possibilities giving either a one molecule (Fig. 7.4d) or a two molecule base (Fig. 7.4e), respectively.

7.3.2 2-D Polymers

To build up the simplest of the plane polymerized structures, it suffices to bring the 2 + 2 cycloaddition chains closer together. In the 2-D polymer formed in this way (Fig. 7.2b) each molecule is linked covalently to four other molecules forming a pseudo-square lattice (according to the theoretical calculations there is a small difference between the two polymerization directions, x and y, of Table 7.4). This effect explains that the best simulation of the observed X-ray patterns contain a mixture of both orientations [10, 11].

Table 7.4. Binding energies (E) with respect to C_{60}, intermolecular center-center distance d, intermolecular bond-lengths R of 2 + 2 cycloaddition, and electronic band gaps E_g for the C_{60} phases studied

	E (eV)	d (Å)	R (Å)	E_g (eV)
rhombohedral	2.1	9.17	1.64	1.0
tetragonal	0.9	9.06(x)	1.64(x)	1.2
		9.13(y)	1.64(y)	1.2
orthorhombic (linear chain)	0.4	9.11	1.64	1.2

The structure that contains this type of C_{60} polymer is a pseudo-tetragonal body-centered one, referred to here as simply "tetragonal". In Fig. 7.5, we can see the X-ray diffraction pattern (solid line) of a sample that is a mixture of tetragonal and rhombohedral phases. The simulated pattern is indicated by the broken line. We have not yet succeeded in preparing a single-phase tetragonal sample. The cell parameters are shown in Table 7.3. The tetragonal structure results from the in-plane compression of (001) fcc planes [87]. In this case we have two reduced interball distances that correspond, as before, to the directions of polymerization. Actually, the structure is not tetragonal, a merohedral disorder between layers increases the apparent molecular symmetry and induces a smearing of a and b, giving rise to a macroscopic tetragonal structure as noted above.

We can construct another 2+2 cycloaddition based 2-D polymer by bringing the C_{60} chains closer together, but in a way slightly different from the tetragonal case. The resulting polymer, Fig. 7.2c, is more compressed than the tetragonal one. In Fig. 7.6 we show X-ray diffraction spectra (data, solid line; simulation, broken line) of a sample that contains this hexagonal polymerized layer in a ABC stacking, whose lattice parameters are shown in Table 7.3. This structure can be viewed as an in-plane compression of the (111) fcc plane that leads to two reduced interball distances, which, as before, correspond to the polymerization directions.

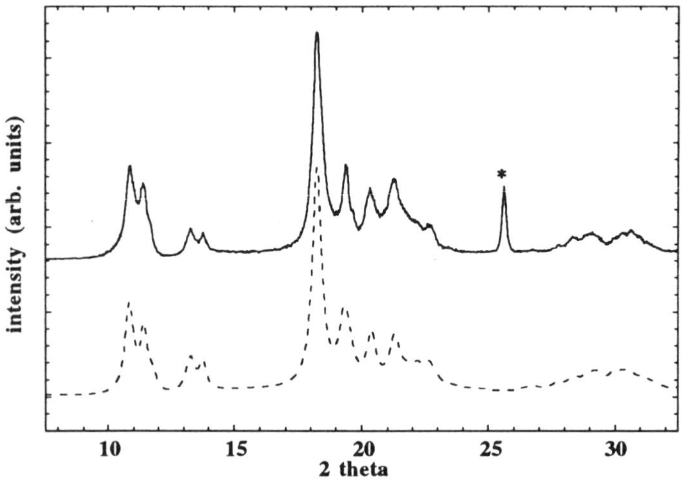

Fig. 7.5. X-ray spectrum from a mixed phase compressed sample tetragonal + rhombohedral (*full line*) together with the simulation using the theoretically derived results (*dashed line*)

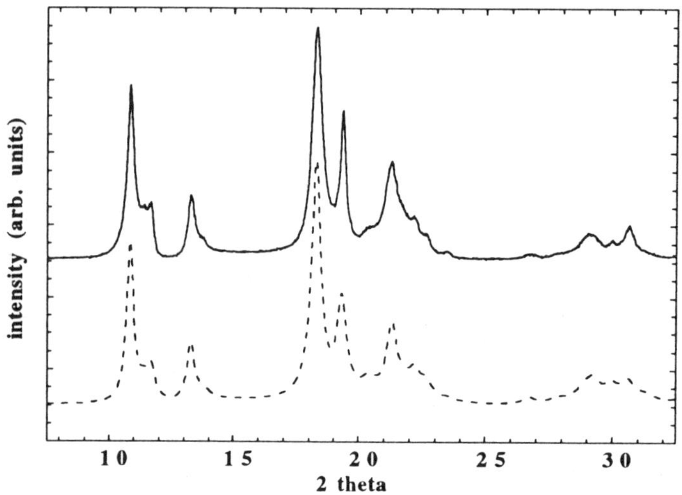

Fig. 7.6. X-ray spectrum from an rhombohedral phase compressed sample (*full line*) together with the simulation using the theoretically derived results (*dashed line*)

Xu and Scuseria [11] have calculated the stability, structure, and electronic bands of the three experimentally observed C_{60} solids. The calculations were carried out using a TB carbon–carbon potential model [90], which predicts that the hexagonal, tetragonal, and linear chain structures proposed are all stable, although they are expected to be slightly endothermic compared to individual C_{60} molecules (Table 7.4).

As shown in Table 7.4, all the structures studied have intermolecular C-C bond lengths of approximately 1.64 Å, characteristic of covalent bonding. The predicted lattice constants (or intermolecular distances) match well with the range of experimental values, 9.1–9.2 Å [64, 10, 66]. The electronic density of states revealed band gaps in the range 1.0–1.2 eV, consistent with the narrowing of HOMO-LUMO gap in a C_{60} dimer due to $2 + 2$ cycloaddition. Additionally, the calculated (two-dimensional) elastic modulus of the hexagonal layer was 1.1 Mbar, comparable to those of many hypothetical schwarzites discussed above, but much larger than the bulk modulus of fcc C_{60} solid (0.14 Mbar [91, 92]).

The best simulations of these two X-ray experimental patterns (see Fig. 7.5 and 7.6) were performed using the TB relaxed atomic positions obtained by computing methods, even if additional improvement was obtained by introducing a merohedral molecular disorder. This clearly indicates that the 2-D phases are more ordered than the 1-D phase. A confirmation of this is the fact that, in contrast to the 2-D phase, in the 1-D experimental X-ray diffraction patttern no longer exhibits peaks at higher angles, but instead exhibits a continuous diffuse scattering (lattice disorder).

If we look now at the chains in the (111) plane of the original fcc structure we see the structure of Fig. 7.3, which has been also observed [66] and its structure analyzed [93] by other groups. The apparent relative ease of preparation stems probably from the fact that their apparatus have only similar working pressures, and not necessarily from a more stable character of this phase.

Chains issued from $1+4$ polymerization (Fig. 7.4a) can also develop plane structures within the (111) plane, that can be reinforced by an additional $2+2$ linking, as shown in Fig. 7.4f. In contrast, the $1 + 4$ chains of Fig. 7.4b do not yield simple plane structures due to their two-molecule base. In the same way, other structures such as the $2 + 4$ phase can be envisaged.

Recent NMR measurements [86] have been performed on ^{13}C enriched rhombohedral samples using the magic angle spinning technique. Six resonances, corresponding to inequivalent carbon atoms, are clearly identified at 149.1, 147.9, 145.2, 139.6, 134.8 and 73.5 ppm. Their intensities agree with the ratio $12 : 12 : 12 : 6 : 6 : 12$ that correspond to the proportions of the inequivalent atom sites for the structure of Fig. 7.2c. This data yield thus the conclusive evidence for the existence of deformed polymerized C_{60} molecules

7.3.3 3-D Polymers

We can now analyze the possibility of bonding between polymerized layers, leading towards a 3-D polymerized phase of C_{60}. Up to now there have been no structural determinations of a three-dimensional fullerite polymerized structure, although several reports have mentioned hard materials issued from C_{60} that apparently conserve the cage structure. Candidates for these hard materials are 3-D fullerite structures with a high degree of polymerization.

A 3-D structure could be derived directly from compression of the high pressure orientational phase, where double bonds face hexagons. At higher pressures (smaller volumes) and relatively low temperatures this structure should be prone to the formation of 1,2+1,4 bonding, and a dense structure where each molecule can be bound to its twelve nearest-neighbour molecules in a very symmetric way could be obtained, though we must remember that this type of bonding is found unstable by the TB model prediction.

Three-dimensional polymerization can be also accomplished by compressing the square or the hexagonal polymerized layers against each other. In each case hexagons of one layer will face those of the nearest sheet. In the first case if the planes are linked by connecting the hexagons through $3(2+2)$ cycloaddditions we can form the structure shown in Fig. 7.7a. If we allow additional $2+2$ bonding, we end up with the dense structure proposed earlier [94], where all 60 atoms are in sp^3 hybridization, suggesting very hard properties but low stability. In this case there are two structures depending on the relative orientation of the succesive tetragonal planes. In a similar way we can obtain a three-dimensional polymer from the rhombohedral structure, which we show in Fig. 7.7b. This phase, with ABC stacking, is the one that is expected experimentally as it implies the smallest layer shift. As shown below, it is energetically stable and could be the stiffest obtainable fullerite.

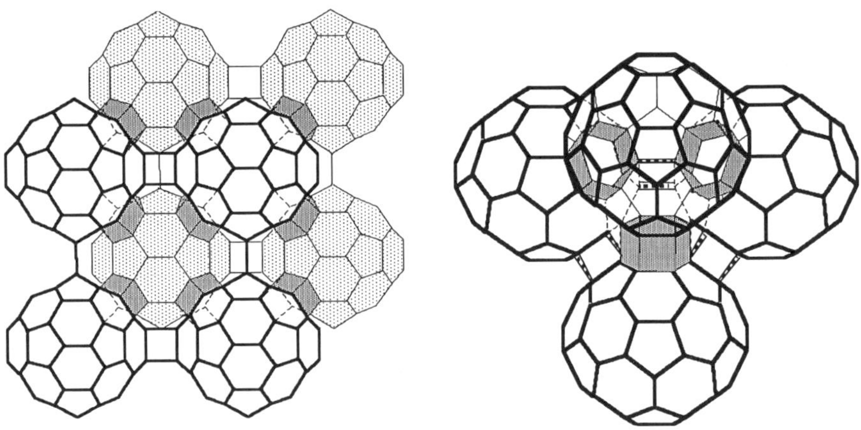

Fig. 7.7. Possible 3-D polymerized structures shown in their simplest expression to clarify the view: (**a**) derived from the tetragonal structure of Fig. 7.2b with an AB stacking and (**b**) derived from the rhombohedral structure of Fig. 7.2c with a ABC stacking

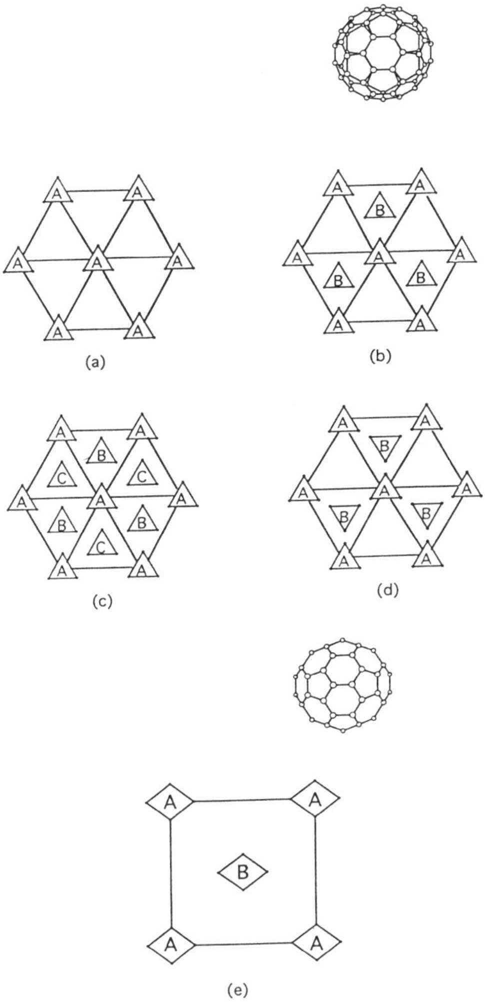

Fig. 7.8. Schemes of the stackings of C_{60} layers used in the theoretical calculations viewed along the stacking direction. The A, B, and C layers are stacked consecutively from the viewer toward the paper The triangles represent C_{60} molecules of two different orientations. An upward triangle denotes a C_{60} molecule as shown in the insert, and a downward triangle is the 180 degree rotation of the upward triangle around its 3-fold axis. (**a**)–(**d**) shows the hexagonal h3d-66, h3d-66-55, h3d-666, and h3d-65 cases as describe in the text. Lines linking the triangles are added to the A layer to indicate a 2D hexagonal lattice. (**e**) ABAB stacking of tetragonal layers viewed along the stacking direction. The rhombs represent C_{60} molecules oriented as shown by the inserted C_{60}. Lines linking the rhombs are added to the A layer to indicate a 2D tetragonal lattice

Several of these models were analyzed by Xu and Scuseria [95] (unpublished) using a TB potential model and performing molecular dynamics (MD) relaxations, and many different solid structures have been found to be stable. A class of these structures is made by stacking the hexagonally packed C_{60} layers in AAA, ABAB, and ABC (the one of Fig. 7.7b) fashion, as schematized in Fig. 7.8. In the academic AAA case (Fig. 7.8a), the top and bottom hexagons of a C_{60} molecule directly face those of the neighbour layers, this structure is called h3d-66. In the ABAB and ABC cases (Figs. 7.8b and 7.8c), molecules embed partially to the lattice space of adjacent layers when the interlayer distances are shortened. The ABAB stacking (Fig. 7.8b) results in hexagon facing hexagon from layer A to layer B, and pentagon–pentagon from layer B to layer A (named h3d-66-55), whereas the ABCABC (Figs. 7.7b and

7.8c) contains only hexagon–hexagon (h3d-666) between adjacent layers, and if the stacking order is reversed (CBACBA), only pentagon facing pentagon (h3d-555). Another possible phase is an ABAB stacking but with the C_{60} molecules in each B layer having a mirror reflection with respect to its layer plane (Fig. 7.8d). This structure gives hexagon-facing-pentagon between all adjacent layers (named h3d-65).

In our simulations, the initial interlayer distances of these stacking cases are chosen such that possible covalent bonds (typically of length 1.5 Å) could be formed between facing polygons (6-6, 6-5, and 5-5) of adjacent layers. Upon relaxation of these 3D periodic structures using TB forces, interlayer bonds of covalent character were found for the 6-6 and 6-5 cases, but not for the 5-5 case, possibly because the facing pairs of 6-6 are all eclipsed, and the facing pairs of 5-5 are all staggered. The resulting stable structures then include h3d-66, h3d-666, h3d-66-55, and h3d-65. The energies and the interlayer distances of these structures are listed in Table 7.5. All these phases are slightly endothermic compared to isolated C_{60} molecules. The interlayer distances are substantially contracted with respect to the value before compression (~10 Å), particularly for the mixed stacking cases (ABAB and ABCABC).

Table 7.5. Energies (eV/C_{60}) with respect to isolated C_{60} molecules, and distance (Å) between adjacent molecular layers of the 3-D phases as described in the text. The distance for h3d-66-55 is that between layers of hexagon facing hexagon

	h3d-66	h3d-666	h3d-66-55	h3d-65	t3d-66
Energy	10.1	20.2	11.8	7.88	24.8
Distance	8.40	6.24	6.21	6.40	4.96
stacking	AAA(H)	ABAB(H)	ABC(H)	ABAB(H) mirror	ABAB(T)

Similar calculations were done for the tetragonal phase. Fig. 7.8e shows schematically the ABAB stacking which results in eight hexagons per molecule to pair with the hexagons in adjacent molecular layers. The stable structure is even more endothermic than other 3D phases (see Table 7.5). The particularly short interlayer distance (4.96 Å) is a result of the strong interlayer embedding of C_{60} molecules. The construction of additional C_{60} solid structures, and the computation of more properties of these phases is currently underway and experimental studies have still to be perform to demonstrate the real existence of these 3-D polymerized fullerites.

7.4 Conclusions

In the search of polymerized fullerites, the number of C atoms in each structural unit is very large and due to the high symmetry of the starting molecule, a large degree of disorder can occur. Consequently, diffraction patterns are, relatively speaking, poor and it is impossible to determine completely the structure from the data. Furthermore, as there are a large number of different C–C bonds, it is also difficult to interpret quantitavely Raman and infrared spectra. However, the combination of all these different techniques and theoretical calculations has allowed the determination of polymerized structures. The study of the behaviour of C_{60} under pressure has thus shown that, as for unsaturated organic molecules, fullerene molecules can polymerize. The three simplest structures that can be obtained by 2+2 cycloaddition reactions have been observed. They constitute new carbon phases, with some expected properties similar to those of schwarzite structures. The theoretical calculations are in excellent agreement with the observed structures. Work is under way to characterize the physical properties of these new phases and to try to pin down the certainly interesting three dimensional polymerized phases, that should be extremely stiff and of possible practical applications.

References

1. H. W. Kroto, J. R. Heath, S. C. O'Brien, R. F. Curl, R. E. Smalley: Nature **318**, 162 (1985).
2. Accounts, Accounts of Chemical Research, 25 (1992).
3. W. Tschmer, L. D. Lamb, K. Fostiropoulos, D. R. Huffman: Nature **347**, 354 (1990).
4. D. J. Klein, T. G. Schmalz, T. G. Hite, W. A. Seitz: J. Am. Chem. Soc. **108**, 1301 (1986).
5. R. Taylor and D. Walton: Nature **368**, 685 (1993).
6. P. A. Heiney, J. E. Fischer, A. R. McGhie, W. J. Romanov, A. M. Denenstein, J. P. McCauley Jr., A. B. Smith, III, D. E. Cox: Phys. Rev. Lett. **67**, 1468 (1991).
7. W. I. F. David, R. M. Ibberson, J. C. Matthewman, K. Prassides, T. J. Dennis, J. P. Hare, H. Kroto, R. Taylor, D. R. M. Walton: Nature **353**, 147 (1991).
8. W. I. F. David, R. M. Ibberson, T. J. Dennis, J. P. Hare, K. Prassides: Europhys. Lett. **18**, 219 (1992).
9. J. P. Lu, X.-P. Li, R. M. Martin: Phys. Rev. Lett. **68**, 1551 (1992); M. Sprik, A. Cheng, M. L. Klein: J. Phys. Chem. **96**, 2027 (1992).
10. M. Núñez-Regueiro, L. Marques, J.-L. Hodeau, O. Béthoux, M. Perroux: Phys. Rev. Lett. **74**, 278 (1995).
11. C. H. Xu and G. Scuseria: Phys. Rev. Lett. **74**, 274 (1995).
12. M. Nicol and G. Z. Yen: J. Phys. **45-C8**, 163 (1984).
13. W. S. Holmes and E. Tyrrall: Trans. Faraday Soc. **52**, 47 (1956); R. Meer, A. L. German, D. Heikens: J. Polym. Sci. **15**, 1765 (1977).
14. J. E. Rice and Y. Okamoto: J. Org. Chem. **46**, 446 (1981).
15. V. V. Yakushev, S. S. Nabatov, O. B. Yakusheva: Dokl. Akad. Nauk. SSSR **214**, 879 (1973).
16. A.N. Dremin and L. V. Babare: J. Phys. **45-C8**, 177 (1984).
17. M. Nicol and G. Z. Yen: J. Phys. **45-C8**, 163 (1984).
18. C.-S. Yoo and M. Nicol: J. Phys. Chem. **90**, 6726 (1986).

19. M. Gauthier, J.C. Chervin, P. Pruzan: in *High Pressure Research*, H.D. Hochheimer (Ed.), (Plenum Press, New York, 1991).
20. R. LeSar: J. Chem. Phys. **86**, 1485 (1987).
21. K. Aoki, Y. Kakudate, S. Usuba, M. Yoshida, K. Tanaka, S. Fujiwara: Sol. Sta. Comm. **64**, 1329 (1987).
22. F.P. Bundy and J.S. Kasper: J. Chem. Phys. **46**, 3437 (1967).
23. W. Utsumi and T. Yagi: Science **252**, 1542 (1991).
24. F.P. Bundy, H.T. Hall, H.M. Strongand R.H. Wentorf: Nature **176**, 51 (1955).
25. A.L. Mackay and H. Terrones: Nature **352**, 762 (1991).
26. D. Vanderbilt and J. Tersoff: Phys. Rev. Lett. **68**, 511 (1992).
27. T. Lenosky, X. Gonze, M. Teter, V. Elser: Nature **355**, 333 (1992).
28. M. O'Keeffe, G.B. Adams, O.F. Sankey: Phys. Rev. Lett. **68**, 2325 (1992).
29. S.J. Townsend, T.J. Lenosky, D.A. Muller, C.S. Nichols, V. Elser: Phys. Rev. Lett. **69**, 921 (1992).
30. R. Phillips, D.A. Drabold, T. Lenosky, G.B. Adams, O.F. Sankey: Phys. Rev. B **46**, 1941 (1992).
31. W.Y. Ching, M.-Z. Huang, Y.-N. Xu: Phys. Rev. B **46**, 9910 (1992).
32. M.-Z. Huang, W.Y. Ching, T. Lenosky: Phys. Rev. B **47**, 1593 (1993).
33. G.E. Scuseria: Chem. Phys. Lett. **195**, 534 (1992).
34. G. Kriza, J.-C. Ameline, D. Jerome, A. Dworkin, H. Szwarc, C. Fabre, D. Schuta, A. Rassat, P. Bernier, J. Phys. I France **1**, 1361 (1991); G.A. Samara, J.E. Schirber, B. Morosin, L.V. hansen, D. Loy, A.P. Sylwester: Phys. Rev. Lett. **67**, 3136 (1991).
35. J.P. Lu, X.-P. Li, R.M. Martin, Phys. Rev. Lett. **68**, 1551 (1992); M. Sprik, A. Cheng, M.L. Klein: J. Phys. Chem. **96**, 2027 (1992); E. Burgos, E. Halac, H. Bonadeo: Phys. Rev. B**49**, 15544 (1994).
36. W.I.F. David and R.M. Ibberson: J. Phys. Cond. Matt. **5**, 7923 (1993).
37. B. Sundqvist, O. Andersson, A. Lundin, A. Soldatov: Sol. Sta. Comm. **93**, 109 (1995); J.A. Wolk, P.J. Horoyski, M.L.W. Thewalt: Phys. Rev. Lett. **74**, 3483 (1995).
38. S.H. Tolbert, A.P. Alivisatos, H.E. Lorenzana, M.B. Kruger, R. Jeanloz: Chem. Phys. Lett. **188**, 163 (1992).
39. S.J. Duclos, K. Brister, R.C. Haddon, A.R. Kortan, F.A. Thiel: Nature **351**, 380 (1991).
40. M. Núñez-Regueiro, P. Monceau, A. Rassat, P. Bernier, A. Zahab: Nature **354**, 289 (1991).
41. M. Núñez-Regueiro, O. Béthoux, J.-M. Mignot, P. Monceau, P. Bernier, C. Fabre, A. Rassat: Europhys. Lett. **21**, 49 (1993).
42. M. Núñez-Regueiro, P. Monceau, J.-L. Hodeau: Nature **355**, 237 (1992).
43. M. Núñez-Regueiro, L. Abello, G. Lucazeau, J.-L. Hodeau: Phys. Rev. B **46**, (1992).
44. H. Hirai, K. Kondo, T. Ohwada: Carbon **31**, 1095 (1993).
45. J.-L. Hodeau, J.-M. Tonnerre, B. Bouchet-Favre, M. Núñez-Regueiro, J.-J. Capponi, M. Perroux: Phys. Rev. B **50**, 10311 (1994).
46. H. Hirai, K. Kondo, N. Yoshizawa, M. Shiraishi: Appl. Phys. Lett. **64**, 1797 (1994).
47. V. Blank, S. Buga, M. Popov, V. Davydov, B. Kulnitsky, E. Tatyanin, V. Agafonov, R. Ceolin, H. Szwarc, A. Rassat, C. Fabre: Mol. Mat. **4**, 149 (1994).
48. M. Núñez-Regueiro, O. Béthoux, J.L. Hodeau, L. Marques, J.M. Tonnerre, B. Bouchet-Fabre, Recent advancements in the chemistry and physics of fullerenes and related materials, K.M. Kadish, R.S. Ruoff (Eds.), **647** (1994).

49. F. Moshary, N.H. Chen, I.F. Silvera, C.A. Brown, H.C. Dorn, M.S. de Vries, D.S. Bethune: Phys. Rev. Lett. **69**, 466 (1992).
50. D.W. Snoke, Y.S. Raptis, K. Syassen: Phys. Rev. B**45**, 14419 (1992).
51. C.S. Yoo and W.J. Nellis: Chem. Phys. Lett. **198**, 379 (1992).
52. B.L. Zhang, C.Z. Wag, K.M. Hoand, C.T. Chan: Europhys. Lett. **28**, 219 (1994).
53. P. Haaland, R. Pachter, M. Pachter, W. Adams: Chem. Phys. Lett. **199**, 379 (1992).
54. M. Jaime and M. Núñez-Regueiro: unpublished.
55. E. Sandré and F. Cyrot-Lackmann: Sol. Sta. Comm. **90**, 431 (1994).
56. C.S. Yoo and W.J. Nellis: Science **254**, 1489 (1991).
57. M. Núñez-Regueiro: Mod. Phys. Lett. B**6**, 1153 (1992).
58. A.M. Rao, P. Zhou, K.A. Wang, G.T. Hager, J.M. Holden, Y. Wang, W.T. Lee, X.X. Bi, P.C. Eklund, D.S. Cornett, M.A. Duncan, I.J. Amster: Science **259**, 95 (1993) and Chap. 4 in this volume.
59. S. Pekker, A. Janossy, L. Mihaly, O. Chauvet, M. Carrard, L. Forro: Science **265**, 1077 (1994) and Chap. 7.
60. N. Takahashi, H. Dock, N. Matzuzawa, M. Ata: J. Appl. Phys. Lett. **74**, 5790 (1993).
61. Y.B. Zhao, D.M. Poirier, R.J. Pechman, J.H. Weaver: Appl. Phys. Lett. **64**, 577 (1994).
62. O. Béthoux, M. Núñez-Regueiro, L. Marques, J.-L. Hodeau, unpublished.
63. H. Yamawaki, M. Yoshida, Y. Kakudate, S. Usuba, H. Yokoi, S. Fujiwara, K. Aoki, R. Ruoff, R. Malhotraand, D. Lorents: J. Phys. Chem. **97**, 11161 (1993).
64. O. Béthoux, M. Núñez-Regueiro, L. Marques, J.-L. Hodeau, M. Perroux, paper presented at the Materials Research Society meeting, Boston MA, 29 November to 3 December 1993.
65. D. Strout, R.L. Murry, C. Xu, W. Eckhoff, G.K. Odom, G.E. Scuseria: Chem. Phys. Lett. **214**, 576 (1993).
66. Y. Iwasa, T. Arima, R.M. Fleming, T. Siegrist, O. Zhou, R.C. Haddon, L.J. Rothberg, K.B. Lyons, H.L. Carter Jr., A.F. Hebard, R. Tycko, G. Dabbagh, J.J. Krajewski, G.A. Thomas, T. Yagi: Science **264**, 1570 (1994).
67. B. Sundqvist, A. Soldatov, O. Anderson, A. Lundin, P.A. Persson: Proceed. 187th Meeting of ECS: Fullerenes, Chemistry, Physics, New Directions VII; ed. K.M. Kadish and R. Ruoff (1995).
68. I.O. Bashkin, V.I. Rashchupkin, A.F. Gurov, A.P. Moravsky, O.G. Rybchenko, N.P. Kobelev, Ya.M. Soifer, E.G. Ponyatovsky: J. Phys. Condens. Matter **6**, 7491 (1994).
69. V.D. Blank, S.G. Buga, N.R. Serebryanaya, V.N. Denisov, G.A. Dubitsky, A.N. Ivlev, B.N. Mavrin, M.Yu. Popov: Phys. Lett. A**205**, 208 (1995).
70. C. Yeretzian, K. Hansen, F. Diederich, R.L. Whetten: Nature **359**, 44 (1992).
71. E.E.B. Campbell, V. Schyja, R. Ehlich, I.V. Hertel: Phys. Rev. Lett. **70**, 263 (1993).
72. D.S. Cornett, I.J. Amster, M.A. Duncan, A.M. Rao, P.C. Eklund: J. Phys. Chem. **97**, 5036 (1993).
73. P. Zhou, Z.-H. Dong, A.M. Rao, P.C. Eklund: Chem. Phys. Lett. **211**, 337 (1993).
74. Y. Wang, J.M. Holden, Z.-H. Dong, X.-X. Bi, P.C. Eklund: Chem. Phys. Lett. **211**, 341 (1993).
75. P.C. Eklund: results presented at the American Chemical Society National Meeting, Denver, Colorado (1993).
76. B.L. Zhang, C.Z. Wang, C.T. Chan, K.M. Ho: J. Phys. Chem. **97**, 3134 (1993).

77. P. Ballone and P. Milani: Z. Phys. D **19**, 439 (1991).
78. G. Seifert and R. Schmidt: Clusters and Fullerenes; V. Kumar, T.P. Martin, E. Tosatti (Eds.) (World Scientific, 1992).
79. X. Long, R.L. Graham, C. Lee, S. Smithline: J. Chem. Phys. **100**, 7223 (1994).
80. M. Menon, K.R. Subbaswamy, M. Sawtarie: Phys. Rev. B **49**, 13966 (1994).
81. N. Matsuzawa, M. Ata, D.A. Dixon, G. Fitzgerald: J. Phys. Chem. **98**, 2555 (1994).
82. M.R. Pederson and A.A. Quong: Phys. Rev. Lett. **74**, 2319 (1995).
83. G.B. Adams, J.B. Page, O.F. Sankey, M. O'Keeffe: Phys. Rev. B **50**, 17471 (1994).
84. W.S. Bacsa and J.S. Lannin: Phys. Rev. B **49**, 14750 (1994).
85. P.W. Stephens, G. Botel, G. Faigel, M. Tegze, A. Janossy, S. Pekker, G. Oszlanyi, L. Forro: Nature **370**, 636 (1994).
86. C. Goze, F. Rachdi, L. Hajji, M. Núñez-Regueiro, L. Marques, J.-L. Hodeau, M. Mehring: Phys. Rev. B **54**, R3676 (1996).
87. L. Marques, J.-L. Hodeau, M. Núñez-Regueiro: in *Proceedings IUTAC 95*, St. Petersburg, in press.
88. I.O. Bashkin: in *Proceedings IUTAC 95*, St. Petersburg, in press.
89. J.M. Thomas, S.E. Morsi, J.P. Desvergne: Adv. Phys. Chem. **15**, 63 (1977).
90. C.H. Xu, C.Z. Wang, C.T. Chan, K.M. Ho: J. Phys. Condens. Matter **4**, 6047 (1992).
91. J.E. Fischer, P.A. Heiney, A.R. McGhie, W.J. Romanow, A.M. Denenstein, J.P. McCauley, Jr., A.B. Smith, III: Science **252**, 1288 (1991).
92. J.M. Schulman and R.L. Disch: J. Chem. Soc. Chem. Commun. **6**, 411 (1991).
93. G. Oszlanyi and L. Forro: Sol. State Comm. **93**, 265 (1995).
94. M. O'Keeffe: Nature **352**, 674 (1991).

8. Polymerization of Charged C_{60} Molecules

H. Kuzmany, J. Winter

8.1 Introduction

A review is given on our present knowledge about polymerization and polymerization products of charged C_{60} in AC_{60} systems where A is K, Rb, or Cs, respectively. It covers initial experiments, conditions for polymerization, structural analysis, morphology, stability, and physical properties of the polymerized material.

The above mentioned compounds are face centered cubic (fcc) at temperatures above about 400 K and undergo a reversible phase transition below this temperature. Reported transition temperatures are between 350 K and 450 K on heating and about 20-30 K lower values on cooling. For RbC_{60} and CsC_{60}, the transition always leads to an orthorhombic phase with an unusual small interfullerene distance of only 9.2 Å along the **a** direction. There is wide evidence that this unusually low value is due to a polymerization by a $2+2$ cycloaddition process. For the potassium compound, either the same type of transition or a phase separation into pristine C_{60} and K_3C_{60} is observed. Which of the two processes is preferred depends on the cooling rate. For slow cooling the fcc compound separates into the two phases whereas fast cooling leads to the polymeric phase.

In the temperature range $\approx 50\,\mathrm{K} < T < T_c$ the polymers are metallic. For lower temperatures RbC_{60} and CsC_{60} undergo a magnetic ordering transition and become insulators. In this chapter, evidence for the metallic state and for the various phase transitions is collected from mainly structural, vibrational, electric, magnetic and thermodynamic investigations. Whereas KC_{60} turns out to be a normal 3-dimensional metal for RbC_{60} and CsC_{60} a quasi-one dimensional electronic structure seems to be a more likely description from the experiments. To date, band structure calculations turn out to be in favor of a 3-dimensional electronic structure.

Quenching the high temperature phase to liquid nitrogen temperature yields a conducting simple cubic phase which transforms into a monoclinic dimer phase on heating. As the sample is warmed further, the dimer phase is dissolved and around 270 K the orthorhombic polymeric phase is generated.

Coevaporation of alkali metal and C_{60} leads to metallic ribbons with highly oriented polymeric chains.

C_{60} fullerenes with intrinsic polymeric character have recently attracted considerable interest. Three routes are known, currently to make such compounds. Photoinduced transformation leads to finite clusters of covalently bonded C_{60} molecules. Similarly, the simultaneous application of high pressure and high temperature results in polymeric structures of different crystalline forms, depending on the experimental conditions for the preparation. Finally, and rather surprising, charging the C_{60} molecules with one electron to -1 can lead to a polymerized material with several unexpected physical properties like a metallic ground state, a reduced dimensionality in the electronic structure, magnetic ordering transitions at low temperatures, etc. The subject of this chapter is concerned with the latter type of material. As it will be shown, these compounds represent a new class of materials in the family of fullerene conductors.

After a short introduction into the various techniques for the doping process, early investigations on the phases AC_{60} (A=K, Rb, Cs) and the detailed structure of the polymeric phases will be reviewed. This description is followed by a summary of the physical properties of the polymeric phases including lattice dynamics, magnetic response and transport. Particular attention is paid to the various structural and magnetic phase transitions. The differences between the various polymeric compounds will be elucidated. The experimental results are supported by a theoretical description of the electronic structure. The last three sections describe another set of phase transitions after quenching the high temperature phase of AC_{60}, the unusual and unexpected stability of AC_{60} versus ambient conditions, and special morphologies of the polymeric compounds. This includes polymeric phases from C_{60} in an anionic state with a charge higher than -1.

8.2 Phases of AC_{60}, A = K, Rb, Cs

One of the many exciting properties of C_{60} fullerenes is the possibility to obtain doped compounds by intercalation of strong electron donors into the pristine C_{60} material. Two structural features of the fullerenes support such a process. The π-orbitals allow for an easy addition of electrons, as in most of the π-electron systems, and the large size of the C_{60} molecule provides ample interstitial lattice space to accommodate the donor atoms. Since C_{60} crystallizes in a fcc structure, one large octahedral and two smaller tetrahedral interstitial sites are available per C_{60} molecule. Filling these sites with a K, Rb, or Cs (or combinations of these metals) has led to the famous family of superconductors A_3C_{60} with transition temperatures as high as 33 K. Any further increase of the concentration of alkali metals is accompanied by changes in the crystal structure. A tetragonally distorted fcc lattice allows the accommodation of four alkali metals, and finally six alkali metals can be accommodated in the six tetrahedral lattice sites of a bcc structure.

The members of the family A_xC_{60} discovered most recently were the compounds with $x = 1$. The delayed observation of these systems may be due to the lack of access to them with the intercalation processes performed at ambient or moderately elevated temperatures. The compounds can only be prepared by doping at high temperatures and exist at room temperature provided that a subsequent cooling was carried out as described in detail below.

Various procedures are used for the doping. The classical way is to equilibrate the fullerene material at a well defined temperature and to expose it to alkali metal vapor [1]. The doping process can be monitored in-situ in this case by watching the change in sample resistance or any other physical property which is significant for the change of the electron concentration. Alternative methods are the sealing of stoichiometric quantities of alkali metal and C_{60} into a quartz tube and equilibrating the system for several days or weeks at elevated temperatures. Instead of pure metals, pristine C_{60} can be sealed into the tube together with an appropriate amount of A_6C_{60}. The subsequent equilibration process yields a single phase material if the concentrations were chosen properly.

The various phases obtained in this way are determined by the structure of the fullerene lattice. This implies that only phases with a well defined stoichiometry are possible (line phases). A generic diagram for the various phases versus alkali metal concentration and temperature was discussed recently for the special case of K_xC_{60} [2]. These types of phase diagrams do not include the polymeric phases as the polymers are not directly accessible by doping.

Details for the description of the doping process and of the resulting structures are summarized in the book *Science of Fullerenes and Carbon Nanotubes* by Dresselhaus et al. [3].

8.2.1 Phase Separation Versus Structural Phase Transition to the Orthorhombic Phase for AC_{60}

Besides direct structural analysis from x-ray diffraction, Raman scattering proved to be an excellent tool to study the various phases of A_xC_{60}. This is possible because the C=C stretch mode of the carbons on the fullerene cage (pentagonal pinch mode) observed at $1469\,cm^{-1}$ in pristine C_{60} exhibits a nearly linear shift of about $7\,cm^{-1}$ per additional electron charge on the cage. Accordingly, early Raman investigations on the doping process gave direct evidence for the existence of the phase KC_{60} if the doping was performed at elevated temperatures. This is demonstrated in Fig. 8.1. In part (a) of the figure, a thin film of C_{60} is exposed to K vapor at 363 K for several hours. The change of the Raman signal for the pinch mode is detected in-situ. With progressive doping, the response from the undoped molecules decreases and a line downshifted by $7\,cm^{-1}$ appears. This line becomes the dominating feature after 4 h of doping. The downshift of the new line by $7\,cm^{-1}$ was good evidence that a phase KC_{60} had developed. The small line on the low frequency side of the response from KC_{60} is downshifted by $22\,cm^{-1}$ from the original position and is thus easily identified as the response from the phase K_3C_{60}. The response of the material on cooling to room temperature is demonstrated in Fig. 8.1b,c for a thin film and for a single crystal single

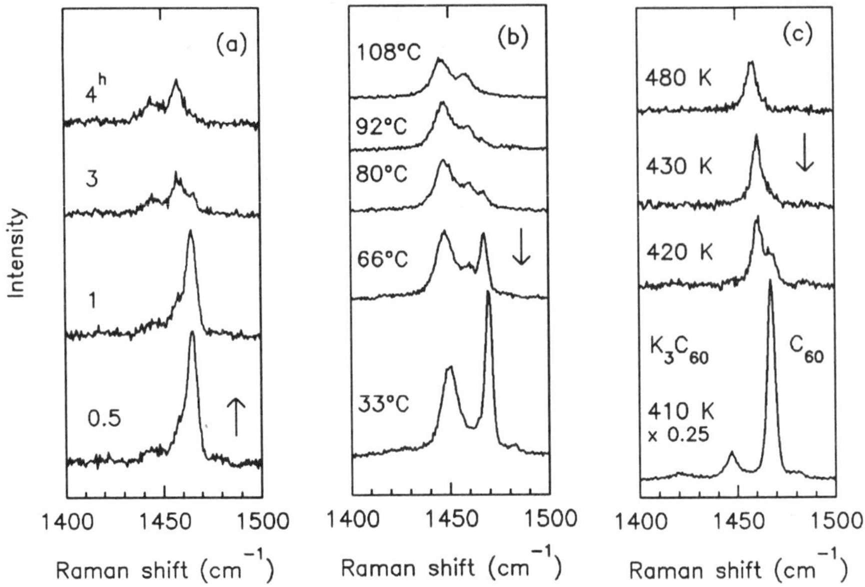

Fig. 8.1. Raman band identified with the pinch mode of C_{60} as observed in-situ during a doping process of a thin film with potassium at 363 K (**a**), change of the response during cooling to room temperature after termination of the doping (**b**); after [4]. Fig. (**c**) is similar to (**b**), but the data are for a single crystal doped to single phase KC_{60}; after [5]

phase KC_{60}, respectively. By approaching a critical temperature, which is slightly higher for the single crystal as compared to the thin film the phase, KC_{60} decomposes and the observed Raman signal is typical for pristine C_{60} with the pinch mode at 1467 cm^{-1} (at 410 K) and metallic K_3C_{60} with the pinch mode at 1448 cm^{-1}. As will be shown below, the situation is very different if the cooling process is performed without interruption at temperatures intermediate between the preparation temperature and room temperature.

Further evidence for the existence of the KC_{60} phase and its instability versus cooling to room temperature was obtained from core level photoemission of the K 2p electrons [6]. For a thin film doped with K to a concentration of $x = 0.7$, a doublet structure with binding energies at 295.3 and 297.8 eV is observed for temperatures of 453 K. The doublet was assigned to electrons with $j = 1/2$ and $j = 3/2$ of the potassium atom in an octahedral lattice site. After cooling to room temperature a pair of doublets is observed which means the potassium is now distributed to the octahedral and to the tetrahedral lattice sites. Accordingly, the KC_{60} phase has decomposed into pristine C_{60} and K_3C_{60}.

RbC_{60} and CsC_{60} turned out to behave different. Early NMR experiments for ^{87}Rb, ^{133}Cs, and ^{13}C in the corresponding compounds gave evidence for

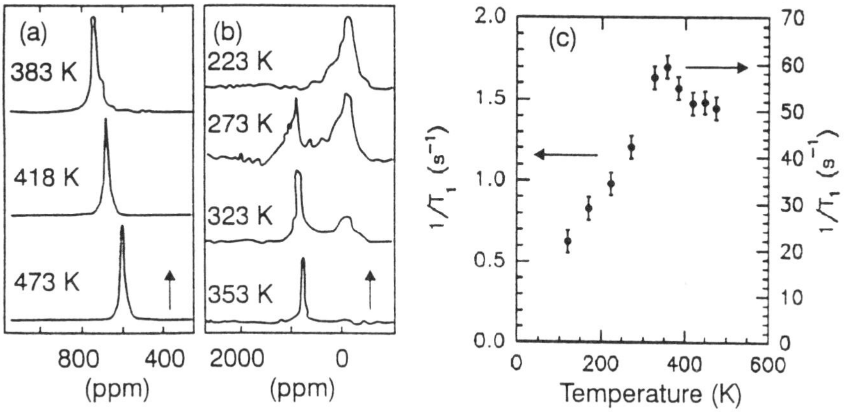

Fig. 8.2. ^{133}Cs NMR spectra for CsC_{60} at the temperatures indicated (**a**) and nuclear-spin-lattice relaxation rate versus temperature for the same compound (**b**). The reference in (**a**) was 1-M $CsCl_{(aq)}$; after [7]

a phase transition between a high temperature paramagnetic state with a rather large chemical shift for the ^{87}Rb and ^{133}Cs nuclei and a low temperature state with a small shift and a broad line. The situation is demonstrated in Fig. 8.2a. The strong increase of the line shift with decreasing temperature is characteristic for localized electron spins in a paramagnetic and nonmetallic state. The spins are located at the C_{60}^- molecules and coupled by $2\,cm^{-1}$ exchange interaction. The extension of the corresponding wave functions to the alkali metals is responsible for the line shift (Fermi contact hyperfine coupling). The small line width indicates the occupancy of a single type of lattice site with rapidly rotating C_{60} molecules.

For ^{13}C NMR, a very narrow line was also observed, but it shifted with the temperature only between 177 and 180 ppm. The orientational correlation was determined from the line width to be less than 10 μs. In the low temperature phase, the line became very broad as for the alkali metals, but showed an unusually small shift of only about 100 ppm.

The existence of a high and a low temperature phase is confirmed from the temperature dependence of the spin-lattice relaxation rate shown in Fig. 8.2b for CsC_{60}. The temperature dependence of this rate is reversed at the phase transition and the spin-lattice relaxation was reduced by a factor of 50 for the low temperature state. For ^{13}C, the relaxation rate was nearly temperature independent and reduced only by a factor of 5.

As a possible explanation for the dramatic changes at T_c, a metallic state was suggested for the low temperature phase where the spin susceptibility is strongly reduced to the Pauli susceptibility for free carriers. This suggestion is consistent with the nearly linear temperature dependence for the relaxation rate shown in Fig. 8.2b for CsC_{60}. From the nearly constant value for $TT_1 \approx 210\,$Ks and the assumption of a Korringa relaxation for the nuclear spins, a density of states at the Fermi level $N(\epsilon_F) = 1/12\hbar A_M \sqrt{\hbar\pi/k_B T T_1} \approx 7.0\,\mathrm{eV}^{-1}$ was determined. A_M is the Fermi contact hyperfine coupling constant which has the value of 0.664 MHz for Cs. The spin-lattice relaxation for the ^{13}C nuclei showed only a smaller reduction at the phase transition. As a tentative explanation for this effect, antiferromagnetic fluctuations were assumed in the metallic state.

A similar evidence for a phase transition around 400 K and for a metallic state at temperatures higher than 50 K was obtained for the RbC_{60} compound from early ESR experiments by Janossy et al. [8]. Actually, this work was written to report on conduction electron spin resonance in Rb_3C_{60} but an additional signal was observed which was later identified as originating from an impurity phase with stoichiometry RbC_{60}. The signal from this phase was found to dependend on temperature treatment. This is now known to be a characteristic property of the phases AC_{60} and will be discussed in Sect. 8.6. The phase transition is evident from the discontinuous change of the line width and occurs at 400 K with a characteristic hysteresis for the cooling and heating cycle of about 40 K as shown in Fig. 8.3. The lower curve in the figure represents the spin susceptibility as derived from the ESR signal. Its nearly temperature independent value is an indication of a metallic behavior. As in metallic systems the linewidth is directly related to the spin-lattice relaxation rate T_1^{-1} as $\Delta H = (2/\gamma)T_1^{-1}$ where γ is the magnetogyric ratio of the electron was determined to be $6 \times 10^7\,\mathrm{s}^{-1}$. The value of T_1^{-1} is similar to Na and thus supported the idea of the metallic state for the low temperature

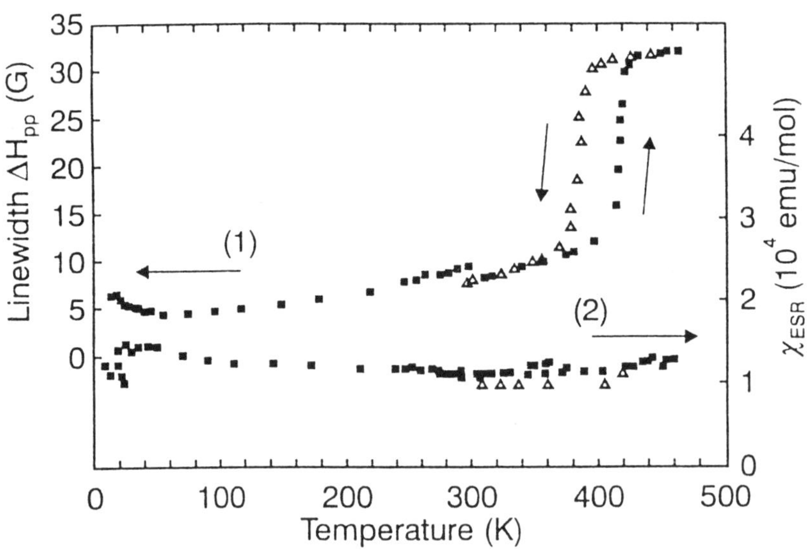

Fig. 8.3. ESR line width ΔH_{pp} (1) and ESR susceptibility χ_{ESR} (2) for an impurity phase Rb_xC_{60} in Rb_3C_{60}. Plotted is a heating and a cooling cycle; adapted from [8]

phase. In fact, from the proportionality of the spin relaxation rate T_1^{-1} to the momentum relaxation rate for the free carriers, a metallic conductivity was derived which turned out to be much higher for RbC_{60} as compared to Rb_3C_{60}.

At the lowest temperatures, another phase transition is indicated (Fig. 8.3 which will be discussed in detail in Sect. 8.3.2.

8.2.2 Structure of the Polymeric Phase

A better understanding of the phase transitions and structural details was obtained from several x-ray investigations [9–12]. The high temperature phase of AC_{60} was identified as a fcc rocksalt structure with the alkali metal ions located only at octahedral interstitial sites. Lattice constants of 14.07 Å, 14.08 Å, 14.13 Å at 473 K for K, Rb, and Cs dopants were observed, respectively [9]. All these lattice constants are smaller than the lattice constant of 14.23 Å for pristine C_{60}. (See also Table 8.1).

A thermal analysis revealed a phase transition for all three compounds with an endothermic peak on heating at around 12 J/g [9]. The transition width was rather broad extending from about 330 K to 430 K, in agreement with the rather wide range of observed transition temperatures described in the NMR experiments. No clear trend could be seen for a scaling of the transition temperature with the size of the alkali metal ion.

Table 8.1. Lattice parameters and bond lengths for polymeric AC_{60}, A = K, Rb, Cs

KC_{60}, r_{ion} = 1.33 Å			Ref.
Lattice parameters (Å)	$a \times b \times c$	9.109 × 9.953 × 14.321	[11]
	fcc cell, 473 K	14.07	[9]
Bond lengths (Å)	C1–C1′	1.65	[11]
	C1–C1	1.74	
	C1–C2	1.50	
RbC_{60}, r_{ion} = 1.48 Å			Ref.
Lattice parameters (Å)	$a \times b \times c$	9.138 × 10.107 × 14.233	[11]
	fcc cell, 473 K	14.08	[9]
Bond lengths (Å)	C1–C1′	1.44	[11]
	C1–C1	1.90	
	C1–C2	1.51	
CsC_{60}, r_{ion} = 1.69 Å			Ref.
Lattice parameters (Å)	$a \times b \times c$	9.1 × 10.21 × 14.17	[43]
	fcc cell, 473 K	14.13	[9]
C_{60}, high temperature, reference cell (for Rb)			
Lattice parameters (Å)	$a \times b \times c$	9.95 × 9.95 × 14.08	

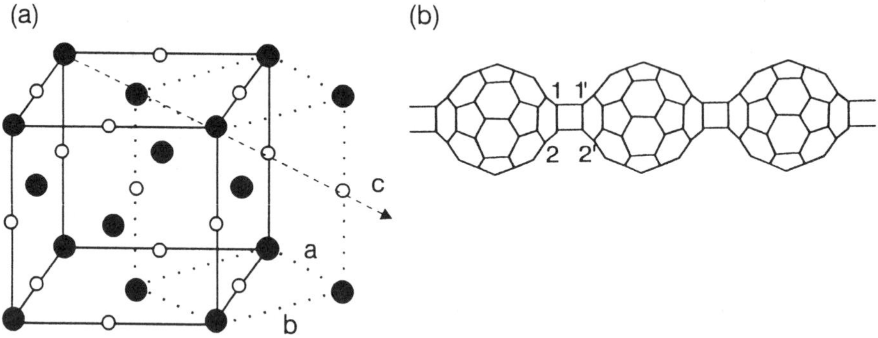

Fig. 8.4. Fcc rocksalt structure (full lines) and derived bco structure (dotted lines) of AC_{60} (**a**) and the polymerized C_{60} chain as obtained from a 2 + 2 cycloaddition (**b**). The dashed line in (**a**) is the chain direction. The numbers on the chain in (**b**) assign the carbon atoms used to describe the bonds in Tab. 8.1

The first x-ray analysis of the low temperature structure in nearly stoichiometric AC_{60} with A = Rb was given by Chauvet et al. [12] and revealed an orthorhombic unit cell with lattice parameters $a = 9.12\,\text{Å}$, $b = 10.10\,\text{Å}$, $c = 14.26\,\text{Å}$. The orthorhombic lattice can be derived from a bc tetragonal representation of the high temperature fcc phase with lattice constants $a = b = 9.95\,\text{Å}$ and $c = 14.07\,\text{Å}$ as shown in Fig. 8.4a. The observed lattice constant in the new a-direction is only 9.12 Å which is unusually short as compared to the center to center distances known for other C_{60} compounds. This unusually short distance motivated Pekker et al. [10] to suggest a polymeric chain structure for the C_{60} molecules in the orthorhombic phase as shown in Fig. 8.4a and b. The connection between the molecules is by a four membered ring established from breaking of two parallel oriented double bonds between adjacent hexagons and reclosing them as σ-bonds between the molecules. This process is known as a 2+2 cycloaddition and is very similar to the one suggested for the photopolymerization of C_{60} by Rao et al. [13]. The chemical reaction is thermally allowed since the conduction band in AC_{60} is derived from a triply degenerate t_{1u} orbital of the neutral molecule. Since from the six available states only one is occupied the reaction can be considered as occurring between an empty and an occupied π^* orbital.

The strong reduction of the center to center distance along the chain in the [110] direction of the original fcc lattice leads to a dramatic volume reduction and thus to large internal strain. The degree of polymerization was estimated to be at least 20 from the width of the x-ray diffraction lines. Also, since the polymerization can start in 6 different [110] directions a multidomain structure with nanosize clusters can be expected. Such clusters have indeed been observed recently by electron microscopy [14].

From the linear arrangement of the molecules in the crystal, a low-dimensional nature for the electronic structure might be expected and was indeed anticipated in the work of Chauvet et al. [12]. The linear character of the electronic structure is not immediately evident since the four membered rings represent an interruption of the next neighbor conjugation. All four carbons are in a highly sp^3 hybridized state. The π-electron overlap relies on second next neighbor transfer integrals. The consequences of this configuration will be discussed in detail in Sect. 8.4.

It is important to stress already at this point that AC_{60} with A = Rb, Cs behaves differently from KC_{60} in several aspects. Intermediate phases during cooling and phase separation as described above have been reported only for the latter compound. On the other hand, if the cooling is fast enough, which means of the order of several K/min, the same orthorhombic structure and the same vibrational pattern are observed for KC_{60} and for RbC_{60}. Whether there is a fundamental difference between the ground state of the two compounds or simply any cooling rate is too fast for the larger diameter Rb or Cs atoms is also an open question.

Activation energies for the two competing processes of phase separation and polymerization were reported recently to be 1.1 and 0.17 eV, respectively [15].

A detailed structural analysis including a determination of the bond lengths was reported by Stephens et al. [11] for KC_{60} and RbC_{60}. The relevant diffraction pattern as obtained from a synchrotron diffraction experiment with 1.14871 Å radiation are shown for the high temperature rocksalt phase and for the low temperature orthorhombic phase in Fig. 8.5. The rocksalt pattern is for KC_{60} and the orthorhombic pattern is for RbC_{60}. The Rietveld fit with an unweighted R factor of 3.8 % fully coincides with the experimental points. The large number of sharp peaks observed even for large angles 2θ is good evidence for the high crystallinity of the polymeric phase. The most plausible orthorhombic space group obtained from the diffraction pattern is Pnnm. This means a simple cubic Bravais lattice with a basis counting of two C_{60} molecules and two alkali metal ions. In other words, two chains, but only one molecule per chain belongs to the primitive unit cell and the cell has D_{2h} point symmetry. This is the same point symmetry as for the C_{60} molecules on the chain.

Without relaxation of the carbon bonds the nearest interfullerene carbon–carbon distance would be 2.23 Å which is too long for any C–C bond. Relaxing the bond lengths around the four membered ring improved the refinement. The obtained lattice parameters and bond lengths are summarized in Table. 8.1. The table contains several important structural data for the AC_{60} phases. The bond lengths between the fullerene cages are in agreement with typical σ bonds between carbon atoms. The intrafullerene C1–C1 bond is unusually large and indicates a very weak σ bonding. It is also important to note that the interfullerene distance C1–C1′ is considerably shorter for RbC_{60} as compared to KC_{60}. The changes of the bond lengths and the new interfullerene bonds reduce the I_h symmetry to D_{2h}.

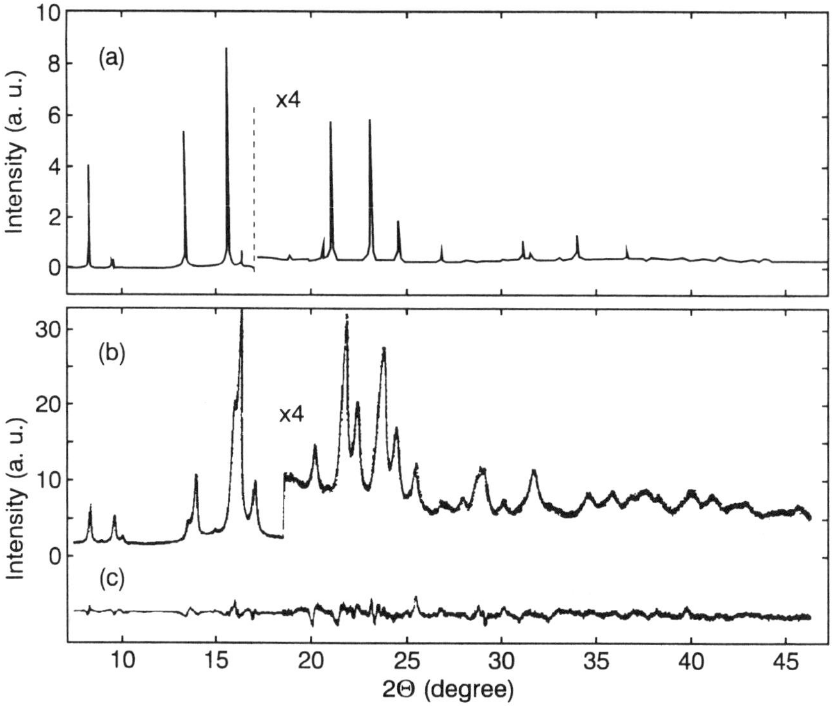

Fig. 8.5. X-ray diffraction pattern for fcc KC_{60} (**a**), orthorhombic RbC_{60} (o-RbC_{60}) (**b**) and the difference between observed and calculated intensity from a Rietveld fit for the orthorhombic pattern (**c**); (**a**) after [16], (**b**) and (**c**) after [11]

8.3 Physical Properties of the Polymeric Phases

Numerous experiments have been carried out to investigate the polymeric character of the orthorhombic phase of AC_{60} and to analyze its physical properties and phase transitions. The following gives selected examples and refers to others by citation. Concentration will be on lattice dynamics, magnetic properties and transport.

8.3.1 Vibrational Spectra and Optical Transitions

Vibrational spectra are most directly related to structural details because of the high symmetry of the C_{60} molecule. Any relaxation from the I_h symmetry of the cage will lead to a splitting of the highly degenerated modes. This splitting has been demonstrated in several papers on Raman [17–20] and IR [28–31] spectroscopy.

The isolated C_{60} cage has 10 Raman active modes, 2 with A_g and 8 with H_g symmetry. The 2 A_g modes appear at 495 cm^{-1} (breathing mode) and

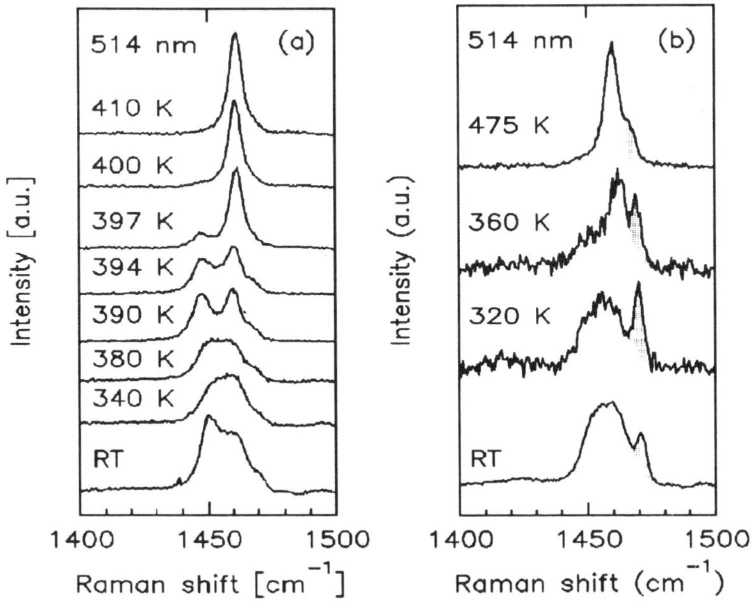

Fig. 8.6. Temperature dependence for the Raman response of the pinch mode on cooling from the high temperature phase for RbC_{60} (**a**), after [18] and for moderate fast cooling of KC_{60} (**b**). The hatched line in (**b**) is from the undoped phase; after [21]

1469 cm^{-1} (pentagonal pinch mode) in pristine solid C_{60}. The H_g modes will be discussed in connection with Table 8.2.

The Raman spectrum showing the temperature dependence of the pentagonal pinch mode during the transition from the high temperature state of RbC_{60} is shown in Fig. 8.6. The behavior is characteristically different from the one shown in Fig. 8.1 for KC_{60}. The example is given in Fig. 8.6a for a single crystal. The pinch mode is downshifted from 1461 cm^{-1} and splits into two components at 1459 and 1453 cm^{-1}, respectively. The splitting starts just below 400 K and is more or less completed at 390 K. A similar behavior is observed for the other Raman lines [18]. This is good evidence for a structural phase transition. Cycling the crystal through the phase transition reveals a hysteresis of several tens of degrees, similar to the hysteresis shown in Fig. 8.3. The width of the hysteresis decreases with increasing number of cycles. The same type of spectra were observed for KC_{60} crystals if the temperature was reduced with a moderate rate [20]. A selected result is shown in Fig. 8.6b. Comparison of this result with the spectra of Fig. 8.1c clearly demonstrates the influence of the cooling speed on the final product. The two spectra taken for 360 K and 320 K exhibit lower signal to noise ratio because they were recorded with a reduced scanning time.

Table 8.2. Raman active vibrational modes observed for AC_{60} phases and orthorhombic pressure polymerized C_{60}. $\Delta\omega_{po}$ and $\Delta\omega_{mo}$ are the frequency shifts between the undoped and doped polymer and the undoped and doped monomer at 450 K, respectively; after [18, 23]

Mode	Pressure polymerized C_{60}			Polymerized RbC_{60}				
	ω cm^{-1}	Int. (a.u.)	γ cm^{-1}	ω cm^{-1}	Int. (a.u.)	γ cm^{-1}	$\Delta\omega_{po}$ cm^{-1}	$\Delta\omega_{mo}$
$H_g(1)$	249	26	8	249	9	7	0	
(266)	256	51	6	259	98	5	+3	−4
(271)	269	27	9	269	10	9	0	−2
	276	38	5	275	17	6	−1	
	344	39	6	344	71	5	0	
$H_g(2)$	416	7	6	402	4	5	−14	
(430)	425	35	5	412	6	9	−13	
	431	41	5	431	18	5	0	−9
	452	112	5	452	10	5	0	
$A_g(1)$ (495)	489	100	5	490	100	5	+1	+1
$H_g(3)$	638	67	5	629	17	8	−9	
(709)	698	33	5	684	7	5	−14	
	708	97	7	696	12	14	−12	
$H_g(4)$	754	57	5	731	5	6	−23	
(770)				738	5	6		
	768	33	7	756	10	5	−12	−4
	775		5	762	8	8	−13	
	952	28	14					
	967	71	14	967	13	10	0	
	1041	13	18	1038	4	8	−3	
				1049	3	7	+8	
$H_g(5)$	1087	39	9	1092	5	20	−5	−5
(1099)	1109	69	9	1109	5	19	0	
	1194	39	12	1191	14	7	−3	
$H_g(6)$ (1246)	1261	30	8	1251	5	9	−10	
	1310	27	8	1300	7	10	−10	
$H_g(7)$	1398	22	9	1389	12	11	−9	
(1421)	1411	60	9	1407	7	16	−4	−15
	1426	75	14					
	1434	149	6	1424	8	20	−10	
	1446	59	10					
$A_g(2)$	1458	878	6	1450	94	9	−8	−7
(1467)	1464	299	10	1460	81	11	−4	
				1493	9	12		
$H_g(8)$	1563	110	18	1530	14	34	−33	−21
(1572)	1577	122	8	1559	8	20	−18	

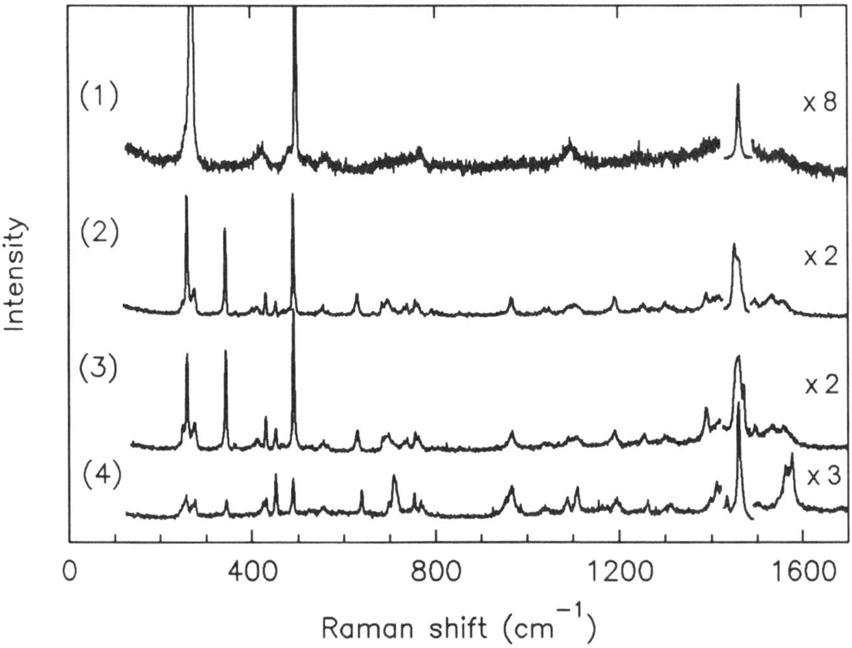

Fig. 8.7. Raman spectra for various phases of AC_{60}; (1): fcc-RbC_{60}, (2): o-RbC_{60}, (3): o-KC_{60}, (4): pressure polymerized C_{60} (orthorhombic phase) after [18] and [23]

The equivalency of the low temperature RbC_{60} phase (o-RbC_{60}) and the low temperature KC_{60} phase (o-KC_{60}) after moderately fast cooling becomes evident from an inspection of the whole range of internal vibrational modes. Interestingly, these spectra are strongly correlated to spectra obtained after pressure polymerization performed at low pressure and rather low temperatures. Such polymers were found also to exhibit an orthorhombic structure (op-C_{60}) [22]. The correlation between the two types of spectra is so close that the o-RbC_{60} material and the o-KC_{60} material can be considered as the doped version of op-C_{60} [23]. The relation between the spectra and their correlation to the high temperature phase of AC_{60} is demonstrated in Fig. 8.7. Spectra (2) and (3) are identical even with respect to small details, which is good evidence for the equivalency of the two phases. As compared to spectrum (1) for the high temperature phase, the H_g modes for both spectra are widely split, most of them into five components, as expected for the reduction of the symmetry from I_h to D_{2h}. The splitting can be as high as $95\,\mathrm{cm}^{-1}$ as for example for the mode $H_g(1)$. This is evidence for a dramatic distortion of the cages, as can be expected from a covalent bonding between them. The Raman response for the orthorhombic pressure polymerized C_{60} is shown as graph (4) in the figure. The correlation between the response from the undoped and the doped phases is evident. An explicit comparison was possible and is shown in Table 8.2 where the detected lines are compiled as they are derived from the fundamental modes for fcc-RbC_{60}, o-RbC_{60} and pressure polymerized orthorhombic C_{60}. The first column gives the assignment of the lines and their position in the high temperature phase. The next two triples of columns give the line parameters and the last two columns the lineshift between the two polymeric phases and between singly charged and neutral monomers.

Table 8.3. Correlation table of symmetry groups I_h and D_{2h}

I_h	D_{2h}
2 A_g	$2A_g$
3 T_{1g}	$3B_{1g}+3B_{2g}+3B_{3g}$
4 T_{2g}	$4B_{1g}+4B_{2g}+4B_{3g}$
6 G_g	$6A_g+6B_{1g}+6B_{2g}+6B_{3g}$
8 H_g	$16A_g+8B_{1g}+8B_{2g}+8B_{3g}$
1 A_u	$1A_u$
4 T_{1u}	$4B_{1u}+4B_{2u}+4B_{3u}$
5 T_{2u}	$5B_{1u}+5B_{2u}+5B_{3u}$
6 G_u	$6A_u+6B_{1u}+6B_{2u}+6B_{3u}$
7 H_u	$14A_u+7B_{1u}+7B_{2u}+7B_{3u}$

Most of the new lines observed in the spectra for the polymer can be assigned to the splitting of H_g modes. However, since in D_{2h} all gerade modes are Raman active additional lines can show up. A correlation table of the symmetry groups I_h and D_{2h} is shown in Table 8.3.

The difference in linewidth between modes on the neutral and on the charged chain deserves attention. The linewidth is expected to be broader for the charged chain since electron–phonon coupling is expected to broaden the lines in a system with free carriers. This effect was explicitly demonstrated for the system A_3C_{60} in several papers [24–26]. From the nearly identical linewidths between o-RbC_{60} and op-C_{60} the coupling to the free carriers is small in general. The mode $H_g(8)$ is an exception, as it shows a very strong lineshift and line broadening upon doping. Line shapes for this mode are explicitly shown in Fig. 8.8. The broadening of the line is consistent with a metallic state of the o-RbC_{60} phase. From Allen's formula $\lambda = g_i\gamma_i/2\pi N(\epsilon_F)\omega_i^2$ the dimensionless electron–phonon coupling constant λ is determined as 0.001 for a density of states at the Fermi level $N(\epsilon_{\mathrm{F}})$ as derived from the NMR experiments of Tycko. This is a factor of 3 smaller than in the case of Rb_3C_{60}. Since the other modes exhibit an even smaller electron–phonon coupling this explains the lack of superconductivity even for KC_{60}.

Infrared spectroscopy can give similar results as Raman scattering with respect to phase separation and phase transitions from lineshift and line splitting. The infrared active modes of the free C_{60} molecule are of T_{1u} symmetry and observed at 527, 576, 1183, and 1429 cm^{-1} for pristine solid C_{60}.

From early experiments on thin films, where a line splitting was not resolved, the four T_{1u} lines were observed for RbC_{60} at room temperature at 525, 574, 1182, and 1392 cm^{-1} [28]. For a fit of the reflectivity spectra a Drude term was required with a plasma frequency half that off the plasma frequency for K_3C_{60}. This indicated already metallic behavior [28]. For KC_{60}, phase separation on slow cooling from the high temperature fcc phase was reported [30] similar to the results from Raman scattering.

Single crystal reflectivity experiments revealed more details about the behavior of the fundamental modes and the phase transition. Figure 8.9 shows the reflectivity from a single crystal [111] face in the spectral range of the 4 fundamental modes for various temperatures. The response for the mode $T_{1u}(4)$ is shown for several closely spaced values of temperature. Between 405 and 403 K, a characteristic change in the spectra is observed. All three lines split into three components each as expected for a reduction of the symmetry from I_h to D_{2h} (See also Table 8.3). These lines must be nondegenerate with B_{u} symmetry. The transition occurs within a temperature interval of 2 K. The relevant line positions are summarized in Table 8.4. The splitting is again very high, indicating a strong deformation of the molecules in the o-RbC_{60} phase. Additional IR modes have been observed, since in D_{2h} symmetry, all ungerade modes from the I_h group become nondegenerate and IR active except the A_u modes [39].

External modes can be expected to be another excellent tool to reveal covalent bonding between molecules. They have not yet been observed by optical methods but were reported recently from neutron scattering [32, 33, 35]. For the high temperature phase only acoustic modes and optical modes from

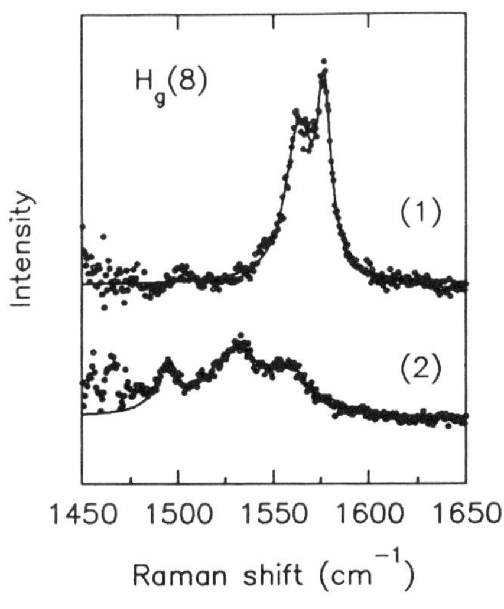

Fig. 8.8. Raman lineshape for the $H_g(8)$ derived mode of pressure polymerized orthorhombic C_{60} (1) and o-RbC_{60} (2); after [23]

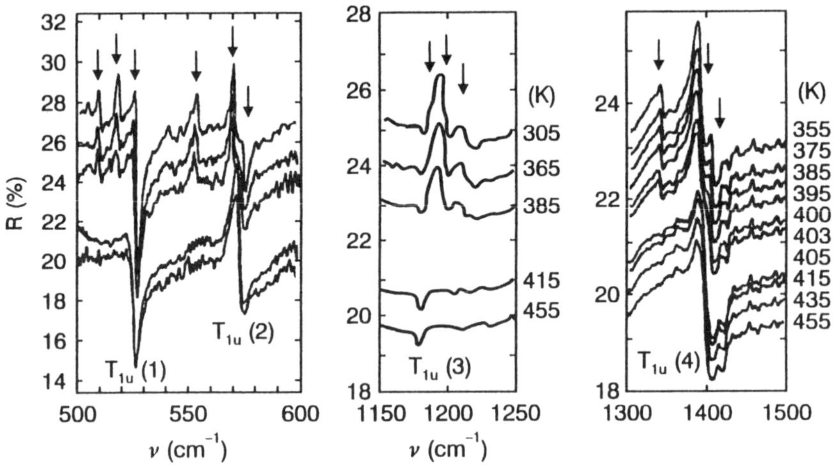

Fig. 8.9. Infrared reflectivity from single crystal RbC_{60} in the spectral range of the T_{1u} modes for various temperatures. The temperatures are indicated on the side of the individual graphs. The down arrows indicate the position of the modes in the low temperature phase; after [27]

Table 8.4. Line positions in cm^{-1} for the IR active modes in RbC_{60}

Mode	1	2	3	4
C_{60} reference (RT)	525	576	1183	1429
fcc- RbC_{60}	526	574	1180	1395
o-RbC_{60}	526	554	1190	1345
	517	571	1198	1390
	503	574	1209	1410

Rb are expected. These modes are in the frequency region below about $50\,cm^{-1}$ as in the conventional alkali fullerides. Below the fcc to orthorhombic phase transition librational modes exist which extend to $130\,cm^{-1}$ as calculated for an infinite uncharged single chain by Adams et al. [34]. Using cold neutrons of only 4.8 meV energy and a time of flight spectrometer, high resolution inelastic scattering revealed the phonon density of states for both phases. Results are shown in Fig. 8.10. Part (a) of the figure is for the high temperature phase. The low frequency peak is in the expected frequency range for the acoustic modes and includes also the Rb modes. The high frequency peak is from the lowest energy internal $H_g(1)$ mode. The two graphs in Fig. 8.10b correspond to the orthorhombic phase and to the dimer phase. For the orthorhombic phase new excitations in the frequency range between 90 and $170\,cm^{-1}$ appear which is characteristic for librations in covalently bonded chains. Thus, these experiments give explicit evidence for the polymeric nature of o-RbC_{60}. The lower energy for the librations in the dimers is evident from their special bonding, as will be discussed below.

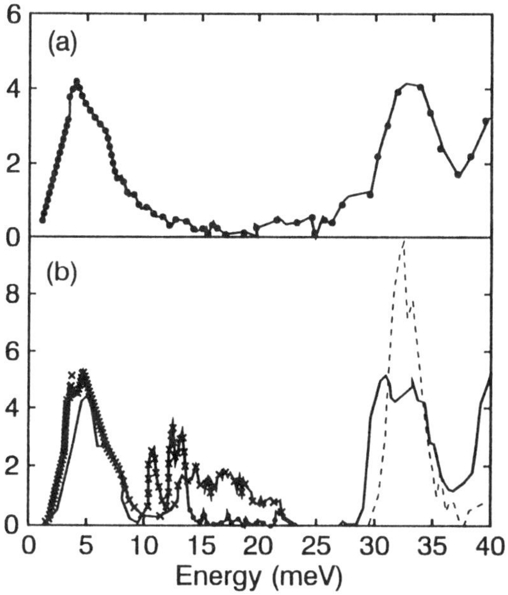

Fig. 8.10. Phonon density of states from inelastic neutron scattering for fcc-RbC_{60} (**a**), and o-RbC_{60} (points), and dimerized RbC_{60} (crosses) (**b**); after [32]

An interesting detail for the temperature behavior of the low energy modes was reported in [32]. It is shown in Fig. 8.11, where the scattering intensity is expressed by the dynamical structure factor $S(q,\omega)/T \propto \omega^{-1}\chi''$. The excitations soften with decreasing temperature but do not really go to zero for $T \to 0$. This unusual effect may have to do with a renormalization of the phonon energy by free electrons. As discussed below, at low temperatures RbC_{60} undergoes a metal–insulator transition which could significantly change the renormalization.

Only preliminary data are available for the optical properties of the polymeric chains. Both early work and more recent work on photoemission revealed densities of states for the lower components of the π-band similar to undoped or metallic A_3C_{60} [17, 36, 37]. Figure 8.12 shows recent results from UPS which also include the photoemission after quenching, as will be discussed in Sect. 8.6. A finite emission at the Fermi level is observed for the orthorhombic phase and for the rocksalt phase but not for the dimer. The result supports the metallic character for the orthorhombic phase, but it is in disagreement with the results from NMR discussed above for the high temperature phase. This phase is claimed in most experiments to be a semiconductor. Either a broadening of the Fermi edge due to the high temperature or a strong disorder with a generation of localized states at the Fermi level may be responsible for this discrepancy.

According to an alternative interpretation by Kirova et al. [38] for the data obtained from the high temperature phase even the latter is metallic. The Curie-like spin susceptibility is a consequence of the low Fermi energy and

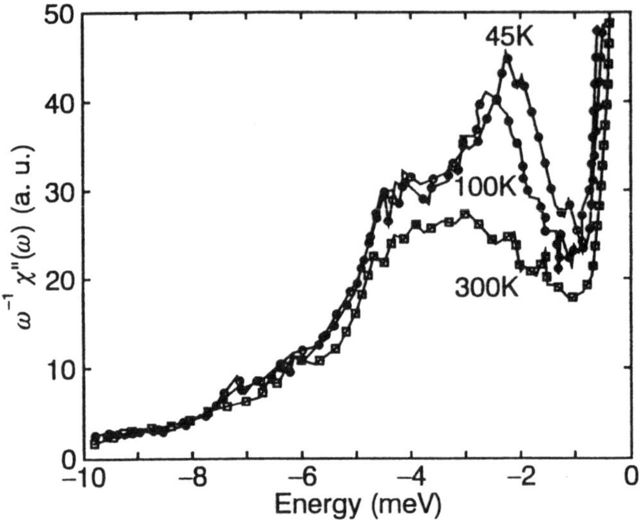

Fig. 8.11. Neutron scattering intensity for o-RbC_{60} in the low energy range of the lattice modes for three different temperatures; after [32]

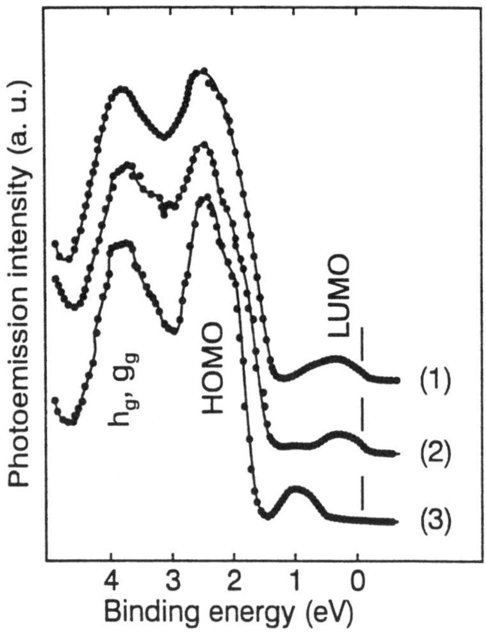

Fig. 8.12. Photoemission for RbC_{60} for three different compounds; (1): rocksalt at 525 K, (2): orthorhombic at 150 K, (3): dimer at 150 K; after [37]

the high temperature. The electron gas loses its degeneracy and Boltzmann statistics applies. Under these conditions the free carrier spin susceptibility is no longer temperature independent.

No optical anisotropy could be observed for the orthorhombic phase [39]. This can be considered as evidence for the distribution of microdomains in the material.

8.3.2 Magnetic Resonance

The early experiments on ESR have been confirmed and extended to the other AC_{60} compounds. The nearly temperature-independent spin susceptibility and a linewidth increasing with T confirmed the metallic character of the polymeric chains in the temperature range between 50 K and 400 K for RbC_{60} [12]. Results are shown in Fig. 8.13(a) for the spin susceptibility, the linewidths and the g-factors as a function of temperature. The small observed line width of only 2 to 6 G was considered as being due to a quasi-onedimensional character of the free electrons. The reduced dimensionality was derived from the chain structure in the orthorhombic phase. Even though this conclusion is not so straightforward, it is consistent with several details of the experimental results. In conventional metallic systems the width of the ESR line is determined by the Elliott mechanism based on a static or dynamic disorder of the lattice coupled to the free electrons by the spin-orbit interaction. This mechanism is not effective in 1-dimensional systems. In contrast, the linewidth is much broader in the 3-dimensional metal Rb_3C_{60} where the Elliott mechanism is fully active.

The observed drop of the spin susceptibility below 50 K and the simultaneous increase of the linewidth with decreasing temperature is also consistent with a quasi-one-dimensional character of the electronic system. For strong electron–electron interaction such systems are known to undergo a transition to a spin-density wave state [41]. For the spin-density wave system, a gap opens at the Fermi energy (without any lattice distortion) and the system becomes an insulator. In a very recent analysis it was demonstrated that the nature of the spin-density wave is based on an antiferromagnetic ordering of the C_{60} spins [42]. Interestingly, the K-polymer behaves differently in this point as shown in Fig. 8.13b. The figure compares the spin susceptibilities as a function of temperature for KC_{60}, RbC_{60}, and CsC_{60}. In the high temperature phase for $T > T_c$ the susceptibilities are Curie type. Below this temperature, the susceptibilities are more or less independent of temperature. Around 50 K the Rb and the Cs compound are subjected to the metal–insulator transition, whereas the K compound stays metallic even at very low temperatures. This suggests that the electronic system in KC_{60} remains 3-dimensional in the paramagnetic o-phase. Since the width of the ESR absorption is still only of the order of several G in this compound, other reasons must be found why the Elliott mechanism is not effective. As the atomic number is known to have a considerable influence on the effectiveness

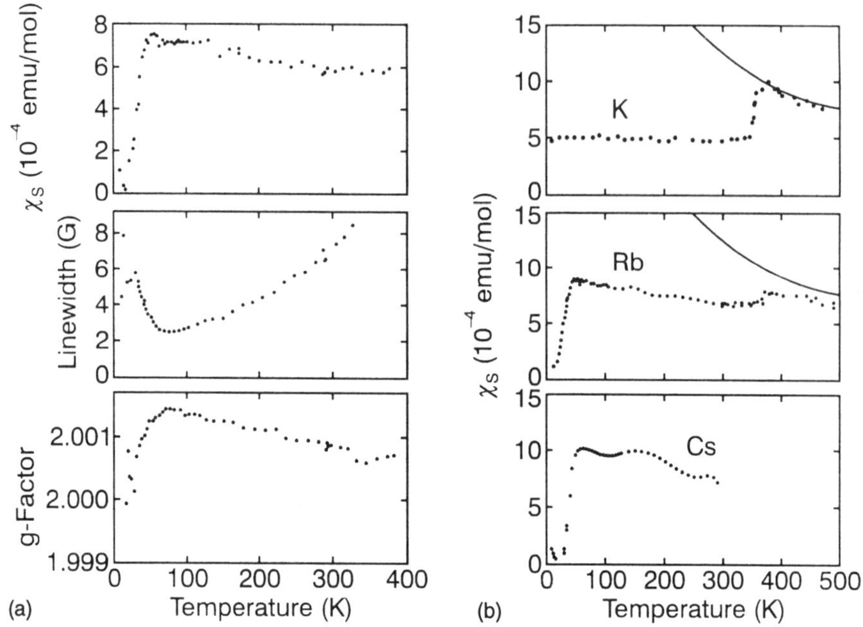

Fig. 8.13. Spin susceptibility, linewidth and g factor for polymeric RbC_{60} versus temperature, after [12] (**a**) and spin susceptibility for three different polymeric C_{60} compounds versus temperature, after [43]. The result for Cs is from [44] (**b**)

of the latter, the smaller atomic number of K as compared to Rb or Cs was suggested to be the reason for the lack of line broadening in the potassium polymer.

As could be expected from the early work of Tycko et al. [7] NMR can give additional valuable information on the structures of the polymeric phases. This information comes from NMR line position, line intensity, linewidth and relaxation rates. Both, the ^{13}C and the alkali metal nuclear spins can be used as probes. Table 8.5 compiles the line positions for ^{13}C NMR in the various phases, as will be discussed in the following.

Table 8.5. Positions for ^{13}C NMR resonances for the three phases KC_{60}, RbC_{60}, CsC_{60}. All numbers are in units of ppm versus tetramethylsilan (TMS)

C_{60}, αC_{60}: 143			$^{13}C(sp^3)$: 37 ±5	
AC_{60}	fcc	polymer	sp^3	A_3C_{60}
K	173–175	40–400	40	186
Rb	179	40–400	40	190
Cs	177–180	0–400		

A detailed analysis of the KC_{60} system with respect to temperature, cooling, and heating conditions performed by Kälber et al. [45] confirmed the results from Raman spectroscopy. Slow cooling or equilibrating just below the temperature for phase separation leads to a disintegration of the phase into α-C_{60} (fcc C_{60} with dissolved K metal) and K_3C_{60}. For rapid cooling to room temperature, a large part of the high temperature phase is transformed to the polymeric state. By heating beyond 380 K it transforms back to the fcc phase. The line after rapid cooling becomes very broad and resembles the one of RbC_{60} in [7]. It thus proves that the same polymeric structure is obtained.

The large width for the resonance in the low temperature phase can be expected to originate from isotropic broadening due to the different carbons in the relaxed chain. If the molecule is distorted from a relaxation after a 2+2 cycloaddition, 9 different carbon sites are expected with relative intensities 4:4:4:8:8:8:8:8:8 [47, 49]. On one site (with intensity 4) the carbons should exhibit a sp^3 hybridization. To see the sp^3 carbons was certainly a fundamental check on the suggestion of the polymeric state in the orthorhombic phase. Such experiments can be performed by high resolution magic angle spinning (MAS) NMR. A strong splitting of the broad line for the RbC_{60} phase was indeed reported by Mehring et al. [46] using a spinning speed of 2.1 KHz. To increase the resolution and avoid spinning sidebands the spinning speed was later increased to 10 kHz. With such experiments it was possible to identify a rather weak NMR response around 40 ppm in KC_{60} as originating from the sp^3 carbons [45]. Evidence for the identification of the line was taken from its position and a different relaxation behavior. The total observed shift δ is in general a combination of a Knight shift K and a chemical shift σ. Because of the lack of paramagnetic electrons, sp^3 carbons exhibit only a chemical shift and relax much slower than the rest of the carbons on the cage in the sp^2 hybridization.

In the case of RbC_{60}, 8 of the nine expected lines including the response from the sp^3 carbons could be resolved [47]. Similar results were reported by Alloul et al. [48, 49]. With an even higher spinning speed of 18 kHz all 9 carbon sites could be resolved for RbC_{60} as shown in Fig. 8.14. To identify the response from the sp^3 carbons, the same technique as described above was used. The observed chemical shift was $\sigma = 35 \pm 5$ ppm. The relative intensities of the lines $1-8$ did not follow the expected relation mentioned above. This was explained by an influence of the interchain transfer integrals on the electron densities at the various sites.

The NMR response for KC_{60} is related to that for RbC_{60} but is different in detail. Because of the larger spin susceptibility for the K-polymer the spectrum is plotted in Fig. 8.14 rescaled by a factor of 1.36 on the x-axis for the purpose of comparison. The differences are particularly strong for low lineshifts which corresponds to carbons in the low spin density region. Since these are likewise the carbons which are most strongly influenced by interchain transfer integrals the carbons in the electron-rich regions are as-

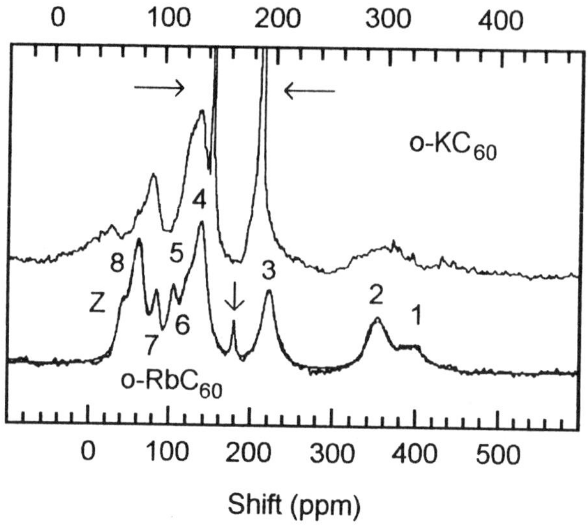

Fig. 8.14. Response of high resolution MAS NMR for KC_{60} and RbC_{60}. The full drawn line for RbC_{60} is an oscillator fit. Z assigns the response from the sp^3 carbons, the numbers 1–8 the response from the other carbons. The spectrum for KC_{60} is scaled by a factor 1.36 as described in the text. Arrows point to lines from pristine C_{60} and metallic A_3C_{60}, respectively; after [49]

sumed to participate more strongly in the intrachain transfer integrals. This assignment provides an explanation for the 1-dimensional character of the electronic system in RbC_{60} and CsC_{60}, but not in KC_{60}.

Besides the difference in the spin susceptibility, also the spin-lattice relaxation rate T_1^{-1} is as much as a factor 4 larger in RbC_{60} as compared to KC_{60}. Moreover, whereas KC_{60} exhibits a Korringa behavior as expected for 3-dimensional metals, T_1 is nearly independent of the temperature in RbC_{60}. Thus, $(T_1T)^{-1}$ increases strongly with decreasing temperature. The situation is demonstrated in Fig. 8.15. The fact that T_1 and χ_{spin} are both independent of temperature is reminiscent for low-dimensional electronic systems with strong correlations as is observed in organic charge transfer metals [50] and oxidic superconductors [51]. Accordingly, antiferromagnetic fluctuations on the chain were suggested to be responsible for the unusual behavior of the Rb and the Cs polymer [52]. At low enough temperatures such fluctuations can condense to a magnetically ordered state.

A magnetic phase transition was also reported from μSR experiments for CsC_{60} [53] and for RbC_{60} [54, 55]. These results are consistent with a magnetic phase transition at low temperatures but indicate a larger amount of magnetic disorder as compared to a spin-density wave state.

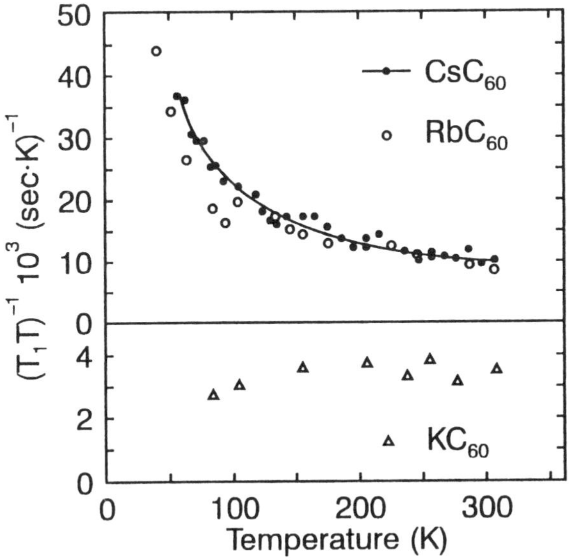

Fig. 8.15. Temperature dependence of $(T_1T)^{-1}$ for ^{13}C NMR in AC_{60} systems. The full drawn line is a fit for $(T_1T)^{-1} = a + bT^{-\kappa}$ with $\kappa \approx 1$; after [52]

8.3.3 Transport

Early experiments on electrical transport were appropriate to reveal the fcc to orthorhombic phase transition as a discontinuity in the resistance [31]. They failed, however, to determine the electrical transport properties because of internal strains and granularity of the material. Recently, more convincing results were obtained, at least for KC_{60}, from free-standing single crystal films [56]. The films were obtained from single crystals doped to a certain penetration depth, typically one to several microns. After this process, the superficially doped crystals were exposed to toluene in which the undoped part of the crystal was dissolved away. The resulting films were well suited for four probe resistivity measurements. Figure 8.16 shows the relative resistivity for KC_{60} in the temperature range from 2 to 450 K. In the intermediate temperature range from 50 K to 350 K $\rho(T)$ has metallic character and can be fitted with a power law in T where the linear term plays an important role. Above 400 K the resistivity increases rapidly in agreement with the proposed transition to a nonmetallic phase. The increase of $\rho(T)$ below 50 K is consistent with the observed magnetic ordering transition observed in the ESR experiments [8, 12], even though this transition was only observed for the Rb and for the Cs compounds. Whether or not the transition occurs seems to be slightly dependent on the sample quality. By applying a hydrostatic pressure the low temperature increase in $\rho(T)$ can be suppressed.

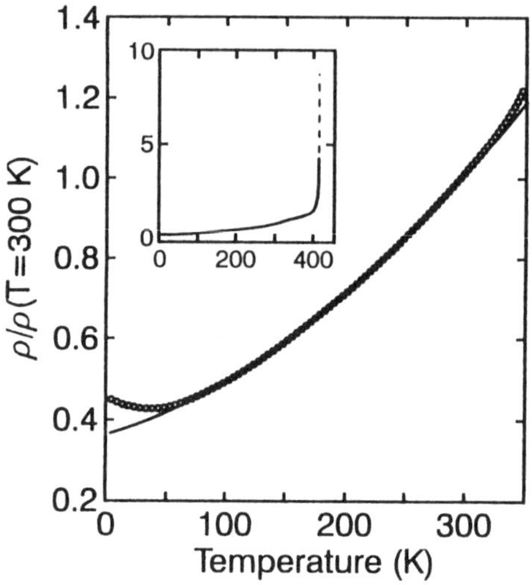

Fig. 8.16. Normalized resistivity for KC_{60} between 2 and 350 K. The insert extends the temperature scale to beyond the orthorhombic to fcc phase transition. The full drawn line is a fit with a polynomial of second order in T (**a**); after [56]. Normalized resistivity versus temperature for RbC_{60} and CsC_{60} (**b**); after [58]

Absolute values for the conductivities for the KC_{60} compounds are about $50\,S\,cm^{-1}$ at 300 K and about $100\,S\,cm^{-1}$ at low temperatures [57]. This compares to $2500\,S\,cm^{-1}$ for K_3C_{60} and $1000\,S\,cm^{-1}$ for Rb_3C_{60} at 300 K.

RbC_{60} and CsC_{60} do not show an explicit metallic dependence of ρ on T. ρ rather strongly increases with decreasing T [58]. This semiconducting behavior is transformed to a metallic temperature dependence under hydrostatic pressure.

Additional information on the conductivity and on the low temperature metal–insulator transition has been obtained from far infrared and microwave reflectivity [43, 59]. Again, the behavior is different for the K-polymer and for the Rb-polymer. In the case of the former, the reflectivity approaches 1 for reducing the frequency to zero for all temperatures below the fcc-o transition. This means a metallic behavior throughout. For the Rb-polymer a similar behavior is observed only for temperatures above 100 K. For temperatures as low as 10 K the reflectivity for $\omega \rightarrow 0$ is reduced to below 60 % as for a semiconducting material. From a Kramers–Kronig transformation, the optical conductivity is obtained as shown in Fig. 8.17. For K it increases for all temperatures and reaches $150\,S\,cm^{-1}$ for the lowest measured frequency of $10\,cm^{-1}$. In contrast, RbC_{60} develops an energy gap for temperatures lower than about 100 K. Since the mean field energy gap for a spin-density wave transition is related to the transition temperature T_{SDW} by a BCS relation as $2\Delta = 3.52k_BT_{SDW}$, the gap energy is expected at $122\,cm^{-1}$ for the $T_{SDW} = 50\,K$, in reasonable good agreement with the beginning of the decrease of the optical conductivity for very low temperatures in Fig. 8.17.

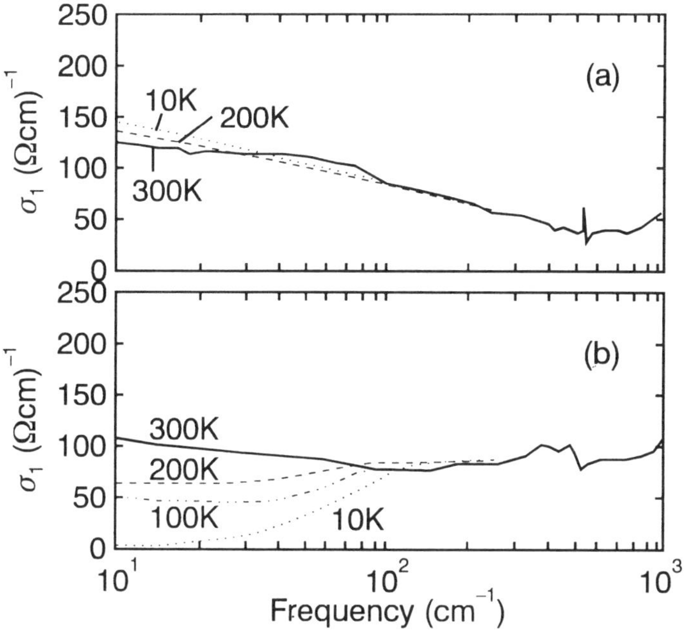

Fig. 8.17. Optical conductivity for KC_{60} (**a**) and RbC_{60} (**b**) versus frequency for the temperatures indicated in K; after [59]

8.4 Electronic Structure of AC_{60}

Calculations concerning the bonding between C_{60} molecules were carried out for various connections and geometries [34, 60, 61] including infinite long chains. Various bonding configurations were found. The bonding by a cycloaddition from two parallel oriented hexagon-hexagon bonds (66/66 bonding) revealed the lowest energy. From several authors, a bonding energy of about 0.5 eV per C_{60} for the dimer as well as for the chain was reported. The electronic structure turned out to exhibit a gap at the Fermi level which is slightly smaller than the gap for the monomer crystal [34]. The electronic structure and optical transitions for the charged dimer as a representation for the charged polymeric chain were calculated by Fagerström et al. [62]. From this, a substantially smaller barrier towards the formation of the polymer was found as compared to the neutral dimer.

The experimental results, for example the shift of the Raman-active pentagonal pinch mode, confirm a full charge transfer between alkali metal and polymeric chain. As a consequence, the three-fold degenerated t_{1u} derived conduction band will be occupied with one electron and render the system in a metallic state in an one electron picture. Even though AC_{60} appears to be indeed metallic at room temperature, the one-electron picture is not really altogether acceptable. For example, it would also predict that the high tem-

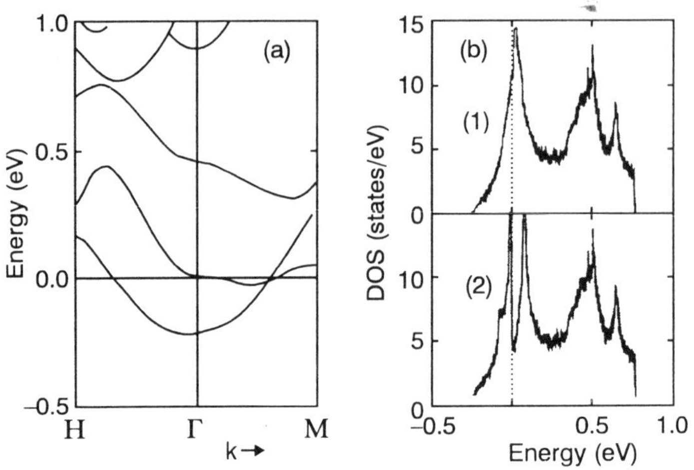

Fig. 8.18. Band structure for polymeric AC_{60} along the chains $(\Gamma - H)$ and perpendicular to the chains $(\Gamma - M)$ (**a**) and density of states for the paramagnetic state (1) and for a magnetically ordered ground state at $T = 0$ (2), (**b**), after [64]

perature phase of AC_{60} is metallic, which is not the case, unless the interpretation of Kirova et al. [38] is anticipated. Also A_4C_{60} which has the conduction band filled up to 2/3 with electrons is not observed to be a metal.

The well defined geometry for the orthorhombic phase allowed for a more detailed evaluation of the electronic structure especially applicable to the AC_{60} family. A simple topological Hamiltonian with a dimerized unit cell on the chain revealed a gap energy of 0.13 eV for the orthorhombic state [63]. More elaborate calculations using the local density approximation in a density functional theory [64] revealed a metallic structure as shown in Fig. 8.18a. The calculation was performed for a simplified unit cell geometry with $Immm$ space group instead of $Pnnm$. This means the translation lattice is bco. The figure gives evidence for a 3-dimensional electronic structure. There is no characteristic difference between the bandwidth along the chains and perpendicular to the chains. It also demonstrates the dramatic lifting of the degeneracy in the t_{1u}-derived band. The orthorhombic symmetry allows only 1-dimensional representations, as was demonstrated for the vibrations. The splitting of the band is as large as the bandwidth itself, and the lower band is about half-filled, as expected from the one electron transferred to the cage. The density of states at the Fermi level for the paramagnetic state is about $13\,\mathrm{eV}^{-1}$ which is nearly a factor of two larger as compared to the results from NMR [7]. Similar results were reported recently by Schulte et al. [65] for the $Pnnm$ geometry.

The stability for the paramagnetic state was checked by allowing for fluctuation of magnetic ordering. A magnetic ordering temperature T_{MO} was found to depend on the electron–electron interaction potential U [64]. For $T_{MO} = 0$ the critical value for U was found to be 250 meV. For $U = 270$ meV, the magnetic ordering temperature was evaluated to be 100 K. In the magnetic ground state a 3-dimensional ordering is obtained. The spin orientation is ferromagnetic along the chains and antiferromagnetic between neighboring chains. The resulting density of states for the paramagnetic ground state and for the magnetically ordered ground state is shown in Fig. 8.18b as graph (1) and (2), respectively. In the magnetically ordered ground state, a pseudogap has developed at the Fermi level, leading to an electrically insulating phase. For U as high as 470 meV, the pseudogap opens to a full gap.

The anomalies observed at low temperatures in the magnetic resonance and transport experiments may be correlated to the opening of the pseudogap. From the calculations presented so far, a reduction in the electronic dimensionality is not obtained unless a change of the interchain distances is provided.

8.5 Stability Versus Atmospheric Conditions

For some time after their discovery, all alkali metal fullerides were observed to be very sensitive to atmospheric conditions. It was therefore quite unexpected when Pekker et al. [66] reported on highly air stable single crystals obtained from a special growth process where K and C_{60} were coevaporated in a sealed quartz tube. In these experiments fibrillar KC_{60} material was obtained embedded in a matrix of undoped C_{60} as will be discussed in detail in the next section. IR spectroscopy was expected to be an excellent tool to check on the air stability since even small amounts of reaction products after air exposure will result in additional absorption lines. Such experiments were reported by Koller et al. [67] from thin film IR transmission and by Winkler et al. [68] from single crystal reflectivity. Both experiments demonstrated the full air stability, as is demonstrated in Fig. 8.19a for the single crystal reflectivity. Spectrum (1) is the reflectivity obtained under vacuum conditions immediately after a RbC_{60} was prepared as a thin film on the surface of the crystal. The response from the T_{1u} modes is indicated by arrows. Spectrum (2) is the reflectivity after the crystal was exposed for one day to air. There is no change observable in the spectrum. This is demonstrated by the difference of the two spectra plotted as spectrum (3). Even after exposure of the crystal for 9 months to air, this difference has hardly changed, as seen from spectrum (4). In contrast, Rb_6C_{60} transforms immediately to the undoped C_{60} plus some reaction products as shown in Fig. 8.19b. The sinusoidal background in the spectrum (1) of the figure is a thin film interference. The numbers indicate the positions of the 4 T_{1u} vibrations for Rb_6C_{60} and C_{60}, respectively. Any time after air exposure the crystals of the orthorhombic phase could be cycled through the orthorhombic to fcc phase transition.

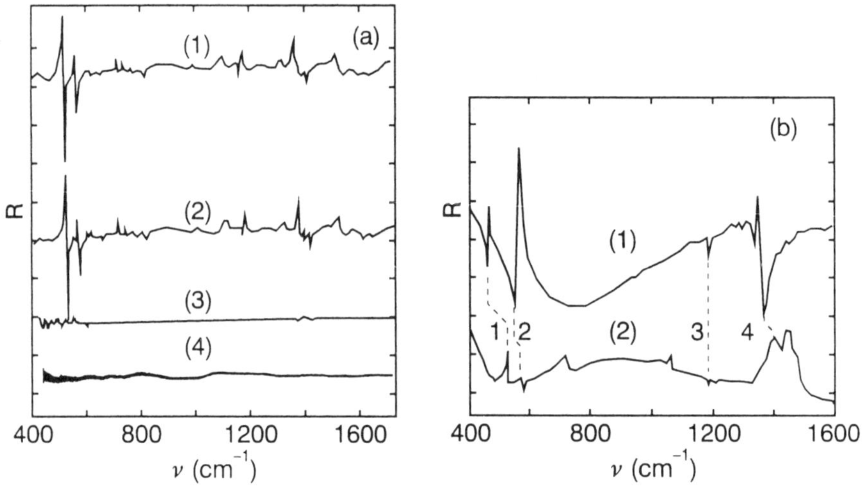

Fig. 8.19. Single crystal reflectivity from a [100] face of a RbC_{60} as measured in-situ in inert conditions (1) and after exposure of one day to air (2). (3) is the difference between (1) and (2) and (4) is the difference after 9 months exposure to air (**a**); adapted from [68]. (**b**) shows similar experiments for Rb_6C_{60}; (1) is the reflectivity for inert conditions and (2) immediately after exposure to air; after [69]

Other experiments like ESR [67], x-ray analysis [67], and NMR [70] fully confirmed the stability of the AC_{60} phase. For RbC_{60} a high stability versus air exposure could be demonstrated, even for the high temperature phase from IR [68] and NMR [70] experiments. Other experiments carried out for KC_{60} showed that the high temperature phase is not air stable [67].

The reason for the unexpected air stability is still under discussion. It might be intrinsic, since other compounds like Ph_4PC_{60} (tetraphenyl phosphonium C_{60}) which exhibits a charge transfer of one electron to the C_{60} cage are known to be air stable. Alternatively, a passivation layer may be generated on the surface like in Al. The fact that even very thin layers of RbC_{60} are fully air stable indicates an intrinsic stability. On the other hand oxygen enrichment in the top layers of the crystal was found from an XPS analysis after sputtering off C_{60} molecules layer by layer from the surface [71].

8.6 Dimeric Phases and Annealing After Quenching of AC_{60}

A large amount of work has been dedicated to rapid cooling, quenching, and annealing after various thermal treatments. The original idea for this type of work was to retain the high temperature fcc phase at low temperatures and to check on special magnetic properties as they are known for some monoanionic C_{60} compounds like TDAE C_{60}. Since cooling rates appear to be a critical experimental detail, quenching, rapid cooling and slow cooling should be discriminated. For slow cooling, phase separation is obtained in KC_{60}. However, it was proven from Raman experiments that a large part of the high temperature phase can indeed be retained after quenching KC_{60} from high temperatures to liquid nitrogen [72]. With annealing, several undefined intermediate states were recorded until finally phase separation was obtained. Similarly Martin et al. [29] found a dramatic increase of the infrared transmission after quenching RbC_{60} from the high temperature phase. The phase obtained was therefore assigned to a nonconducting compound.

More detailed information on the structure and phases obtained after quenching and annealing was obtained from x-ray, calorimetric, and ESR analysis [44, 73, 74].

A refined x-ray analysis for KC_{60} after quenching from the high temperature rocksalt phase was reported by Zhu et al. [73]. For rapid cooling with about 10 K/min to room temperature, the orthorhombic phase was obtained, but quenching from 573 K to ice water and subsequent cooling to 19 K revealed a new phase. The x-ray pattern of this phase showed a superlattice structure which required a doubling of the unit cell and a distortion from a cubic or even from the orthorhombic symmetry. The final assignment was to a monoclinic cell of space group $C2/m$ with lattice constants $a = 17.092$ Å, $b = 9.771$ Å, $c = 19.209$ Å, and $\beta = 124.06°$. The C_{60} molecules were arranged in a dimeric structure with a center-to-center distance of 9.34 Å. This distance is slightly larger than in o-C_{60} but still much shorter than in all other known C_{60} compounds.

The structure was observed to be unstable on heating. At 270 K the superlattice peaks disappeared and finally at room temperature the polymeric phase of KC_{60} (and not the phase separated compound) was obtained. These experiments triggered a new series of quenching and annealing experiments for the AC_{60} phases. The dimerized phase was expected to be a Peierls insulator as a consequence of the doubling of the unit cell. This idea is fully consistent with the photoemission spectra shown in Fig. 8.12. If the samples are quenched from the high temperature phase to liquid nitrogen, a metallic phase with $Pa\overline{3}$ is obtained [75].

The variety of new metastable phases obtained during an annealing process after quenching was demonstrated in experiments of differential thermal analysis (DTA) by [44]. To identify the various phases, ESR and x-ray structural analysis were used simultaneously. Figure 8.20a shows two DTA diagrams, one for quenched and one for slowly cooled CsC_{60}. The figure displays 5 phase transitions for the quenched sample at the indicated temperatures. Two of the transitions are exothermic and three are endothermic. For the slowly cooled compound, only one transition is observed. Similar results with a quantitative thermal analysis have been reported by Granasy et al. [76–78].

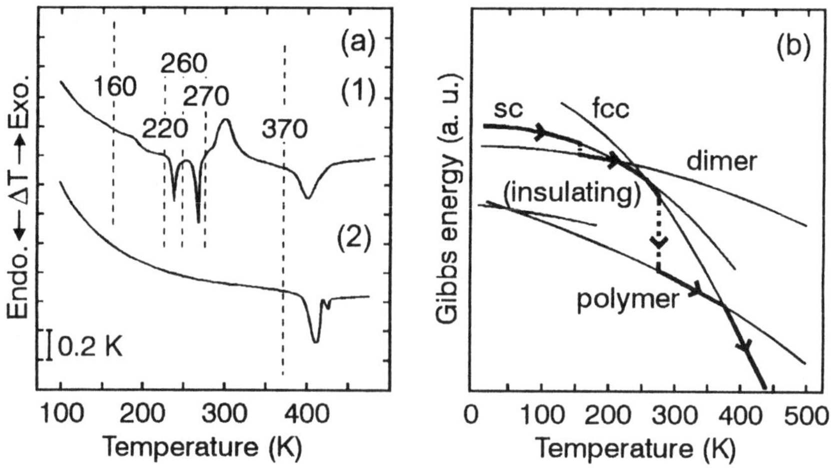

Fig. 8.20. Differential thermal analysis of quenched CsC_{60} (1) and slowly cooled C_{60} (2) (**a**) and generic diagram for the Gibbs free energy of the system (**b**); after [44]

As supported from the simultaneous measurements of ESR and x-ray diffraction, the phase transitions exhibit the following behavior [44]:

- for $T < 160\,\mathrm{K}$ with an exothermic transition at 160 K the phase is 3-dimensional metallic
- for $160\,\mathrm{K} < T < 220\,\mathrm{K}$ with an endothermic transition at 220 K the phase is dimeric and semiconducting
- for $220\,\mathrm{K} < T < 260\,\mathrm{K}$ the phase is sc, 3-dimensional metallic and orientationally ordered
- for $260\,\mathrm{K} < T < 270\,\mathrm{K}$ with an exothermic transition at 270 K the phase is fcc(I), 3-dimensional metallic and disordered
- for $270\,\mathrm{K} < T < 370$ the phase is polymeric, 1-dimensional metallic with an endothermic transition at 370 K
- for $T > 370\,\mathrm{K}$ the phase is fcc(II) and semiconducting (or metallic with the degeneracy of the electron gas lifted).

The temperatures, as given above, are for the Cs-compound. Part (b) of the figure shows a schematic diagram for the Gibbs free energy of the system with the development of the phases indicated by the full and dotted heavy line. The same scenario with similar temperatures holds for the Rb- and the K-compound, except for the latter, no discrimination between a sc and a fcc(I) phase was observed. Also, for KC_{60} another intermediate phase between the polymeric phase and the high temperature fcc phase was reported by Faigel et al. [77, 80]. The heat of transition for the change from the polymeric phase to the high temperature phase was observed to be 35 J/g which is much higher than the value of 12 J/g reported earlier [9].

Very recently, the structure of the dimer phase was optimized by determining the mutual orientation of the two molecules in the dimer [81]. Several possible geometries are compiled in Fig. 8.21. The double bonded interconnection is obtained from $2+2$ cycloaddition and known from quantum-chemical calculations as the lowest energy state for the neutral compound [34, 60]. The larger center-to-center distance observed for the charged dimers and their instability at rather low temperatures suggests the possibility of a single bond.

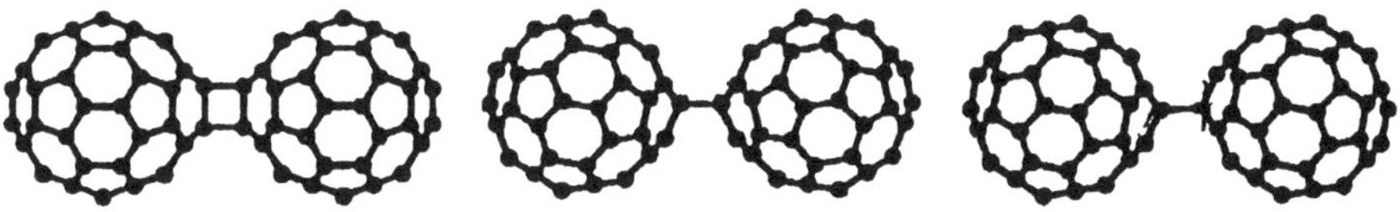

Fig. 8.21. Three selected dimer structures for C_{60}. (1): double bonded with symmetry D_{2h}, (2): single bonded with symmetry C_{2v}, (3): single bonded with symmetry C_{2h}; after [82]

The best fit of the x-ray pattern was indeed obtained for a single bonded dimer in a trans configuration with C_{2h} geometry [81]. The intermolecular C–C distance was determined to be 1.54 Å with a space group $P2_1/a$. This result explains why the dimer is dissolved on heating before the polymer is generated. The two compounds have a completely different bonding structure. As a matter of fact, the dissolution of the dimer is a necessary condition for the generation of the polymer since the latter relies on freely rotating molecules [13].

Ground state optimized geometries and total energies have been reported by Kürti et al. [82] for the neutral and for the double negatively charged dimer of C_{60} for various single bonded and double bonded structures. The semiempirical calculations were carried out using AM1 and PM3 parameterization. The calculation revealed the 66/66 D_{2h} configuration and the single bond configuration with geometry C_{2h} from Fig. 8.21 as the lowest energy states for the neutral pairs and for the dianion $(C_{60})^{-2}$, respectively, both in full agreement with the experiment.

8.7 Other Polymeric Forms of C_{60} Fullerides

To obtain fully doped single crystals of AC_{60} cocrystallization of C_{60} with the alkali metal is required. As mentioned already [66] this was possible for KC_{60}. More recently, it was also demonstrated for RbC_{60} [83, 84]. The RbC_{60} crystals were rather small, of the order of 50 µm, but exhibited a narrower ESR line of only 4.8 G, indicating a higher coherence length for the chains as compared to the conventionally doped material.

The KC_{60} samples showed an interesting morphology. The crystals grew to a larger size of about 100 μm but were only fractional doped to KC_{60}. Large areas remained undoped and could be desolved away with toluene. The remaining insoluble material was pure KC_{60} with a fibrillar structure. A close inspection revealed that the fibers are actually ribbons with typical dimensions 500 μm (length), 10 μm (width), and 0.04–0.14 μm (thickness). The ribbons consist of lamellas with dimensions 0.1–0.5 μm (width) and up to 10 μm (length). The thickness is less than that of the ribbons since the lamellas are arranged in an "accordion" type geometry. This means the lamellas are connected along their longest extension but their surfaces are inclined by an angle of about 70°. The long direction of the lamellas can be either along the direction of the ribbons or perpendicular to them.

Electron diffraction and scanning tunneling microscopy [85] were used to obtain more details about the orientation of the chains in the lamellas. The resolved geometry is shown schematically in Fig. 8.22. Part (a) represents the arrangement of the lamellas as observed from the front of the ribbon. The crystallographic directions below the graph refers to a cubic reference cell. The ribbons and the lamellas extend in [011] direction in this picture. Two lamellas are connected along a twin boundary. The chain directions in the crystals are along **a** and **a′**. This means they are oriented nearly perpendicular to the surface of the lamellas and the coherence length of the chains is only of the order of 100 molecular units. At the twin boundary the C_{60} molecules are connected under an angle of about 60° which allows a continuation of the polymerization across the boundary as indicated by the linked chain in the figure. Part (b) of the figure gives details of the geometry for two connected lamellas in the cubic reference system. The hatched areas are the surface

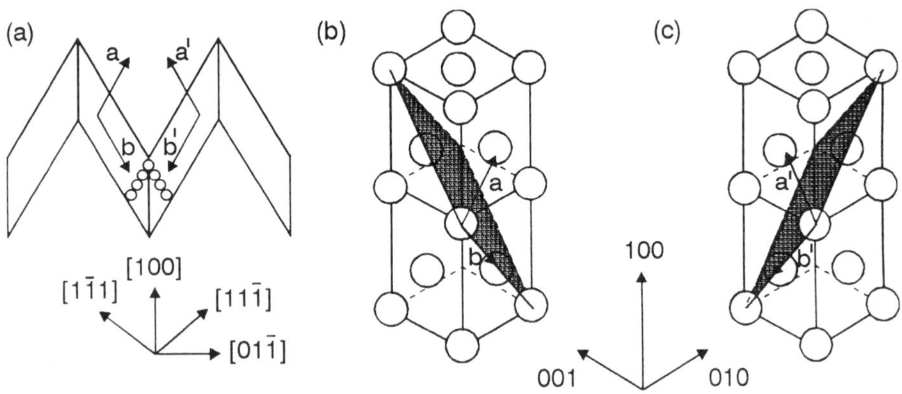

Fig. 8.22. a–c. Geometrical arrangement of the lamellas and of the polymeric chains in the ribbons (**a**) and orientation of the twined lamellas in the cubic reference geometry (**b**); adapted from [84]

of the two lamellas with surface normal pointing in $[11\bar{1}]$ (left) and in $[1\bar{1}1]$ (right). The chains are oriented in **a** direction and thus make an angle of about 35° with the surface of the lamellas.

Even though the observed material has a complicated geometry, it exhibits macrosize single crystal areas with fully ordered chains. This is in contrast to the nanosize domains in diffusion doped single crystals.

Polymerization under moderate pressure in trianionic ternary compounds of C_{60} with alkali metals as counter ions has been reported recently by Zhu [86]. Polymerization was achieved for the superconductor Na_2RbC_{60} and Na_2CsC_{60} for a hydrostatic pressure of 3 kbar and room temperature. The type of compounds investigated are good candidates for a pressure polymerization as the C_{60} molecules are still in a state of rotational diffusive motion at room temperature and the lattice constant is still small, intermediate between C_{60} and AC_{60}. The nearest interball distance at room temperature and ambient pressure is $D = 9.95$ Å for the compounds investigated.

The pressure induced polymerization was deduced from an x-ray analysis. The x-ray pattern obtained could be indexed with a bco unit cell with dimensions $a = 9.347$ Å, $b = 9.898$ Å and $c = 14.268$ Å. The transition was fully reversible with release of the pressure. The observed lattice constant for a is close to that for the monoanionic polymeric compounds and to the two-fold charged dimeric compounds. Improved agreement of the calculated x-ray pattern was obtained for relaxed C1–C1′, C1–C1, and C1–C2 bonds which turned out again to be similar to the values for the known polymers and dimers. These results were taken as arguments for the establishment of covalent bonds. The polymer was found to be superconducting from a recent pressure experiment [87].

8.8 Summary and Open Questions

There is wide evidence that the phases AC_{60}, A = K, Rb, Cs, have a polymeric structure in their orthorhombic phase. Below the transition temperature to this phase the systems are metallic.

K behaves differently from Rb and Cs in several aspects. For slow enough cooling, it is subjected to a phase separation into uncharged C_{60} and K_3C_{60} rather than to the fcc to orthorhombic phase transition. If the cooling is fast enough, the orthorhombic state is obtained as for the Rb and for the Cs compound. It stays metallic to any low temperature observed so far. 'Fast enough' means the K atoms do not have enough time to diffuse over resonable distances to establish phase separation. So, the system rather accepts covalent bonding.

In contrast to K, Rb and Cs always undergo the fcc to orthorhombic transition but they are, in addition, subjected to a process of magnetic ordering at very low temperatures of the order of $T_{cM} = 50$ K. Below this temperature the compounds become nonmetallic. The basic argument refers to the different behavior is the different dimensionality of the electronic system in the three compounds. There are several experiments indicating that o-KC_{60} is a 3-dimensional metal, whereas o-RbC_{60} and o-CsC_{60} are quasi-one-dimensional systems. This difference has not yet been confirmed from theoretical considerations. Some evidence for the different dimensionality of the electronic system in the three compounds may be drawn from structural data. Figure 8.23 shows the interchain distance (1) in the a, b plane, the intermolecular distance along the chains (2) and the nearest intermolecular carbon–carbon distance (3) versus the ionic radius for the compounds. Whereas for increasing ionic radius the interchain distance increases the C1–C1′ distance decreases. This means the transfer integral along the chains will increase relative to the transfer integral between the chains with increasing r and the electronic structure may become anisotropic.

Another solution for the discrepancy between the experimentally observed dimensionality of the electronic system and the calculation was recently suggested in [88]. For a more or less half-filled tight-binding band the Fermi surface was shown to be unstable as in the case of a low dimensional system. Only minimal differences in the anisotropy or in the deviation from half-filling may be responsible for the different behaviour between KC_{60} and RbC_{60}.

Quenching fcc-AC_{60} from high temperatures to liquid nitrogen yields a metastable metallic state with $Pa\overline{3}$ space group. By annealing from this state several metastable phases are created with metallic and nonmetallic character until finally the polymeric o-AC_{60} is established.

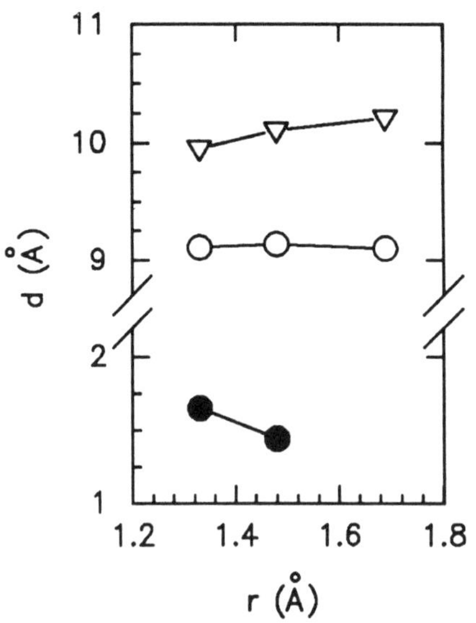

Fig. 8.23. Interchain distance in the a, b-plane (1), intermolecular distance along a (2) and nearest intermolecular carbon carbon distance (3) versus ionic radius r for AC_{60} compounds

The described behavior renders AC_{60} as a new class of metallic and airstable compounds in the family of fulleride conductors.

Even though a fairly good understanding of the system has been obtained there are several open questions.

On the structural side it would be beneficial to have a better (microscopic) understanding of the fcc to orthorhombic phase transition in addition to the very phenomenological theory of Shakhmatov et al. [89]. There seems to be no direct access to the polymeric phases. This means the phases cannot be obtained directly by doping a pristine polymeric phase. Rather a transition from high or low temperatures is needed. However, more experiments are still required. Another problem is NaC_{60}. Why does it not form a polymeric state or under which condition will it form such a state? Also, it does not seem to be air stable, in contrast to the statement that all AC_{60} compounds are stable in air.

On the electronic side, the calculations should be carried out to a level where they can provide a basis for understanding the differences observed for the dopants K, Rb, and Cs. Also, more results on the electron–phonon coupling are needed to get a better understanding of the metallic phase. Why is the coupling constant so much smaller in AC_{60} as compared to A_3C_{60}? A better characterization of the state in the high temperature fcc phase and below the magnetic ordering transition would be helpful, from both a theoretical and an experimental perspective. If the high temperature state is metallic, the discontinuities in the electronic properties need explanation, and the impact of electron–electron correlation must be characterized.

Acknowledgement. We are grateful for valuable discussions with J. Kürti, S. Pekker, A. Zettl, J. Weaver, M. Golden, B. Renker, P. Stephens, and N. Kirova. The work was supported by the FFWF in Austria, project P09741.

References

1. R. C. Haddon et al.: Nature **350**, 320 (1991).
2. D. M. Poirier, D. W. Owens, J. H. Weaver: Phys. Rev. B **51**, 1830 (1995).
3. M. S. Dresselhaus, G. Dresselhaus, P. C. Eklund: *Science of Fullerenes and Carbon Nanotubes*, (Academic Press, San Diego, Boston, New York 1996).
4. J. Winter, H. Kuzmany: Solid State Commun. **84**, 935 (1992).
5. J. Winter, H. Kuzmany: Proc. Int. Winterschool on *Electronic Properties of Novel Materials*, H. Kuzmany, J. Fink, M. Mehring, S. Roth (Eds.) (World Scientific, Singapore 1994), p 271.
6. D. M. Poirier, J. H. Weaver: Phys. Rev. B **47**, 10959 (1993).
7. R. Tycko, G. Dabbagh, D. W. Murphy, Q. Zhu, J. E. Fisher: Phys. Rev. B **48**, 9097 (1993).
8. A. Janossy et al.: Phys. Rev. Lett. **71**, 1091 (1993).
9. Q. Zhu et al.: Phys. Rev. B **47**, 13948 (1993).
10. S. Pekker et al.: Solid State Commun. **90**, 349 (1994).

11. P. W. Stephens et al.: Nature **370**, 636 (1994).
12. O. Chauvet et al.: Phys. Rev. Lett. **72**, 2721 (1994).
13. Rao et al.: Science **259**, 955 (1993).
14. N. G. Chopra, J. Hone, A. Zettl: Phys. Rev. B **53**, 8155 (1996).
15. J. Robert, P. Petit, J. E. Fischer: Synth. Met. **77**, 119 (1996).
16. G. Bortel: private communication.
17. G. P. Lopinski, M. G. Mitch, J. R. Fox, J. S. Lannin: Phys. Rev. B **50**, 16098 (1994).
18. J. Winter, H. Kuzmany: Phys. Rev. B **52**, 7115 (1995).
19. M. C. Martin et al.: Phys. Rev. B **51**, 3212 (1995).
20. J. Winter, H. Kuzmany: J. Raman Spectrosc. **27**, 373 (1996).
21. J. Winter, H. Kuzmany: unpublished.
22. A. Soldatov et al.: Proc. Int. Winterschool on *Electronic Properties of Novel Materials*, H. Kuzmany, J. Fink, M. Mehring, S. Roth (Eds.) (World Scientific, Singapore 1996), p 344.
23. J. Winter, H. Kuzmany: Phys. Rev. B , 17486 (1996).
24. C. H. Varma et al.: Science **245**, 989 (1991).
25. M. G. Mitch, J. S. Lannin: J. Phys. Chem. Solids **54**, 1801 (1993).
26. J. Winter, H. Kuzmany, Phys. Rev. B **53**, 655 (1996).
27. H. Kuzmany, R. Winkler, T. Pichler: J. Phys. Condens. Matter **7**, 6601 (1995).
28. T. Pichler, R. Winkler, H. Kuzmany: Phys. Rev. B **49**, 15879 (1994).
29. M. C. Martin et al.: Phys. Rev. B **49**, 10818 (1994).
30. D. Koller, M. C. Martin, L. Mihaly: Mol. Cryst. Liq. Cryst. **256**, 275 (1994).
31. T. Pichler, R. Winkler, H. Kuzmany: Proc. Int. Winterschool on *Electronic Properties of Novel Materials*, H. Kuzmany, J. Fink, M. Mehring, S. Roth (Eds.) (World Scientific, Singapore 1994), p 261
32. B. Renker, R. Heid, F. Gompf, H. Schober: Proc. Int. Winterschool on *Electronic Properties of Novel Materials*, H. Kuzmany, J. Fink, M. Mehring, S. Roth (Eds.) (World Scientific, Singapore 1996), p 136.
33. B. Renker et al.: Phys. Rev. B **53**, R14701 (1996).
34. G. B. Adams, J. B. Page, O. F. Sankey, M. O'Keefe: Phys. Rev. B **50**, 17471 (1994).
35. M. C. Martin, P. Wochner, L. Mihaly, L. Forro: Proc. Int. Winterschool on *Electronic Properties of Novel Materials*, H. Kuzmany, J. Fink, M. Mehring, S. Roth (Eds.) (World Scientific, Singapore 1996), p 132.
36. P. J. Benning, F. Stepaniak, J. H. Weaver: Phys. Rev. B **48**, 9086 (1993).
37. D. M. Poirier, C. G. Olson, J. H. Weaver: Phys. Rev. B **52**, R11662 (1995).
38. N. Kirova, L. Firlej: Synth. Met. **77**, 63 (1996).
39. K. Kamaras, S. Pekker, D. B. Tanner, L. Forro: Proc. Int. Winterschool on *Electronic Properties of Novel Materials*, H. Kuzmany, J. Fink, M. Mehring, S. Roth (Eds.) (World Scientific, Singapore 1996), p 119.
40. Y. Yafet: Solid State Phys. **14**, 1 (1963).
41. G. Grüner: Rev. Mod. Phys. **66**, 1 (1994).
42. A. Janossy, N. Nemes, T. Feher, G. Oszlanyi, G. Baumgartner, L. Forro: Phys. Rev. Lett. **79**, 2718 (1997).
43. F. Bomelli et al.: Phys. Rev. B **51**, 14794 (1995).
44. M. Kosaka et al.: Phys. Rev. B **51**, 12018 (1995).
45. T. Kälber, G. Zimmer, M. Mehring: Phys. Rev. B **51**, 16471 (1995).
46. T. Kälber, G. Zimmer, M. Mehring: Z. Phys. B **97**, 1 (1995).
47. K. F. Thier, G. Zimmer, M. Mehring, F. Rachdi: Phys. Rev. B **53**, R496 (1996).
48. H. Alloul et al.: Proc. Int. Winterschool on *Electronic Properties of Novel Materials*, H. Kuzmany, J. Fink, M. Mehring, S. Roth (Eds.) (World Scientific, Singapore 1995), p 443.

49. H. Alloul et al.: Phys. Rev. Lett. **76**, 2922 (1996).
50. H. J. Pederson et al.: Solid State Commun. **35**, 207 (1980).
51. G. Grüner: Rev. Mod. Phys. **66**, 1 (1994).
52. V. Brouet et al.: Phys. Rev. Lett. **76**, 3638 (1996).
53. L. Cristofolini et al.: J. Phys.: Cond. Matter **7**, L567 (1995).
54. Y. J. Uemura et al.: Phys. Rev. B **52**, R6991 (1995).
55. W. A. MacFarlane et al.: Phys. Rev. B **52**, R5995 (1995).
56. J. Hone, M. S. Fuhrer, K. Khazeni, A. Zettl: Phys. Rev. B **52**, R8700 (1995).
57. A. Zettl: private communiation.
58. A. Zettl: Proc. Int. Winterschool on *Electronic Properties of Novel Materials*, H. Kuzmany, J. Fink, M. Mehring, S. Roth (Eds.) (World Scientific, Singapore 1996), p 14.
59. F. Bomelli, L. Degiorgi, P. Wachter, L. Forro: Proc. Int. Winterschool on *Electronic Properties of Novel Materials*, H. Kuzmany, J. Fink, M. Mehring, S. Roth (Eds.) (World Scientific, Singapore 1995), p 351.
60. D. Porezag, M. R. Pederson, Th. Frauenheim, Th. Köhler: Phys. Rev. B **52**, 14963 (1995).
61. P. Surjan, K. Nemeth: Proc. Int. Winterschool on *Electronic Properties of Novel Materials*, H. Kuzmany, J. Fink, M. Mehring, S. Roth (Eds.) (World Scientific, Singapore 1995), p 297.
62. J. Fagerström, S. Stafström: Phys. Rev. B **53**, 13150 (1996).
63. P. Surjan, K. Nemeth: Solid State Commun. **92**, 407 (1994).
64. S. C. Erwin, G. V. Krishna, E. J. Mele: Phys. Rev. B **51**, 7345 (1995).
65. J. Schulte, M. Böhm, Chem. Phys. Lett. **252**, 367 (1996).
66. S. Pekker et al.: Science **265**, 1077 (1994).
67. D. Koller et al.: Appl. Phys. Lett. **66**, 1015 (1995).
68. R. Winkler, T. Pichler, H. Kuzmany: Appl. Phys. Lett. **66**, 1221 (1995).
69. H. Kuzmany, R. Winkler, J. Winter: Am. Electrochem. Soc. Proceedings **95-10**, R. S. Ruoff, K. M. Kadish (Eds.) (The Electrochem. Soc. Inc., Pennington 1995), p 823.
70. C. Goze et al.: Synth. Met. **77**, 115 (1996).
71. T. Danno, unpublished.
72. H. Kuzmany, M. Matus, T. Pichler, J. Winter: Proc. Nato Conference on *Physics and Chemistry of Fullerenes*, K. Prassides (Ed.) (Kluwer Acad. Pub., NL 1994).
73. Q. Zhu, D. E. Cox, J. E. Fischer: Phys. Rev. B **51**, 3966 (1995).
74. G. Oszlanyi et al.: Phys. Rev. B **51**, 12228 (1995).
75. A. Lappas, M. Kosaka, K. Tanigaki, K. Prassides: J Am. Chem. Soc. **117**, 7560 (1995).
76. L. Granasy et al.: Solid State Commun. **97**, 573 (1996).
77. L. Granasy, S. Pekker, L. Forro: Phys. Rev. B **53** 5059 (1996).
78. L. Granasy et al.: Proc. Int. Winterschool on *Electronic Properties of Novel Materials*, H. Kuzmany, J. Fink, M. Mehring, S. Roth (Eds.) (World Scientific, Singapore 1995), p 331.
79. P. Petit, J. Robert, J. E. Fischer: Phys. Rev. B **51**, 11924 (1995).
80. G. Faigel et al.: Phys. Rev. B **52**, 3199 (1995).
81. G. Oszlanyi et al.: Proc. Int. Winterschool on *Electronic Properties of Novel Materials*, H. Kuzmany, J. Fink, M. Mehring, S. Roth (Eds.) (World Scientific, Singapore 1996), p 354.
82. J. Kürti, K. Nemeth: Chem. Phys. Lett. **256**, 119 (1996).
83. S. Pekker, M. Carrard, L. Forro, L. Mihaly: Proc. Int. Winterschool on *Electronic Properties of Novel Materials*, H. Kuzmany, J. Fink, M. Mehring, S. Roth (Eds.) (World Scientific, Singapore 1996), p 110.

84. M. Carrard et al.: Synth. Met. **80**, 29 (1996).
85. L. P. Biro, S. Pekker, E. Balazs: Synth. Met. **77**, 123 (1996).
86. Q. Zhu: Phys. Rev. B **52**, R723 (1995).
87. J. E. Schirber, L. Hansen, B. Morosin, J. E. Fischer, J. D. Jorgensen, G. H. Kwei: Physica C **260**, 173 (1996).
88. M. Fally, H. Kuzmany: Phys. Rev. B **56**, 13861 (1997).
89. V. S. Shakhmatov, S. L. Drechsler, T. Galbaatar: Proc. Int. Winterschool on *Electronic Properties of Novel Materials*, H. Kuzmany, J. Fink, M. Mehring, S. Roth (Eds.) (World Scientific, Singapore 1995), p 306.
90. J. Winter, H. Kuzmany: Appl. Phys. **A 64**, 301 (1997).

9. Electronic Structures and Optical Excitations in C_{60}-Polymers

K. Harigaya

9.1 Introduction

Variations in the band structures of C_{60}-polymers are studied, when conjugation conditions and the electron number are changed. We use a semiempirical model with the Su–Schrieffer–Heeger type electron–phonon interactions. In the neutral one-dimensional (and two-dimensional) C_{60}-polymer, electronic structures change between direct-gap insulators and metals, depending on the degree of conjugation. High pressure experiments could observe such pressure-induced metal–insulator transitions. It is found that one-dimensional C_{60}-polymer doped with one electron per one molecule is always a metal. The energy difference between the highest-occupied state and the lowest-unoccupied state of the neutral system becomes smaller upon doping owing to the polaron effects. When the one-dimensional C_{60}-polymer is doped with two electrons per one C_{60}, the system is insulating. When the conjugation in the direction of the polymer chain is smaller, it is a direct-gap insulator. The energy gap becomes indirect when the conjugation is stronger. We also study the antiferromagnetic phase of A_1C_{60} by using a tight-binding model with long-range Coulomb interactions. The antiferromagnetism is well described by the model. The comparison with the photoemission studies shows that the new band around the Fermi energy of the A_1C_{60} phase can be explaind by the extremely large intrusion of an energy level into the gap of the neutral system. This indicates that the interaction effects among electrons are important in doped C_{60}-polymers. Finally, we investigate charge-transfer (CT) exciton effects in optical absorption spectra of crosslinked C_{60} systems by using an exciton theory. We consider C_{60}-dimers and two molecule systems of the C_{60}-polymer. Variations in the optical spectra are studied by changing the conjugation parameter between molecules. We find that the total CT-component increases in small conjugation, and saturates at intermediate conjugation. It decreases in large conjugation. We also find that the CT-component of doped systems is smaller than that of neutral systems, indicating that the electron–hole distance becomes shorter in doped C_{60}-polymers.

Recently, it has been found that linear (one-dimensional) C_{60}-polymers can be realized in alkali-metal doped C_{60} crystals: A_1C_{60} (A = K, Rb, Cs) [1–5]. Their solid state properties are being intensively investigated and are reviewed in Chapters . One electron per A atom is donated to the polymer chain. Metallic conductivity is observed and it seems that Fermi surfaces exist in high temperatures, but the system shows antiferromagnetic correlations

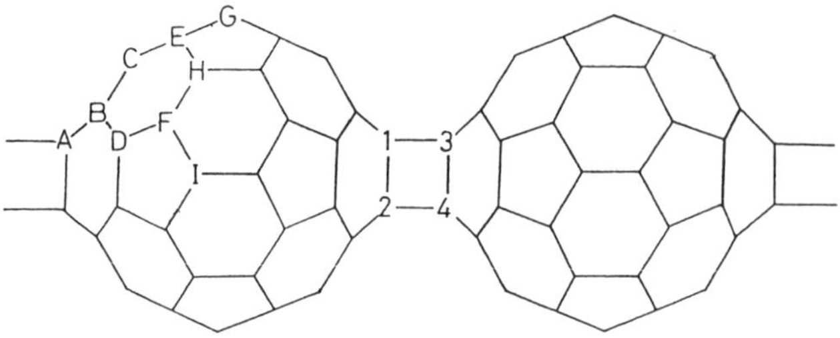

Fig. 9.1. Crystal structure of the one-dimensional C_{60}-polymer. The labels, A–I, indicate carbon atoms whose charge densities are not equivalent due to symmetry. The conjugation along four bonds, which connect carbon atoms with labels, 1–4, is controlled by the parameter a in (9.1)

in low temperatures [1]. X-ray data has been interpreted to indicate that the A_1C_{60}-polymer has a lattice structure in which the C_{60} molecules are formed in a linear chain. The bonds between C_{60} are formed by the [2 + 2] cycloaddition mechanism. The structures of the one-dimsensional polymers are shown in Fig. 9.1. There are four-membered rings between neighboring C_{60} molecules. Quite recently, other new polymer phases of C_{60} systems which are synthesized at high pressures have been reported, and two-dimensional polymer structures have been proposed [6–9]. A tight-binding calculation for these structures has been already reported [10], and the lattice structures can be explained by the calculation. The two kinds of lattice structures – rectangular and triangular – are present in the two-dimensional polymers, and these are shown for the rectangular polymer (Fig. 9.2a) and the triangular polymer (Fig. 9.2b).

In this Chapter, we shall review studies on the variations in the electronic band structure of one- and two-dimensional C_{60}-polymers when the conjugation conditions between molecules are changed. When the conjugation of a certain bond is larger, the π-character of the bond is larger than the σ-character. When the conjugation of a bond is smaller, the σ-character is larger than the π-character. We assume that the bonding states between C_{60} might be changed easily, possibly by applying high pressures. As illustrated in Fig. 9.3 (the idea proposed in [2]), there are three candidates for the classical bonding structures around the four-membered ring. The first case (1) is that there are only weak van der Waals interactions between C_{60}, and the classical double bonds remain as in the isolated C_{60} molecule. The second case (2) is that all the bonds of the four-membered ring are single or σ bonds. All the four-membered rings in Figs. 9.1 and 9.2 are formed by the σ-bonds in this case. Then, in the third case (3), the bonds which connect neighboring molecules are double bonds, and the bonds derived from the isolated C_{60} – the bonds, $\langle 1,2\rangle$, $\langle 3,4\rangle$, $\langle 5,6\rangle$, $\langle 7,8\rangle$, of Fig. 9.2a – are destroyed completely. Here, $\langle i,j\rangle$ indicates the specific pair of neighboring atoms. In this case, the intermolecular bonds, $\langle 1,5\rangle$, $\langle 2,6\rangle$, $\langle 3,8\rangle$, $\langle 4,7\rangle$, of Fig. 9.2a

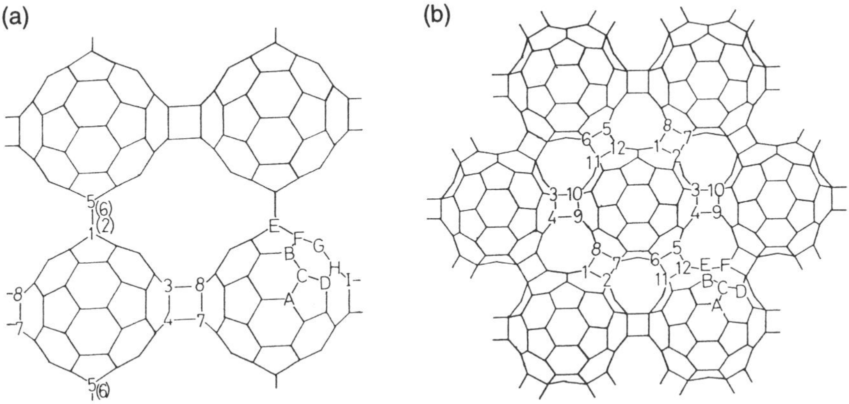

Fig. 9.2. Crystal structures of two-dimensional C_{60}-polymers. The rectangular polymer is shown in (**a**), and the triangular polymer is displayed in (**b**). The labels, A–I, indicate carbon atoms which are not equivalent due to symmetry. The carbon sites which constitute the four-membered rings are indicated by the numbers. In (**a**), the atomic sites 2 (and 6) overlap with the sites 1 (and 5), respectively, and cannot be seen from the front. The parenthesis indicate the overlapped sites

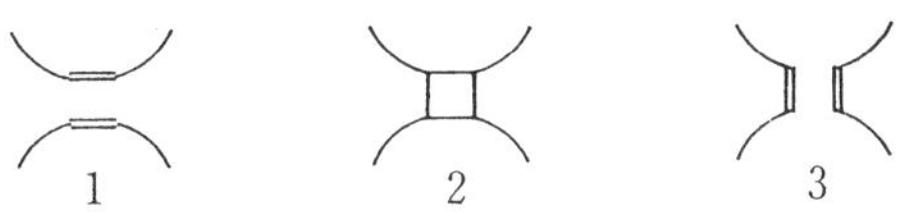

Fig. 9.3. Three possible bonding structures for a four-membered ring

have the σ- as well as π-characters. In other words, the degree of conjugation between the molecules becomes maximum. We have proposed a model which can deal with changes of conjugation conditions among the above three cases, and have examined the electronic band structures of the two types of C_{60}-polymers as a function of the conjugation conditions [11, 12]. We note that the operator at the lattice sites of the four-membered rings is one of the relevant linear combinations of sp^3 orbitals. Similar assumptions regarding σ-conjugation has been used in Si-based polymers, for example, in [13].

In [11, 12], we have proposed a semiempirical tight-binding model analogous to the Su–Schrieffer–Heeger (SSH) model [14] of conjugated polymers. The first purpose of this article is to review the model and the main results. The electronic structures may depend sensitively upon the conjugation conditions even in the neutral polymer, because several bonds connecting neighboring molecules are largely distorted, and the mixing among sp^3 orbitals will change only by slight change of the bond structures [2]. We have studied the effects of the change of the conjugation conditions by introducing a phenomenological parameter in a tight-binding model. The model is an extension of the SSH-type model which has been applied to C_{60} [15, 16] and

C_{70} [16, 17] molecules. The model is solved with the assumption of the adiabatic approximation, and band structures are reported in order to discuss metal–insulator changes and polaron effects by varying conjugations.

We have concluded that the electronic structures change from those of direct-gap (or indirect-gap) insulators to those of metals, depending on the degree of conjugation. In the one-dimensional polymer and the two-dimensional rectangular polymer, with increasing conjugation between molecules, the electronic structures change from a direct gap insulator with the gap at the Γ point, through a metallic structure, to that of an insulator with the direct gap at the Brillouin zone boundary. It is anticipated that high pressure synthesis may be able to change conjugation conditions between C_{60} molecules, and the resulting electronic structure changes could be observed. The reentrant change from the insulator through a metal into an insulator is specific to the present C_{60}-polymer systems and is quite interesting. In the two-dimensional triangular polymer, there is an indirect energy gap at intermediate conjugation, and the system changes into a metal at the maximum conjugation.

We will also review the doping effects in one-dimensional (1D) polymers [18]. We discuss the following properties: (1) The 1D C_{60}-polymer doped with one electron per molecule is always a metal. The energy difference between the highest-occupied state and the lowest-unoccupied state of the neutral system becomes smaller upon electron doping due to polaron effects. (2) When the 1D C_{60}-polymer is doped with two electrons per molecule, the system is found to be insulating. When the conjugation in the direction of the polymer chain is smaller, a direct-gap insulator is found. The energy gap becomes indirect when the conjugation is stronger.

The next purpose of this Chapter is to review a microscopic model for the antiferromagnetism in one-dimensional C_{60}-polymers [19] in order to show that the antiferromagnetic ground state can be actually explained by our model with strong electron–electron interactions. We have pointed out the importance of Coulomb interactions, comparing with photoemission experiments of C_{60}-polymer systems [20]. The model is one dimensional and has two contributions. The first term is the hopping interaction or transfer integral along the polymer backbone structure, and the second term is the Coulomb interaction among electrons. We use the long-range Ohno potential which we have used in a previous paper [21]. In [22], the relevance of the Hubbard model, where the C_{60} molecule is regarded as a site, has been discussed. We will compare the present calculation with the assumptions of the Hubbard model.

Finally, we review optical excitation properties in cross-linked C_{60} systems. Many optical experiments on neutral C_{60} thin films and alkali-metal doped crystals have been performed (see reviews, [23, 24], for example). Along with the experimental developments, we have theoretically analysed the optical spectra of C_{60} and C_{70} molecules [21] and solids [25, 26]. The C_{60} and

C_{70} systems maximally doped with alkali metals have been also studied [27]. We have searched for parameter sets in order to explain the photoexcitation energies, and the relative oscillator strengths, which have been measured in experiments of C_{60} and C_{70} systems. Then, we have studied optical excitation properties of the polymerized C_{60} systems by using the exciton theory [21] used in the above literature. We have considered a C_{60}-dimer and a two-molecule system of the C_{60}-polymer (C_{60}-dimer with periodic boundary condition), as the examples of the cross-linked C_{60}. An interesting problem, special to the polymerized systems, is the possibility of the cross-over between Frenkel (molecular) excitons and Wannier (charge-transfer (CT)) excitons. If interactions are not present between molecules, the exciton is localized on a molecule. As the interactions turn on, some of the optical excitations might have amplitudes over several molecules, and thus the excitations become like CT-excitons. After calculations, we have found the following: (1) The CT-exciton component first increases in smaller conjugation, i.e. in weak intermolecular interactions, and saturates at the intermediate conjugation. The component begins decreasing in the large conjugation. This fact is commonly seen in the C_{60}-dimers and polymers. (2) The CT-component of systems doped with one electron per C_{60} is smaller than that of the neutral systems. The electron–hole separations become shorter upon doping in polymerized C_{60} systems.

This Chapter is organized as follows. In the next section, the electron–phonon model and the tight-binding model with electron–electron interactions are introduced. In Sect. 9.3, polaron effects in C_{60} are reviewed. The electronic band structures and doping effects on the one-dimensional ploymers are discussed in Sect. 9.4. The band structures of the two-dimensional rectangular and triangular polymers are discussed in Sect. 9.5. The magnetization mechanism of the A_1C_{60} phase is discussed in Sect. 9.6. The CT-exciton effects in crosslinked C_{60} are reviewed in Sect. 9.7. The final section is devoted to a summary of the important results.

9.2 Models

In Sects. 9.4 and 9.5, we apply an SSH-type model to the two-dimensional C_{60}-polymers. In C_{60}, all the carbon atoms are equivalent, so it is a good approximation to neglect the mixing between π- and σ-orbitals. The presence of the bond alternation and the energy level structures of the neutral C_{60} molecule can be quantitatively described by the calculations within the adiabatic approximation. In C_{70}, the molecular structure becomes elongated, suggesting that the degree of the mixing between the π- and σ-states is different, depending on the particular carbon site. In this respect, the extended SSH model does not take into account these different mixings. However, it has been found [16, 17] that the qualitative character of the electronic level structure can be reasonably calculated when the extended SSH model is applied to the C_{70}. This is a valid approach because the σ-orbitals can be simulated by the classical harmonic springs in the first approximation.

In this Chapter, we assume the same idea that the lattice structures and the related molecular orbitals of each C_{60} molecule in the C_{60}-polymers can be described by the SSH-type model using a hopping interaction for the π-orbitals and classical springs for the σ-orbitals. However, the mixings between the π- and σ-orbitals near the eight bonds, $\langle i,j \rangle$ $(i,j = 1-8)$, shown in Fig. 9.2a are largely different from those of regions far from the four bonds. We shall shed light on the special character of bonding between the neighboring C_{60}. Electronic structures would be largely affected by changes of conjugation conditions around the four bonds (or local σ-conjugations, as in Si-based polymers [13]). We shall introduce a semiempirical parameter a to describe variable conjugation as shown in the following Hamiltonian:

$$\begin{aligned} H_{\mathrm{pol}} &= a \sum_{l,\sigma} \sum_{\langle i,j \rangle}{}' (-t + \alpha y_{l,\langle i,j \rangle})(c_{l,i,\sigma}^\dagger c_{l+1,j,\sigma} + \mathrm{h.c.}) \\ &+ (1-a) \sum_{l,\sigma} \sum_{\langle i,j \rangle}{}'' (-t + \alpha y_{l,\langle i,j \rangle})(c_{l,i,\sigma}^\dagger c_{l,j,\sigma} + \mathrm{h.c.}) \\ &+ \sum_{l,\sigma} \sum_{\langle i,j \rangle = \mathrm{others}} (-t + \alpha y_{l,\langle i,j \rangle})(c_{l,i,\sigma}^\dagger c_{l,j,\sigma} + \mathrm{h.c.}) \\ &+ \frac{K}{2} \sum_{i} \sum_{\langle i,j \rangle} y_{l,\langle i,j \rangle}^2 , \end{aligned} \tag{9.1}$$

where t is the hopping integral of the system without the bond alternations in the isolated C_{60} molecule; α is the electron–phonon coupling constant which changes the hopping integral linearly with respect to the bond variable $y_{l,\langle i,j \rangle}$, and where l means the lth molecule and $\langle i,j \rangle$ indicates the pair of the neighboring i and jth atoms; the atoms with $i = 1-4$ of the one-dimensional polymer are shown in Fig. 9.1, the atoms with $i = 1-8$ of the rectangular polymer are shown by numbers in Fig. 9.2a, and the atoms with $i = 1-12$ of the triangular polymer are in Fig. 9.2b; the i and j in the third term of (1) label the nonnumbered atoms in the same molecule; $c_{l,i,\sigma}$ is the annihilation operator for a π-electron at the ith site of the lth molecule with spin σ; and the sum is taken over the pairs of neighboring atoms. The last term with the spring constant K is the harmonic energy of the classical spring simulating the σ-bond effects. Note that the sum with the prime is performed over $\langle i,j \rangle = \langle 1,5 \rangle, \langle 2,6 \rangle, \langle 3,8 \rangle, \langle 4,7 \rangle$ for the rectangular polymer, and over $\langle i,j \rangle = \langle 1,8 \rangle, \langle 2,7 \rangle, \langle 3,10 \rangle, \langle 4,9 \rangle, \langle 5,12 \rangle, \langle 6,11 \rangle$ for the triangular polymer. The sum with the double prime is performed over $\langle i,j \rangle = \langle 1,2 \rangle, \langle 3,4 \rangle, \langle 5,6 \rangle, \langle 7,8 \rangle$ for the rectangular polymer, and over $\langle i,j \rangle = \langle 1,2 \rangle, \langle 3,4 \rangle, \langle 5,6 \rangle, \langle 7,8 \rangle, \langle 9,10 \rangle, \langle 11,12 \rangle$ for the triangular polymer. The same convention is used for the one-dimensional polymers.

As stated before, the parameter a controls the strength of conjugation between neighboring molecules. When $a = 1$, the σ-bondings between atoms,

1 and 2, 3 and 4, 5 and 6, 7 and 8, in Fig. 9.2a are completely broken and the orbitals would become like π-orbitals. The bond between the atoms 1 and 5 and the equivalent other bonds become double bonds. This is the case (3) discussed above (see Fig. 9.3. As the parameter a becomes smaller, the conjugation between the neighboring molecules decreases, and the C_{60} molecules become mutually independent. For case (2), the operator $c_{l,i,\sigma}$ at the lattice sites of the four-membered rings is associated with the relevant linear combinations of the sp^3 orbitals. Here, we include the possibility of local σ-conjugations at the four-membered rings. When $a = 0$ (case (1)), the C_{60} molecules are completely isolated from each other. As we shall see in Sects. 9.4 and 9.5, the band structure of the C_{60}-polymers will depend strongly on the conjugation conditions a.

In Sect. 9.6, we discuss the mechanism of the antiferromagnetism in one-dimensional C_{60}-polymers doped with one electron per one C_{60} molecule. In Sect. 9.7, we consider optical excitation properties in cross-linked C_{60} systems. The Hamiltonian is given by the following:

$$H = H_0 + H_{\text{int}}, \tag{9.2}$$

$$\begin{aligned} H_0 = &-at\sum_{l,\sigma}\sum_{\langle i,j\rangle=\langle 1,3\rangle,\langle 2,4\rangle}(c^\dagger_{l,i,\sigma}c_{l+1,j,\sigma}+\text{h.c.}) \\ &-(1-a)t\sum_{l,\sigma}\sum_{\langle i,j\rangle=\langle 1,2\rangle,\langle 3,4\rangle}(c^\dagger_{l,i,\sigma}c_{l,j,\sigma}+\text{h.c.}) \\ &-t\sum_{l,\sigma}\sum_{\langle i,j\rangle=\text{others}}(c^\dagger_{l,i,\sigma}c_{l,j,\sigma}+\text{h.c.}), \end{aligned} \tag{9.3}$$

$$\begin{aligned} H_{\text{int}} = &\, U\sum_{l,i}(c^\dagger_{l,i,\uparrow}c_{l,i,\uparrow}-\frac{n_{\text{el}}}{2})(c^\dagger_{l,i,\downarrow}c_{l,i,\downarrow}-\frac{n_{\text{el}}}{2}) \\ &+\sum_{l,l',i,j}W(r_{l,l',i,j})(\sum_\sigma c^\dagger_{l,i,\sigma}c_{l,i,\sigma}-n_{\text{el}})(\sum_\tau c^\dagger_{l',j,\tau}c_{l',j,\tau}-n_{\text{el}}). \end{aligned} \tag{9.4}$$

In (9.2), the first term is the tight binding part of the C_{60}-polymer, and the second term is the Coulomb interaction potential among electrons. Equation (9.3) is equivalent to (9.1), if $\alpha = 0$ and $K = 0$ in (9.1). The electron–phonon interactions are not taken into account, because they are not necessary in order to show the presence of the antiferromagnetism and to discuss optical excitations. The numbers in (9.3) indicate the carbon atoms displayed in Fig. 9.1. Equation (9.4) describes the Coulomb interactions among electrons. Here, n_{el} is the average number of electrons per carbon site; $r_{l,l',i,j}$ is the distance between the ith site of the lth C_{60} and jth site of the l'th C_{60}; and

$$W(r) = \frac{1}{\sqrt{(1/U)^2+(r/r_0V)^2}} \tag{9.5}$$

has the parametrized form of the Ohno potential [21]. The quantity $W(0) = U$ is the strength of the onsite interaction; V is the strength of the long-range part ($W(r) \sim r_0 V/r$ in the long distance limit $r \gg r_0$); and $r_0 = 1.433$ Å is the mean bond length of the single C_{60} molecule.

9.3 Polarons in an Isolated C_{60} Molecule

In this section, we review lattice and electronic structures of the C_{60} molecule doped with one or two electrons (holes) [16], and discuss polaron effects in an isolated molecule. Polarons in C_{60}-polymers will be discussed in the next section.

First, we discuss structural distortions and electronic structures of C_{60} [16]. The bonding (lattice) configurations of the doped molecular systems are shown in Figs. 9.4a, b. The site and bond labels are displayed in Fig. 9.4a. We show three kinds of short bonds. The shortest bonds, d, are represented by the thick lines. The next shortest ones, b, are shown by the usual double lines. Finally, the dashed lines indicate the longest of the short bonds. They are the bonds f in Fig. 9.4a and bonds g in Fig. 9.4b. Other longer bonds are not shown. The figures are the same for the electron and hole dopings. When the change in the number of electrons is one, the change in the electron density is the largest at the sites at the ends of dashed lines, namely, points D. The dashed lines are mostly located along an equatorial line of C_{60}. The absolute value of the length of the bonds g is the smallest of the four kinds of bonds with negative bond variables. The bond alternation becomes the weakest along this equatorial line. The distortion of the intramolecular lattice is similar to that of a polaron [28] in conjugated polymers. When the change in the electron number is two, configurations of dashed lines along the equatorial line change, as shown in Fig. 9.4b. The ordering of bonds, f and g, with respect to the bond variable is now reversed. Other configurations are the same,

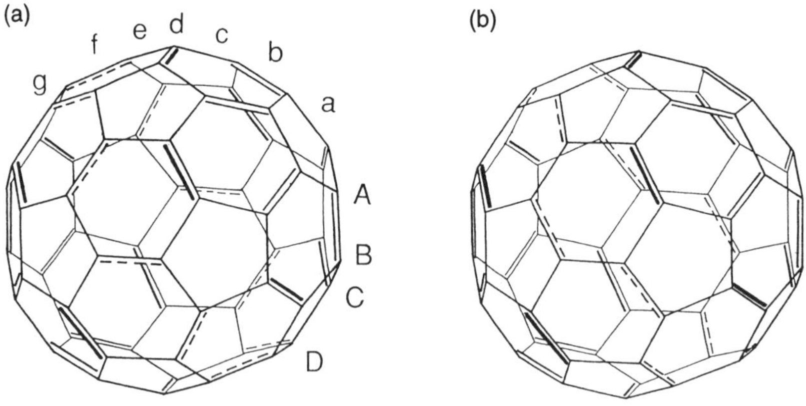

Fig. 9.4. Lattice structures of doped C_{60} ((**a**) $|N_{\text{ex}}| = 1$ and (**b**) $|N_{\text{ex}}| = 2$]. Here, N_{ex} means the excess electron number per C_{60}

however. The change in the electron density is also the largest at points D. Therefore, polaronic distortion persists when the doping proceeds from one to two electrons (or holes).

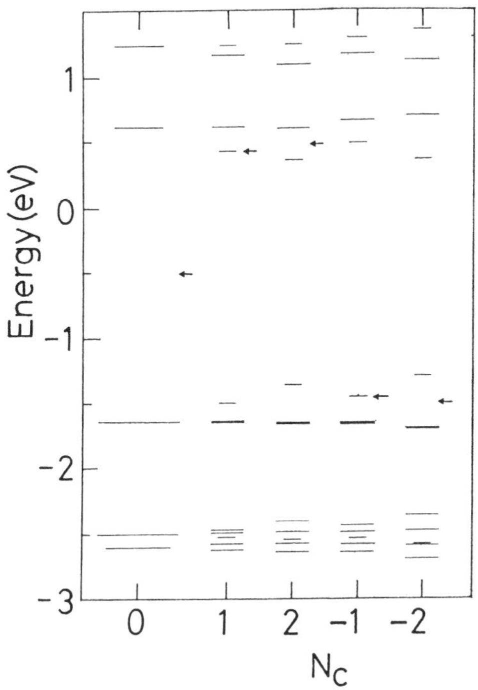

Fig. 9.5. Energy-level structures of C_{60} with $-2 \le N_{ex} \le 2$. The line length is proportional to the degeneracy of the energy level. The shortest line is for the nondegenerate level. The arrow shows the position of the Fermi level

Next, we look at changes in the electronic level structures. They are shown in Fig. 9.5. When the system is doped, the high degeneracy associated with the icosahedral symmetry decreases due to the reduced symmetry stemming from the additional short bonds and the associated deformation of the lattice, or so-called Jahn–Teller distortion. When the excess electron (hole) number is one and two, the highest level is nondegenerate and upshifts in energy relative to the highest occupied molecular orbital (HOMO) of the neutral system. In contrast, the other two levels shift only slightly. Similarly, the lowest unoccupied molecular orbital (LUMO) of the neutral system splits into two levels. Of these two levels, the energy of the nondegenerate level shifts downward, while the change of the energy of the doubly degenerate level is small. This change in the level structure near the HOMO-LUMO gap is common to both the case of electron and hole doping. Furthermore, the change is similar to that in the polaron formation [28] in conjugated polymers.

9.4 Doping Effects and Metal-Insulator Transition in One-Dimensional C_{60}-Polymers

To obtain the following results, we use the parameter values, $t = 2.1\,\mathrm{eV}$, $\alpha = 6.0\,\mathrm{eV/Å}$, and $K = 52.5\,\mathrm{eV/Å^2}$, which yield the energy gap 1.904 eV and a bond length difference of 0.04557 Å for an isolated C_{60} molecule.

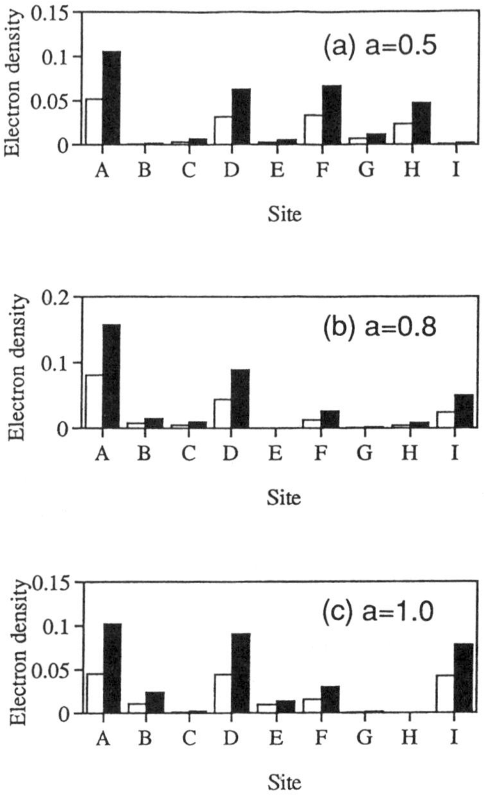

Fig. 9.6. Excess electron distribution for the three conjugation conditions, (**a**) $a = 0.5$, (**b**) 0.8, and (**c**) 1.0. The white bars are for the case $N_{\mathrm{ex}} = 1$ and the black bars are for $N_{\mathrm{ex}} = 2$

Figure 9.6 shows the excess electron distribution for the three conjugation conditions, $a = 0.5$, 0.8, and 1.0. The site labels, A–I, are shown in Fig. 9.1. The symmetry of the polymer chain decreases from the icosahedral symmetry of the isolated molecule, and mutually symmetry equivalent sites have the same electron density. The various labels represent the sites with different electron density. The white bars are for the case of the excess electron number per C_{60}, $N_{\mathrm{ex}} = 1$, and the black bars are for $N_{\mathrm{ex}} = 2$. The excess electron density at the sites A is the largest for all the displayed cases. The bond alternation patterns are largely distorted near these sites, so the electron density change is the largest too. In Fig. 9.6a, the densities at sites, D, Φ, and H, are relatively larger. In Figs. 9.6b, c, the densities are larger at sites D and I. In this way, the sites, where excess electrons are prone to accumulate and thus the bond alternation patterns are highly distorted, are spatially localized in the molecular surface. This is one of the polaron effects which we have discussed in the previous section. Here, we for simplicity do not show bond alternation patterns, and we only note that the distortion of the bond alternation is larger where the change of the electron density is larger. The polaronic distortion pattern is different from that in the isolated C_{60}, and this is due to the difference in the symmetry group. Tanaka et al. [22] have drawn a schematic figure where the electron density change is the largest at the molecule center, but the present result does not agree with this feature. Numerical calculations are necessary in order to derive the actual distributions.

Next, we discuss band structures of electrons in detail. Figures 9.7–9.9 display the band structures for the conjugation parameters, $a = 0.5$, 0.8, and 1.0, respectively. Figures (a), (b), and (c) are for $N_{\text{ex}} = 0$, 1, and 2, respectively. In each figure, the unit cell is taken as unity, so the first Brillouin zone extends from $-\pi$ to π. Due to the inversion symmetry, only the wavenumber region, $0 \leq k \leq \pi$, is shown in the figures. The dashed line indicates the position of the Fermi energy E_{F}.

Figures 9.7a–c show the band structures of the polymer for $a = 0.5$ and with $N_{\text{ex}} = 0$, 1, and 2, respectively. A HOMO-LUMO energy gap of about 0.8 eV is found at the zone center (Fig. 9.7a). So, the system is a direct gap insulator. When doped with one electron per C_{60}, the system is a metal as shown by the presence of the Fermi surface in Fig. 9.7b. The system is an insulator again when $N_{\text{ex}} = 2$, as shown in Fig. 9.7c. Here, the energy gap is at the boundary of the Brilloune zone, i.e., at $k = \pi$.

Increasing the parameter a generates an overlap of the HOMO and LUMO bands in the neutral system. This is shown for $a = 0.8$ in Fig. 9.8a. A Fermi surface must develop, so the system becomes metallic. If a increases further, the positions of the previous HOMO band and LUMO band are reversed as shown for $a = 1.0$ in Fig. 9.9a, and the system becomes a direct gap insulator again, but at $k = \pi$.

When $N_{\text{ex}} = 1$, the system is always a metal for all values of the parameter a between 0 and 1. Representative cases $a = 0.8$ and 1.0 are displayed in Figs. 9.8b and 9.9b. The number of the Fermi surfaces is two or four, depending upon the parameter a. The number is four in Fig. 9.8b: two of the Fermi surfaces are in the region $k > 0$, and the other two are in $k < 0$, due to the symmmetry of the Brillouin zone. The number of the Fermi surfaces is two in Fig. 9.9b. However, the metallic property is obtained for all the a we have considered. We also find that the HOMO band and LUMO band of the neutral system move into the energy gap upon doping, and the positions of the other energy bands do not change as much. This is due to the polaronic

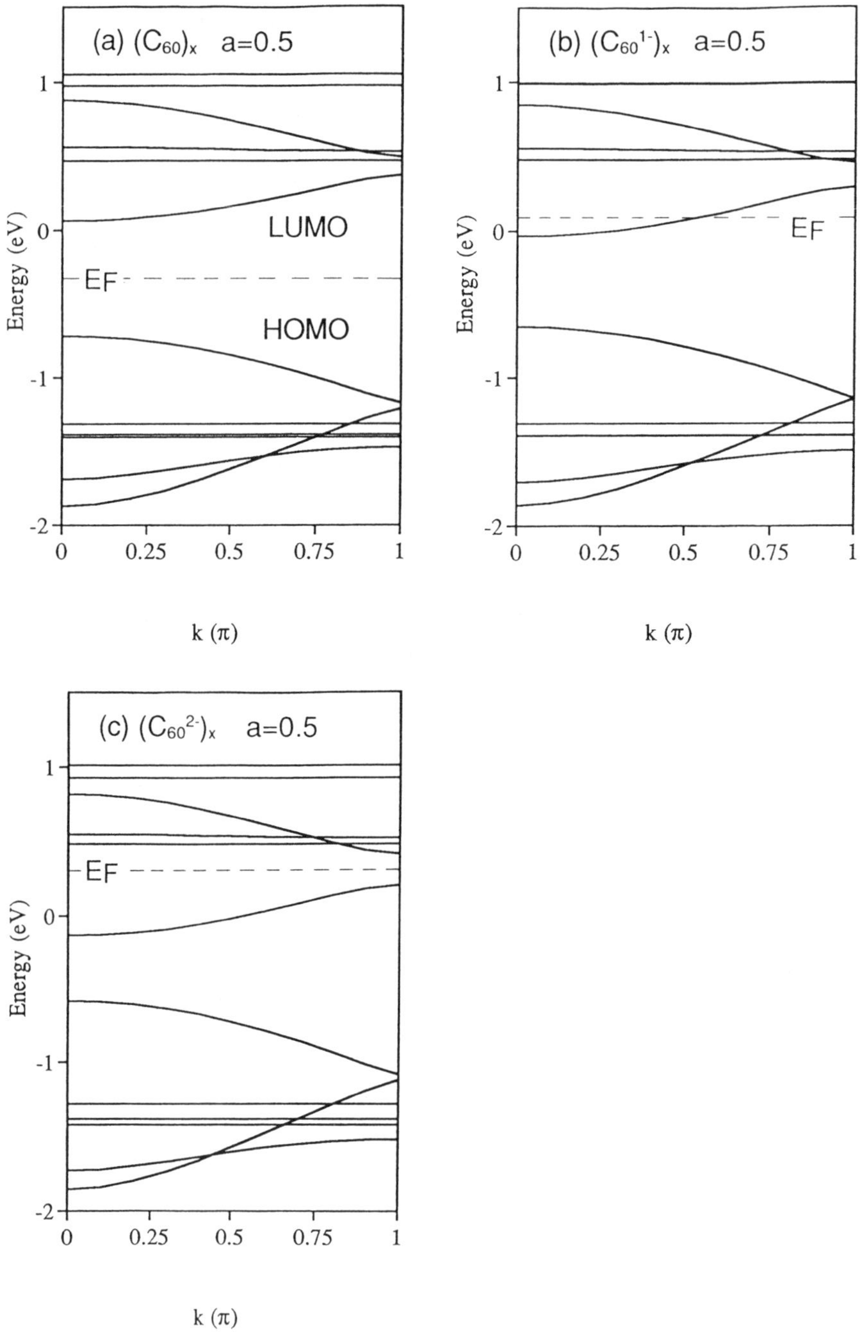

Fig. 9.7. Band structures of the one-dimensional C_{60}-polymer for the conjugation $a = 0.5$. The excess electron number N_{ex} per one C_{60} is (**a**) 0, (**b**) 1, and (**c**) 2, respectively. The lattice constant of the unit cell is taken as unity

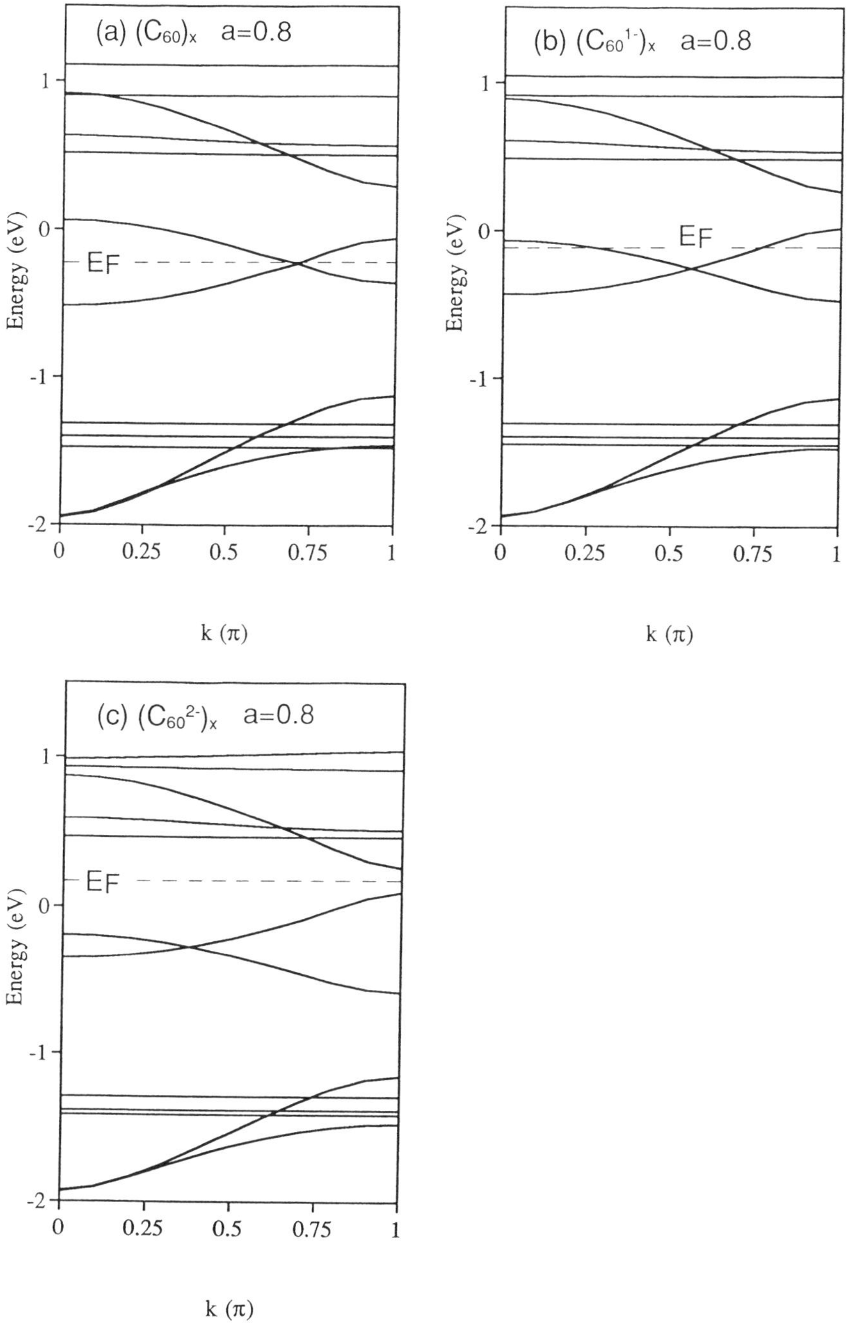

Fig. 9.8. Band structures of the C_{60}-polymer for the conjugation $a = 0.8$. The excess electron number N_{ex} per one C_{60} is (**a**) 0, (**b**) 1, and (**c**) 2, respectively. The lattice constant of the unit cell is taken as unity

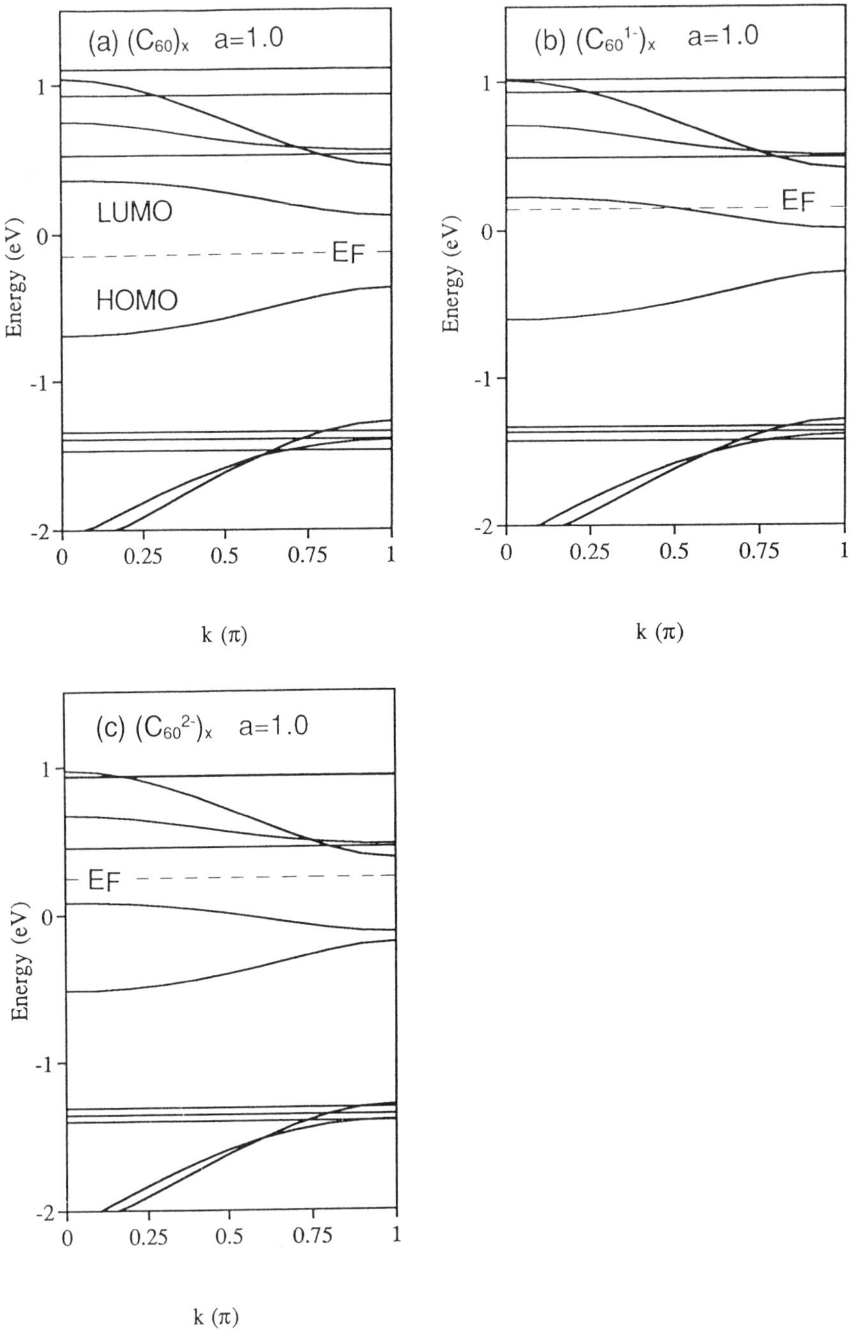

Fig. 9.9. Band structures of the C_{60}-polymer for the conjugation $a = 1.0$. The excess electron number N_{ex} per one C_{60} is (**a**) 0, (**b**) 1, and (**c**) 2, respectively. The lattice constant of the unit cell is taken as unity

distortion of the lattice, which we have discussed in the calculation of an isolated molecule in Sect. 3.

When $N_{ex} = 2$, the system is always found to be an insulator. For smaller a, for example, $a = 0.5$ and 0.8, the energy gap appears at $k = \pi$. For larger a, for example, $a = 1.0$ (Fig. 9.9c), the energy gap becomes indirect. The polaronic distortion becomes larger as the doping concentration increases. Thus, the motions of the HOMO and LUMO bands from those of the neutral system become larger, too.

The above variations of the energy gap are summarized for the cases $N_{ex} = 0$ and 2, where a finite energy gap appears for certain a values. The results are shown in Figs. 9.10a, b. The white (black) squares indicate that the system is a direct gap insulator where there is a energy gap at $k = 0$ (π). The squares with the plus mean that the system is an indirect gap insulator. The crosses are for metals. In the neutral system $N_{ex} = 0$ (Fig. 9.10a), the energy gap decreases almost linearly for smaller a. The system changes into a metal as a increases, and finally an energy gap appears again. For $N_{ex} = 2$ shown in Fig. 9.10b, the system is a direct gap insulator with the energy gap at $k = \pi$ up to $a \sim 0.9$. The system turns into an indirect gap insulator near $a = 1.0$.

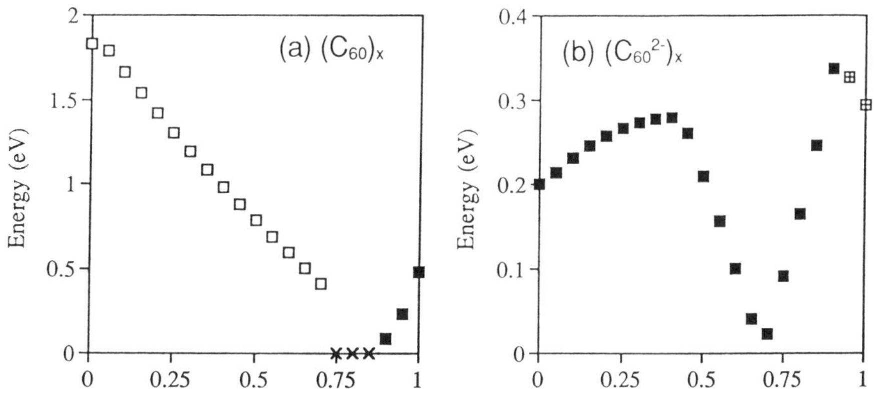

Fig. 9.10. Variations of the energy gap plotted against a. The cases $N_{ex} = 0$ and 2 are shown in (**a**) and (**b**), respectively. The white (black) squares indicate that the system is a direct gap insulator where there is a energy gap at $k = 0$ (π). The squares with the plus symbol indicate that the system is an indirect gap insulator. The crosses are for metallic cases

9.5 Metallic and Insulating States in Two-Dimensional C_{60}-Polymers

We next discuss band structures of the rectangular polymers for the conjugation conditions, $a = 0.5$, 0.7, 0.8, 0.9, and 1.0. The results are shown in Fig. 9.11. At $a = 0.5$ (Fig. 9.11a), there is a direct energy gap at the Γ point ($k = 0$), and the system is an insulator. As discussed in [10], the density of states of the system at ambient pressures is large as the bandwidths are quite narrow. The parameter value $a = 0.5$ may be too large for the rectangular polymer at room temperature and pressure [8]. However, we shall report the relatively large a cases, because our central interest is in the behavior of the system for large conjugation which could be realized at high pressures. As the parameter a increases, the HOMO-LUMO gap decreases, as shown for $a = 0.7$ in Fig. 9.11b. Bands further removed from E_F do not move as much. This is because their wave functions have large amplitudes in the central parts of the C_{60} molecules. In Fig. 9.11c ($a = 0.8$), the crossing of the HOMO and LUMO bands at smaller a occurs. There are Fermi surfaces at about -0.3 eV, and the system is metallic. The crossing becomes more apparent for $a = 0.9$ as shown in Fig. 9.11d. At the maximum conjugation condition ($a = 1.0$), a direct energy gap appears again at the K point, as shown in Fig. 9.11e, and the system changes into an insulator again.

Why does such reentrant behavior take place? In order to discuss the reason, we show the magnitudes of wave functions at the Γ point for (a) $a = 0.5$ and (b) $a = 1.0$, in Fig. 9.12. The inequivalent sites are labelled as A–I, as shown in Fig. 9.2a. The wave functions can be taken as real, so we use this convention. The HOMO is shown by white bars, and the LUMO is displayed by the black bars. In Fig. 9.12a, the HOMO has the negligible amplitude at site E, and the LUMO is near zero at site I. In contrast, as shown in Fig. 9.12b, the LUMO has negligible amplitude at site E and the HOMO amplitude is near zero at site I. This fact indicates that the symmetries of the HOMO and LUMO are reversed in the two insulators which are derived from small and large values of the conjugation parameter a. Therefore, the crossing of the HOMO band and LUMO band should occur at an intermediate value of a in the present model, and this is therefore the origin of the metallic band structures. Applying high pressures might decrease the distances between C_{60} molecules, and thus with increasing a, an insulating system evolves into a metal and then to an insulator again. This behavior is specific to the rectangular C_{60}-polymers, and it seems a quite interesting result.

Next, we shall discuss band structures of the triangular C_{60}-polymer shown in Fig. 9.2b. There are six inequivalent sites, and they are labeled as A–F. Qualitatively, the band structure changes with conjugation are similar to those of the rectangular polymer, so we show band structures only for (a) $a = 0.5$ and (b) $a = 1.0$, in Fig. 9.13. For $a = 0.5$, the HOMO band has the maximum at the Γ point, and the LUMO band has the minimum at the

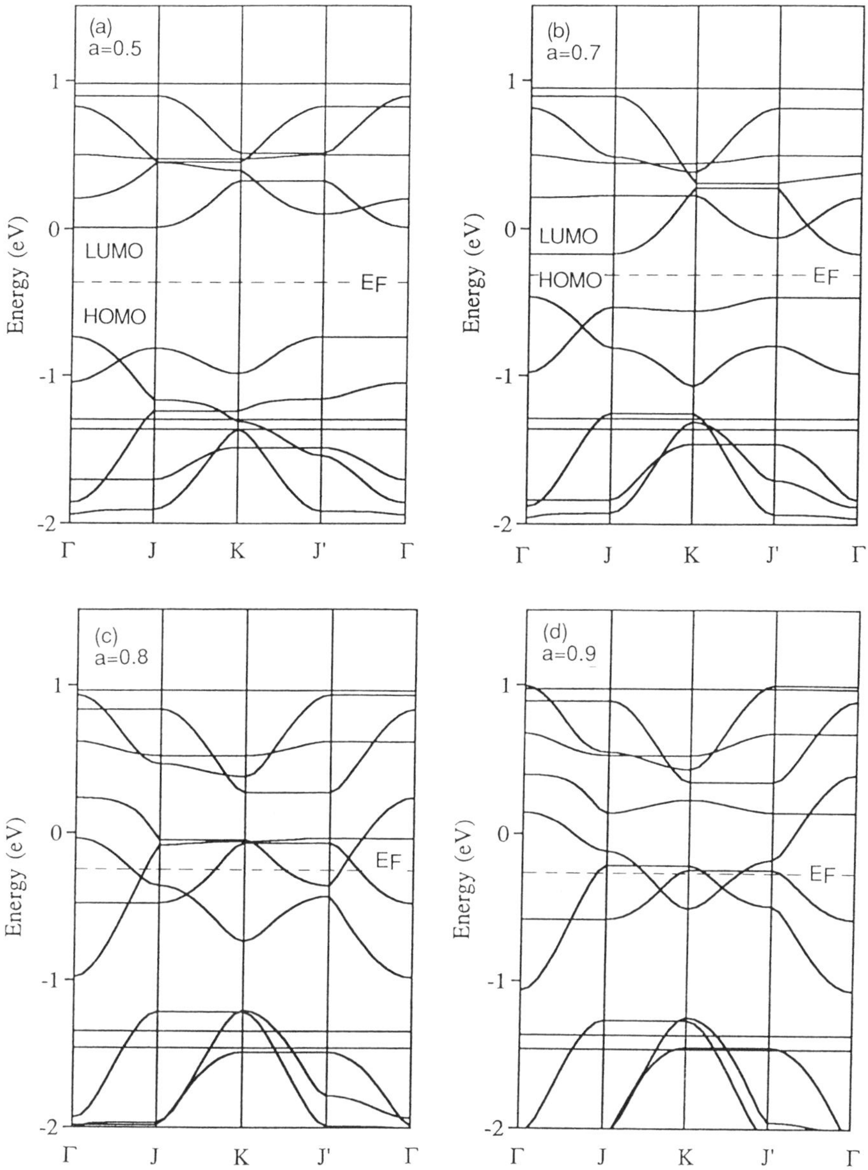
(a)
a=0.5
LUMO
HOMO
EF
Energy (eV)
1
0
-1
-2
Γ J K J' Γ
(b)
a=0.7
LUMO
HOMO
EF
Energy (eV)
1
0
-1
-2
Γ J K J' Γ
(c)
a=0.8
EF
Energy (eV)
1
0
-1
-2
Γ J K J' Γ
(d)
a=0.9
EF
Energy (eV)
1
0
-1
-2
Γ J K J' Γ

Fig. 9.11 a–d.

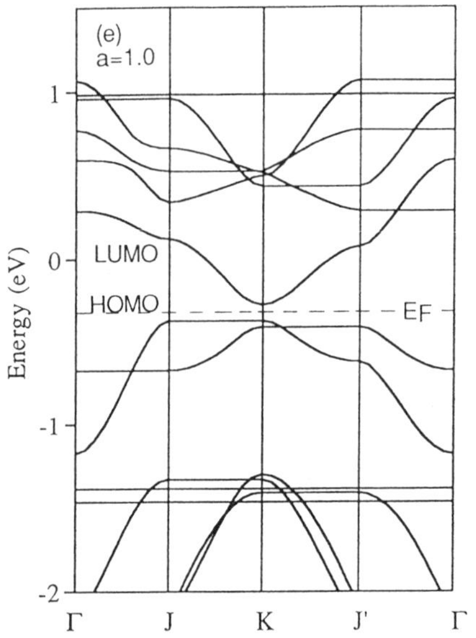

Fig. 9.11. Band structures of the rectangular C_{60}-polymer for various values of the conjugation parameter (**a**) $a = 0.5$, (**b**) 0.7, (**c**) 0.8, (**d**) 0.9, and (**e**) 1.0

K point. The system is an indirect gap insulator. For $a = 1.0$, there are Fermi surfaces at about -0.1 eV. Therefore, a metallic transition could be observed in the triangular polymer, too. In this system, a reentrant behavior is not found in the large a region. This might come from the geometry difference between the rectangular and triangular polymers.

9.6 Magnetism in One-Dimensional C_{60}-Polymers

The model Hamiltonian (2) is solved with the assumptions of the unrestricted Hartree–Fock approximation. We assume that one electron is doped per C_{60} molecule and $V = U/2$. Results are calculated for various values of the parameters a and U.

The magnitude of the ordered spin per one C_{60} is calculated, and is shown in Fig. 9.14 as a function of a and U. If the magnitude is zero, there is no antiferromagnetic order. The magnitude is finite when the magnetic ordering appears. As U becomes stronger, the magnetization appears and becomes larger, as expected. When a is smaller, the itinerancy of electrons in the chain direction decreases, and thus the antiferromagnetism appears more easily. A general feature is realized in the present calculation: the critical value of U, where a finite magnetization begins to appear, decreases as a becomes smaller.

The calculated results are summarized as a phase diagram in Fig. 9.15. The metallic phase region is labeled as M, and the antiferromagnetic phase is labeled as AF. Antiferromagnetism appears in the small a region when U is

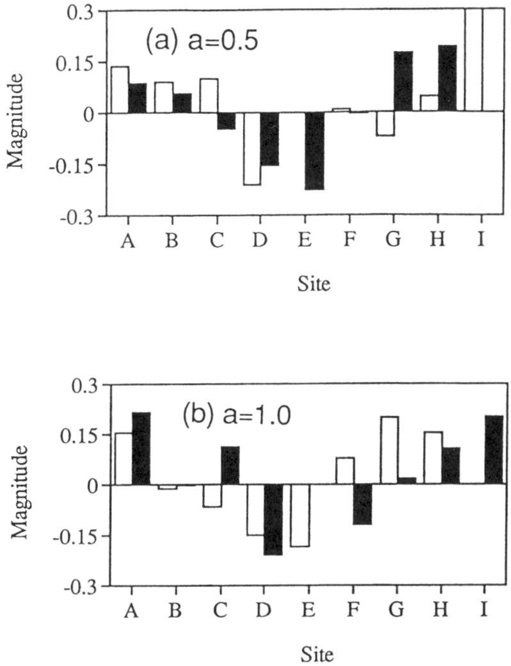

Fig. 9.12. Magnitudes of the wavefunctions of the HOMO and the LUMO at the Γ point for (**a**) $a = 0.5$ and (**b**) $a = 1.0$, in the rectangular polymer. The nonequivalent sites are labelled as A–I, as shown in Fig. 9.1a. The HOMO is shown by the white bars, and the LUMO is displayed by the black bars

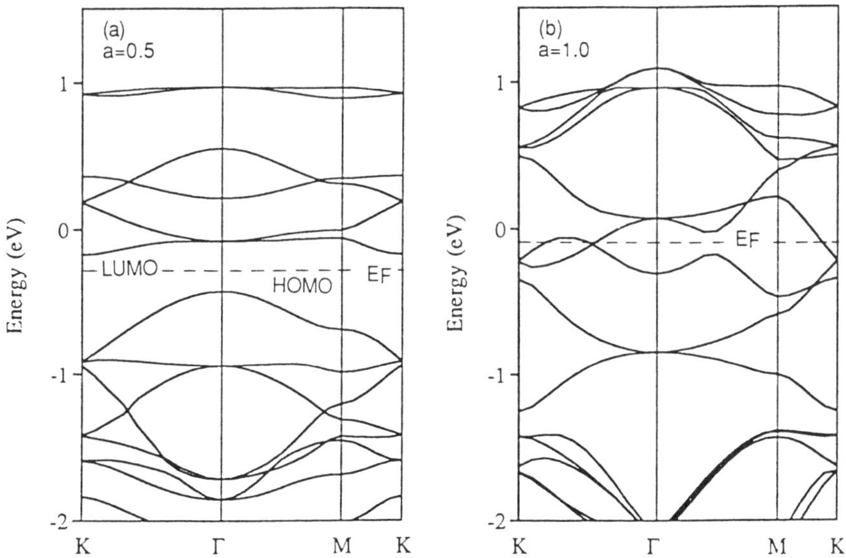

Fig. 9.13. Band structures of the triangular C_{60}-polymer of the cases (**a**) $a = 0.5$ and (**b**) $a = 1.0$

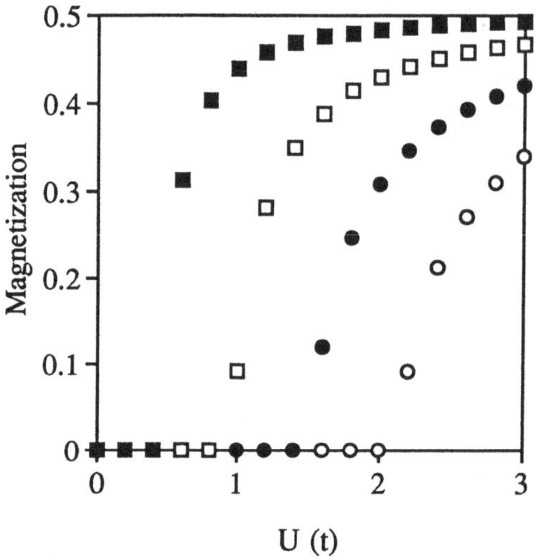

Fig. 9.14. Magnitude of the magnetization per one C_{60} shown as a function of a and U. We use $V = U/2$ in the Ohno potential (see text). The closed and open squares are for $a = 0.1$ and 0.2, respectively. The closed and open circles are for $a = 0.3$ and 0.4, respectively

taken constant. The boundary between the AF and M phases is almost linear. In [22], the applicability of the one-dimensional Hubbard model is discussed. When this discussion is adopted to the present model (2), the model can be mapped to the following Hamiltonian:

$$H = -t_{\text{eff}} \sum_{\langle l,l' \rangle,\sigma} (a_{l,\sigma}^\dagger a_{l',\sigma} + \text{h.c.}) + U_{\text{eff}} \sum_l n_{l,\uparrow} n_{l,\downarrow}, \tag{9.6}$$

where t_{eff} is an effective hopping integral between neighboring C_{60} molecules; U_{eff} is the on-ball Coulomb repulsion; $a_{l,\sigma}$ is an annihilation operator of an effective orbital of the doped electron at the lth molecule; and $n_{l,\sigma} = a_{l,\sigma}^\dagger a_{l,\sigma}$. In the one-dimensional Hubbard model, it is known that the antiferromagnetism appears even when U_{eff} is small but positive. However, there is a region of the phase diagram (Fig. 9.15) where magnetic order is not present. Thus, we find that the mapping onto the Hubbard model is not always relevant, and that it can be done only for the case of small a and large U. In [22], the antiferromagnetism has been discussed in connection with the polaron in the doped C_{60} [16]. In the present section, electron–phonon interactions are not considered, and only electron–electron interactions are taken into account. Therefore, the antiferromagnetism found is not directly related with polaron formation due to the Jahn–Teller effects [16]. Rather, we could regard the antiferromagnetism due to electron–electron interactions and as one of the spin density wave states in an itinerant electron system. It is, of course, found that the extra (doped) electrons tend to have large amplitudes at certain sites on the surface of the C_{60} molecules. However, the main origin of the localization would be strong electron–electron interactions rather than the Jahn–Teller effect.

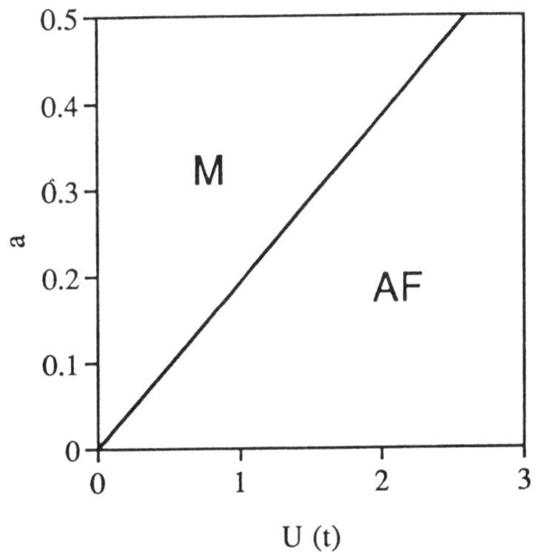

Fig. 9.15. Phase diagram shown against a and U. The region "M" is the metallic phase, and the region "AF" is the antiferromagnetic phase

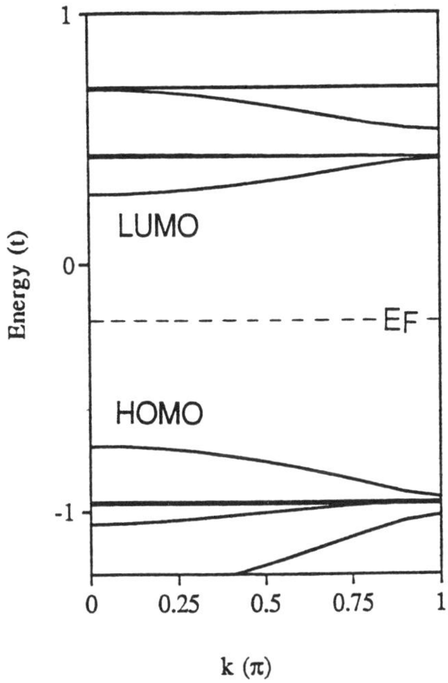

Fig. 9.16. Band structures of the neutral C_{60}-polymer of the case $(a, U, V) = (0.3, 2.4t, 1.2t)$. The unit cell consists of one C_{60}. The length of the unit cell is taken as unity

It is useful to look at how the band structures change upon the formation of the antiferromagnetic order. The dispersion of the bands of the system before the electron doping and without the magnetic order are shown for $(a, U, V) = (0.3, 2.4t, 1.2t)$ in Fig. 9.16. Here, the electron number is 60 per one C_{60}, and the unit cell consists of one C_{60}. The band structures of the

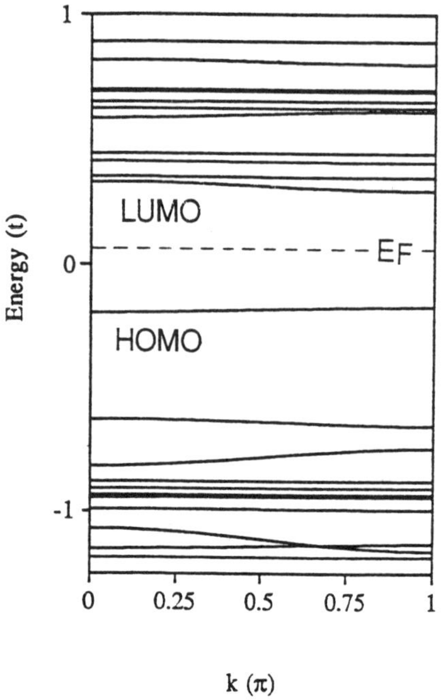

Fig. 9.17. Band structures of the C_{60}-polymer doped with one electron per one C_{60} of the case $(a, U, V) = (0.3, 2.4t, 1.2t)$. The unit cell consists of two C_{60} molecules. The length of the unit cell is taken as unity

antiferromagnetic system are shown for the same (a, U, V) in Fig. 9.17. The unit cell becomes doubled, so the first Brillouin zone is half that shown in Fig. 9.16. The energy bands in Fig. 9.17 become narrower. The band named "HOMO" in Fig. 9.16 changes into the HOMO-1 band and the HOMO-2 band in Fig. 9.17. There is a small energy gap between the HOMO-1 band and HOMO-2 band at the boundary of the first Brillouin zone in Fig. 9.17. In the same way, the band named "LUMO" of Fig. 9.16 is the origin of both the "HOMO" and "LUMO" of Fig. 9.17. There is a large energy gap at the zone boundary as well. These energy gaps are opened by the antiferromagnetic states.

It should be noted that, in the presence of antiferromagnetism, the HOMO-derived energy band rises in energy and is located near the center of the original energy gap for the neutral system. This intrusion is much larger than that in the polaron formation due to the Jahn–Teller interactions in the doped C_{60} [16], and cannot be obtained from models with electron–phonon interactions only. Actually, the new band has been observed in the photoemission studies of the orthorhombic phase of Rb_1C_{60} [20]. Such large changes in band structure might be evidence for a strong electron–electron interaction in the C_{60}-polymers, rather than the result of a Jahn–Teller distortion due to the electron–phonon interactions. The positions of the HOMO band and HOMO-1 band of Fig. 9.16 do not shift so much upon the formation of the antiferromagnetism. This observation is also consistent with the photoemission studies [20].

In this section, we have not considered possible interchain interactions which might effect on the dispersion of bands perpendicular to the polymer direction [29, 30]. In the tight-binding type models as in the present calculations, there might be many candidates for the form of the interchain interaction. Even though they have not been taken into account in the present calculations, we might expect that the nature of one-dimensional chains with strong electron–electron interactions is still effective in establishing an antiferromagnetic state in the real A_1C_{60} system. The low dimensionality of the polymer-chain might be favorable for the AF ordering to occur due to interactions among itinerant electrons.

9.7 CT-Excitons in C_{60}-Dimers and Polymers

The exciton formalism applied to isolated C_{60} and C_{70} molecules [21] is extended to cross-linked C_{60} systems in this section. The ideas of the Frenkel (molecular) exciton and the charge-transfer (CT) exciton are shown in Fig. 9.18. The Frenkel exciton is an excitation of an electron–hole pair in a single molecule, and the CT-exciton is an excitation of an electron from one molecule to another molecule. The parameters in the Ohno potential are $U = 4t$, $V = 2t$, and $t = 1.8\,\mathrm{eV}$. They have been determined in order to explain the experimental optical spectra of isolated molecules [21].

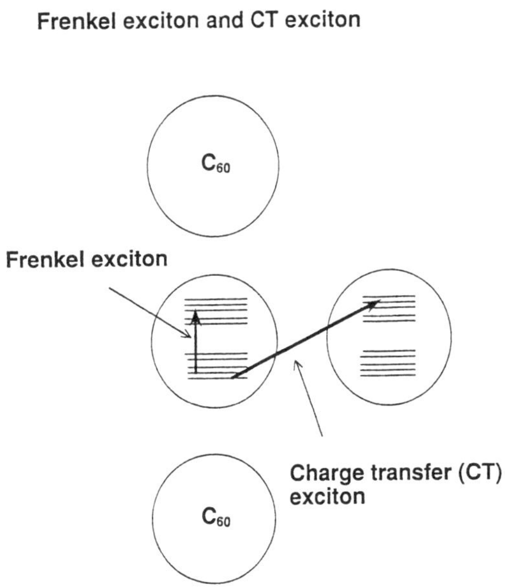

Fig. 9.18. Frenkel exciton charge-transfer (CT) exciton

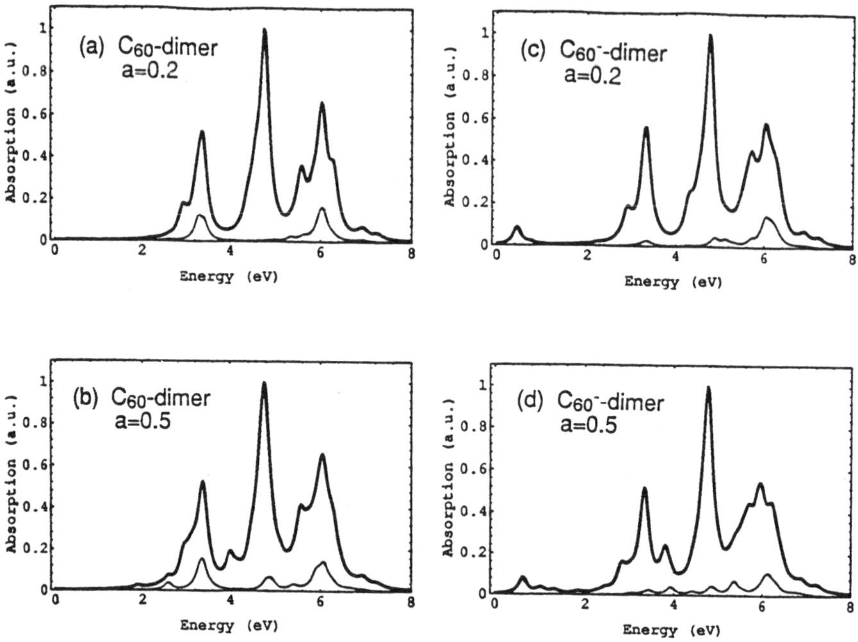

Fig. 9.19. Optical absorption spectra of C_{60}-dimers for the neutral case with (**a**) $a = 0.2$ and (**b**) $a = 0.5$, and for the one-electron (per C_{60}) doped case with (**c**) $a = 0.2$ and (**d**) $a = 0.5$. The data are shown in arbitrary units. The bold line represents the total absorption, and the thin line shows the contribution from the charge-transfer excitons. The Lorentzian broadening 0.108 eV is used

First, we look at theoretical optical spectra of C_{60} dimers. Figures 9.19a, b show the optical absorption spectra of the neutral systems, and Figs. 9.19c, d show the absorption of the systems doped with one electron per C_{60}. There are three main peak structures which are typical of a single molecule in Figs. 9.19a, b [21]. In Figs. 9.19c, d, a small feature appears around 0.4–0.6 eV. This is due to the filling of electrons in t_{1u} orbitals derived from the C_{60} molecule. Optical transition from t_{1u} to t_{1g} orbitals is dipole allowed, and a new peak appears at low energies. The similar fact has been recently reported in the optical spectra by the quantum chemical calculations [31].

In order to characterize the CT-excitons, we calculate the probability that the excited electron and hole exist on different molecules. If the probability is larger than 0.5, we regard the excitation as CT-like. If it is smaller than 0.5, the excitation tends to localize on a single molecule and is Frenkel-like. We calculate contributions from the CT-excitons to the optical absorption, by retaining only the CT-like part. The results are shown by the thin lines in Fig. 9.19.

Even though molecular exciton character is dominant, a small CT-like component develops in the optical spectra. In Figs. 9.19a, b, relatively larger oscillator strengths of the CT-excitons exist at the higher energy sides of the main three features centered around 3.2 eV, 4.8 eV, and 6.0 eV. In the doped case of Figs. 9.19c, d, the absorption of the CT-excitons become somewhat broader, but a certain amount of CT-like oscillator strengths exist. We also find that the small features around 0.4–0.6 eV, which appear upon doping, are mainly Frenkel-like.

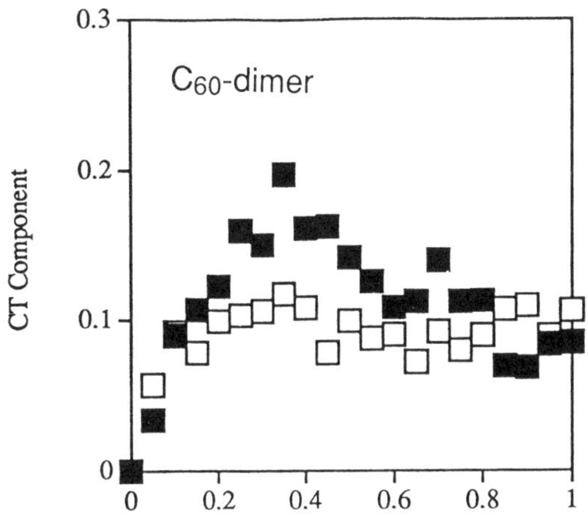

Fig. 9.20. Ratio of the oscillator strengths (CT-component) of the charge-transfer excitons with respect to the total oscillator strengths in the C_{60}-dimers. The closed squares are for the neutral cases and the open squares are for the one-electon (per C_{60}) doped cases

Now, we summarize development of the CT-component as increasing the conjugation parameter. The ratio of the sum of the oscillator strengths of the CT-excitons with respect to that of the total absorption is calculated and is shown as a function of the conjugation parameter in Fig. 9.20. In other words, we have calculated the ratio of the area below the thin line with respect to the area below the bold line in Fig. 9.19. The filled and open squares are the data of the neutral and doped systems, respectively. We find that the CT-components first increase as the intermolecular interactions switch on, and saturate at about $a = 0.3$, or so. It is somewhat surprising that the CT-component gradually decreases at the larger a than about 0.5, i.e. in the large conjugation region. Without calculations, one may expect that the value will continue increasing as the intermolecular interactions become stronger. We also find that the CT-component of the doped system is smaller than that of the neutral system. Thus, the doping effects slightly suppress separations between the excited electron and the hole.

Next, calculations are performed for the C_{60}-polymers. The two-molecule system is considered. The two-molecule system is equivalent to the dimer system with periodic boundary conditions. The characterization of the CT-excitons has been done similarly as in the dimer system. Figures 9.21a, b show the optical absorption spectra of the neutral system for the two conju-

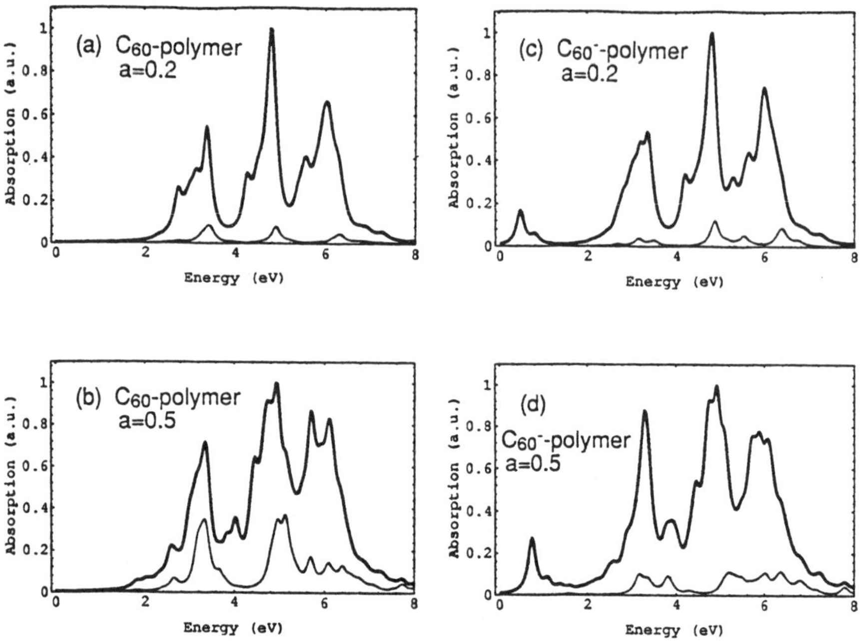

Fig. 9.21. Optical absorption spectra of C_{60}-polymers for the neutral case with (**a**) $a = 0.2$ and (**b**) $a = 0.5$, and for the one-electron (per C_{60}) doped case with (**c**) $a = 0.2$ and (**d**) $a = 0.5$. The data are shown in arbitrary units. The bold line represents the total absorption, and the thin line shows the contribution from the charge-transfer excitons. The Lorentzian broadening 0.108 eV is used

gation parameters. The thin line represents the optical spectra due to the CT-excitons. The CT-excitons exist at the higher energy side of each of the three main features. This property is clearly seen for stronger conjugation, $a = 0.5$, in Fig. 9.21b. The broadening of the three main features is larger when the intermolecular interactions are stronger. A small shoulder appears in the energy region lower than 2.8 eV in Fig. 9.21b. There are dipole-forbidden transitions in these energy region for the single molecule. They become partially allowed when there are intermolecular interactions. This is the origin of the broad low energy structure. Such broad structures have been measured in the optical absorption of C_{60}-polymers synthesized under high pressures [29], and have been ascribed to the symmetry lowering effects.

We look at the optical spectra of the doped system with one electron per C_{60}. The total absorptions and the CT-like component are shown for $a = 0.2$ and 0.5 in Figs. 9.21c, d, respectively. The appearance of the small feature at 0.4–0.6 eV is more distinct than in C_{60}-dimers. As we are treating the two-molecule system, the small feature is a peak. However, if we could do calculations for an infinitely long polymer, the feature will develop into a Drude tail which is typical of metallic systems. In the present exciton formalism, we cannot reproduce the Drude structure. This is a limitation of calculations on small systems. By comparing the neutral and doped systems, we find that the CT-component is suppressed upon doping.

The ratio of the oscillator strengths of the CT-excitons with respect to the total absorption is calculated, and is shown as a function of the conjugation parameter in Fig. 9.22. The filled and open squares are for the neutral and doped systems. We find that the CT-component first increases

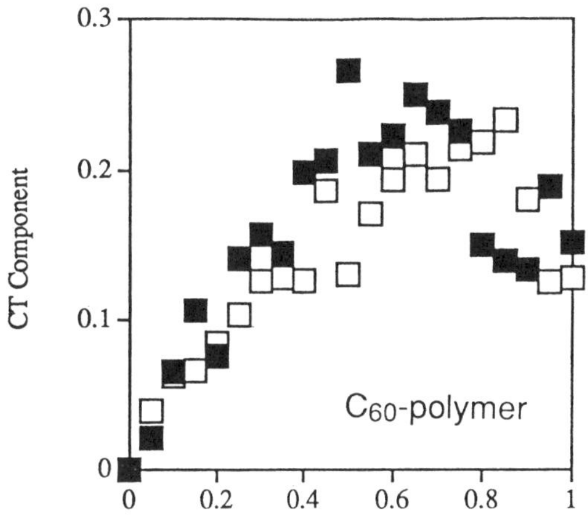

Fig. 9.22. Ratio of the oscillator strengths (CT-component) of the charge-transfer excitons with respect to the total oscillator strengths in the C_{60}-polymers. The closed squares are for the neutral cases and the open squares are for the one-electon (per C_{60}) doped cases

in smaller conjugation, and saturates at the intermediate conjugation. The dramatic decrease is seen in the large conjugation. The CT-component of the C_{60}-polymers is larger than that of the C_{60}-dimers in Fig. 9.20. This is due to the enhanced intermolecular interactions in polymers. We also note that the CT-component decreases when the neutral system is doped with electrons. The same qualitative feature has been seen in the calculations of the dimers.

9.8 Summary

In the neutral one-dimensional and two-dimensional C_{60}-polymers, the electronic structure can be that of a direct-gap insulator or a metal, depending on the degree of conjugations. High pressure experiments may be able to change conjugation conditions in the chain direction, and the electronic structure changes discussed here could be observed.

Within the model calculations performed here, the one-dimensional C_{60}-polymer doped with electron per one molecule is always found to be a metal.

The energy difference between the highest-occupied state and the lowest-unoccupied state of the neutral system becomes smaller upon doping owing to the polaron effects. When the one-dimensional C_{60}-polymer is doped with two electrons per one C_{60}, the system is found to be insulating, and for decreasing conjugation in the direction of the polymer chain, a direct-gap insulator is obtained. The energy gap becomes indirect for strong conjugation.

Next, we have studied the antiferromagnetic phase of the C_{60}-polymers by microscopic calculations. The formation of the antiferromagnetic order is closely related with strong itinerant electron interactions rather than by a Jahn–Teller mechanism. Similar to photoemission studies on A_1C_{60}, we find that a new metallic band can be explained by the fact that the HOMO of the neutral system rises into the HOMO-LUMO gap by virtue of the electron doping. This is another indication that the interaction effects among electrons play a crucial role in doped C_{60}-polymers.

Finally, we have discussed properties of the CT-like excitons in optical absorption of the neutral and doped C_{60}-dimers (and polymers). We have looked at changes of the optical spectra as a function of the conjugation parameter. We have found that the total CT-component first increases in smaller conjugation, as expected. However, it saturates at the intermediate conjugation, and a dramatic decrease follows in the large conjugation. These qualitative features are commonly seen in the calculations of C_{60}-dimers and polymers, though the strengths of CT-like oscillator strengths are quantitatively different. We have also found that the CT-component of the doped systems is smaller than that of the neutral systems, and thus the electron–hole separations are suppressed upon doping in polymerized C_{60} systems.

References

1. O. Chauvet, G. Oszlànyi, L. Forró, P. W. Stephens, M. Tegze, G. Faigel, A. Jànossy: Phys. Rev. Lett. **72**, 2721 (1994).
2. P. W. Stephens, G. Bortel, G. Faigel, M. Tegze, A. Jànossy, S. Pekker, G. Oszlanyi, L. Forró: Nature **370**, 636 (1994).
3. S. Pekker, L. Forró, L. Mihaly, A. Jànossy: Solid State Commun. **90**, 349 (1994).
4. S. Pekker, A. Jànossy, L. Mihaly, O. Chauvet, M. Carrard, L. Forró: Science **265**, 1077 (1994).
5. M. Kosaka, K. Tanigaki, T. Tanaka, T. Atake, A. Lappas, K. Prassides: Phys. Rev. B **51**, 12018 (1995).
6. O. Béthoux, M. Núñez-Regueiro, L. Marques, J. L. Hodeau, M. Perroux: Paper presented at the *Materials Research Society Meeting*, Boston, USA, November 29-December 3 (1993).
7. Y. Iwasa, T. Arima, R. M. Fleming, T. Siegrist, O. Zhou, R. C. Haddon, L. J. Rothberg, K. B. Lyons, H. L. Carter Jr., A. F. Hebard, R. Tycko, G. Dabbagh, J. J. Krajewski, G. A. Thomas, T. Yagi: Science **264**, 1570 (1994).
8. M. Núñez-Regueiro, L. Marques, J. L. Hodeau, O. Béthoux, M. Perroux: Phys. Rev. Lett. **74**, 278 (1995).
9. G. Oszlanyi and L. Forró: Solid State Commun. **93**, 265 (1995).

10. C.H. Xu and G.E. Scuseria: Phys. Rev. Lett. **74**, 274 (1995).
11. K. Harigaya: Phys. Rev. B **52**, 7968 (1995).
12. K. Harigaya: Chem. Phys. Lett. **242**, 585 (1995).
13. T. Hasegawa, Y. Iwasa, H. Sunamura, T. Koda, Y. Tokura, H. Tachibana, M. Matsumoto, S. Abe: Phys. Rev. Lett. **69**, 668 (1992).
14. W.P. Su, J.R. Schrieffer, A.J. Heeger: Phys. Rev. B **22**, 2099 (1980).
15. K. Harigaya: J. Phys. Soc. Jpn. **60**, 4001 (1991).
16. K. Harigaya: Phys. Rev. B **45**, 13676 (1992).
17. K. Harigaya: Chem. Phys. Lett. **189**, 79 (1992).
18. K. Harigaya: Chem. Phys. Lett. **253**, 420 (1996).
19. K. Harigaya: Phys. Rev. B **53**, R4197 (1996).
20. G.P. Lopinski, M.G. Mitch, J.R. Fox, J.S. Lannin: Phys. Rev. B **50**, 16098 (1994).
21. K. Harigaya and S. Abe: Phys. Rev. B **49**, 16746 (1994).
22. K. Tanaka, Y. Matsuura, Y. Oshima, T. Yamabe, Y. Asai, M. Tokumoto: Solid State Commun. **93**, 163 (1995).
23. M.S. Dresselhaus, G. Dresselhaus, P.C. Eklund: *Science of Fullerenes and Carbon Nanotubes* (Academic Press, San Diego, 1996).
24. Y. Iwasa: *Optical Properties of Low-Dimensional Materials*, T. Ogawa and Y. Kanemitsu (Eds.) (World Scientific, Singapore, 1995) Chap. 7.
25. K. Harigaya and S. Abe: Mol. Cryst. Liq. Cryst. **256**, 825 (1994).
26. K. Harigaya and S. Abe: *22nd International Conference on the Physics of Semiconductors* (World Scientific, Singapore, 1995), p. 2101.
27. K. Harigaya: Phys. Rev. B **50**, 17606 (1994).
28. A.J. Heeger, S. Kivelson, J.R. Schrieffer, W.P. Su: Rev. Mod. Phys. **60**, 781 (1988).
29. S.C. Erwin, G.V. Krishna, E.J. Mele: Phys. Rev. B **51**, 7345 (1995).
30. K. Tanaka, Y. Matsuura, Y. Oshima, T. Yamabe, H. Kobayashi, Y. Asai: Chem. Phys. Lett. **241**, 149 (1995).
31. J. Fagerström and S. Stafström: Phys. Rev. B **53**, (1996) 13150.

10. Superconductivity of "Fullerene Conducting Polymer" Composites Doped by Alkali Metals

A.A. Zakhidov, H. Araki, K. Yoshino

10.1 Introduction

The great interest in buckminsterfullerene C_{60} was accelerated by the discovery of superconductivity in alkali metal doped C_{60} crystals [1–2]. One of the most amazing features of this fascinating molecule is its ability to provide both of the cooperative phenomena: superconductivity and ferromagnetism [3–6] in its solid state complexes.

When embedded into the tetrahedral and octahedral interstitial positions of the fcc lattice of the A_3C_{60} crystal (A being an alkali metal, e.g. K, Rb) each C_{60} is triply charged to C_{60}^{3-} and a narrow conduction band is formed due to $\pi - \pi$ wave function overlap between adjacent molecules. This band is half filled due to the degeneracy of t_{1u} LUMO of C_{60} and the electrons undergo a superconducting transition below a critical temperature T_c, which ranges from 19 K in K_3C_{60} to 33 K in $Cs_2Rb_1C_{60}$, The latter is higher than the T_c of any conventional inorganic superconductors except high T_c SC oxide ceramics.

On the other hand, when the C_{60} molecules are in a 1:1 solid state complex with the organic donor TDAE (tetrakis (dimethylamino)ethylene), the molecules are in the C_{60}^{1-} charged state, and the solid exhibits ferromagnetism below $T_c = 17$–24 K [3–6]. This is the highest Curie temperature for an organic magnet. Theories for the mechanisms of both phenomena have been proposed [9–17], and it looks like all other higher fullerenes C_n ($n > 70$) are not able to show these cooperative phenomena.

The absence of superconductivity in K_xC_{70} and in all higher fullerides K_xC_n ($n = 76$, 78, 86, 92) [7] (at least above 4 K), and the systematic rapid decrease of T_c with x in molecular alloys of $A_3(C_{60})_{1-x}(C_{70})_x$ [7] strongly suggest that the C_{60} molecule is exceptional in the fullerene family C_n. The unique ability of C_{60} to exhibit cooperative phenomena is supported by the absence of ferromagnetism in TDAE-C_n ($n = 70$–78) [4, 5] and by the weak suppression of ferromagnetism (and Curie temperature) in the molecular alloys TDAE-$(C_{60})_{1-x}(C_{70})_x$ [6]. This evidence indicates that either due to the extremely high symmetry and degeneracy of the C_{60} electronic states, or due to other reasons, C_{60} is superior among other C_n fullerenes for the cooperative behavior of its additional π-electrons. Furthermore, if the symmetry of C_{60} is disturbed by the chemical addition of side groups as in C_{60}-CH_3A, superconductivity and ferromagnetic behavior will disappear [8].

A question arises as to what extent this special behavior is connected with intramolecular properties of C_{60} and to what extent does the contribution from the specific crystal structure and intermolecular interactions play a role? Regarding superconductivity, it is believed that the electron pairing interactions are provided by intramolecular forces, either vibrational (i.e. the exchange of intramolecular A_g and H_g symmetry type vibrons [9–11]), or of purely electronic origin. For the electronic mechanisms various models have been considered: negative U-centers [12, 13], parity doublets [14] and bipolarons [15].

Intermolecular or "lattice" phonons appear to be not effective for electron pairing, consistent with the absence of an alkali metal isotope effect [16], and recently justified theoretically [17].

We must also ask if the superior property of C_{60} for SC pairing can be effective if C_{60} is embedded into a lattice other than the fcc lattice of A_3C_{60}, or even into an amorphous matrix, in which it can be charged to an appropriate anionic state. If such a matrix can provide the mobility of charges at the Fermi level, will it support superconductivity in the composite? In fact, this question was the primary motivation for our present study. There were some indications that the type of lattice was not crucially important for SC. For example, the existence of superconductivity in alkaline-earth-metal-fullerides Ca_5C_{60} [18], Sr_6C_{60} and Ba_6C_{60} [19] in which C_{60} molecules form a bcc rather than a fcc lattice, and the number of electrons donated to C_{60} is 5 or 6 rather than 3. It has recently been claimed that in quasi-two-dimensional C_{60}-graphite intercalation compound doped by K, in $C_{32}K_4C_{60}$, superconductivity occurs at $T_c = 19.5\,K$ [20]. However, this latter material, and its SC, has not been confirmed.

One of the approaches to this interesting problem is connected with the alkali metal (A) doping of the quasi-two-dimensional charge transfer (CT)-complexes with planar organic donors like OM-TTF-C_{60}. Preliminary results in the material showed that a ternary compound OM-TTF-C_{60}-A_x (A = K, Rb) exhibits multi-phase SC behavior, with several superconducting transition temperatures T_c [21].

In the present paper, we discuss a distinct system, the composite in which C_{60} is intercalated between the chains of a conducting π-conjugated polymer (CP), and is also doped by A (A = K, Rb) with the aim of filling electrons into conduction states of the system. CP are chosen as a matrix for C_{60}, because this class of π-conjugated quasi-one-dimensional solids, like polyacetylene, undergoes a phase transition from an insulating to a metallic state upon exposure to n-type (e.g. Na, K, TBA^+, etc.) or p-type dopants (I_2, SbF_6, etc.) [22]. Charges in the CPs are known to be accumulated in self-trapped, mobile solitons, polarons and bipolarons, which also have interesting optical properties. Superconductivity is expected in doped CP states and even postulated theoretically at intermediate doping level [23], but it has never been observed experimentally.

We have recently initiated the search for SC phases that may exist in such $CP(C_{60})_yA_x$ composites [24–26], where both y and x are defined with respect to a monomer unit of the CP. In this chapter, we review briefly the experimental results [24–30] on these new CP based C_{60} composites, where OO-PPV (poly(2,5-dialkoxy-1,4-phenylene vinylene) or PAT (poly(3-alkylthiophene)) are the conducting polymers and A=Rb,K are used as the dopants. A general picture of a novel, granular SC has been found and is discussed below, which shows unusual multi-phase behavior and orbital supercurrents in the ground state. Further strategies for the search of C_{60}-induced superconductivity and its separation from granular SC phases are also suggested.

10.2 Charge Transfer Processes in Pristine CP-C_{60} Composites

We recently proposed that C_{60} or C_{70} can be used as a weak acceptor dopant in conducting polymers (CP); and in a series of experiments we have found quenching of photoluminescence (PL) [31] and enhancement of photoconductivity (PC) [32, 33] due to photoinduced charge transfer (PCT) between these fullerenes and a number of CPs. The results emphasized in this chapter follow our earlier work for weak dopants in CP [34]. This photoinduced charge transfer (PCT) is shown schematically in Fig. 10.1, and has been discussed in detail in our reviews [35–37]. PCT from C_{60} to CP has been independently found also in other CPs, like MEH-PPV and also demonstrated by complementary methods, such as photoinduced absorption (PA), light induced ESR [38] and transient spectroscopy which clarified the ultrafast speed of the PCT process [39, 40]. This phenomenon is also discussed in detail in Chap. 3 of this book.

It has also been found that a sizable PCT exists at the interface between a C_{60} film and various CP films, and this charge transfer across the interface leads to a sizable photovoltaic effect, which is being investigated for novel solar cell applications [38, 41–43].

All these studies have proved that C_{60} is well hybridized with the CP chains. Forward CT (electron transfer from the CP to C_{60}) indicates strong π–π overlap, while backward recombinative CT is observed to be strongly suppressed. Such a suppression may arise for several reasons; e.g. an overlap between self-trapped states on the CP and C_{60} that can be orientationally modulated due to the string-type geometry of the polarons on C_{60} [37, 43]. We have shown that the π-electrons of the CP chains are important for good hybridization. Coincident with this view, is the observation that if polysilane chains without aromatic side groups (with non π-electrons) were used as the matrix, PCT was strongly suppressed [44].

It has also been found that upon injection of carriers from electrodes into such composites, electroluminescence is quenched even more strongly

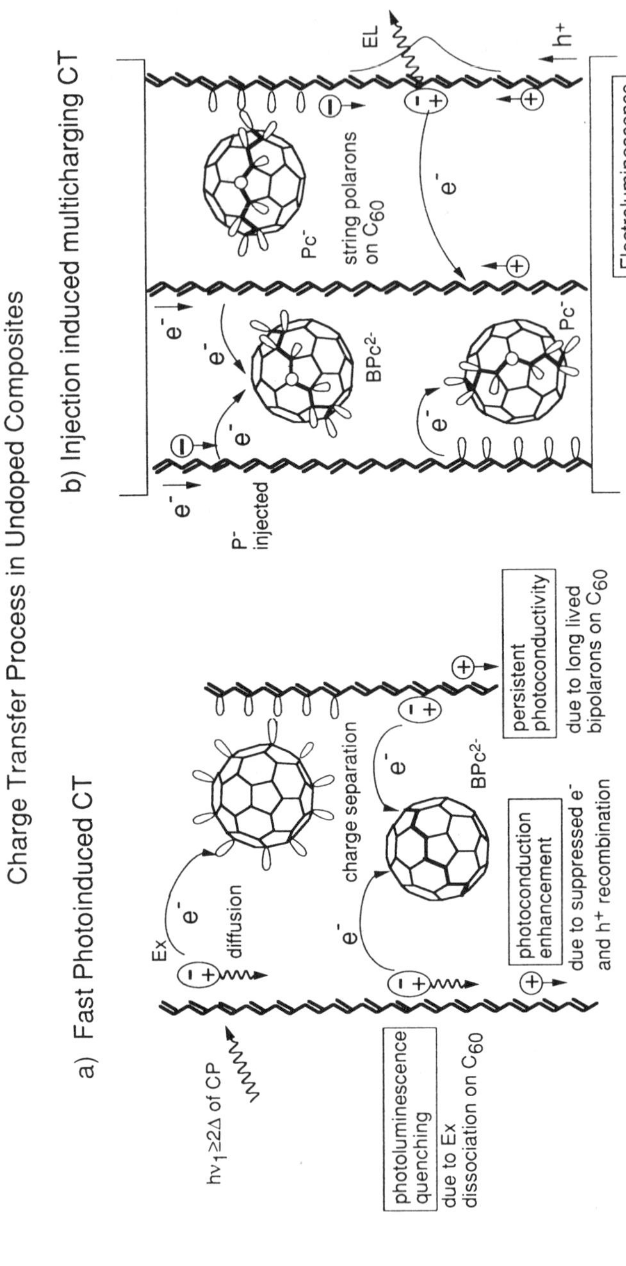

Fig. 10.1. Charge transfer processes in undoped C_{60}/CP composite: (**a**) forward photoinduced CT by exciton dissociation on C_{60}, which is fast due to good $\pi - \pi$ overlapping. (**b**) electrode injection followed by capture of electron by C_{60} in string polarons Pc and bipolarons BPc, which strongly quenches electroluminescence

than PL, indicating that the C_{60} molecule may act as such a multiple trapping and/or a recombination center for carriers [35–37]. The effect of persistent photoconductivity at room temperature has also been found in C_{60}-CP nanocomposites [45]. Both these latter effects strongly suggest that multi-charging of C_{60}, e.g. to a bipolaronic state C_{60}^{2-} may take place, and that these bipolarons are long lived.

In all the above studies, charge carriers are generated in CP-C_{60} composites due only to photopumping or electrode injection. Under such conditions (i.e. neutral C_{60} in a neutral polymeric matrix) the remarkable solid state properties of C_{60} anions in cooperative phenomena at low temperature, like superconductivity, can obviously not take place, simply due to the lack of equilibrium charge carriers. Of course, under extreme situations such as high density photoexcitation, the concentration of charged C_{60}^{n-} ($n = 1, 2, \ldots$) may become so high that cooperative phenomena like Bose condensation of C_{60} bipolarons, nonequilibrium superconductivity or excited state ferromagnetism might appear. We will not discuss these questions here, since a more realistic possibility exists to provide charges in the ground state. Fortunately, both C_{60} and CP can be doped by the same type of n-dopants, e.g. by alkali metals. Such CT doping should actually create a novel type of π-electron system in which charged C_{60}^{n-} anions coexist with a sea of quasi-1-D mobile electrons in the CP chains. The optical and conductive properties of such a system are very intriguing, while at low temperatures the appearance of superconductivity might be expected. In this sense, for all other higher fullerenes, e.g. C_{70} (which shows PCT quite similar to C_{60} [46]), one should not expect any special cooperative behavior upon A-doping in composites with CP. On the other hand SC phases may be expected in C_{60}-CPs provided the C_{60} molecules are properly charged and possess continuity throughout the matrix.

10.3 Models of C_{60}-Induced Superconductivity and Role of Granularity

Upon doping a CP-C_{60} composite with a n-dopants such as A-atoms, the electrons should be transferred from the A- atoms to both CP chains and C_{60} molecules filling their empty bands and levels, depending on their relative positions (i.e. on the δ_1 and δ_2 energies shown schematically at Fig. 10.2). This CP$(C_{60})_yA_x$ system should be rich in possible conducting and SC phases due to many choices for the CP matrix, and for independent variation of y and x. In our view, the most exciting possibility is to obtain a truly microscopic molecular level SC phase in which formation of Cooper pairs in the conduction band of CP chains may be induced by electron–electron attraction on C_{60} molecules because both electronic systems are strongly hybridized via strong $\pi - \pi$ overlap. This is shown schematically in Fig. 10.2, which, at first glance,

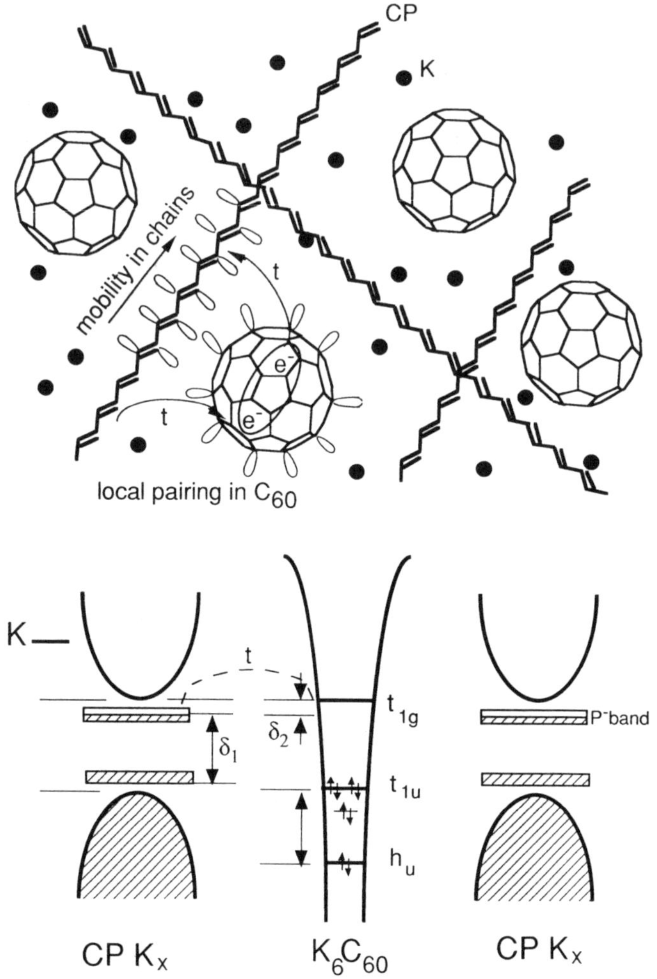

Fig. 10.2. Schematics of two-component or C_{60}-induced superconductivity in ideal CP-C_{60} system, in which C_{60} is intercalated as isolated molecules: C_{60} provides SC pairing of electrons in its local level, while CP chains provide their mobility due to good hybridization. Inset shows hypothetical optimal energy matching with resonating states and doubly degenerate regime of negative-U level in C_{60}

resembles the idea of excitonic superconductivity proposed much earlier by Little [46]: it has similar components like conductive polymeric chains with large molecules attached on the sides.

One cannot exclude the possibility that the polarization of the C_{60}^{n-} anions can contribute to better electronic pairing in the CP chains via exchange of excitons in C_{60}, as proposed by Little [46], (particularly C_{60}^{1-} is claimed to have unusually high polarizability [14]). However, we have in mind another picture, similar in spirit to "induced pairing models" or models involving mixture of local pairs and itinerant electrons.

In this qualitative picture, the mobility of the Cooper pairs should be provided by the CP chains, while C_{60} molecules should induce SC pairing for electrons partially localized on C_{60} in π-electron chains. The essential idea is that if electrons may spend a relatively long time on quasi-stationary (due to CP-C_{60} hybridization) orbitals of C_{60}, they may attractively interact more effectively than in a hypothetical pristine CP-A_x superconductor. In the case of ceramic high T_c SC, the pairing of mobile holes in a wide electronic band can be provided via hybridization with localized electrons of negative U-centers which have been associated with Cu ions, or with oxygen vacancies. Since it is believed that the SC of the fullerides is due to intramolecular phonons within each C_{60} molecule (as discussed above), the C_{60} molecules might be expected to play the role of "pair inducing centers".

One can estimate qualitatively the best conditions for a C_{60}-induced SC phase:

1) C_{60} molecules should have a strong tendency for pair formation. In the picture of negative U-center, it was proposed that there is a non-zero pair-binding energy in the electronic pairing mechanism [13] which is expressed as $E_{\text{pair}}N = 2\,E_N - (E_{N+1} + E_{N-1})$, where E_N is the energy of C_{60} molecule with N additional electrons. This shows that when adding N electrons per C_{60}, there is a tendency for disproportionation, e.g. at $N = 1$, the energy of the state with a pair: C_{60}^{2-} + C_{60} is lower than that of C_{60}^{1-} + C_{60}^{1-}. Moreover, for the best contribution to superconductivity, such a negative U-center should be in a degenerate double valence regime: $E_{N+1} = E_{N-1}$ [51].

2) Good hybridization is needed, i.e. a large enough π–π overlap integral t. This is actually proved in CP-C_{60} by PCT and t, can be roughly estimated as $t = 0.05$–$0.1\,\text{eV}$, slightly smaller than the C_{60}-C_{60} overlap integral in fullerides [2].

3) Favorable matching of the energy levels, which should nearly resonate in the ideal case. The mismatch δ_2 between the Fermi level in a polaronic band of the CP and the t_{1u} or t_{1g} LUMO levels of C_{60} should be small, i.e. $\delta_2 < t$. This condition may occur, for example, between the P-level of PAT and the t_{1g} level of C_{60} (as sketched in Fig. 10.2). In general, the CP provides a wide possibility for the tuning of their conductive levels by adding various side groups, e.g. a cyano group added to poly(para)phenylene-vinylene makes its energy spectrum more "acceptor-like". [54], increasing its electron affinity and its dopability. Choosing an appropriate CP matrix, which can meet this resonance condition with C_{60} is therefore important. For this reason, we checked experimentally several CPs as matrices for the CP$(C_{60})_y$ A_x composites.

4) Charge carriers in the CP chains should be sufficiently delocalized to provide large mobility and conductivity and better delocalization in the SC state [23]. This will require rather high doping x levels of the CP$(A)_x$,

generating a carrier concentration and conductivity high enough to be termed "metallic". These conditions can be met in various heavily doped CPs, e.g. $(CH)_y(Na)_x$ or $PPP(Na)_x$, [22]. It can be expected that in CP composites formed with C_{60}, a good delocalization can be similarly achieved, although, as we will show below, experimentally this is a hard task, since higher values overall x are needed.

So the requirements 2)–4) can be, in principle, achieved in real systems by control of the matrix parameters and the doping level x. Thus, the requirement 1) for the role of C_{60} as a strong pairing center is a most important one.

The work of Chakravarty et al. [13] spurred many studies on the value of $E_{\text{pair}}N$, but there is still uncertainty whether $E_{\text{pair}}N$ can really be large enough. It has also been found theoretically that the electron–lattice interaction can also lead to sizable $E_{\text{pair}}N$, or, in other words, bipolarons on C_{60} can be rather stable [15]. From the experimental test on the disproportionation ($N = 1$) reaction, however, it has been clarified that only 3% of spins in C_{60}^{1-} anion form C_{60}^{2-} dianions exist in solids [55, 56]. On the other hand, as we have discussed above, C_{60} in a CP matrix acts as a multicharging center upon electrode injection, suggesting that bipolarons are indeed stable on the C_{60} molecule.

We should mention here that the pairing is not necessarily connected only with a nonzero $E_{\text{pair}}N$ binding energy, or the existence of stable C_{60}^{2-} states. If the pairing in A_3C_{60} is of vibronic origin [9–11] rather than electronic, C_{60} molecules can still play the role of pairing centers. Similar to the observation of the chemisorption enhancement of the SC T_c in metallic films [53], electrons from CP chain can tunnel to the localized t_{1g} level of C_{60}, and being localized there, can strongly interact with the A_g and H_g vibrons of C_{60}. If the local level is narrow enough and resonating with the Fermi level of the conduction electrons, this can make a significant contribution to the effective pairing in CP chains. Estimates for the chemisorption case show that T_c may increase by factor of 2 or 3 due to pairing on the adatoms via the interaction of localized electrons with a surface lattice phonon mode. A theory should be developed for the particular case of the interaction with internal vibrons of C_{60}, but the results of [53] should suffice for qualitative estimates.

In this case, C_{60} may offer a first example of such a pair inducing center, i.e. the C_{60}^{3-} center. In all previously studied cases [47–52], superconductivity in a lattice containing negative U-centers has not been observed. Therefore, it is worthwhile to study experimentally how a random distribution of C_{60}^{n-} centers in a CP matrix will behave at low temperatures.

Hence, this was a primary motivation for our study and a working hypothesis. However, since C_{60} is known (from our earlier experimental results [30, 57]) to aggregate into clusters in $CP(C_{60})_y$ composites, at least at large C_{60} content $y = 0.05$; it is more realistic, although less exciting, to obtain SC from the C_{60} clusters upon A-doping of the composite. In this system

A_3C_{60} granules will, of course, be responsible for the superconductivity. We show below that such SC grains are indeed observed, at least at large y, and are coupled either via the proximity effect through conducting CP-A_x, or by Josephson-type, weak links through insulating CP barriers and give an interesting example of disordered granular SC [58, 59]. This is an interesting problem in its own right, because the CP matrix may play an active role via the nonlinear charge and/or spin excitations in the CP chains which may induce new effects.

Granular high T_c SC ceramics have recently attracted much attention [58, 59] due to unusual mesoscopic effects, and we demonstrate below that fulleride-polymer composites also exhibit some new unusual phenomena which cannot be found in conventional granular superconductivity with transmissive intergrain barriers.

"High T_c SC ceramic-polymer" composites have been recently studied [60] due to better mechanical properties compared to brittle pristine ceramics. The CP-C_{60} molecular composites considered here are qualitatively different from these composites made with insulating polymers.

10.4 Low Field Microwave Absorption (LFMA) in Superconducting $CP(C_{60})_yA_x$

We have used the very sensitive low field microwave absorption (LFMA) test to search for SC $CP(C_{60})_yA_x$ phases. Since the observed LFMA signals demonstrated rather unusual and unexpected features not seen previously in other SC fullerides, we provide a short introduction to the LFMA method.

10.4.1 Basics of LFMA as a Sensitive Test for Superconductivity

The measurement of the nonresonant low magnetic field microwave absorption (LFMA) has proved to be a sensitive method for the search and study of novel SC phases in HTSCs [61–65], organic materials [66–68] and, particularly, fullerides [1, 7, 69–73]. This method has higher sensitivity than resistivity measurements or even SQUID magnetometry for detecting small fractions of SC material. It has the following advantages:

1) It can detect submicrogram quantities of a SC phase, (in both SC and non-SC matrices), determine T_c, and detect multiple T_c-phases in the same sample [74]. 2) It is a contactless probe, which can be employed in a high vacuum, an important feature for air-sensitive fullerides. 3) It therefore can track the evolution of the SC fractions upon various sample treatments, such as doping [69], annealing [73] or alloying [7]. 4) It can provide information about the quality of the SC state (e.g. granular nature, glassy behavior, critical fields, [61, 62] etc.), which is particularly important for samples with a composite structure, as in the present case.

Using LFMA, several new C_{60}-based superconductor have been discovered. $Na_3N_xC_{60}$ has been found with $T_c = 10$–$17\,K$, prepared from the reaction of C_{60} with the sodium azide (NaN_3) [71, 72]. Recently another new superconductor with a composition of $Li_2Na_x(N_y)C_{60}$ has been also found by LFMA [70]. Furthermore LFMA was used to detect superconductivity in a 50 layer Langmuir–Blodgett film of C_{60} doped by K with a $T_c = 8\,K$ [75].

However, it is important to note that the appearance of a LFMA does not directly imply the existence of SC phase. A LFMA absorption not associated with SC may appear in doped CPs, but it has a different character compared to a SC LFMA: it exhibits no hysteresis, and exhibits a different temperature dependence [76]. The non-SC LFMA alluded to above originates from a negative magnetoresistance effect caused by a spin-selective polaron–bipolaron transformation during electron hopping [76]. Another type of non-SC LFMA has been reported in spin-glass systems such as Gd_2CuO_4 [77]. This type of LFMA has a distinct form of hysteresis with an "anticlockwise" loop opposite to that of a SC LFMA hysteresis loop. It is therefore believed that the observation of a "clockwise" hysteresis loop, at low H_{mod} can provide unequivocal evidence for a SC state [77].

Let us recall here that a magnetic field H_{ext} at the sample loaded into an ESR spectrometer cavity is comprised of three components:

$$H_{ext}(t) = H(t) + H_{mod}\sin(\omega_{mod}t) + H_{mw}\cos(\omega_{mw}t) \tag{10.1}$$

where $H(t)$ is the slowly varying DC magnetic field of the spectrometer magnet, H_{mod} is a modulation field amplitude changing with frequency f_{mod} (10–100 kHz, and H_{mw} is a third magnetic field amplitude at the microwave frequency f_{mw} (10 GHz). To carry out the LFMA method, the ESR spectrometer needs a minor modification. Additional DC Helmholtz coils must be attached to the spectrometer DC magnet poles to both compensate for a residual field as well as to allow the DC magnetic field to be scanned around zero field within the range of (50–200 G). The sample is cooled down to some chosen temperature ($T < T_c$) in a zero DC magnetic field (zero field cooling, ZFC), followed by reversible H scanning within the range allowed by the Helmholz coils. ZFC is necessary to eliminate effects from trapped magnetic flux.

In principle, the microwave power P absorbed by the sample is a function of the applied DC magnetic field and temperature, i.e. $P = P(H,T)$ for fixed values of H_{mod} and H_{mw}. The signal $I(H)$ recorded by an ESR spectrometer is proportional to $\mathrm{d}P/\mathrm{d}H$. In the case of a SC sample, the LFMA signal depends on the amplitude of the modulation field, and can be expressed as $I(H)\delta P/\delta H)H_{mod}$. To make sure that the LFMA found in our composites is a true signature of SC, the hysteresis behavior at various H_{mod} must be obtained. Moreover, SQUID magnetometry was also used in optimally doped samples with a sufficiently large SC fraction.

10.4.2 Features of Superconducting LFMA in PAT-C_{60}-K and OO-PPV-C_{60}-K

The details of the sample preparation, doping procedure etc. are given in [24–30], so they are not described here. We next discuss the important general features of the LFMA results for various (x, y) regimes of $CP(C_{60})_yK_x$.

10.4.3 Low C_{60} concentration $y < 0.005$

Characteristic SC LFMA signals appeared after several hours of doping, even in dilute composites containing rather small concentrations of C_{60} molecules, i.e. with $y = 0.0005$, or 0.05 mol % relative to the monomer. Typical LFMA spectra taken at various temperatures $T < T_c$ ($T_c = 12.4$ K) is shown in Fig. 10.3a for optimally doped $PAT(C_{60})_yK_x$ composite with y = 0.005. Sim-

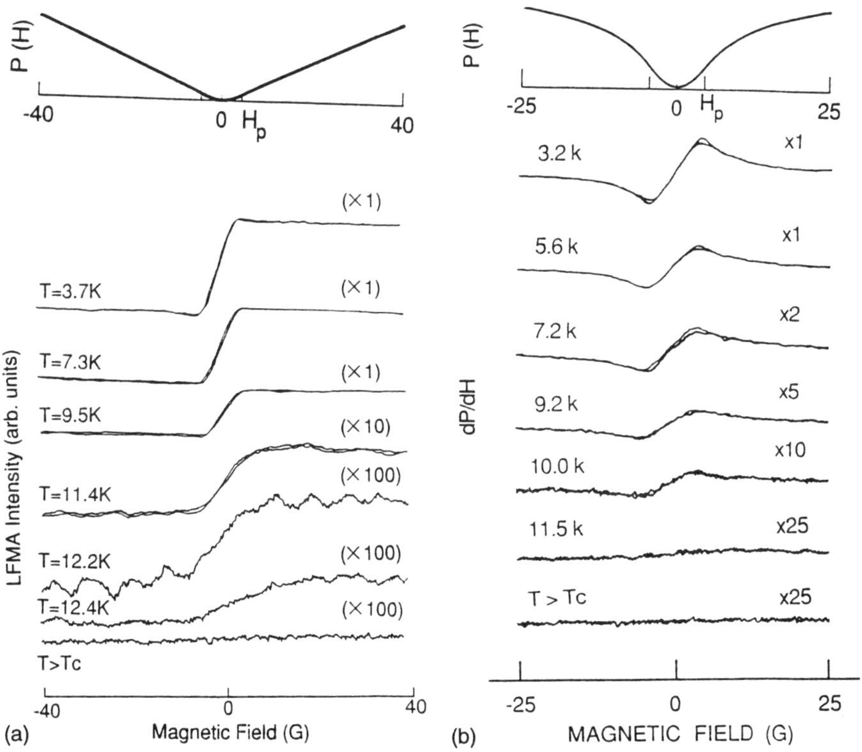

Fig. 10.3. LFMA spectra of SC $CP(C_{60})_yK_x$ composites with small $y = 0.005$ measured at various T below T_c, with normal-phase shape of derivative dP/dH, corresponding to microwave absorption $P(H)$ with minimum at $H = 0$, as shown in the inset: (**a**) CP = PAT, $T_c = 12$ K, and (**b**) CP = OO-PPV, $T_c = 14$ K

ilar to the LFMA data in bulk alkali metal-doped fullerides, this composite exhibits the normal positive signal. The DC field hysterisis is negligibly small, which is probably due to the very small volume fraction of SC phase. Figure 10.3b shows the T-evolution of a similar LFMA signal of another composite OO-PPV$(C_{60})_yK_x$ also at $y = 0.005$ which exhibits a $T_c = 11.5$ K. We show them together to stress that the LFMA behavior at small y is qualitatively similar for these two different CPs. The structure in the LFMA abruptly declines near T_c, typical of LFMA of SC origin, and contrary to a magnetoresistive LFMA signature which has an exponentially decreasing signal with increasing temperature. This magnetoresistive LFMA signature has been observed in some CPs such as Na-doped polyparaphenylene [76]. Note that T_c is significantly lower than $T_c = 19$ K reported for bulk K_3C_{60}. Furthermore, T_c was observed to slightly increase with increasing doping time t_d for each y composition. The tendency of T_c to decrease is found with the decrease of the C_{60} content y. For the lowest $y = 0.0005$ composite studied, only traces of a LFMA signal were found at a $T_c = 4.7$ K, though it was not possible to prove the SC nature of that LFMA, since it exhibits little or no hysteresis, and the T-dependence was difficult to obtain due to very low T_c.

Let us briefly discuss how the origin of the SC phase can be understood from the observed LFMA behavior. Although there are many models for the microscopic origin of LFMA [75–80] (each of them being valid in a certain range of H and T), it is commonly believed that at low $H = 0$ and low T ($T = T_c$) that the LFMA originates mainly from trapped magnetization (either in the bulk and/or in Josephson-junctions (JJ), or in other words, by damped fluxon dynamics in the microwave field. This view is consistent with the observations above, where the LFMA intensity $P(H)$ increases with increasing fluxon density with increasing magnetic field H. The LFMA signal ($\mathrm{d}P/\mathrm{d}H$), for the case of the normal phase is negative ($\mathrm{d}P/\mathrm{d}H < 0$) at $H < 0$, and positive ($\mathrm{d}P/\mathrm{d}H > 0$) at $H > 0$ (Fig. 10.4), which is opposite to the phase, or H-dependence, of conventional resonant ESR, so it is frequently called the "out-of-phase" signal.

We should also recall the contribution to LFMA from the damped motion of vortices. In type II superconductors Abrikosov vortices start to penetrate into the bulk in a field below H_{c1}, which is typically in the range of 30–50 G for alkali fullerides [2]. However in JJ the vortices can enter below H^*_{c1}, which is much lower than H_{c1} and, that is apparently the case in the CP$(C_{60})_yA_x$ compounds. The equation of motion of the vortices trapped in the microwave field is described by the equation:

$$\mu \frac{\mathrm{d}^2 x}{\mathrm{d}t^2} + \eta \frac{\mathrm{d}x}{\mathrm{d}t} + kx = \frac{1}{c} J_{mw} \Phi_0 \tag{10.2}$$

here J_{mw} is the microwave current density, Φ_0 is the flux quantum. The first term in (10.2) represents the inertia, the second the viscosity damping, and the third the restoring force related to the pinning potential. Typical values of J_{mw} can be estimated following [65]. The empty cavity microwave field

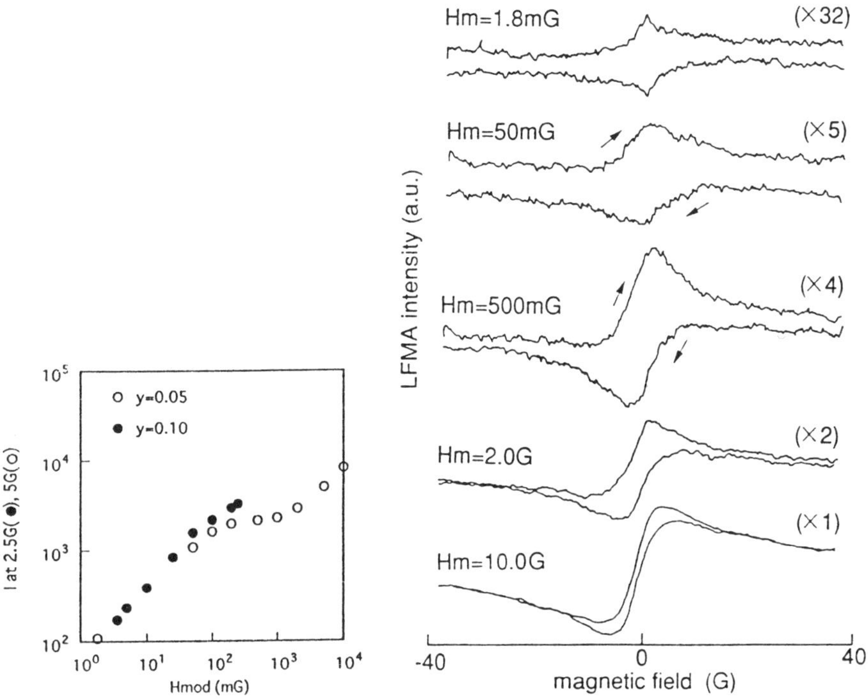

Fig. 10.4. Hysteresis behavior of LFMA in composite with $y = 0.05$ measured at low $T = 4.2\,\mathrm{K}$ for various H_{mod}. The width of hysteresis increases with decrease of H_{mod}, showing sign change at backward scan for small H_{mod}, typically for damped vortex pinning-depinning contribution to LFMA. Inset shows the dependence of intensity on H_{mod}, from which $H^*_{\mathrm{c1}} = 0.5\,\mathrm{G}$ is estimated

power P_0 is given by $H_{\mathrm{mw}} = (32\pi P_0 Q/\omega V_{\mathrm{c}})^{1/2}$, where the cavity volume is V_{c} ($10\,\mathrm{cm}^3$, and a typical quality factor is $Q = 8000$. For $P_0 = 20\,\mathrm{mW}$, at $\omega/2\pi = 9.4\,\mathrm{GHz}$ one arrives at $H_{\mathrm{mw}} = 0.52\,\mathrm{G}$. This field penetrates the composite only to a depth of lL. The corresponding microwave surface current density is then $(1/c)\ J_{\mathrm{mw}} = (1/4\pi)H_{\mathrm{mw}}/\lambda_L$, which corresponds to $J_{\mathrm{mw}} = 10^4\,\mathrm{A/cm}^2$ for $\lambda_L = 1000\,\mathrm{A}$. This is really large compared with a typical tunneling current density. It has been shown in [65] that due to the extremely low viscosity of JJ vortices their velocity in a typical microwave field ($H_{\mathrm{mw}} = 0.56$) is five orders of magnitude larger than observed for bulk fluxons. Consequently the energy relaxation rate for them is ten orders of magnitude larger than the bulk rate, which makes the damped motion of JJ fluxons the main mechanism of MA even near $H = 0$. This is due solely to H_{mw} and nonzero H_{mod} for non-zero $P(H)$ at near $H = 0$. However, increasing the external static field H and for $H_{\mathrm{mod}} > 2H^*_{\mathrm{c1}}$, a larger density of fluxons starts to be created, thus this type of LFMA always has a minimum $P(H)$ at $H = 0$. Later, we will demonstrate that another mechanism, connected with the critical state phase slippage in loops of JJ arrays, and also known to contribute to LFMA, is important usually at higher fields H, and may give a maximum in $P(H)$ at $H = 0$, i.e. contribute to an "in-phase" LFMA signal. Since in our case $P(H)$ has a minimum at $H = 0$, and increases linearly with H, this can be attributed to the dissipation of fluxons pinned at JJs connecting SC grains. Thus we conclude that the SC phase of the composite with low y contains a large fraction of JJs and has a rather small H^*_{c1}. Note also that the LFMA data in PAT12 and OO-PPV matrices show different H-field dependencies, as can be seen from their direct absorption curves $P(H)$ shown in the inset of Fig. 10.3. In PAT12, $P(H)$ exhibits a linear H behavior (Fig. 10.3a), while in OO-PPV it tends to saturate with increasing H, having maximal derivative at $H_{\mathrm{p}} = \pm 1\,\mathrm{G}$ (Fig. 10.3b). This behavior may reflect a different geometry of JJs in these two matrices due to a different morphology. Above H_{p}, the number of fluxons is not increasing in OO-PPV which might be connected with a very small size of JJ, or, on the other hand, indicate a rather low H^*_{c2} field. On the contrary, in PAT12, the density of vortices increases with H, although other mechanisms may start to contribute at large H.

The intensity of the LFMA in the composites is higher than in bulk A_3C_{60} which also suggests that the LFMA comes not from the bulk, but from the magnetization trapped in JJ. This observation explains why LFMA is more sensitive than SQUID for the present composites: it can sense the JJs between the SC grains, rather than the net volume of the SC phase.

10.4.4 Intermediate to High C_{60} Concentration $0.01 < y < 0.1$

The LFMA for $CP(C_{60})_yA_x$ composites with intermediate to high y exhibits several interesting features which, on the one hand, prove the SC origin of LFMA (particularly the correct hysteresis behavior (Fig. 10.4)) and, on the other hand, reveal rich spectra with multi-peak behavior. The latter implies a granular SC [25], and the coexistence of several SC phases [28–30]. The system must also exhibit unusual supercurrent, loops flowing through network of JJs in the ground state [25, 27, 30]. It is important to mention here that the evolution of these features with doping depends now on the type of CP matrix and A dopant. The data become more complicated as more n-dopable CP matrices are taken as the host for a composite. These CPs have better electron withdrawing ability, i.e. smaller δ_2 mismatch (c.f. Fig. 10.2) with C_{60} LUMOs.

In $PAT12(C_{60})_yK_x$, the LFMA has a normal-phase signal at low $T > T_{\mathrm{c}}$. As T approaches T_{c} from above, the LFMA exhibits phase reversal (Fig. 10.5a). In some doped and not well annealed samples where the CP chains are probably not fully doped, we found the coexistence of a central normal-phase (CN) narrow signal superposed onto an anomalous reverse-phase (AP) broad signal existing down to the lowest T ($T < T_{\mathrm{c}}$). The narrow CN signal disappeared at a certain $T^* < T_{\mathrm{c}}$ (Fig. 10.5b).

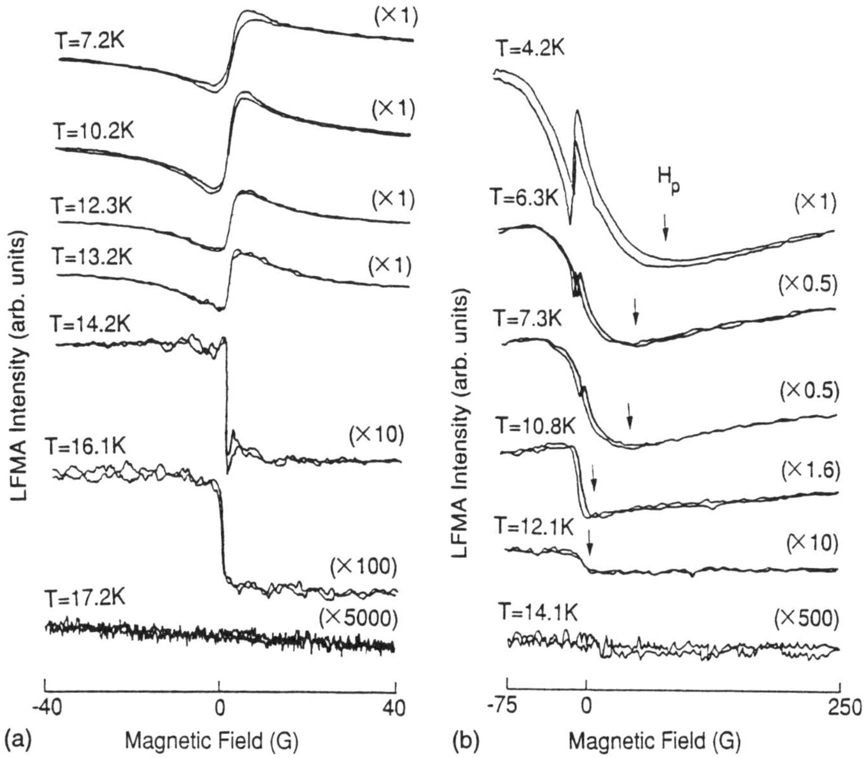

Fig. 10.5. LFMA in PAT12$(C_{60})_yK_x$ with large $y = 0.05$: (**a**) well equilibrated composite, showing temperature dependent phase reversal at T close to T_c: The shape of LFMA becomes of reverse-phase type at $T = 12\,K$, which corresponds to $P(H)$ with maximum at $H = 0$. (**b**) in freshly doped (nonequilibrated) SC composite, the broad AP-LFMA, superposed on the CN-LFMA which disappears above T^*. H_p indicates the field at maximum of AP component

In OO-PPV$(C_{60})_yK_x$, a multi-component LFMA signal was observed in all samples, independent of the annealing or aging time and appeared even in a composite with intermediate $y = 0.01$. The "clockwise" hysteresis CN type LFMA was found in both components, indicative of SC, and the width of the hysteresis loop was found to increase with increasing y (Fig. 10.6). The T-dependence of the peak-to-peak intensity of both components exhibits typical multiphase behavior, indicating either the existence of two spatially separated SC phases, with different T_c [28] or contributions from two distinct mechanisms to LFMA.

We next describe the main LFMA features common to a variety of CP hosts in more detail, trying to understand the origin of the multi-component behavior.

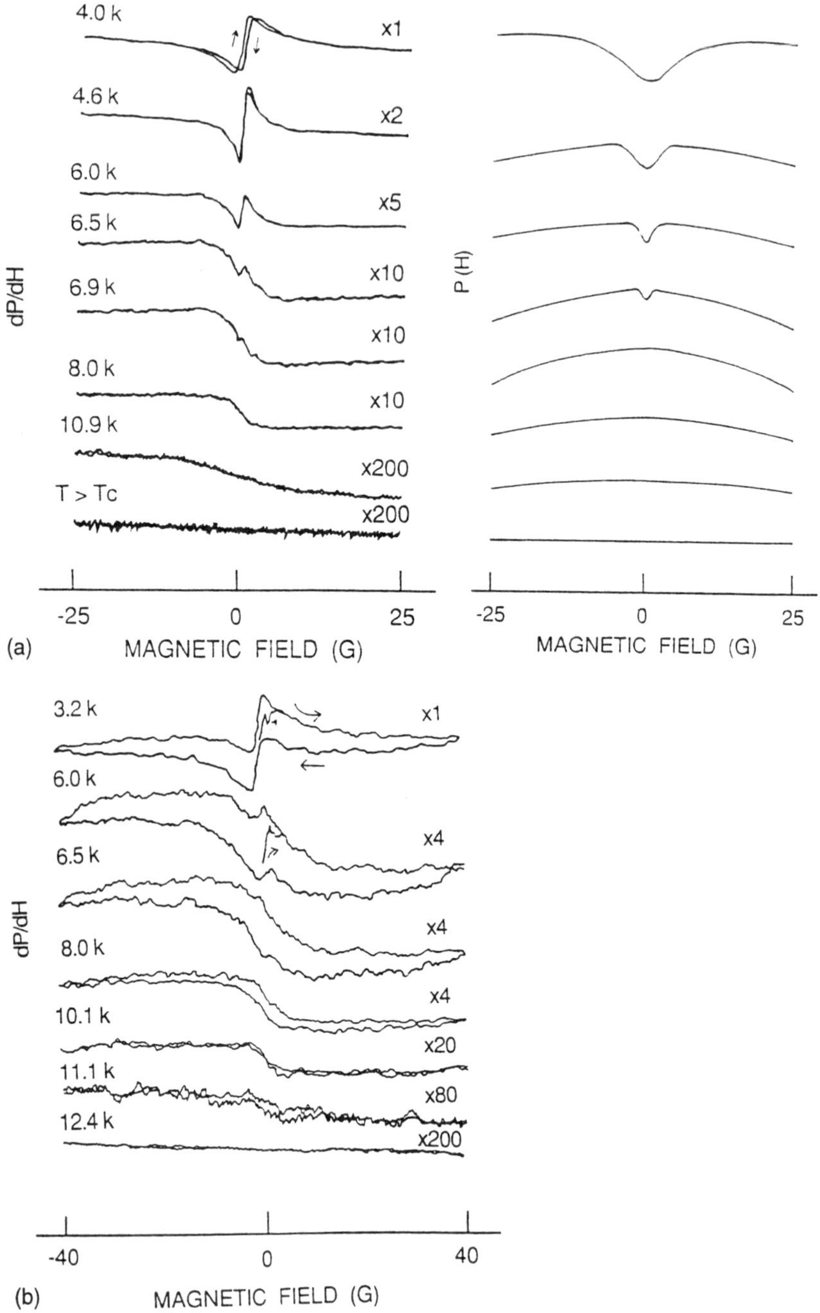

Fig. 10.6. Evolution of two-compo- nent $\mathrm{d}P/\mathrm{d}H$ of OO-PPV$(C_{60})_yK_x$ with T: (**a**) medium $y = 0.01$. CN LFMA appears below $T_{c2} = 7\,K$ coexisting with AP wing. Inset shows schematics of direct absorption $P(H)$ behavior with T. (**b**) Large $y = 0.1$, The broad hysteresis, exhibiting clockwise loop typical to SC phase. $T_{c1} = 12\,K$, $T_{c2} = 7\,K$

Superconducting Hysteresis of LFMA In all $CP(C_{60})_yA_x$ composites, the LFMA exhibits the typical CN component in low H ($H < 0$) at low $T < T_c$. Most importantly, it has a correct hysteresis observable at low H_{mod}. Hysteresis can be clearly found at larger y due to the correspondingly larger SC volume fraction. This results from the larger fraction of magnetization trapped at JJs between SC grain. Figure 10.4 shows the hysteresis behavior of the CN component which is enhanced in low H_{mod} and is typical of granular superconductors with Josephson vortices.

From the hysteresis dependence on H_{mod}, one can obtain the JJ-critical field H^*_{c1}. The LFMA for small H_{mod} changes sign at the turning points of the scanning DC magnetic field. Furthermore, the relative width of the hysteresis loop is enhanced with decreasing H_{mod}. The JJ allows the flux to enter at a very low effective H^*_{c1} [65]. The change of sign of the LFS (low field signal) at the reverse sweep is related to the change of the critical state at the surface [63], i.e. to the pinning and depinning of fluxons during each modulation cycle. Since H_{mw} and H_{mod} are always applied to the sample in a cavity, the critical current J_{cs} already flows at the surface of the grain at these low fields and contributes to the microwave absorption. With the reversal of the scanning direction, J_{cs} reverses only after a certain field interval equal to twice H^*_{c1} [63]

$$2H^*_{c1} = (4\pi/c)\lambda_J J_{cs}, \tag{10.3}$$

where λ_J is Josephson penetration depth. The same change of J_{cs} occurs over a modulation field cycle. As seen in the inset of Fig. 10.4, dependence of LFMA intensity on H_{mod} exhibits two linear parts. These two regions involve two different contributions [62]: the signal induced by J_{cs} for $H_{mod} < H^*_{c1}$ and the signal caused by the change in the fluxon density for $H_{mod} > 2H^*_{c1}$(the MA is proportional to the concentration of fluxons) [64]. The minimum point between these two regions corresponds to $H_{mod} = 2H^*_{c1}$. On increasing the temperature, the minimum point shifts to lower field. The minimum point can not be observed clearly, but still it is possible to estimate the value of H^*_{c1} to 0.2–0.4 G, which is smaller than that in high T_c SC cuprates. Since the flux is known to penetrate into JJs at the critical field [64], derived as

$$H^*_{c1} = \Phi_0/(4\pi\lambda_L\lambda_J), \tag{10.4}$$

the J_{cs} can be estimated from (10.3) and (10.4). Using $H^*_{c1} = 0.3$ G found experimentally and taking representative $\lambda_L = 200$ nm, we can get $\lambda_J = 20$ mm, and $J_{cs} = 10^2$ A/cm^2.

Temperature-Dependent Reverse-Phase LFMA The LFMA data for larger $y = 0.025$ and 0.05 in the PAT12 matrix showed an interesting phenomena of phase reversal in a narrow temperature region below T_c (Fig. 10.5a). At T close to T_c LFMA shows the anomalous "reverse-phase" dP/dH form, which corresponds to the case having $P(H)$ maximum at $H = 0$. In this case, the phase change happens abruptly between $T = 11$ and 12 K, i.e. close to $T_c = 14$ K.

The reversal of the LFMA phase means that $P(H)$ corresponds to a zero-field maximum which cannot be understood in terms of fluxon density (which always increases with H as explained above). Phenomena similar to AP-LFMA have been observed near T_c in rf absorption in ceramic high T_c SC, and it has been suggested that the mechanism is absorption in a random network of SC grains weakly coupled through normal (N) metal links [78], i.e. S-N-S type contacts are created near T_c as the SC fraction strongly decreases with increasing T and H. Such a mechanism can be applied to a granular SC phase which is in the percolation regime and for T close to T_c. The MA is due to conductivity of the normal fraction, exhibiting a reverse-phase LFMA as a result of nonmonotonous dependence of SC fraction in a percolative network. Our LFMA data can therefore be interpreted to indicate that the SC phase is a granular one, composed of K_3C_{60} clusters separated by conductive PAT-K_x barriers (more details can be found on MW power and T-dependence of reverse-phase LFMA in [29, 30]).

Multi-Phase LFMA with an Anomalous Reverse-Phase Component (π-Junctions) In some PAT12$(C_{60})_yK_x$ samples, it is clearly found that the LFMA is of the anomalous reverse-phase type (AP) even at $T \ll T_c$ (Fig. 10.5b). Similar LFMA has been seen in all composites with CP matrices which exhibit strong n-dopability, i.e. OO-PPV and PAT6. To investigate this phenomenon in more detail, we have measured the LFMA at small modulation amplitudes H_{mod}, and have found that at $T < T^*$ the reverse phase AP signal splits into two components: the CN component, which can be revealed only at a small H_{mod} (< 1 G), exhibiting a normal-phase signature down to the lowest T, and the AP type component existing over whole T range below T_c which can be seen as wings in higher H regions. The peak-to-peak width of this component is rather broad and the field associated with this peak (H_p) shifts to lower H with increasing T.

The peak-to-peak width of the CN-component depends on H_{mod} and it cannot be resolved if H_{mod} is too large, which is just a technical problem of over-modulation. In PAT12$(C_{60})_yK_x$, this component appears at $T \leq T^* =$ 6.5–8 K, much below T_c = 12–14 K, and in Rb-doped PAT6$(C_{60})_y$ $T^* =$ 13.5 K and T_c = 24.5 K. So the value of T^* depends on y and the type of CP and is nearly one half of the T_{c1} value. We discuss below that T^* may be interpreted as the T_{c2} of a separate SC2 phase. The appearance of a CN component in OO-PPV-C_{60}-K appears for y as low as 0.001 (Fig. 10.6). The multi-peak derivative LFMA corresponds to a direct absorption, which has narrow minimum at $H = 0$, on the top of a broad maximum, and this local minimum disappears above T^*, as shown in Fig. 10.6b. The relative intensities of CN- and AP-component for $y = 0.01$ are plotted as a function of T in Fig. 10.7. It appears that this normal LFMA originates from a different SC

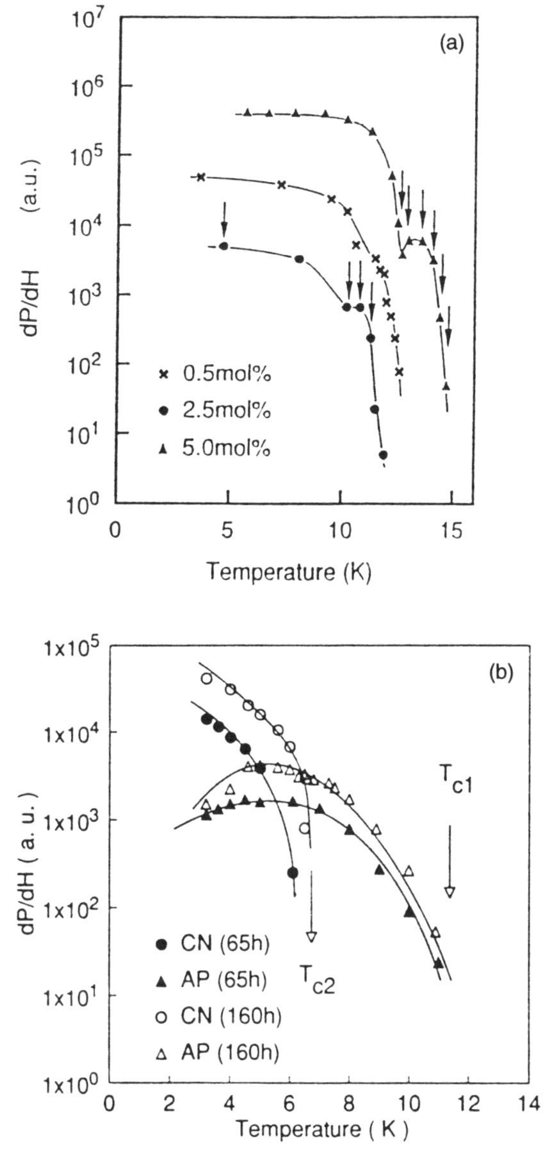

Fig. 10.7. Temperature dependence of LFMA peak-to-peak intensity in (**a**) PAT12 $(C_{60})_yK_x$, and (**b**) OO-PPV $(C_{60})_yK_x$ ($y = 0.01$) at different doping times (t_d = 65 h and 160 h). AP-component abruptly appears at T_{c1}, while CN-component at T_{c2}. Note that CN and AP coexist below T_{c2}

phase with lower T_c, since the T-dependence of the net LFMA intensity shows a saddle point at T^*, similar to multi-phase superconductors with several T_cs [74].

The intensity of the AP-component has a tendency to increase with increasing y. For the largest y, the AP-component becomes very intense and broad, exhibiting a maximum at H_p which shifts to the lower field side with increasing T (i.e. H_p (12 K) = 8 G, H_p (4 K) = 88 G), while the H_p of the

CN-component does not shift with T. So from the different $H_P(T)$ behavior of the CN an AP components, we speculate that they originate from different SC phases SC1 and SC2, each having distinct T_{c1} and T_{c2}. SQUID data also supports such an assignment (also, see [29, 30] for details of mechanisms of CN and AP LFMA).

With the aim of explaining these H_p-shifts and to understand the origin of the AP-component, let us recall another microscopic mechanism of LFMA in more detail, known to contribute mainly at large static H when loops of weak linked JJs start to play a role in the MA, in addition to the damped fluxon motion [79, 80].

If there is a SC current loop with the radius r, the flux starts to slip into the loop at the field of $H_o = \Phi_0/2\pi r^2$, where $\Phi_0 = \mathrm{ch}/2\mathrm{e} = 2 \times 10^{-7}\,\mathrm{G\,cm}^2$ is the fundamental flux quantum. The size of the JJ loops in a granular superconductor varies over a wide range. Most groups associate the maximum of the LFMA at H_p with the maximum in the size distribution of the JJ loops and use it to deduce an average projected area $\langle S \rangle = \pi \langle r^2 \rangle$ of the SC loop as

$$\langle S \rangle = \Phi_0/(2H_p). \tag{10.5}$$

It should be pointed out that the loop may either encompass only a single grain (if it has defects like twin boundaries, etc.) or multiple grains connected to each other and making a closed loop. According to Xia and Stroud [79], when the applied magnetic field induces a current greater than the critical current, the junction breaks down, and emits a flux quantum accompanied by MA.

It can be explained in another way, following [80]. If the SC loop is oriented perpendicular to H in the microwave field, a series of equally spaced, sharp MA peaks should be observed with increasing H. The current density J across the JJ depends on the phase difference of Cooper pairs Q on each sides of junction as $J = J_s \sin(Q_1 - Q_2)$. When an external field is applied and the flux slips through the loop, the phase difference $(Q_1 - Q_2)$ changes by

$$\Delta\Theta = 2\pi(\Phi/\Phi_0). \tag{10.6}$$

Due to quantization of flux, there should be phase jumps with increasing H which occur when Φ $(= HS)$ becomes an integer multiple n of Φ_0, i.e. when $H = n\ (\Phi_0/S)$. When this happens, the value of loop current exceeds the critical current J_c and normal current flows across the barrier with the resistance R_n of JJ. These jumps may happen in a time on the order of 10^{-12} s, the voltage jumps $V(t) = -(h/4\pi\mathrm{e})\mathrm{d}Q/\mathrm{d}t$ can therefore be very large. The pulse of normal current $I_n = V(t)/R_n$ absorbs microwave energy P, which can be roughly estimated from

$$P = \frac{V(t)^2}{R_n} = \frac{h^2}{16\pi^2\mathrm{e}^2} R_n^{-1} \left(\frac{\mathrm{d}\Theta}{\mathrm{d}t}\right)^2 \tag{10.7}$$

This mechanism is believed to give rise to a broad MA in granular SC due to the superposition of sharp contributions from many loops with areas

S. Model calculations for a Boltzmann type distribution of $\langle S \rangle$ showed that H_p for the normal-phase type of LFMA signal can thus be explained [80]. If we assume this model to be correct, even for the anomalous reverse-phase type, then using an experimental value of $H_p(T)$, we can arrive at an $\langle S \rangle$ value in our composites, as $\langle S_1 \rangle = \Phi_0/(2H_p) = 1.3 \times 10^{-9}\,\mathrm{cm}^2$, i.e. typical cluster radius of $r = 2 \times 10^{-4}\,\mathrm{cm}$, in good agreement with TEM images of the intragrain C_{60} networks, (discussion below and Fig. 10.10). Note that for the narrow CN component (with $H_p = 0.5\,\mathrm{G}$), we get unreasonably large $S_{cp} = 2 \times 10^{-7}\,\mathrm{cm}^2$, or loop size of 2.5 mm, which is too large, and implies that CN originates from another microscopic microwave (MW) mechanism, e.g. damped fluxon motion, as discussed above for the case of small y. This also suggests that normal phase LFMA at small y and CN at large y may have the same origin.

We analyzed the temperature dependence of $H_p(T)$ in a resistively shunted JJ model [30] (in which $H_p(T) = cp/\langle A \rangle$, with $cp = a + bI_{co}$, with zero field critical current $I_{co} = (1-t)n$, $t = T/T_c$, and $\langle A \rangle$ is the average junction area (λ_d), λ is the penetration depth, and d is average grain size). The best fits to the experimental data yielded, $n = 2$ in the expression for I_{co}. The value $n = 2$ corresponds to an S-N-S type junction in our system (N is the normal metallic state, S is SC state; $n = 1$ corresponds to an S-I-S type junction in which I is the insulating state).

The most important question connected with the anomalous component is to explain the reverse-phase character, i.e. the maximum in $P(H)$ at $H = 0$. The only possible explanation for such reverse-phase LFMA at $T \ll T_c$, is the existence of orbital supercurrents in the ground state of our SC composite which are induced by so-called Josephson π-junctions, i.e. junctions containing localized magnetic moments (unpaired spins). This problem has been studied theoretically in 1977 by Bulayevskii et al. [81] and it has been predicted that due to the spin-flipping component the sign of the pair tunneling contribution to the Josephson coupling energy can be reversed. As a result, the ground state of a SC loop formed by several weakly coupled SC grains containing π-junctions in the barrier should have a phase slip $Q = \pi$ in the current across the junction. It can be viewed as the permanent orbital magnetic moment whose alignment in the external H gives a paramagnetic response.

The idea of the π-junction has been used for the interpretation of recently observed paramagnetic Meissner effect (PME) in high T_c SC ceramics [82, 83]. Moreover, it has been shown experimentally that PME is correlated with anomalous reverse-phase LFMA [84], very similar to the data observed here.

The following arguments can explain the phase reversal of LFMA by π-junctions: The microwave loss expressed in (10.7) is related with the total flux Φ inside the loop through the relation (10.6) between ΔQ and the flux, as $\mathrm{d}Q/\mathrm{d}t \propto \mathrm{d}\Phi/\mathrm{d}t = (\mathrm{d}\Phi/\mathrm{d}\Phi_{\mathrm{ext}})(\mathrm{d}\Phi_{\mathrm{ext}}/\mathrm{d}t) \propto \mathrm{d}\Phi/\mathrm{d}\Phi_{\mathrm{ext}}$, so that

$$P(H) = P\Phi_{\mathrm{ext}} \propto (\mathrm{d}\Phi/\mathrm{d}\Phi_{\mathrm{ext}})^2. \tag{10.8}$$

Now for a SC loop containing a π-junction the total flux Φ through the loop is related to the external flux Φ_{ext} [81] as

$$\Phi_{ext} = \Phi - \beta \sin(2\pi\Phi/\Phi_0), \tag{10.9}$$

where $\beta = LJ_c/c$ is the measure of self-generated field, L is inductance, and J_c is the critical current of the π-junction. It is known [81] that when $\beta > 1$ (hysteretic junction) the ground state is current carrying and that when $\beta \gg 1$ (as expected for mezoscopic defect structures), $\Phi = \Phi_0/2$ (i.e. $Q = 2\pi\Phi/\Phi_0 = \pi$) in the ground state. The latter condition then shifts the operating point of the JJ from $Q = 0$ to $Q = \pi$. From (10.8) and (10.9) one arrives at the following expression for the relative power absorption p as

$$p = P(\Phi_{ext})/P(0) \tag{10.10}$$

$$= [\{1 - (2\pi\beta/\Phi_0)\cos 2\pi\Phi(0)/\Phi_0\}/ \\ \{1 - (2\pi\beta/\Phi_0)\cos 2\pi\Phi(\Phi_{ext})/\Phi_0\}]^2 \tag{10.11}$$

where $\Phi(\Phi_{ext})$ is a function of Φ_{ext}, given implicitly by (10.9). The prefactor in the expression for $P(H)$ depends on the specific mechanism of the microwave dissipation. In a low loss junction the expression (10.7) can be used for an instantaneous loss connected with a normal current pulse across R_n. The important qualitative point of this treatment, according to [84], is that the change of the operating point from $Q = 0$ to $Q = \pi$, or from a normal to a π-junction, can change the minimum in $P(H)$ from $H = 0$ (CN-LFMA) to maximum at $H = 0$ (AP-LFMA). The behavior of P as function of Φ_{ext} (in units of $\Phi_0/2\pi$) as expressed in (10.10) is plotted in [84] for typical β, and the curve shows a maximum of p at $\Phi_{ext} = 0$ and two minima around $\Phi_{ext} = \pi$. To our knowledge, this is the only reasonable explanation for the AP-LFMA in SC. Previously, when this "in-phase" LFMA was observed in low quality high TC SC films [78], they were assigned to negative magnetoresistance in small SC loops [86]. We do not discuss this possibility, since in our case the PME has also been observed in SQUID experiments, supporting the π-junction origin. For the origin of the local spins in the π-junctions, we propose spin carrying polarons, either P^+ or P^-, as sketched in Fig. 10.11. Tunneling with spin flipping via the P level should be large enough if d is small, giving a strong π-junction. This assignment is supported by the observation of the AP component in CP matrices with strong n-dopability, i.e. OO-PPV and PAT6, while in freshly doped PAT12, only transient polarons exist and the AP-LFMA and PME were observed, and disappeared after (t =? days at 240°C) long sample aging this phenomena disappeared. Our spin assignment is uncertain at this time, since imperfections within each grain (twin boundaries, dislocations, etc.) also can act as Josephson junctions. The unpaired spins such as on C_{60}^{1-} and C_{60}^{3-} radicals whose existence is consistent with the evaluation of ESR spectra upon K-doping [24], might be another origin of the spins in our π-junctions. TEM images (see Fig. 10.10b) and further discussion below shows that intergrain links in large y composites are formed by disordered C_{60} bridges which might contain C_{60} and C_{60}-radicals, the latter serving in π-junctions.

Returning to the origin of two components in LFMA spectra appearing at two temperatures, we conclude that T^* can be interpreted as T_{c2} of the second SC2 phase which is spatially separated from the phase SC1. Since SC2 seems to make no contribution to LFMA from phase slippage in loops, one may speculate that it originates from single K_xC_{60} clusters which have an intragrain JJ but are not connected into loops with other clusters. Indeed, in TEM image of small $y = 0.005$ composites, we have seen much smaller clusters which are either isolated or loosely connected in a fractal type agglomerates (see Fig. 10.9b). Rather low T_{c2} values might be connected with nonstoichiometric composition of A_xC_{60}, with x distinct from $x = 3$ of bulk fullerides.

An alternative explanation for multi-phase LFMA behavior is possible with the existence of only one SC phase (i.e. K_3C_{60} grains), but with two different microscopic mechanisms for microwave loss. So, at small H, the CN component may arise due to damped fluxon motion of vortices penetrating into the JJs at small H_{c1}, giving a small T_{c2} of Josephson media which is known to be smaller than bulk T_{c1}. Then at temperatures above T_{c2}, these weak JJs are decoupled and this contribution to LFMA disappears. However, the contribution from damped bulk fluxons, and/or quantized flux penetration into JJ loops in the bulk still persists, giving a phase slippage contribution which survives up to T_{c1} and shows up as an AP component at larger H. In this case, JJs corresponding to T_{c2} need to be weaker than those responsible for the higher T_{c1}. In conflict with this explanation is that a reverse-phase LFMA appears at higher T_{c1} which requires the existence of π-junctions inside each grain. These π-junctions can be C_{60}^{1-} and/or C_{60}^{3-} spins localized at the twin boundaries in the grains (instead of the suggested spin polarons in CP chains). The question is then "why should they appear only for composites with large y"? It does not seem natural that with increasing y, intragrain junctions should become stronger and π-junctions appear. In TEM, we have observed intergrain loops at large y, and therefore an intergrain origin for π-junctions, and T_{c2} originating from a separate SC phase, seems a more realistic explanation. More study is needed to clarify the origin of the two different T_cs in our system. Our recent studies by the heating of SCs upon applying larger MW-power [29] supports the existance of π-junctions, which are observed only at small power and at lowest $T \ll T_c$. On the other hand, the behavior of the peak position of the broad AP signal with T [30] and MW power [29] has been explained by the resistively shunted Josephson junction (RSJ) model which suggests that the broad AP signal originates from S-N-S junctions.

10.5 SQUID Magnetometry

Magnetization studies were successful only in samples with larger $y = 0.05$ [24, 27]. These samples showed two types of behavior below T_c. In "equilibrated" (after long aging at about 2 weeks at 240°C) PAT12$(C_{60})_yK_x$ samples, a normal diamagnetic Meissner effect can be observed. A small SC shielding fraction was observed, i.e. only about 0.1% of the ideal diamagnetic response which corresponds to 0.8% with respect to C_{60}.

On the other hand, in freshly doped samples, with a nonequilibrium distribution of K-dopant within the composite, the PME has been observed in a FC run in fields from 5 to 50 G, above which PME disappeared (Fig. 10.8a). The SC volume fraction was higher, about 1.8% with respect to C_{60} clusters, and $T_c = 14$ K similar to the T_c determined by LFMA [27]. As noted in [27], it was not proved that PME correlates with anomalous LFMA. Possibly this was due to sample aging effects. Recent SQUID studies on OO-PPV$(C_{60})_yK_x$ composites allowed us to find a better correlation of PME with reverse-phase LFMA [29, 30]. This was observed in the same sample ($y = 0.1$). Figure 10.8 shows the typical DC magnetic susceptibility of the optimally K-doped $y = 0.1$ composite. The sample was cooled to 5 K in zero magnetic field and the zero

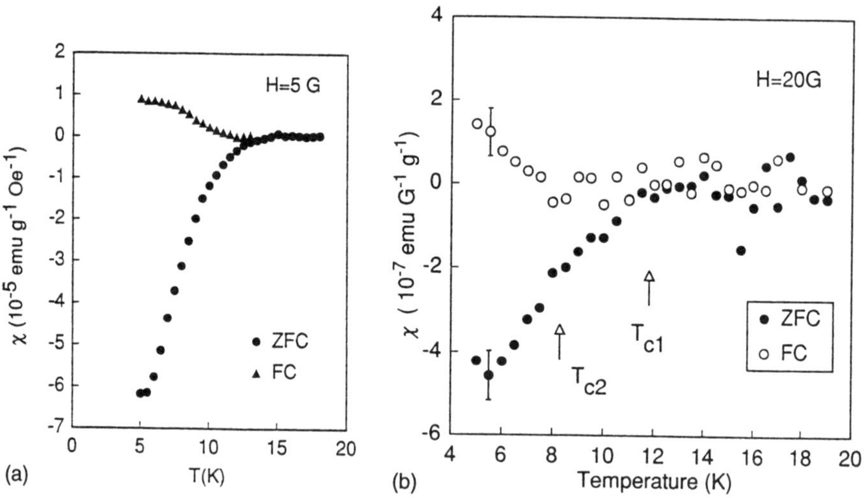

Fig. 10.8. (a) SQUID magnetization of PAT$(C_{60})_yK_x$ composite ($y = 0.05$). (**a**) Temperature dependence of χ in freshly doped sample, showing 1.8% diamagnetizm with $T_c = 17$ K at zero field cooling (ZFC), in $H = 5$ G field. Field cool (FC) run shows PME assigned to π-junction. (**b**) SQUID magnetization of OO-PPV$(C_{60})_yK_x$ composite: ($y = 0.1$, $t_d = 180$ h): Temperature dependence of χ, showing 0.02% diamagnetizm with $H_c = 12$ K at zero field cooling (ZFC), in $H = 10$ G field. At field cooled (FC) run $\chi(T)$ is paramagnetic with same H_c. Note that $\chi(T)$ has a structure at $T^* = 7$–8 K which is close to onset of CN appearance in LFMA spectra

field cooled (ZFC) magnetization was then measured up to 20 K in external field of $H = 20\,\mathrm{G}$ (Fig. 10.8b). Measurements were repeated in a field cooled (FC) run, and the hysteresis was clearly seen between the ZFC and FC data. This result is typical for type II superconductors due to flux trapping in FC. Moreover, the FC run clearly showed the PME effect, i.e. paramagnetism, which disappeared at T_c, instead of the diamagnetic response observed in ZFC. It should be noted that the value for T_c found in low $H = 20\,\mathrm{G}$ was 12 K, which is the same as T_c determined by the LFMA onset for AP-LFMA. From Fig. 10.8a, it is evident that at low T a paramagnetic tail appears in the $\chi(T)$ data and that the SC transition is broad, typical of a multi-phase sample. Moreover, $\chi(T)$ shows a cusp (both in the ZFC and FC runs) around $T^* = 7$–8 K where the CN component appears in the LFMA spectra. Also, some systematic fluctuations are seen around $T = 16$–17 K. The SC shielding fraction corresponds to 0.02% measured relative to the C_{60} content only which is much lower than we observed in PAT12$(C_{60})_yK_x$ composites. This is consistent with lower T_c values and lower LFMA intensities observed in OO-PPV$(C_{60})_yK_x$ composites, compared to PAT12 case. For $y = 0.01$, and 0.05 composites which showed clear SC phases with $T_c = 12$–13 K by LFMA method, we were not able to detect a SC phase by the SQUID magnetization, proving once again that LFMA is the much more sensitive method.

10.6 General Discussion

10.6.1 Paths of Electron Transfer and SC Phases

Before we start a discussion of CT processes in composites, it is important to emphasize that we are dealing with a very complicated system, the properties of which depend on several independent parameters: the type of CP matrix, and specific y and x. The CP matrix may exhibit various electronic behaviors. Most important here are the parameters δ_1 and δ_2 (Fig. 10.2): which are the energy level mismatches between the conduction state in the CP and LUMO levels of C_{60} molecule. Also of importance are the (overlap integrals. Implicitly, the CP will influence the nature of the composite through its morphology, or macromolecular structure which can affect the formation and/or size of the C_{60} clusters, and through parameters like its melting temperature which determines the practical upper limit for the doping temperature.

We studied the morphology of the composites by TEM microscopy and found that it depends on y and the type of CP. Figure 10.9 shows a TEM image of a $y = 0.05$ composite. It revealed that the C_{60} are aggregated in mainly two type of clusters: medium size, isolated clusters ($d = 600$–800 Å) or agglomerates of smaller clusters ($d = 50$–150 Å) which appear to form fractal structures, like trees, similar to electrochemically grown CPs. Small clusters seem to be not strongly coupled to closed loops. Isolated C_{60} molecules also can be found randomly distributed in CP matrix. At large y, the morphology

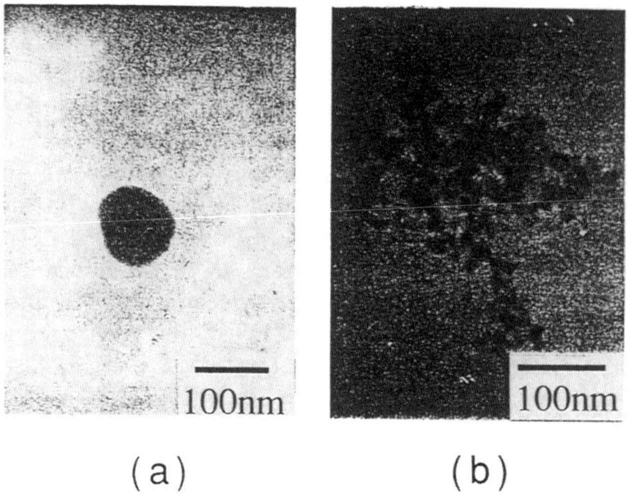

Fig. 10.9. TEM image of PAT6(C_{60})$_y$ composite at small $y = 0.005$. Two typical regions can be found: (**a**) Isolated C_{60} clusters of medium size $d = 800$, sometimes coexisting with a network of small size ($d = 50$–150 Å) clusters, (**b**) Fractal networks of loosely bound small size C_{60} clusters ($d = 50$–150 Å)

of the composite is different from that observed for small y as can be seen in Fig. 10.10. Larger size C_{60} clusters are formed ($d = 1000$ Å) that are strongly coupled into a dense network. In Fig. 10.10b large magnification clearly shows that intergrain links are composed of C_{60} bridges which are less dense and have a disordered structure, in contrast to the polycrystalline structure observed within each grain. Within these grains, it was possible to see regular crystalline planes and defects, like twin boundaries.

The CT processes from A dopants in these CP composites are much more complicated than found in pristine CP, where they can be qualitatively understood, as shown in the schematic diagram of Fig. 10.11b. At $y = 0$, electrons from the A atomic level are well known to be transferred to the conduction band (CB) of the PAT chains and then relax to polaronic P-levels. At large enough A-doping level above a critical $x_c = 0.65$–0.7 (see [22]), the insulator–metal transition usually takes place with the formation of a half-filled "polaronic band". At $y = 0$, however, the electrons from P-levels should be transferred to t_{1u}-derived CB of C_{60} clusters (or to a t_{1u} level of an isolated C_{60} molecule). This process can be kinetically hindered by a slow diffusion of K atoms which can lead to a slow-kinetics aging effect as observed in the PAT12 case. Crystalline C_{60} clusters can be transformed upon A-doping to ionic salts with stoichiometries A_3C_{60} (metallic and SC) and A_6C_{60} (insulating). Other phases, like A_1C_{60} and A_4C_{60} are normally metastable and more difficult to obtain.

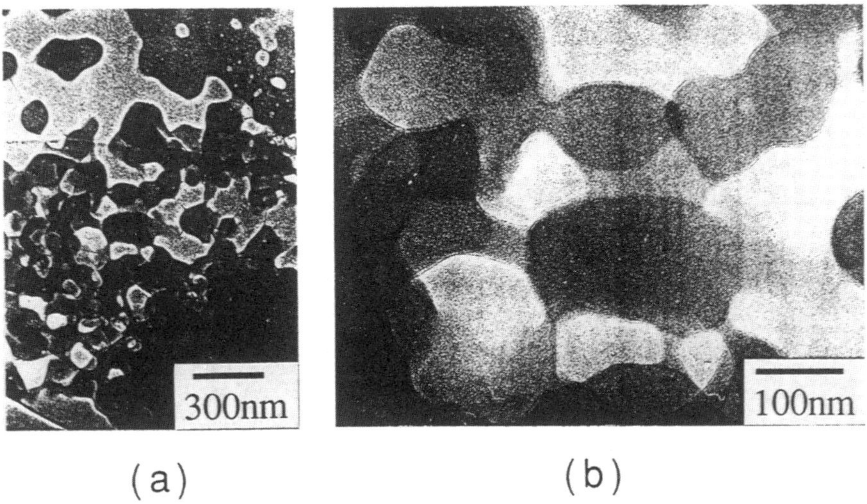

Fig. 10.10. TEM image of PAT6$(C_{60})_y$ composite at large $y = 0.05$: (**a**) the dense network of strongly coupled C_{60} grains of medium or large size ($d = 800$–2000 Å). (**b**) enlarged image showing that intergrain links are in fact filled with less transparent (i.e. thinner) C_{60} solid, creating strong intergrain coupling

The morphology of the $CP(C_{60})_yK_x$ composite is therefore very complicated, as observed by TEM, and shown schematically in Fig. 10.11a. Note first of all that two qualitatively different regions (A and B) may exist (or coexist) depending on y and the degree of homogeneity. The A-region of Fig. 10.11a corresponds to that shown in Fig. 10.10 at large y, where we have proposed that a network of JJ coupled SC loops form at low T. The B-region describes the dilute regions as in small y case of Fig. 10.9, where it is thought that single C_{60} molecules and small C_{60} grains are randomly distributed within the PAT12 host.

Charge transfer processes in dilute B-regions: At a low A-doping level (x), electrons are transffered from A atoms to CP chains, compensating residual P^+ states and charging an isolated neutral C_{60} to C_{60}^{1-}. Since the position of the t_{1u} LUMO of C_{60} is rather low compared to the upper level of the P^- in PAT12, the formation of P^- states in this system is energetically unfavorable in the presence of C_{60}. However, for kinetic reasons, P^- may be formed in chains far from C_{60} due to the slow diffusion of A atoms through the composite. In contrast to this behavior in PAT12, in OO-PPV and PAT6, the P^- state is more stable due to smaller δ_2. At higher A doping level separate C_{60} molecules will be charged to C_{60}^{2-}, C_{60}^{3-} up to C_{60}^{6-} until the filling of the t_{1u} levels is complete with 6 electrons. The role of the Madelung energy in the stabilization of higher C_{60}^{n-} radicals is not clear, but radicals even higher than $n = 6$ can be expected due to filling of t_{1g} second LUMO levels. Indeed, we have observed at low y up to 4 lines in ESR of dilute composites ($y = 0.005$)

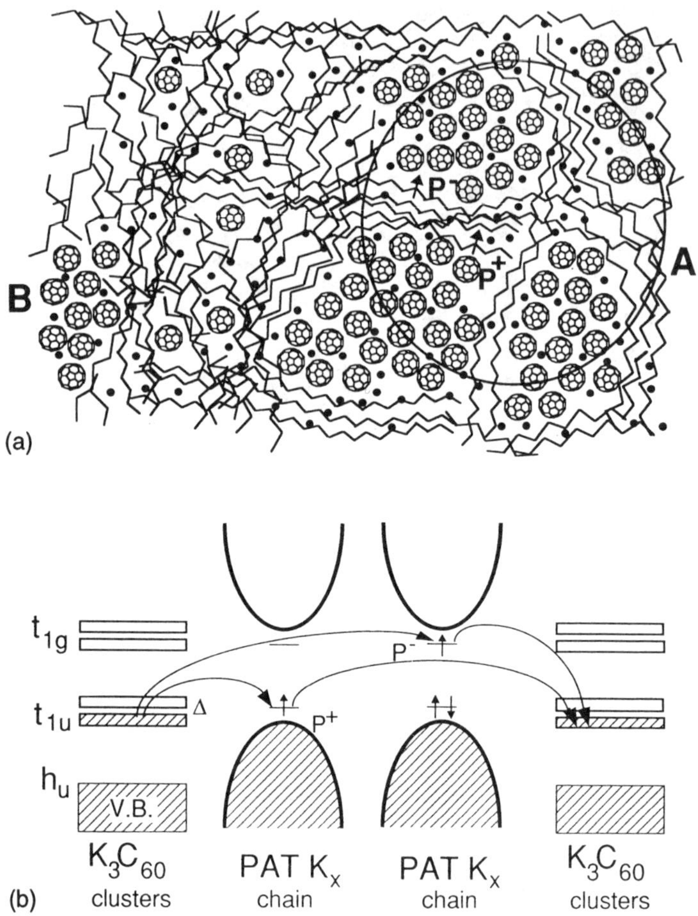

Fig. 10.11. (**a**) Schematic morphology of realistic $CP(C_{60})_yA_x$ composite showing two qualitatively different regions: A-region, containing dense network of K_3C_{60} clusters, weakly coupled into JJ arrays (sketched by circle loop), polaronic π-junction (marked by arrow of its spin); and B-region, composed of isolated C_{60} molecules and dilute, isolated small K_xC_{60} clusters in CP matrix, (**b**) energetics of A-phase with spin-flipping tunneling via P^- and P^+ π-junction

[24, 26]. Small disordered A_xC_{60} clusters should give a SC phase with a low value for T_c (i.e. 12 K), which is probably observed as normal phase LFMA in low y case and as a lower T_{c2} (CN-component) in large y composites. T_{c2} should be low either due to disorder effects of the grains, i.e. nonstoichiometric (K_xC_{60}, $x = 3$) and/or intragrain imperfections which act as JJs and are responsible for LFMA, similar to the case of SC LB (Langmuir–Blodgett) films [75 a]. Also, a low T_{c2} may be due to the phase-locking transition which can be induced in a critical regime of SC granular systems, distinct from quasiordering within the grains [75 b].

An insulator-metal transition in the host PAT should occur upon further doping when A_6C_{60} clusters cannot absorb any more A-atoms. In this phase, the two component C_{60}-induced SC phase may be expected. Electrons in the polaronic upper band (shown in Fig. 10.2) should be hybridized with the LUMO of C_{60} molecules, and so their pairing can be induced, as discussed above.

Charge transfer processes in dense A-region: In this case, the C_{60} clusters can play the role of "electronic sponges" and absorb all dopant (although this process may be slowed down by kinetics). As a result, the CP matrix can be insulating, even at large x. However, the conductivity of the composite at this stage should increase due to metallic A_3C_{60} islands, as we have indeed observed [26]. At this stage, a granular SC phase can be found. Polarons in CP chains can play the role of intermediates in tunneling between the SC grains, creating π-junctions. Further doping of A_xC_{60} clusters then occurs, filling them until insulating K_6C_{60} grains are formed.. At this x-level, conductivity should show decrease and the granular SC phase should disappear. We were able to detect the beginning of overdoping in OO-PPV$(C_{60})_yK_x$ at doping times of 300–400 h. Upon even further doping, A will finally dope the CP chains and induce an insulator-metal transition at a much higher critical value $x_c(y) > x_c(0)$. Conductivity should then show a dramatic increase and the SC phase can be expected in the host CP states, possibly enhanced by hybridization with molecular prestine C_{60} and $K6C_{60}$.

Experimentally, we have evidently found SC phases in several composites, as summarized in Table 10.1. From the evolution of the LFMA spectra and SQUID data in various CP matrices, we can conclude that the general tendencies of doping described above are indeed observed for several CP matrices. Choosing a CP matrix with lower δ and better n-dopability due to a higher melting temperature, we observed more pronounced two-phase SC behavior, and more complicated LFMA with increasing y.

10.6.2 Prospects for C_{60}-Induced Superconductivity

Returning to our original motivation to search for a SC phase with the CP playing an active role in the electron pairing; can we decide whether the LFMA with normal phase signal in small y sample, or alternatively a small T_{c2} SC phase in large y composites; originates from this hypothetical C_{60}-induced SC phase?

We believe that the experimental evidence suggests that disordered C_{60} grains with smaller size, observed even in small y composites by TEM microscopy, are the most reasonable explanation for the lower T_{c2} phase and normal LFMA with small H^*_{c1}. Moreover, both ESR and conductivity [26], demonstrated that the CP matrix is nearly insulating (at least in the PAT12 case), so that the concentration of electrons in polymeric chains is not high enough for SC. However, in the OO-PPV and PAT12 cases, the conditions

Table 10.1. Summary of LFMA/SQUID data for various $CP(C_{60})_yA_x$ superconducting composites

Type of $CP(C_{60})_yA_x$ composite and its main characteristics	T_d $t_d(h)$	y	LFMA number of peaks	LFMA phase	LFMA hysteresis	SQUID type	SQUID volume fract.	T_c (K)	Origin of SC phase	Ref.
PAT12(C_{60})$_y$K$_x$ large $\delta_c \sim 1.8\,eV$ low $T_{melt} = 130°C$	120 ↘ 110	0.0005	1	N(normal)	—	—	—	4.5(?)	non SC LFMA (?), non hysteresis	[30–32]
		0.005	1	N	—	—	—	13	single K_xC_{60} ($x \neq 3$) grains or C_{60}–induced paring	[30–32]
		0.025	CN/AP	N+AP(T_{dep})	clockwise	M	–	12	granular percolated	[31–33]
		0.05	CN/AP	N+AP	CN:clockwise	M+PME	1.8%	10/15	granular width π-junction	[31–33]
	160 h	0.1	CN/AP	N+AP(T_{dep})	clockwise	M	0.8%	17	granular percolated	[31–33]
OOPPV(C_{60})$_y$K$_x$ lower $\delta_c \sim 1.2\,eV$ $T_{melt} = 140°C$	135 ↘ 130	0.005	N	N	—	—	—	11.5	single K_xC_{60} ($x \neq 3$) grains or C_{60}–induced paring	[34]
		0.01	CN/AP	N+AP	—	—	—	7/11.5	first SC1: large K_3C_{60} grains with intergrain π-junctions	[34]
	180–	0.05	CN/AP	N+AP	CN:clockwise	—	—	8.6/12	second SC2: small defected	
	200 h	0.1	CN/AP	N+AP	clockwise based in both CN and AP	M(ZFC) PME(FC)	0.02%	7.5/12	K_xC_{60} ($x \neq 3$) grains or C_{60}–induced paring	
PAT16(C_{60})$_y$Rb$_x$ low $\delta_c \sim 1.3\,eV$ high $T_{melt} = 200°C$	165 ↘ 150 325 h	0.05	2CN AP	N+AP	CN: clockwise AP: anti-clockwise	M+strong background diamagnet above T_c	very small	13.5 25	first SC1: large K_3C_{60} grains with intergrain π-junctions second SC2: small defected K_xC_{60} ($x \neq 3$) grains or C_{60}–induced paring	[35]
PAT6(C_{60})$_y$K$_x$	165 ↘ 155 200 h	0.05	CN AP	N+AP	CN: clockwise AP: anti-clockwise	M+PME	small	12 15	granular with π-junction	[35]
PAT6(C_{60})$_y$Na$_x$	165 ↘ 155 70 h	0.05	1	N only at large HM	—	—	—	9(?)	C_{60}–induced SC or non SC LFMA of the magnetoresist LFMA type	[35]

for C_{60}-induced SC are better. We cannot exclude the possibility of a microscopic SC phase not directly associated with K_xC_{60} granules associated with SC pairing in the CP chains enhanced somehow by C_{60}^{n-} anions and, of course, there is no clear evidence for this C_{60}^{n-} mediated SC yet. ESR vs conductivity studies are in progress to determine the concentration of carriers in the polymeric part.

However, the strategy for the search of two-component SC, under doping conditions when it can be masked by granular K_xC_{60} SC phases, is becoming clearer after having studied composites prepared from several polymeric matrices. There are two types of problems on the way to new SC: technical ones, connected with the possible masking effects of A_xC_{60} cluster SC, and physical ones, connected with the redistribution of A dopant in favor of A_xC_{60} clusters, since C_{60} grains can absorb almost all A atoms like "black holes" due to their strong electron affinity, therefore leaving polymeric conducting bands empty. The best solution to avoid these masking effects will be to eliminate C_{60} clusters. We have always observed C_{60} clusters in the optical spectra at large y, and proved that simple grinding of composite films allows to make clusters smaller [58]. We found by TEM that even at small y clusters exist, although x-ray analysis does not show C_{60} crystallites even at large y. It has been realized recently that C_{60} shows a strong tendency to make clusters even in the best solvents like toluene and benzene [86], where such clusters containing 10–20 C_{60} molecules show up in light scattering [87] and are responsible for diffusion anomalies [88]. In general, macromolecules have a very small entropy of mixing and therefore the tendency for C_{60} to aggregate in solution or in the composite probably cannot be avoided altogether.

From the above discussion, we suggest the following strategy for the search for polymeric, two-band superconductivity, stimulated by the presence of C_{60} molecules:

1. In large y composites, one should try to achieve the overdoping state at large x, when the SC phase of A_xC_{60} should be suppressed. Residual single C_{60} molecules may then induce SC pairing of electrons in metallic CP chains. This type of experiment is in progress now, but it seems that there is some mechanism inhibiting the overdoping due to the CP environment, and this shifts the insulator-metal transition of the CP host to very large critical $x_c(y)$-values.

2. In small y composites, it is important to choose a CP matrix with small energy mismatch δ_2, so that electrons can be favorably placed into conductive states in the CP chains and provide delocalization and metallicity of the host CP. The resonance between CB states of the polymer and the levels of C_{60} can be possibly improved choosing a high electron affinity polymer, like cyano-PPV [54], as a host CP. Much higher levels of x will be needed to overdope the residual single A_xC_{60} grains, and rule out the masking effects.

3. Another promising way is to rule out the effect of cluster SC by using the A dopants which do not create SC phases in A_3C_{60} solids. For example, Na,

Cs and Li are the candidates for this. Na_3C_{60} at low temperature segregates into two non-SC phases, Na_2C_{60} and Na_6C_{60}, Cs has too large a radius to accommodate itself in the tetragonal sites of the C_{60} fcc lattice, while Li is too small and, moreover, seems to make a covalent bond with C_{60} [70]. Hence, observation of SC in CP-C_{60} composites doped by these dopants would clearly prove a novel type of fulleride SC. However, due to low vapor pressure of Na and Li there are difficulties for vapor doping which we are trying to overcome by using other doping methods. Also, we are investigating CP with higher melting point, like PAT6, which allows us to increase the doping temperature T_d. We already proved by ESR that PAT6$(C_{60})_y$ can be Na-doped and we found LFMA, at $y = 0.05$, below $T = 9\,K$ [29]. But the LFMA is very weak, and can be observed only at large H_{mod}, so that a hysteresis behavior cannot be observed, and thus it might not be a SC LFMA, but rather a magnetoresistance type of LFMA as in the PPP-Na_x case [76].

10.7 Conclusions

In summary, we have found the appearance of superconductivity upon A-doping of CP$(C_{60})_y$ composites (at $y > 0.005$) by means of LFMA and SQUID magnetometry. We have studied the evolution of the SC phase upon variation of CP matrix, the C_{60} content y, and the A-doping level x:

1. The SC phase at small C_{60} content ($y < 0.005$) shows a normal LFMA signature without phase reversal; also a low $T_c = 12\,K$ is found. Small size-disordered A_xC_{60} SC grains with intragrain Josephson junctions are proposed to be the origin of this SC phase, although C_{60}-induced SC also may contribute to this signal. Traces of LFMA found at the lowest $y = 0.0005$ at $T = 4.7\,K$ encourage us to study low y composites more carefully, since monomolecular C_{60} may induce SC pairing in CP chains at large doping levels x when the CP chains will become metallic and the A_xC_{60} grains are overdoped to suppress SC in the A_3C_{60} grains.

2. The SC phase found at high C_{60} content ($y > 0.01$) in OO-PPV and PAT 12, PAT6 matrices show two-component LFMA spectra, arising from two spatially separated SC phases with distinct T_c. The anomalous phase (AP) LFMA component, appearing below T_{c1}, is supposed to originate from π-junctions carrying ground state supercurrents in JJ loops. Normal phase narrow LFMA component identified with the onset at T_{c2} is probably connected with the SC phase found at dilute regions of the composite, which contain (as in the low y case) small A_xC_{60} clusters and single C_{60} molecules.

3. PME which is found in SQUID in a FC run in small magnetic field support the existence of π-junctions. Either spin carrying polarons in PAT chains (P^+ or P^-) or C_{60}^{n-} radicals in intergrain barriers are suggested as the possible origin of such π-junctions.

So, in the search for C_{60}-induced superconductivity in polymeric matrices, we have found curious granular SC phases with distinct T_cs which showed rich LFMA spectra with anomalous LFMA phase behavior. These samples also demonstrated PME in SQUID. So these composites are a first example of a molecular type granular SC in which π-junctions are found coexisting with other phenomena, like periodic oscillations of the magnetization with magnetic field (which is presently under study). These granular SC phases, however, mask the observation of possible C_{60}-induced SC, and it is necessary to find a way to avoid such masking effects, e.g. by using an n-dopant which would not create SC in C_{60} grains, e.g. like Na, Li or Cs.

We believe, that $CP(C_{60})_yA_x$ composites offer many possibilities for the study of both mesoscopic SC phenomena and new superconductors based on C_{60}-induced pairing. For these purposes, the various doping methods, particularly electrochemical doping by large organic donors, and favorable structures (like electrode injection in inversed MISFET structures) may be suggested. They are currently under investigation in our laboratories [89].

All these results call for more study. We realize that we are still at the very early stage of research, and the desire to attract more attention to this promising field has encouraged the authors to present the embryonic results in this book.

Acknowledgement. We sincerely thank our principle collaborators, who contributed to different parts of this work, particularly Professor K. Yakushi and Dr K. Tada, Dr. I. Khayrullin H. Kajii, K. Yamasaki and J. Saiki. The valuable assistance of Professor K. Niihara and Dr T. Sekino of Osaka University with TEM microscopy is highly appreciated. Thanks are due to Low Temperature Center of IMS, Okazaki for providing SQUID facilities. A.A. Zakhidov is particularly thankful for the hospitality of IMS, Okazaki, and Osaka University.

References

1. A.F. Hebard, M.J. Rosseinsky, R.C. Haddon, D.W. Murphy, S.H. Glarum, T.T.T. Palstra, A.P. Ramirez, A.R. Kortan: Nature **350**, 600 (1991).
2. For a review on A_3C_{60} see e.g. A.F.Hebard: in Annum. Rev. Mater. Sci. **23**, 139 (1993) and A.P. Ramirez: in *Superconductivity Review*, Vol.1., P. Kumar (Ed.), (Gordon and Breach) pp.1, **102**, 1994, M.P. Gelfand: ibid pp. 103, 150, (1994).
3. P.M. Allemand, K.C. Khemani, A. Koch, F.Wudl, K. Holczer, S. Donovan, G. Gruner, J.D. Thompson: Science **253**, 301 (1991).
4. K. Tanaka, A.A. Zakhidov, K. Yoshizawa, K. Okahara, T. Yamabe, K. Yakushi, et al.: Phys. Lett. A **164**, 221 (1992), and also in Phys. Rev. B **47**, 7554 (1993).
5. K. Tanaka, A.A. Zakhidov, K. Yoshizawa, K. Okahara, T. Yamabe, K. Kikuchi, S. Suzuki, I. Ikemoto, Y. Achiba: Solid State Commun. **85**, 69 (1993).
6. K. Tanaka, K. Yoshizawa, T. Sato, T. Yamabe, K. Okahara, A.A. Zakhidov: Solid State Commun. **87**, 1055 (1993), and also in Phys. Rev. B **51**, 990 (1995).
7. A.A. Zakhidov, K. Imaeda, A. Ugawa, K. Yakushi, H. Inokuchi, Z. Iqbal, R.H. Baughman, B.L. Ramakrishna, Y. Achiba: Physica C **185–189**, 411 (1991), also in Mol. Cryst. Liq. Cryst. **218**, 299 (1992).
8. A.A. Zakhidov, R. Whitehead, T. Suzuki, Yu. Maruyama, (unpublished).

9. C.M. Varma, et al.: Science **245**, 989 (1991).
10. R.A. Jishi and M.S. Dresselhaus: Phys. Rev. B **45**, 2597 (1992).
11. M. Schluter et al.: Phys. Rev. Lett. **68**, 526 (1992).
12. G. Baskaran and E. Tossati: Curr. Sci. **61** 3452 (1991).
13. a) S. Chakravarty, M.P. Gelfand, S. Kivelson: Science **254**, 970 (1991), b) S. Chakravarty and S. Kivelson: Europhys. Lett. **16**, 751 (1991).
14. R. Friedberg, T.D. Lee, H.C. Ren: Phys. Rev. B **46**, 14150 (1992).
15. G.P. Zhang, Y.S. Ma, X. Sun, K.H. Lee, T.Y. Park: Phys. Rev. B **52**, 6081 (1995).
16. a) B. Burk, V.H. Crespi, A. Zettl, et al.: Phys. Rev. Lett. **72**, 3706 (1994), b) T.W. Ebbesen, T.S. Tsai, K. Tanigaki, H. Hiura, Y. Shimakawa, Y. Kubo, I. Hirosawa, J. Mizuki: Physica C **203**, 163 (1992).
17. V. Ivanov and Y. Maruyama: Physica C **247**, 147 (1995).
18. A.R. Kortan, et al.: Nature **355**, 529 (1992), H. Araki, N. Yamasaki, A.A. Zakhidov, K. Yoshino: Physica C **233**, 242 (1994).
19. A.R. Kortan, et al.: Nature **360**, 566 (1992).
20. M.S. Fuhrer, J.G. Hou, X.-D. Xiang, A. Zettl: Solid State Commun. **90**, 357 (1994)
21. a) A. Otsuka, G. Saito, A. Zakhidov, K. Yakushi, M. Kusunoki, K. Sakaguchi, Mol. Cryst. Liq. Cryst. **285**, 187 (1996), b) A. Otsuka, G. Saito, A. Zakhidov, K. Yakushi, Synth. Metals **85**, 1459 (1997).
22. For a review on conducting polymers see: A. J. Heeger, S. Kivelson, J.R. Schrieffer, W.P. Su: Rev. Mod. Phys. **60**, 781 (1988), V. Vardeny, Ch. 2 of this book.
23. S.A. Kivelson and V.J. Emery: Synth. Met. **65**, 2494 (1994).
24. H. Araki, A.A. Zakhidov, E. Saiki, N. Yamasaki, K. Yakushi, K. Yoshino: Jpn. J. Appl. Phys. **34**, L1041 (1995).
25. A.A. Zakhidov, H. Araki K. Tada, K. Yakushi, K. Yoshino: Phys. Lett. A **205**, 317 (1995).
26. H. Araki, A. Zakhidov, K. Tada, K. Yakushi, K. Yoshino: Synth. Met. (in Proc. EMRS, Spring Meeting, 1995).
27. A.A. Zakhidov, H. Araki, K. Tada, K. Yoshino: Synth. Met. (in Proc. EMRS Spring Meeting, 1995)
28. K. Yoshino, A.A. Zakhidov, H. Kajii, H. Araki, K. Tada, T. Noguchi, T. Ohnishi, K. Yakushi, Physica C **264** (1996).
29. H. Kajii, H. Araki, A. Zakhidov, K. Tada, K. Yakushi, K. Yoshino: Physica C **277**, 107 (1997).
30. H. Araki, H. Kajii, K. Tada, A. Zakhidov, K. Yakushi, T. Sekino, K. Niihara, K. Yoshino: Physica C (in press).
31. a) S. Morita, A.A. Zakhidov, K. Yoshino: Solid State Commun. **82**, 249 (1992). b) S. Morita, A.A. Zakhidov, T. Kawai, H. Araki, K. Yoshino: Jpn. J. Appl. Phys. **31**, L890 (1992), also b) J. Phys: Cond. Matter **5**, 2103 (1993).
32. K. Yoshino, X.H. Yin, S. Morita, T. Kawai, A.A. Zakhidov: Chem. Express **7**, 817 (1992), also in Solid State Commun. **85**, 85 (1993).
33. K. Yoshino, X.H. Yin, A.A. Zakhidov, T. Noguchi, T. Ohnishi: Jpn. J. Appl. Phys. **32**, L357 (1993).
34. A.A. Zakhidov: Proc. Int. Conf. ELORMA-87, Moscow, 1987, p 256, also in Synth. Met. **41–43**, 3393 (1991).
35. K. Yoshino, X.H. Yin, T. Akashi, K. Yoshimoto, S. Morita, A.A. Zakhidov: Mol. Cryst. Liq. Cryst. **225**, 197–211 (1994).
36. K. Yoshino, T. Akashi, K. Yoshimoto, M. Yoshida, S. Morita, A.A. Zakhidov: Mol. Cryst. Liq. Cryst. **256**, 343 (1994).
37. A.A. Zakhidov, T. Akashi, K. Yoshino: Synth. Met. **70**, 1519 (1995).

38. N. S. Sariciftci, D. Braun, C. Zhang, V. Srdranov, A. J. Heeger, F. Wudl: Science **258**, 1474 (1992) and also Appl. Phys. Lett. **62**, 585 (1993), see also a review: N. S. Saricifty and A. J. Heeger: Int. J. Mod. Phys. B **8**(3), 237 (1994).
39. L. Smilovitz, N. S. Sariciftci, R. Wu, C. Gettinger, A. J. Heeger, F. Wudl: Phys. Rev. B **47**, 13835 (1993).
40. C. H. Lee, G. Yu, D. Moses, K. Pakbaz: C. Zhang, N. S. Sariciftci, A. J. Heeger, F. Wudl: Phys. Rev. B **48**, (1993), and also in N. S. Saricifti and A. J. Heeger: Synth. Met. **70**, 1349 (1995).
41. S. Morita, A. A. Zakhidov, K. Yoshino: Jpn. J. Appl. Phys. **32**, L873 (1993), also S. Morita, S. B. Lee, A. A. Zakhidov, K. Yoshino: Mol. Cryst. Liq. Cryst. **256**, 839 (1994).
42. A. A. Zakhidov, K. Tada, K. Yoshino: Synth. Met. **71**, 2113 (1995).
43. A. A. Zakhidov and K. Yoshino: Synth. Metals **71**, 1875 (1994), A. A. Zakhidov and K. Yoshino: Synth. Met. **64**, 155 (1994).
44. K. Yoshino, M. Hamaguchi, T. Kawai, A. A. Zalhidov, H. Ueno, M. Kakimoto, H. Kojima: Jpn. Appl. Phys. **34**, L141 (1995).
45. K. Yoshino, T. Akashi, K. Yoshimoto, S. Morita, T. Kawai, A. A. Zakhidov: Jpn. J. Appl. Phys. **34**, L127 (1995).
46. W. A. Little: Phys. Rev. A **135**, 1416 (1964).
47. S.P. Ionov: Izv. Acad. Nauk. SSR **49**, 310 (1985) (Engl. Trans.: Bull. Acad. Sci. USSR, Phys. Ser. (USA) **49 A**, 90 (1985)).
48. Robaszkiewicz, R. Minas, Ranninger: Phys. Rev. B **36**, 180 (1987); ibid. Rev. Mod. Phys. **63**, 113 (1990).
49. Y. Bar-Yam: Phys. Rev. B **43**, 359, 2601 (1991), D. Eagles: Solid State Commun. **80**, 823, 831 (1991).
50. E. Simanek: Sol. State. Commun. **32**, 731 (1979), C.S. Ting, D.N. Talwar, K.L. Ngai: Phys. Rev. Lett. **45**, 1213 (1980).
51. A. G. Malshukov: Solid State Commun. **77**, 57 (1991).
52. Yu. N. Gartstein and A. G. Malshukov: Solid State Commun. **83**, 989 (1992), Physica C **219**, 39 (1994).
53. V. M. Agranovich, V. E. Kravtsov, A. G. Malshukov: Solid State Commun. **33**, 137 (1980).
54. N.C. Greenham, S.C. Moratti, D. D. C. Bradley, R. H. Friend, A. B. Holmes: Nature **365**, 628 (1993).
55. P. M. Allemand, et al.: J. Am. Chem. Soc. **113**, 2780 (1991), P. J. Krusic, et al.: ibid **113**, 6274 (1991), P. N. Keizer, et al: J. Phys. Chem.**95**, 7117 (1991).
56. F. Mehran et al.: Phys. Rev. B **46**, 8579 (1992).
57. S. Luzzati, F. Speroni, A. Zakhidov, S. Morita, K. Yoshino: Mol. Cryst. Liq. Cryst. **256**, 927 (1994).
58. T. V. Ramakrishna: Phys. Scr. T **27**, 24 (1989).
59. S. John and T. C. Lubensky: Phys. Rev. Lett. **55**, 1014 (1985), Phys. Rev. B **34**, 4815 (1986).
60. J. R. Benlhachemi, J. Gavari, C. Musso, J. Alfred-Duplaan, Marfaing: Physica C **230**, 246 (1994), A. Benlhchemi, J. R. Gavari, Y. Massiani, S. Aityazza: Physica C **235**, 240 (1994).
61. See for review on the LFMA: K. W. Blazey, in *Earlier and Recent Aspects of Superconductivity* Springer Series in Solid-State Sciences, Vol. 90, J. G. Bednorz, K. A. Muller (Eds.), (Springer-Verlag Berlin, Heidelberg, 1990) 262.
62. S. V. Bhat, P. Ganguly, T. V. Ramakrishnan, C. N. R. Rao: J. Phys. C **20**, L559 (1987), also for review: K. N. Shrivastava: Phys. Repts. **200**, 51 (1991).
63. K. W. Blazey, A. M. Portis, J. G. Bednorz: Solid. State. Commun. **65**, 1153 (1988).

64. A. M. Portis, K. W. Blazey, K. A. Muller, J. G. Bednorz: Europhys. Lett. **5**, 467 (1988).
65. A. M. Portis and K. W. Blazey: Solid State Commun. **68**, 1097 (1988).
66. J. Bohandy, B.F. Kim, F.J. Adrian, K. Moorjani, S. D'Arcangelis, D. O. Cowan: Phys. Rev. B **43**, 3724 (1991).
67. R. C. Haddon, S. H. Glarum, S. V. Chichester, A. P. Ramirez, N. M. Zimmerman: Phys. Rev. B **43**, 2642 (1991).
68. A. A. Zakhidov, A. Ugawa, K. Yakushi, K. Imaeda, H. Inokuchi, I. I. Khairullin, P. K. Khabibullaev: Physica C **185**, 2669 (1991).
69. A. A. Zakhidov, A. Ugawa, K. Imaeda, K. Yakushi, H. Inokuchi, K. Kikuchi, I. Ikemoto, S. Suzuki, Y. Achiba: Solid State Commun. **79**, 939 (1991).
70. H. Araki, N. Yamasaki, A. A. Zakhidov, K. Yoshino: Physica C **233**, 242 (1994).
71. K. Imaeda, I. I. Khairullin, K. Yakushi, H. Inokuchi: Solid State Commun. **87**, 375 (1994).
72. I. I. Khairullin, K. Imaeda, K. Yakushi, H. Inokuchi: Physica C **231**, 26 (1994).
73. N. Yamasaki, H. Araki, A. A. Zakhidov, K. Yoshino: Solid State Commun. **92**, 547 (1994).
74. I. I. Khairullin, A. A. Zakhidov, P. K. Khabibullaev, Z. Iqbal, R. H. Baughman: Synth. Met. **33**, 243 (1989).
75. a) P. Wang, R. Metzger, S. Badow, Y. Maruyama: J. Phys. Chem. **97**, 2926 (1993), b) G. Dentscher, Y. Imry, L. Gunther: Phys. Rev. B **10**, 4578 (1974)
76. A. A. Zakhidov, I. I. Khairullin, V. Yu. Sokolov, R. H. Baughman, Z. Iqbal, M. Maxfield, B. L. Ramakrishna: Synth. Met. **41–43**, 3717 (1991).
77. M. D. Sastry, K. S. Ajayakumar, R. M. Kadam, G. M. Phatak, R. M. Iyer: Physica C **170**, 41 (1990).
78. A. Dulcic, R. H. Crepeau, J. H. Freed, L. F. Schneemeyer, J. V. Waszczak: Phys. Rev. B **42**, 2155 (1990).
79. T. Xia and D. Stroud: Phys. Rev. B **39**, 4772 (1989).
80. F. J. Owens: Phys. Status Solidi B **162**, 565 (1990).
81. L. N. Bulaevskii, V. V. Kuzii, A. A. Sobyanin: JETP Lett. **25**, 290 (1977) (Pisma Zh. Eksp. Teor. Fiz. **25**, 314 (1977)).
82. F. V. Kusmartsev: Phys. Rev. Lett. **69**, 2268 (1992).
83. L. N. Brausnich, N. Nauf, V. Kataev, S. Neuhausen, A. Grutz, A. Kock, B. Roden, D. Khomski, D. Wohlleben: Phys. Rev. Lett. **68**, 403 (1992), see also in W. Brausnich et al.: Phys. Rev. **48**, 4030 (1993).
84. S. V. Bhat, A. Rastogi, N. Kumar, R. Nagarajan, C. N. R. Rao: Physica C **219**, 87 (1994).
85. P. Santhaham, C. P. Umbach, C. C. Chi: Phys. Rev. B **40**, 889 (1989).
86. V. N. Bezmelnitsin, A. V. Eletskii, E. V. Stepanov: J. Phys. Chem. **98**, 6665 (1994).
87. Q. Ying, J. Marecek, B. Chu: Chem. Phys. Lett. **219**, 214 (1994).
88. V. N. Bezmelnitsin, A. V. Eletskii, M. V. Okun, E. V. Stepanov: Himicheskaya Fiz. **13**, 156 (1994).
89. K. Tada, T. Kawai, H. Harada, A. A. Zakhidov, K. Yoshino: Tech. Repts. of Osaka Univ. **45**, 161 (1995).

11. Solution-State Syntheses of Fullerene-Based Pendant, Pearl Chain, and Dendritic Polymers

M.S. Meier

11.1 Introduction

Polymeric materials are macromolecular solids or liquids formed of chains of monomer units. These macromolecules can be formed, for example, as straight chains, as long chains with sidechains branching off the main chain, or as cross-linked solids in which different chains are linked together. The incorporation of fullerenes into organic polymers has been seen as a way to improve the notoriously poor processibility of fullerenes [1] or to improve upon the desirable properties of fullerenes. Alternatively, the incorporation of fullerenes into known polymers may improve the properties of the polymeric material.

The polymers which are discussed in this chapter [2] have strong, covalent bonds between the fullerene moiety and the rest of the polymer. This is distinct from composites or mixtures in which a fullerene and a polymer are mixed together but not chemically bonded together. Several different structural motifs can be imagined for fullerene-based polymers. A polymer formed by linking fullerene units by a single bond is conceivable, and such materials are discussed elsewhere in this book. A second motif involves linking fullerenes through *pairs* of single bonds, typified by the photopolymer reported by Rao et al. [3] (Fig. 11.1). These are discussed in detail in several chapters of this volume.

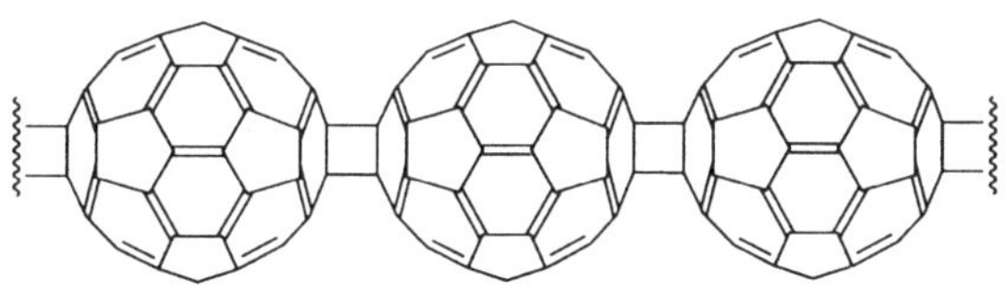

Fig. 11.1. Idealized structure of the photopolymer formed by [2+2] cycloaddition reactions between fullerene monomers [3]

This chapter focuses on the chemistry of fullerene-containing polymers prepared by *solution-state* syntheses. We will make a distinction between polymers in which the fullerene is part of the main chain of the polymer and those in which the fullerene is a pendant group, dangling from the main chain. The former motif is also known a "pearl necklace" polymer, since the fullerene units appear as balls on a string [4]. The latter is the "charm bracelet" motif [5]. We will also discuss some heavily cross-linked materials formed by copolymerization of C_{60} with other monomers.

11.2 Attachment of Fullerenes to Existing Polymers: Charm Bracelet Polymers

This approach relies on chemical reactions between C_{60} and functional groups on an existing polymer. Numerous reactions between C_{60} and nucleophiles (reactive species that are electron rich such as anions, nucleophiles, radicals, and other species) are known and can be found in any of several excellent reviews of the organic chemistry of fullerenes [6–10]. In principle, all reactions that form covalent bonds to C_{60} could be used to link C_{60} to a polymer, given a polymer with the appropriate functional group. Covalent bonds form between the reactive group and the fullerene, leading to pendant C_{60} groups along the chain. This approach has been referred to as "buckyball fishing" (Fig. 11.2) [11].

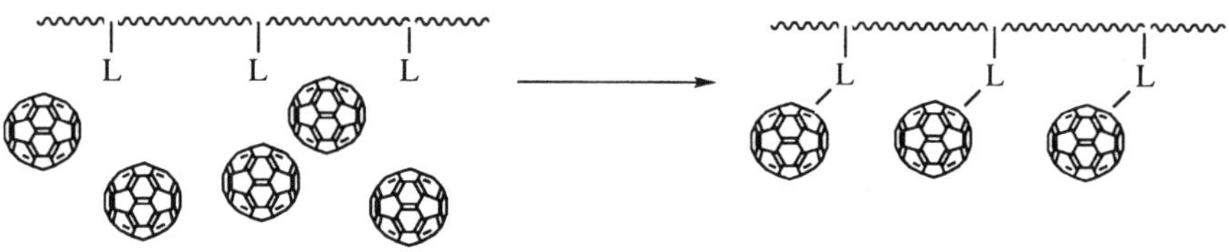

Fig. 11.2. "Buckyball fishing" in which fullerenes are caught by a reactive functional group L on a polymer

The nucleophilic addition of amine groups [6, 7, 12] to C_{60} has been exploited in the preparation of a number of different fullerene-modified polymers. This reaction results in the formation of a C_{60}-nitrogen bond and a C_{60}-hydrogen bond.

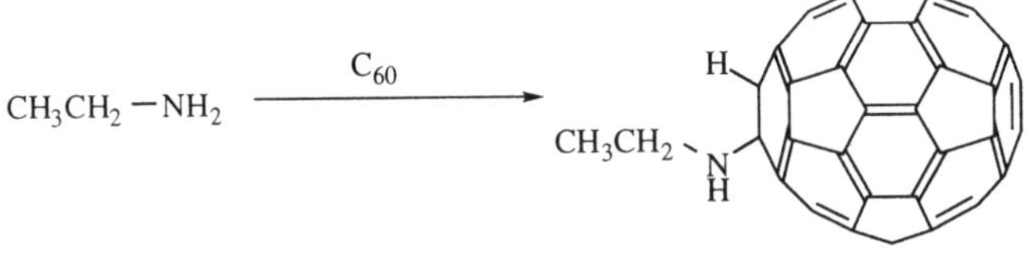

Amines do not add across most carbon–carbon double bonds. With C_{60}, amines undergo electron transfer to form a transient radical ion pair (shown in brackets in Fig. 11.3). This undergoes a coupling reaction and a proton transfer to produce the C_{60}-amine adduct. Simple alkenes (such as ethylene) do not react with amines under comparable conditions.

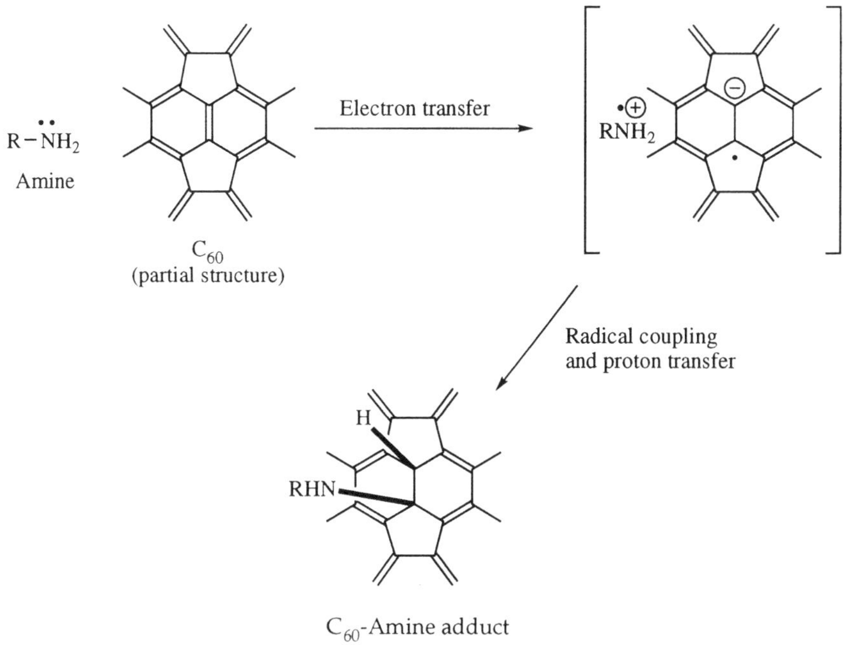

Fig. 11.3. Attack of an amine (RNH_2) on C_{60}

In the following examples, preformed polymers bearing primary amine groups react with C_{60} to produce materials bearing fullerene groups that are covalently bound to the polymer through the amine groups. Amine-terminated polystyrene **1**, shown schematically below, undergoes addition to C_{60} to produce polymer product **2** that bears fullerene groups at the ends of the chain [13].

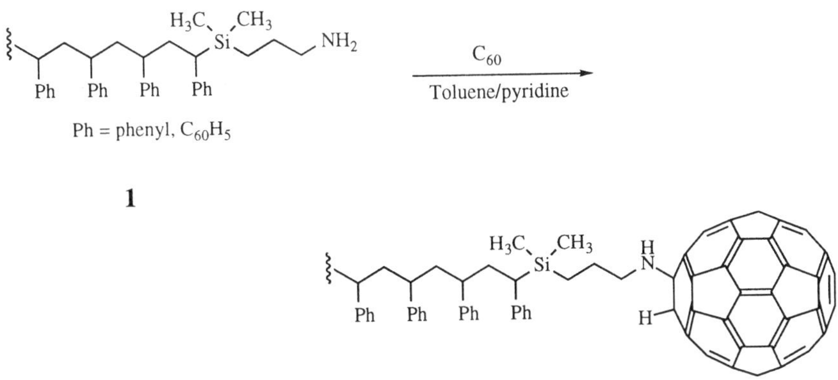

1

2

As many as twelve propylamine molecules can add to a single C_{60} molecule [6], so a significant amount of cross-linking could occur in the reaction of polymeric amine **1** with C_{60}. A 2:1 molar ratio of C_{60} to polymer **1** was used in an effort to minimize the amount of cross-linking from the addition of more than one polymer segment to the fullerene. The resulting polymer is soluble in THF, facilitating removal of unreacted C_{60} which is insoluble in THF. Glass transition temperatures are essentially the same as in **1**, indicating that the presence of the fullerene does not significantly alter this property.

Similar chemistry has been used to prepare a series of poly(ethylene glycol) and poly(propylene glycol) polymers that are endcapped with fullerenes [14]. Reaction of the starting polymer (neat or in toluene) with C_{60} leads to products that are soluble in aqueous acid and in neutral water in some cases. While the entire structure of this polymer may be incompletely understood, the essential reaction involved is shown below in Fig. 11.4. Elemental analysis indicates that in the actual materials there is an average of three polymer chains per fullerene unit, suggesting a cross-linked structure.

This common theme was also used to prepare polymer **4**, in which the fullerene unit is connected to an interesting saturated hydrocarbon backbone (**3**) [15].

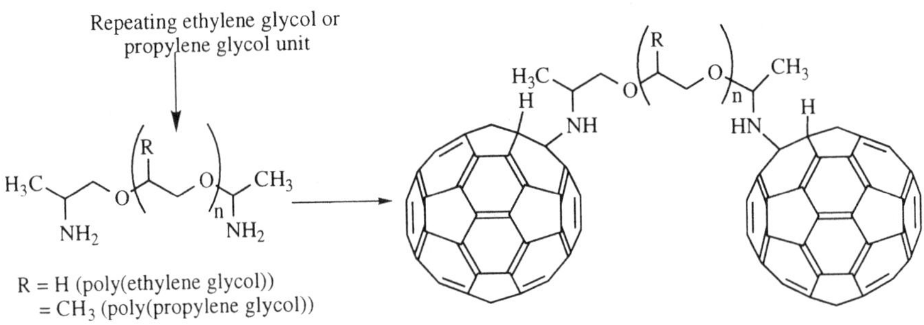

Fig. 11.4. A fullerene end-capped poly(propylene glycol). On average there are 3 polymer chains attached to each fullerene

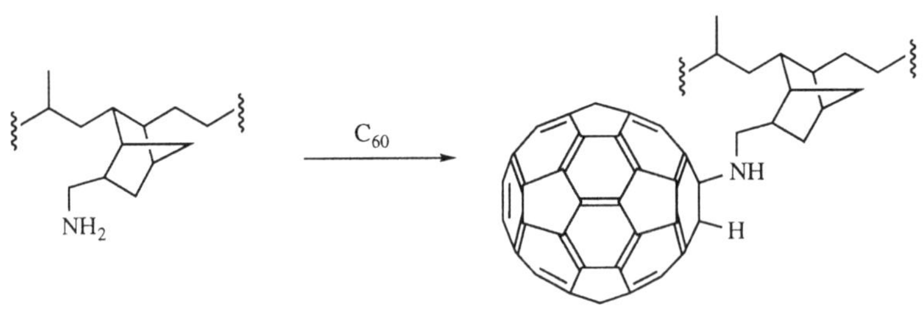

3 **4**

Polymer **4**, a reddish material, is soluble in THF and heptane, solvents in which C_{60} is very poorly soluble [16]. Viscosity evidence was presented to support the idea that more than one amine reacts with each molecule of C_{60} leading, in effect, to cross-linking.

The authors of this work point out that the hydrogen atom bound to the fullerene in **4** is assumed: it has not been directly observed but is necessary to satisfy the valency of carbon. Work in nonpolymeric systems has shown that in certain C_{60}-amine adducts, the fullerene-bound hydrogens can be lost, presumably by oxidation, sometimes resulting in fullerene–fullerene bonds [8, 17]. These bonds are generally formed across a six-membered ring from the amine substituent. Such reactions are potential cross-linking processes for all fullerene-containing polymers generated by addition of amines to fullerenes.

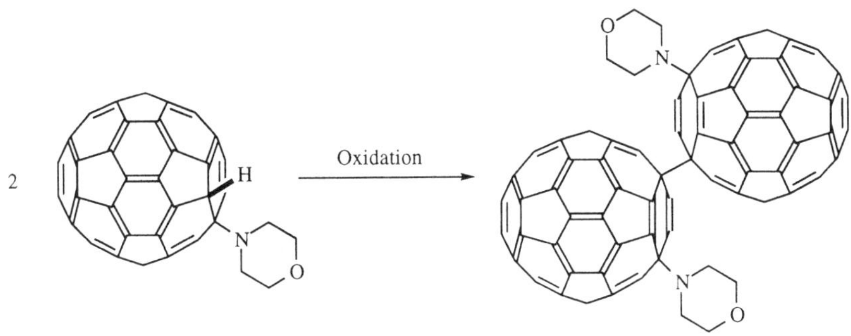

Adding soluble poly(ethylene imine) or a soluble 1:1 co-polymer of styrene and 4-(2-aminoethyl)iminomethyl styrene to C_{60} leads to the formation of polymers **5** and **6**, respectively, with pendant fullerene groups [11]. The secondary amine groups in the polymer backbone are unreactive under these conditions, as shown by a lack of reaction between C_{60} and poly(propylene imine).

These materials were prepared by adding the soluble polymer in toluene to a toluene solution of C_{60}. The addition was stopped when 60% of the C_{60} had been consumed, leading to polymers in which the fullerenes have reacted with only one amino group. The ratio of polymer to C_{60} in these materials is 18–19 to 1. Evidence for monosubstitution includes optical absorption maxima at 326 and 435 nm, which are typical for monofunctionalized fullerenes. These polymers are soluble in toluene, giving golden brown solutions, and they are soluble in CS_2 but not in hexane.

Polymeric amines can be used to tether fullerenes through reactions on fullerene derivatives, as demonstrated in the attachment of **7** to poly(propionylethylenimine-co-ethylenimine) [18].

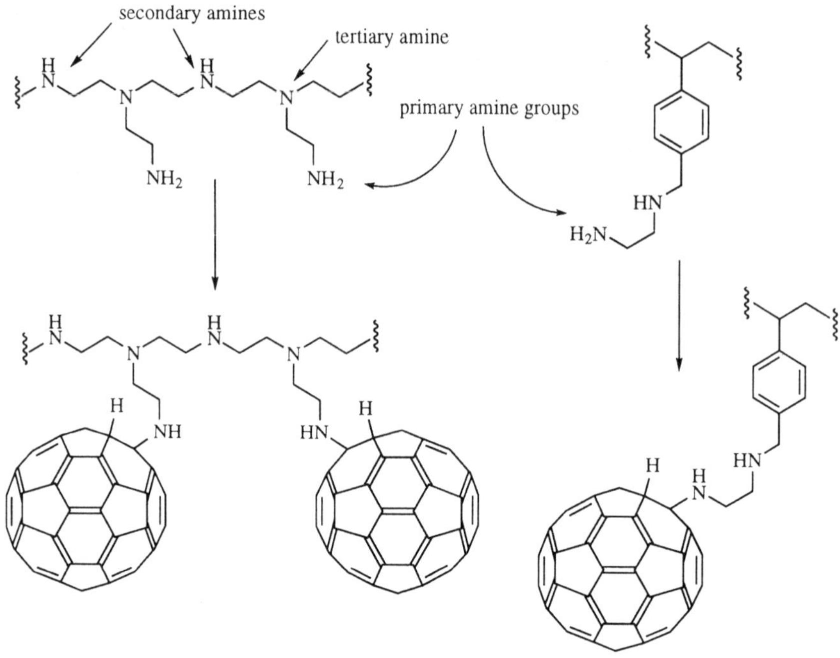

5 6

The resulting polymer **8** has a fullerene loading of about 15% and is highly soluble in water. The measured solubility of **8** is over 90 mg/mL in water, a remarkable change from the complete insolubility of C_{60} in water.

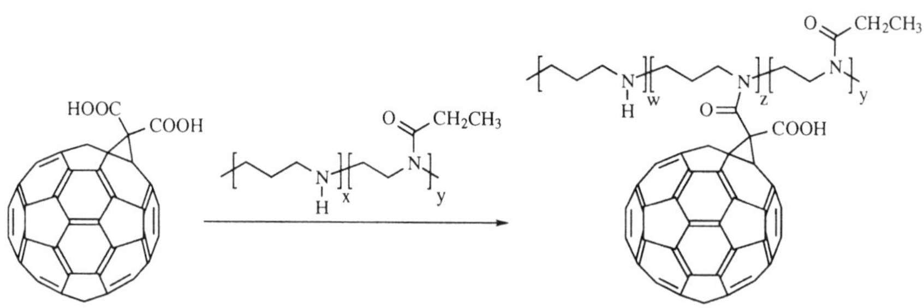

7 8

This poly(propionylethylenimine-co-ethylenimine) copolymer has also been decorated with C_{60} directly by irradiation of a mixture of the copolymer and C_{60} [19]. Fullerene incorporation of 13% was achieved. The aqueous solubility of the resulting polymers were between 30 mg/mL and 74 mg/mL, depending on specific nature of the sample. Again, this degree of solubility is much higher than that of C_{60} alone.

Carbon anions, such as CH_3^-, are highly nucleophilic and readily add to fullerenes [6, 20, 21]. This reaction has been used to attach C_{60} to poly(isoprene), poly(butadiene) anions and poly(styrene) [22] and poly(ethylene) [23] anions. In a typical case, poly(styrene) was treated [1] with a strong base (sec-butyl lithium) and tetramethylethylenediamine (TMEDA) in dry hydrocarbon solvent, resulting in lithiation of a few percent of the polymer chains, then treated with a solution of C_{60} in toluene.

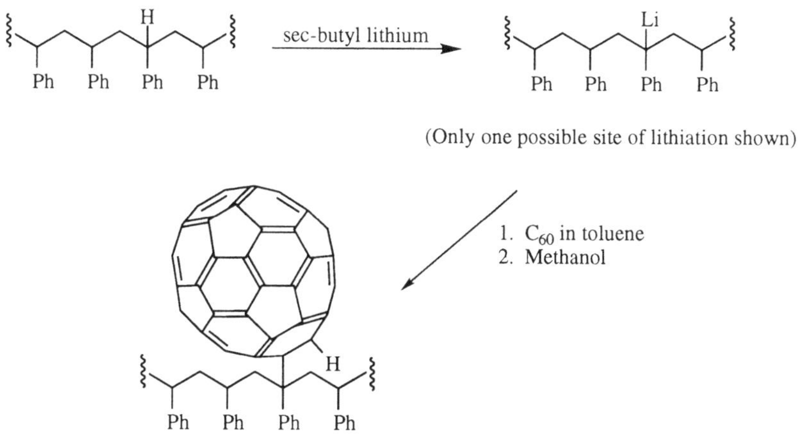

The resulting polymer can be precipitated from solution with methanol, but is significantly more soluble than is C_{60}. This reaction leads to an increase in the polydispersity from 1.1 to 1.8, suggesting that not all chains reacted equally. Significant cross-linking did not occur. Bulk samples were determined to be roughly 10% (w/w) C_{60}. Grafting C_{60} on to these polymers results in some changes in the properties of the polymer. C_{60}-derivatized poly(butadiene) [1] shows a somewhat higher T_g (−90°C, up from −100°C) than the unreacted polymer, a result that suggests that chain rigidity is higher in the grafted polymer, and the result is much more striking in the poly(isoprene) case (−65°C to +10°C).

Reaction of C_{60} with anionic living poly(styrene) **9** followed by methyl iodide results in the formation of polymers (**10**) in which fullerene units are specifically attached at the ends of the polymer chains [22]. Poly(styrene) anions were used in excess, leading to structures in which more than one polymer chain is attached to each fullerene. This type of structure has been termed a "flagellene" due to its resemblance to protozoans.

As expected, the flagellene polymer **10**, which bears a fullerene chromophore, has much stronger absorbance at 600 nm than does unmodified poly(styrene), which contains a chromophore (the phenyl, C_6H_5 group) that

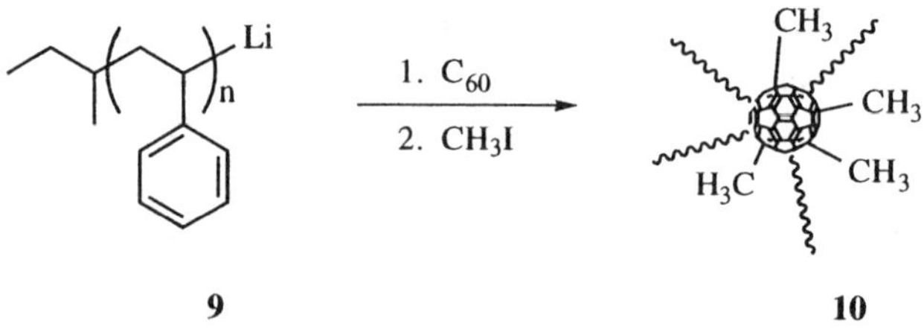

absorbs only below 300 nm. The molecular weights of the polymer appears (by gel-permeation chromatography, GPC) to reach a maximum of a few thousand, although the authors point out that the hydrodynamic volume, the effective volume occupied by the molecule tumbling in solution, is probably insensitive to increasing substitution and therefore GPC-based determination of molecular weights may not be reliable.

Fullerenes will react with strong electrophiles (electron deficient species) [24], and these reactions can be used to attach fullerenes to existing polymers. For example, reaction of poly(styrene) with C_{60} in the presence of $AlCl_3$ leads to a fullerene-derivatized polystyrene in which the fullerene is attached to the polymer through the rings rather than directly to the polymer backbone [25]. There must be two new bonds formed to the fullerene, and the second bond (to X) has not been described in this case.

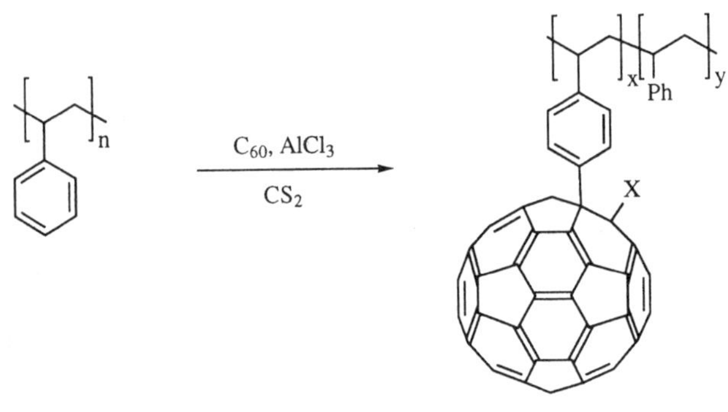

Treatment of C_{60} with alkyl azides results in the formation of azafulleroids [26], in which a nitrogen atom has been inserted into a C–C *single* bond of the fullerene. The product has a nitrogen atom bridging across two carbon atoms, and is a $C_{60}N$ cage, a 61-atom cluster rather than a 60-atom cluster. These 61-atom spheres are known as fulleroids. Another distinctive characteristic of this reaction is that the addition of azides to fullerenes results in addition across a 6,5 ring fusion (1,6 addition) with cleavage of a fullerene carbon–carbon bond (the 1,6 bond) and regeneration of the double bond. The starting fullerene (C_{60} in this case) has 30 double bonds and the product here also contains 30 double bonds.

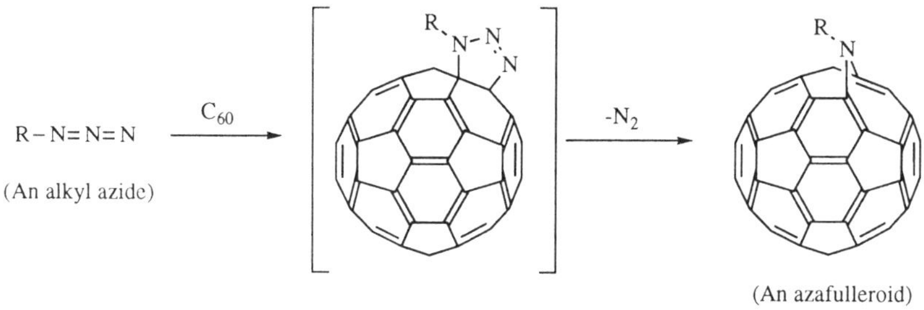

This reaction has been used to prepare an azafulleroid-containing polymer, using an azidomethyl substituted polystyrene as the polymer backbone [27]. This is "buckyball fishing" again, although a different "lure" is being used.

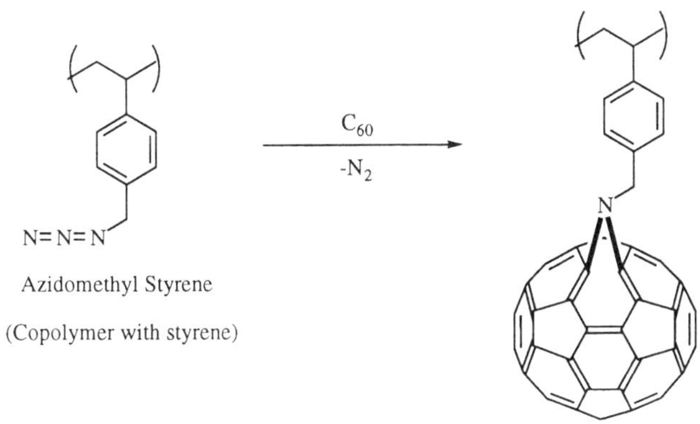

11

Weight percentages of C_{60} in polymer **11** were as high as 29%. A major advantage of the azide "lure" over amine lures is that monoaddition is highly favored over multiple addition. Multiple addition can lead to cross-linking and to poor solubility (and processibility) in the final polymer, but the material obtained here is soluble in solvents such as THF and chloroform ($CHCl_3$). Reaction of a dendrimer (highly branched) azide with C_{60} results in the formation of compound **12** (Fig. 11.5) [28]. This material is significantly more soluble in organic solvents than unmodified C_{60}, presumably the highly soluble branched unit dominates the overall structure. The electrochemical behavior of **12** differs from simple azafulleroids [26] in that the reduction potentials are shifted to more negative values in **12**. The authors suggest that the insulating effect of the large dendrimer unit is responsible for this shift.

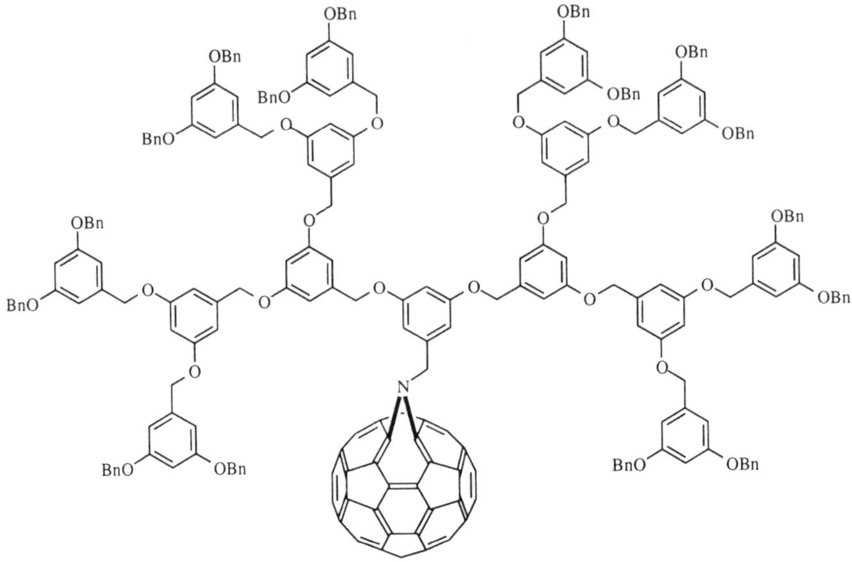

12

Fig. 11.5. A "dendritic" azafulleroid. BnO is an abbreviation for benzyloxy, $C_6H_5CH_2O$

A final example of a method for attaching C_{60} to an existing polymer backbone utilizes the Diels–Alder reaction of a polystyrene-tethered cyclopentadiene with C_{60} [29]. A Diels–Alder reaction involves the formation of a six-membered ring from a diene and another alkene known as a dienophile. Fullerenes in general are highly reactive dienophiles and participate in the Diels–Alder reaction with a variety of dienes [30].

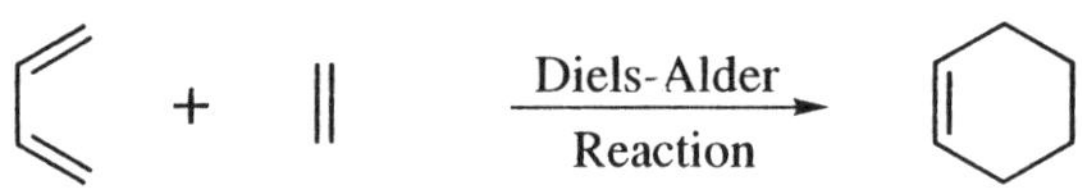

(A diene) (Dienophile)

Starting with commercial "Merrifield" resin, approximately 2% cross-linked chloromethylated polystyrene developed for the solid-phase synthesis of peptides, cyclopentadiene-bearing polymer **13** was prepared. The maximum amount of C_{60} that could be bound to the resin was 23 mg C_{60} per gram of resin.

The Diels–Alder reaction of cyclopentadiene with C_{60} is reversible, as are a number of other Diels–Alder additions to C_{60}.[30] The fullerene content of **13** can be liberated by heating the polymer, or more efficiently, heating in the

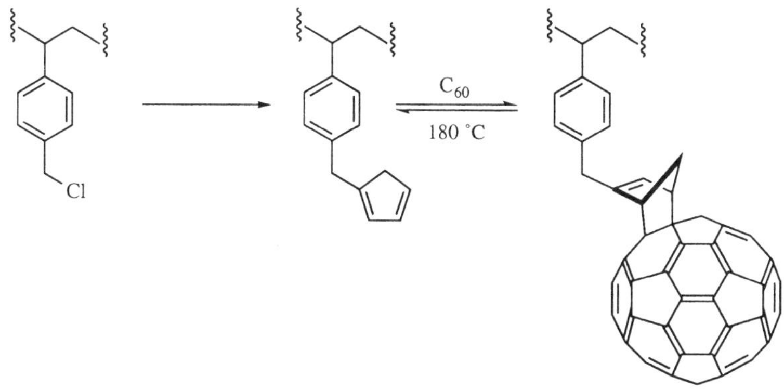

13

presence of maleic anhydride. The authors point out that the reversibility of this reaction might have applications in the nonchromatographic purification of fullerenes. In principle, fullerenes in a sample would react with the polymer, while non-fullerene material would not. The polymer could be isolated and washed, then heated to release the fullerenes.

A soluble fullerene-containing polyester has been prepared by condensation of bisphenol **14** with a diacid chloride [31]. Heating **14** with sebacoyl chloride results in the formation of **15** which is soluble in nitrobenzene and benzonitrile. The electrochemical behavior of **15** is very similar to that of **14**, showing that the properties of the fullerene monomer are preserved in **15**. Interestingly, when hexamethylene diisocyanate is substituted for sebacoyl chloride, the resulting polyurethane is insoluble.

11.3 Pearl Necklace Polymers

In principle, any chemical reaction that forms covalent bonds to a fullerene can be used to create a bridged fullerene dimer, oligomer, or polymer. A bifunctional monomer reacts with two fullerene molecules to form a bridged dimer. Further reactions leads to the growth of the chain. This approach is shown schematically in Fig. 11.6.

The most serious limitation of this approach is that the solubility of compounds of this type tends to be very low, resulting in the precipitation of short oligomers. The diacid **16** is an interesting starting material for polymer synthesis. Treatment of **16** with isophthalic acid and 4,4'-diaminodiphenyl ether in the presence of triphenylphosphite leads to the formation of polyamide **17** [32]. The overall yield in this process is good (90–95%). The weight-average molecular weight of the product is between 50 000 and 60 000.

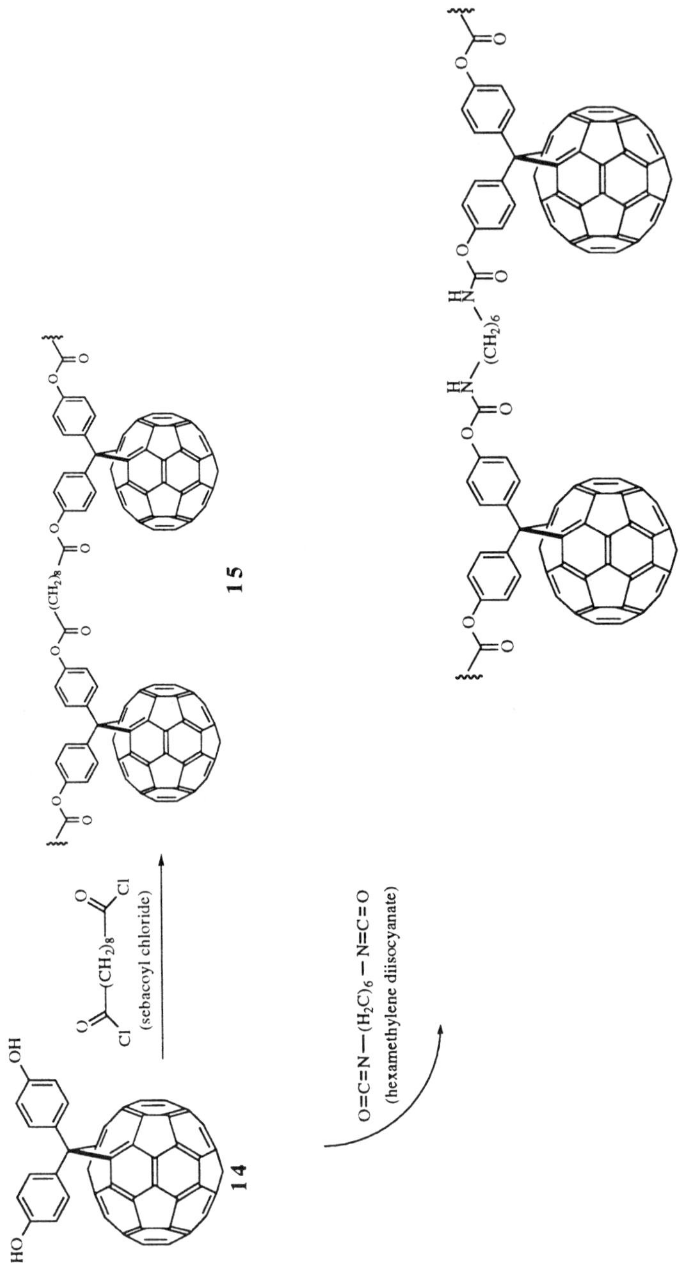
14
(sebacoyl chloride)
15
O=C=N—(H$_2$C)$_6$—N=C=O
(hexamethylene diisocyanate)
(a polyurethane)

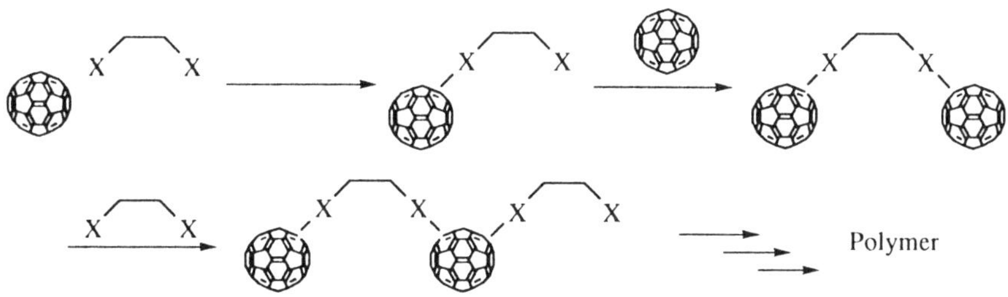

Fig. 11.6. Sequential growth of a bridged "pearl necklace" polymer

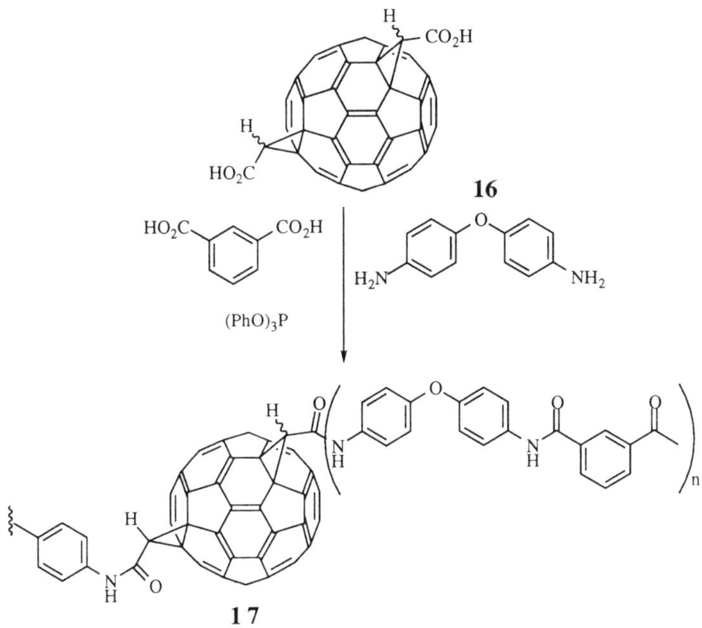

Diazoalkanes undergo 1,3-dipolar addition with a wide variety of alkenes to form pyrazolines, five-membered rings, and fullerenes are no exception.

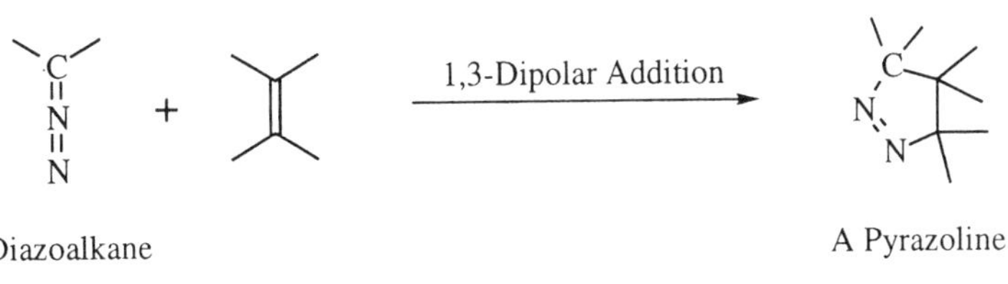

Reaction of diazoalkanes with C_{60} to forms pyrazolines (**18**) which break down to form either methanofullerenes (**19**) or fulleroids (**20**) [6, 33–35].

18

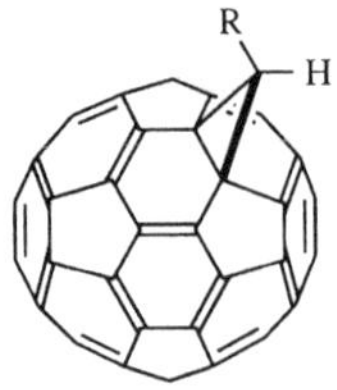

19

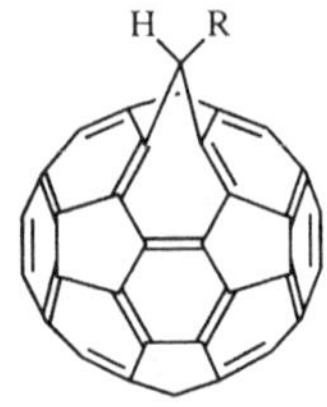

20

Diazoalkane chemistry has been used to link two fullerenes together via a phenyl bridge [4]. Treatment of C_{60} with either of two bis(diphenyldiazo) compounds results in the formation of a bridged fullerene dimer. Larger structures, involving more than two fullerene molecules, were not formed. The solubility of these bridged compounds is very low and they precipitate from solution and hence do not react further.

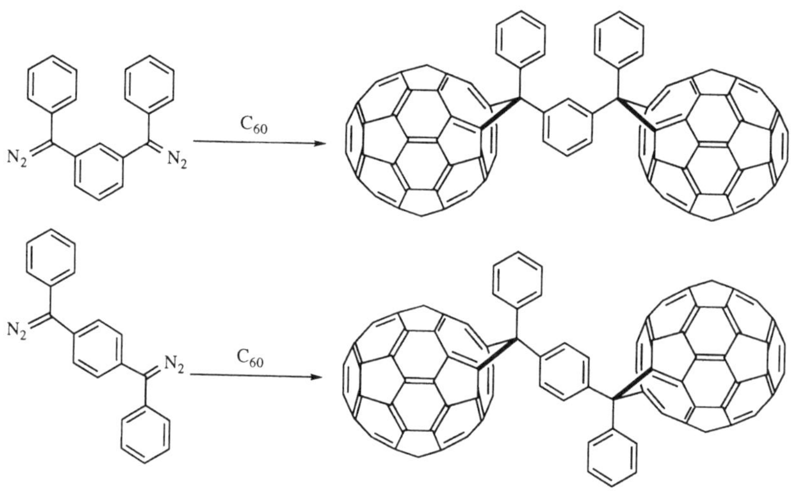

The poor solubility of most bridged fullerene dimers is a major limitation to this approach to the preparation of fullerene-containing polymers. Increasing solubility through derivatization is an approach which has been used successfully by Diederich and co-workers [36, 37] in the preparation of compounds **22** and **23** (Fig. 11.7), which are hybrids between the fullerene family of carbon allotropes and the "cyclocarbon" family [36]. Compounds **22** and **23**[37] were prepared by oligomerization of monomer **21** which bears several groups to improve the solubility. These substituents are crucial for keeping the compounds in solution [37], making it possible for these larger

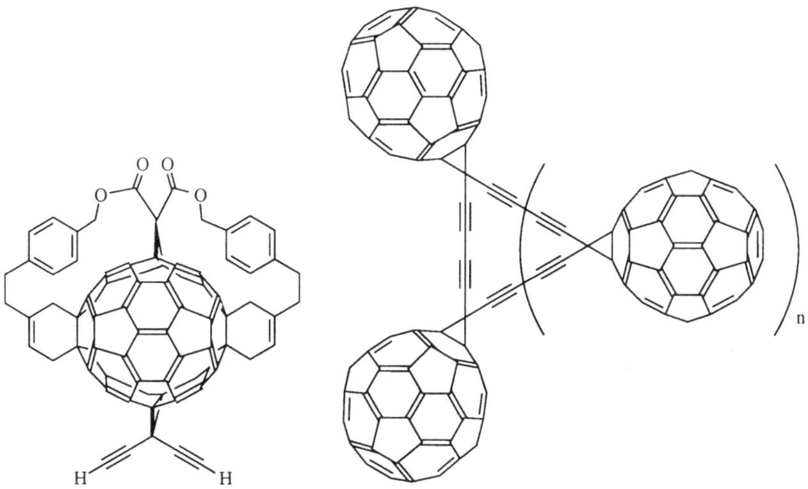

21 **22** (n=1) **23** (n=2)

Fig. 11.7. Compounds **21**, **22**, and **23**. The twin benzylester chains are essential for solubility. For clarity, these solubility-enhancing substituents are omitted in the drawings of **22** and **23**

structures to form. The acute angle between the diyne units (–C≡C–C≡C– groups) favors the formation of cyclic structures as shown here, but extended chains and three-dimensional networks are conceivable and would be very interesting structures.

Radical copolymerization of styrene with C_{60} [38–40] or with C70 [39] has lead to a series of copolymers composed of roughly 0.2% to 50% C_{60}, including one containing about 17% C_{60} [38], and one with 0.25–5% C70 [39]. These copolymers are soluble in hot toluene, but can be precipitated with methanol. The 0.4% C_{60} polymer has a polydispersity of 4 and a weight-average molecular weight (Mw) of 62 000, while the 17% C_{60} material has a polydispersity of 2.6 and Mw of 17 000 [38]. Styrene-C_{60} copolymers have lower intrinsic viscosities than linear styrene homopolymers, consistent with the proposed branched structure [41].

The fullerene content of these materials is responsible for broad, featureless absorption in the long-wavelength end of the UV/Vis spectrum, a region where the absorption of pure polystyrene is minimal [39]. It is also clear from this broad absorption and from the increase in fluorescence of these materials that the fullerene units have been covalently modified, and the authors point out that breaking the symmetry of C_{60} is an important step on the way toward nonlinear optical materials based on fullerenes. However, there is evidence that these polymerizations lead to microclusters or microgels, and that UV/Vis spectra are due to scattering phenomena [42]. The presence of 30 double bonds in C_{60} make extensive cross-linking possible.

The miscibility of poly(styrene)-C_{60} copolymers with other polymers (poly(methyl vinyl ether), PMVE) and poly(2,6-dimethyl-1,4-phenylene oxide), PPO) has been studied using differential scanning calorimetry (DSC) [43]. Three different poly(styrene)-C_{60} copolymer samples were found to be miscible with PPO, while miscibility of poly(styrene)-C_{60} samples with PMVE decreases as the fullerene content of the copolymer increases.

11.4 Highly-Crosslinked Polymers

Flash vacuum pyrolysis (650°C) of 2,2-paracyclophane (**24**) leads to the formation of p-xylylene (**25**, shown below in the diradical form). A polymer results when the gas flow containing **25** is swept over a chilled solution of C_{60} in toluene [44].

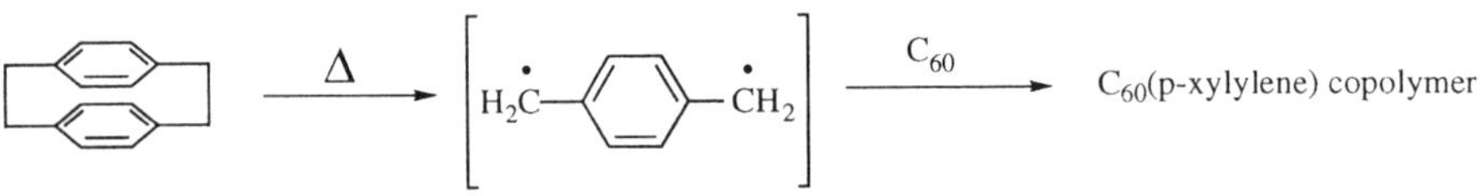

24 **25**

The brown, solid C_{60}-xylylene copolymer produced here is insoluble and difficult to characterize structurally. It is not likely to have a regular, ordered structure. Thermal gravimetric analysis (TGA) shows a steady loss of mass (loss of 44% of the starting mass) over a temperature range up to 1000°C, well above the temperature where pure C_{60} has either sublimed or decomposed. The polymer has been characterized by solid-state ^{13}C NMR, in which a set of overlapping resonances, spanning the range from 129 to 150 ppm, were observed and assigned to fullerene carbons. Resonances that could be assigned to the p-xylylene moiety appeared at 38, 129, and 139 ppm.

Addition of excess 2-methylaziridine to C_{60} results in the formation of amine adducts of the general form **26**, in which ten (on average) methylaziridine groups are added to each C_{60} [45]. This compound is highly soluble in a number of solvents, including chloroform ($CHCl_3$) and tetrahydrofuran (THF).

Heat curing mixtures of **22** with phenols such as bisphenol A (shown below), or with epoxides such as Epon 828, leads to soluble polymers. Given the ten aziridine groups on **26**, these polymers are likely to be significantly cross-

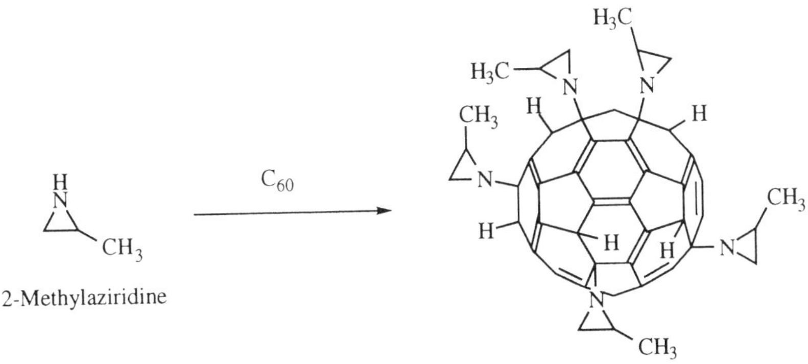

26

linked, yet are sufficiently soluble to be spin-coated onto surfaces. Friction testing of spin-coated quartz disks showed that the lubrication behavior of these polymers is similar to wet graphite, and the polymers show some of the same film-regeneration behavior known in some of the best solid lubricants.

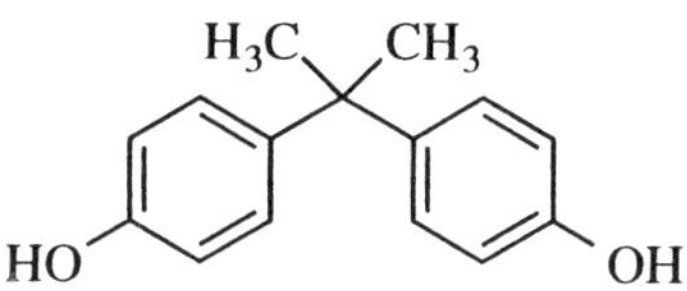

Bisphenol A

11.5 Others

There are a number of reports of fullerene-containing polymers that do not fall into the categories above. These are materials where the material is either highly cross-linked or where the structure is not well understood. Electrochemical reduction of $C_{60}O$ using a gold electrode at potentials of at least $-1.2\,V$ (vs Ag/Ag^+), sufficient to produce $C_{60}O^{2-}$, results in a polymeric film on the electrode surface [46]. The rate of polymerization is solvent dependent, and at slow sweep rates there is significant decomposition of $C_{60}O^{\bullet -}$ to C_{60} rather than further reduction and polymerization. The polymer, whose structure is not well established, is yellowish and at least partially soluble in o-dichlorobenzene.

A clever route to 'poly(fullerenes)' has been reported recently, in which fullerene (C_{60} or C_{70}) clusters are prepared in mixed solvent systems, then photochemically polymerized [47]. The structure of this polymer is likely to similar to the [2 + 2] photopolymer that has been prepared in the solid state (Fig. 11.1) [3], although the material prepared from clusters is soluble in DMSO, despite the high molecular weight.

11.6 Future Directions

There has been relatively little work on higher fullerenes such as C_{70}, C_{76}, C_{78}, and C84, probably due to the higher cost and poor availability of those compounds. The lower symmetry of the higher fullerenes also complicates structural determinations. Endohedral fullerenes, compounds in which metal atoms are trapped inside fullerene shells, could produce very interesting polymeric materials, but at this time these compounds are very precious and not widely available. Much of the research effort into fullerene containing polymers of the types discussed in this chapter has centered on the development of methodology for the preparation of different fullerene-containing materials. The literature base of the physical, chemical, electronic, and optical data on these materials is slowly growing. Sufficient data is in for a number of companies and individuals to have applied for patents (see [2] for a short list to 1996).

References

1. L. Dai, A. W. H. Mau, H. J. Griesser, T. H. Spurling, J. W. White: J. Phys. Chem. **99**, 17302–17304 (1995).
2. A. O. Patil: *Fullerene-Functionalized Polymers*, (CRC Press, Boca Raton, 1996) Vol. 4, pp. 2597–2605.
3. A. M. Rao, P. Zhou, K.-A. Wang, G. T. Hager, J. M. Holden, Y. Wang, W.-T. Lee, X.-X. Bi, P. C. Eklund, D. S. Cornett, M. A. Duncan, I. J. Amster: Science **259**, 955–957 (1993).
4. T. Suzuki, Q. Li, K. C. Khemani, F. Wudl, Ö. Almarsson: J. Am. Chem. Soc. **114**, 7300–7301 (1992).
5. I. Amato: Science **252**, 646 (1991).
6. F. Wudl, A. Hirsch, K. C. Khemani, T. Suzuki, P.-M. Allemand, A. Koch, H. Eckert, G. Srdanov, H. M. Webb: *Survey of Chemical Reactivity of C_{60}, Electrophile and Dieno-Polarophile Par Excellence*, G. S. Hammond and V. J. Kuck (Ed.), (ACS, Washington, DC, 1992) Vol. 481, pp. 161–175.
7. A. Hirsch: Angew. Chem. Int. Ed. Engl. **32**, 1138–1141 (1993).
8. A. Hirsch: Synthesis 895–913 (1995).
9. H. Schwarz: Angew. Chem. Int. Ed. Engl. **31**, 293–298 (1992).
10. T. Taylor, M., W. D. R. Nature **363**, 685–693 (1993).
11. K. E. Geckeler, A. Hirsch: J. Am. Chem. Soc. **115**, 3850–3851 (1993).
12. R. Taylor: *The Chemistry of Fullerenes*, T. Taylor (Ed.) (World Scientific, River Edge, NJ, 1995).
13. C. Weis, C. Friederich, R. Mülhaupt, H. Frey: Macromolecules **28**, 403–405 (1995).

14. N. Manolova, I. Rashkov, F. Beguin, H. J. van Damme: Chem. Soc. Chem. Commun. 1725–1727 (1993).
15. A. O. Patil, G. W. Schriver, B. Carstensen, R. D. Lundberg: Polym. Bull. **30**, 187–190 (1993).
16. R. S. Ruoff, D. S. Tse, R. Malhotra, D. C. Lorents: J. Phys. Chem. **97**, 3379–3383 (1993).
17. G. Schick, K.-D. Kampe, A. Hirsch: J. Chem. Soc. Chem. Commun. 2023–2024 (1995).
18. Y. P. Sun, B. Liu, D.K. Moton: J. Chem. Soc. Chem. Commun. 2699–2700 (1996).
19. Y.-P. Sun, B. Liu, G. E. Lawson: Photochem. Photobiol. **66**, 301–308 (1997).
20. A. Hirsch, A. Soi, H. R. Karfunkel: Angew. Chem. **104**, 808–810 (1992).
21. P. J. Fagan, P. J. Krusic, D. H. Evans, S. A. Lerke, E. Johnston: J. Am. Chem. Soc. **114**, 9697–9699 (1992).
22. E. T. Samulski, J. M. DeSimone, M. O. Hunt, Y. Z. Menceloglu, R. C. Jarnagin, G. A. York, K. B. Labat, H. Wang: Chem. Mater. **4**, 1153–1157 (1992).
23. D. E. Bergbreiter, N. H. Gray: J. Chem. Soc., Chem. Commun. 645–645 (1993).
24. G. A. Olah, I. Bucsi, C. Lambert, R. Aniszfeld, N. J. Trivedi, D. K. Sensharma, G. K. S. Prakash: J. Am. Chem. Soc. **113**, 9387–9388 (1991).
25. B. Liu, C. E. Bunker, Y.-P. Sun: J. Chem. Soc., Chem. Commun. 1241–1242 (1996).
26. N. Prato, Q. C. Li, F. Wudl, V. Lucchini: J. Am. Chem. Soc. **115**, 1148–1150 (1993).
27. C. J. Hawker: Macromolecules **27**, 4836–4837 (1994).
28. C. J. Hawker, K. L. Wooley, J. M. Fréchet: J. Chem. Soc. Chem. Commun. 925–926 (1994).
29. K. Guhr, M. D. Greaves, V. M. Rotello: J. Am. Chem. Soc. **116**, 5997–5998 (1994).
30. M. S. Meier: *Cycloadditions to* C_{60}, R. Taylor (Ed.), (World Scientific Books, 1994) Vol. 4, pp. 174–194.
31. S. Shi, K. C. Khemani, Q. C. Li, F. Wudl: J. Am. Chem. Soc. **114**, 10656–10657 (1992).
32. J. Li, T. Yoshizawa, M. Ikuta, M. Ozawa, K. Nakahara, T. Hasegawa, K. Kitazawa, M. Hayashi, K. Kinbara, M. Nohara, K. Saigo: Chem. Lett. 1037–1038 (1997).
33. M. Prato, F. Wudl: *Fulleroids, Methanofullerenes, and Oxa-, Aza-, and Sila-Homologues*, R. Taylor (Ed.), (World Scientific, New York, 1995) Vol. 4, pp. 151–173.
34. A. B. I. Smith, R. M. Strongin, L. Brard, G. T. Furst, W. J. Romanow, K. G. Owens, R. J. Goldschmidt, R. C. King: J. Am. Chem. Soc. **117**, 5492–5502 (1995).
35. T. Suzuki, Q. Li, K. C. Khemani, F. Wudl: J. Am. Chem. Soc. **114**, 7301–7302 (1992).
36. F. Diederich, Y. Rubin: Angew. Chem. Int. Ed. Engl. **31**, 1101–1123 (1992).
37. L. Issacs, P. Seiler, F. Diederich: Angew. Chem. Int. Ed. Engl. **34**, 1466–1469 (1995).
38. C. E. Bunker, G. E. Lawson, Y.-P. Sun: Macromolecules **28**, 3744–3746 (1995).
39. T. Cao, S. E. Webber: Macromolecules **28**, 3741–3743 (1995).
40. Y.-P. Sun, G. E. Lawson, C. E. Bunker, R. A. Johnson, B. Ma, C. Farmer, J. E. Riggs, A. Kitaygorodskiy: Macromolecules **29**, 8441–8448 (1996).
41. W. T. Ford, T. D. Graham, T. H. Mourey: Macromolecules **30**, 6422–6429 (1997).

42. P. L. Nayak, S. Alva, K. Yang, P. K. Dhal, J. Kumar, S. K. Tripathy: Macromolecules **30**, 7351–7354 (1997).
43. J. Zheng, S. H. Goh, S. Y. Lee: Macromolecules **30**, 8069–8071 (1997).
44. D. A. Loy, R. A. Assink: J. Am. Chem. Soc. **114**, 3977–3978 (1992).
45. A. Nigam, T. Shekharam, T. Bharadwaj, J. Giovanola, S. Narang, R. Malhotra: J. Chem. Soc. Chem. Commun. 1547–1548 (1995).
46. K. Winkler, D. A. Costa, A. L. Balsh, W. R. Fawcett: J. Phys. Chem. **99**, 17431–17436 (1995).
47. Y.-P. Sun, C. E. Bunker, B. Liu: J. Am. Chem. Soc. **117**, 12705–12711 (1996).

Index

Springer Series in

MATERIALS SCIENCE

Editors: R. Hull R. M. Osgood, Jr. H. Sakaki A. Zunger

1 **Chemical Processing with Lasers***
By D. Bäuerle

2 **Laser-Beam Interactions with Materials**
Physical Principles and Applications
By M. von Allmen and A. Blatter
2nd Edition

3 **Laser Processing of Thin Films and Microstructures**
Oxidation, Deposition and Etching of Insulators
By. I. W. Boyd

4 **Microclusters**
Editors: S. Sugano, Y. Nishina, and S. Ohnishi

5 **Graphite Fibers and Filaments**
By M. S. Dresselhaus, G. Dresselhaus, K. Sugihara, I. L. Spain, and H. A. Goldberg

6 **Elemental and Molecular Clusters**
Editors: G. Benedek, T. P. Martin, and G. Pacchioni

7 **Molecular Beam Epitaxy**
Fundamentals and Current Status
By M. A. Herman and H. Sitter 2nd Edition

8 **Physical Chemistry of, in and on Silicon**
By G. F. Cerofolini and L. Meda

9 **Tritium and Helium-3 in Metals**
By R. Lässer

10 **Computer Simulation of Ion-Solid Interactions**
By W. Eckstein

11 **Mechanisms of High Temperature Superconductivity**
Editors: H. Kamimura and A. Oshiyama

12 **Dislocation Dynamics and Plasticity**
By T. Suzuki, S. Takeuchi, and H. Yoshinaga

13 **Semiconductor Silicon**
Materials Science and Technology
Editors: G. Harbeke and M. J. Schulz

14 **Graphite Intercalation Compounds I**
Structure and Dynamics
Editors: H. Zabel and S. A. Solin

15 **Crystal Chemistry of High-T_c Superconducting Copper Oxides**
By B. Raveau, C. Michel, M. Hervieu, and D. Groult

16 **Hydrogen in Semiconductors**
By S. J. Pearton, M. Stavola, and J. W. Corbett

17 **Ordering at Surfaces and Interfaces**
Editors: A. Yoshimori, T. Shinjo, and H. Watanabe

18 **Graphite Intercalation Compounds II**
Editors: S. A. Solin and H. Zabel

19 **Laser-Assisted Microtechnology**
By S. M. Metev and V. P. Veiko
2nd Edition

20 **Microcluster Physics**
By S. Sugano and H. Koizumi
2nd Edition

21 **The Metal-Hydrogen System**
By Y. Fukai

22 **Ion Implantation in Diamond, Graphite and Related Materials**
By M. S. Dresselhaus and R. Kalish

23 **The Real Structure of High-T_c Superconductors**
Editor: V. Sh. Shekhtman

24 **Metal Impurities in Silicon-Device Fabrication**
By K. Graff 2nd Edition

25 **Optical Properties of Metal Clusters**
By U. Kreibig and M. Vollmer

* The 2nd edition is available as a textbook with the title: *Laser Processing and Chemistry*

Printing: Saladruck, Berlin
Binding: Buchbinderei Lüderitz & Bauer, Berlin